CHOOSING AN EXPERIMENTAL DESIGN

KISS → Keep It Simple Statistically

PF → Process flow diagram

CE → Cause-and-effect diagram

inputs

outputs

C → inputs on cause-and-effect to be held constant

N → inputs on cause-and-effect that are noise or uncontrolled

X → inputs (factors) on cause-and-effect identified for experimentation

SOPs → standard operating procedures to insure all Cs are held constant

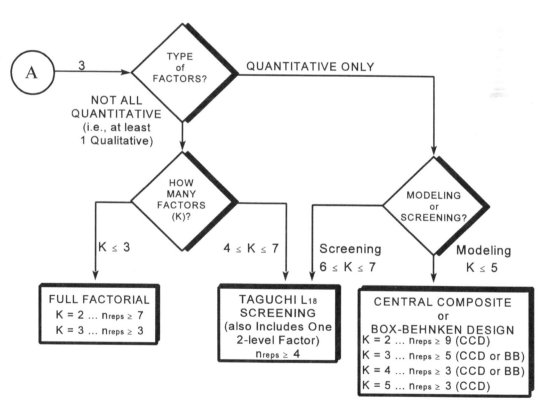

Source: Steve Schmidt
Mark Kiemele
Cass Grandone

ABOUT THE BOOK

The primary motivation for this book is to provide material that will bridge the gap between texts too mathematical and those that attain simplicity by omitting important concepts. This is an applications oriented text with an emphasis on Keeping It Simple Statistically (KISS) and "Statistical Thinking." Instead of focusing on a specific approach to designing experiments, this book blends the competing Taguchi, Shainin, and classical approaches. Several Rules of Thumb are presented to reduce statistical complexity. Over 275 pages of actual industrial case studies plus software for design, analysis, and simulated processes are included.

ABOUT THE AUTHORS

Stephen R. Schmidt, Ph.D., founder of Air Academy Associates, a multi-million dollar international consulting business, has more than 20 years of teaching and consulting experience in process improvement methodology. His "Keep It Simple Statistically" (KISS) approach has gained him widespread popularity and an impressive list of clients: General Electric, Sony, Motorola, Boeing, Texas Instruments, Lockheed Martin, AlliedSignal, Levi Strauss, Advanced Microdevices, Abbott Labs, Citicorp, Ford Motor Company, Dannon, Delco Electronics (Delnosa-Mexico), National Semiconductor, plus many others. Because of his vast experience, he has served as an adjunct faculty member to the Motorola Six Sigma Research Institute where he trained the first wave of Black Belts at Motorola and Texas Instruments. In addition, he served on the Board of Directors for the Statistical Applications Institute. Dr. Schmidt is also an adjunct professor at The University of Texas at Austin where he provides management training and serves as a principal instructor for the Accelerated Six Sigma program. He has a B.S. in Math from the U.S. Air Force Academy, an M.S. in Operations Research from the University of Texas, and a Ph.D. in Applied Statistics from the University of Northern Colorado. He served for 20 years in the U.S. Air Force as an instructor pilot and tenured professor at the U.S. Air Force Academy. Dr. Schmidt is the co-author of three other recent texts: *Basic Statistics: Tools for Continuous Improvement*; *Total Quality: A Textbook of Strategic Quality, Leadership and Planning*; and *Knowledge Based Management (KBM)*.

Robert G. Launsby, M.S. in Engineering, is President of Launsby Consulting. He has over 20 years of industrial experience in management, engineering, training, and applications of problem solving approaches. Bob rose to the rank of Consultant Engineer at Digital Equipment Corporation because of his innovative work in the application of statistical tools to engineering problems. Over the last 14 years, he has led over 500 applications of statistical methods in a wide variety of industries. Possessing strong presentation and facilitation skills, Bob has instructed over 3,000 people in the use of statistical problem solving tools. Bob has taught seminars for Drexel University, Washington State University, University of Phoenix, University of Colorado, University of Nebraska, and the University of Northern Colorado. Bob is co-author of the texts: *Straight Talk on Designing Experiments* and *Process Validation for Business Success*.

OTHER TEXTS BY AIR ACADEMY PRESS

Basic Statistics: Tools for Continuous Improvement, 4th Edition (1997)
by Mark J. Kiemele, Stephen R. Schmidt, and Ronald J. Berdine.
ISBN 1-880156-06-7

This text provides a refreshingly new approach to applying statistical tools for moving up the quality improvement ladder beyond the low-hanging fruits. Emphasis is on "statistical thinking" for transforming data into information, plus applications. Topics include: *The Role of Statistics in Process Improvement; Steps Before Collecting Data; Descriptive Statistics; Probability Distributions; Confidence Intervals; Hypothesis Testing; Analysis of Variance; Regression; Design of Experiments; Statistical Process Control; Gage Capability; Multivariate Charts; Reliability; and Quality Function Deployment.* More than 65 examples and case studies, contributed by more than 10 industrial practitioners, span manufacturing, service, software, government, and the health care industries. Recommended for use at the college level and within industry, this is the only single book on the market that integrates Knowledge Based Management (KBM), basic graphical and statistical tools, Statistical Process Control (SPC), Design of Experiments (DOE), Quality Function Deployment (QFD), and reliability. Included with the book is the software program SPC KISS student version.

Knowledge Based Management (KBM), (1996)
by Stephen R. Schmidt, Mark J. Kiemele, and Ronald J. Berdine.
ISBN 1-880156-05-9

Why do some companies succeed in quality improvement and others see little or no return on their investment? Many managers are frustrated or sometimes intimidated by the complexity of modern-day processes. This text provides a common-sense, easy-to-follow approach to gaining process knowledge while showing how a proper knowledge base can provide greater return on investment and expanded profit margins. Our experience has indicated that the major barrier to success in most organizations is a lack of knowledge about processes, products, and people. Without an adequate knowledge base, it is difficult to communicate and to make good decisions. Without knowledge, we develop a culture of diversity — diverse opinions.

UNDERSTANDING INDUSTRIAL DESIGNED EXPERIMENTS

Fourth Edition

Stephen R. Schmidt, Ph.D.
Robert G. Launsby, M.S.

Edited By
Mark J. Kiemele, Ph.D.

Colorado Springs, Colorado

Library of Congress Catalog Card Number: 88-92885

ISBN 1-880156-03-2

Printed in the United States of America
16   15   -12   13

The authors recognize that perfection is unattainable without continuous improvement. Therefore, we solicit comments as to how to improve this text. To relay your comments or to obtain further information, contact:

AIR ACADEMY ASSOCIATES
1650 Telstar Drive, Suite 110
Colorado Springs, CO   80920
(719) 531-0777 ◆ FAX (719) 531-0778
e-mail: aaa@airacad.com
website: www.airacad.com

Preface

The time is quickly approaching when one will not be considered a competent engineer, scientist, or technical manager without the ability to "think statistically" and implement proper experimental design techniques. The demands for increased efficiency of processes, lower product cost, and shortened development cycles will dictate that we use simple, but powerful tools to get the most out of our experiments. No longer do we have the luxury of running one-factor-at-a-time experiments or experiments with excessive sample sizes. In competitive environments, only those groups who **apply** experimental design approaches efficiently and effectively to maximize knowledge will survive.

Once experimenters decide to start using orthogonal or nearly orthogonal designs, they are quickly faced with a plethora of competing design and analysis strategies. Full factorial, Plackett-Burman, Taguchi, and CCD are just some of the competing design types. What is the strength and weakness of each? When should I choose one design over another? How should I analyze the data? Should I use "Pick the Winner", plots of marginal means, ANOVA, Normal Probability Plots, or Regression Analysis? What is the objective of my experiment: Screening, Modeling, or Robust Design? How does the experimental objective affect the choice of design and analysis techniques? Is the brainstorming session critical? Who should be involved? What is a logical step-by-step process to conduct a successful experiment? What are some Rules of Thumb that will simplify and inject meaning and insight into the statistics?

Academia and industry need a book that will answer the above questions in a way that practioners can understand! The purpose of this text is to answer as many of your questions about experimentation as possible and to guide you through your own success story. Rather than presenting only one specific philosophy of experimentation (such as Taguchi, Classical, etc.), the authors have selected the best from competing experimental strategies and then blended them into a more useful and powerful approach. The end result is a text that represents a revolutionary change in how we communicate statistics and how we approach experimentation in order to maximize the amount of knowledge gained from experimental data.

Acknowledgements

Writing a book this size is a major undertaking and it requires a team effort to be successful. Our customers are a vital part of our team and thanks to their feedback we have been able to put together this exciting new fourth edition. A special thanks to all the professors, scientists, engineers, and practitioners using our text and supplying suggestions for improvements.

Thanks to Debbie and Derek Brown, Diane Launsby and Ronda Churchill for their typing and editing of the first three editions. Also, a very special thanks to Dr. Mark J. Kiemele, our chief editor, and to Beatriz Orozco, our technical layout and graphics specialist; without their outstanding work this edition would not have made it to print. The following is a partial list of those who have helped over the years to make this text a success.

Willis Adcock
Alan Arnholt
Mark Bailey
Maureen Beaty
Ron Berdine
Maurice Berryman
Tom Bingham
Mike Bishop
Susan Blake
Robin Bloom
James Boudot
Randy Boudreaux
James Brickell
Tom Cheek
Sandy Claudell
Karen Cornwell
Richard Counts
Thomas Curry
David Daniel
Sue Darby
Rodney Davis
Steve Dziuban

Larry Erwin
Gary Fenton
David Fisk
Robert Frost
Tom Gardner
Jan Gaudin
Jay Gould
Jim Grayson
Rafael Gutierrez
Dave Hallowell
Jim Hecker
Phil Herre
Suellen Hill
John Hopkinson
Mike Jackson
Omer Jenkins
Bruce Johnson
Kelly Johnson
Bradley Jones
S.E. Jones
Joe Juarez
Ragu Kacker

Robert Kaliski
Carol Kiemele
Peter Knepell
Kenneth Knox
James Kogler
Tara Kroger
Jayme Lahey
Robert Lawson
David LeBlond
William Lesso
Dan Litwhiler
Jim Merritt
Douglas Montgomery
Ken Mooney
William Motley
Raymond Myers
Terry Newton
Dale Owens
Cliff Polo
Lisa Reagan
Jack Reece

James Riggs
C. B. (Kip) Rogers
Bill Russell
Jim Rutledge
Ewald Schelert
Vicki Schmidt
Doug Sheldon
Tamara Sheldon
Steve Smith
Gary Steele
David Swint
Tom Tanis
John Tomick
Barbara Warsavage
Dan Warwick
Dan Watson
Karen Welch
Rita Whiteley
Don Wilson
William Woodard
Barbara Yost

Cover by Bernard Sandoval, Muir Agency, Colorado Springs, CO

Contents

Case Study and Example Topics   xi

Chapter 1   Foundations

1.1	Intended Audiences for this Text	1-1
1.2	Quality and the Role of Statistical Techniques	1-2
1.3	What is Experimental Design	1-3
1.4	Why Use Experimental Design	1-7

- Cases 1, 2, 3, and 4
- Japanese success
- Deming philosophy
- Barriers to U. S. implementation of experimental design

1.5	Variation and Its Impact on Quality	1-16
1.6	Measurements of Quality	1-19
1.7	Organizing the Experiment	1-23

- Guidelines for brainstorming
- Steps for experimental design

1.8	Selecting the Quality Characteristic(s)	1-29
1.9	Selecting the Design Matrix (Set of Test Conditions)	1-30
1.10	Conducting the Experiment	1-33
1.11	Causality versus Correlation	1-34
1.12	A Quality Improvement Example	1-36
1.13	Taguchi Approach to Designed Experiments	1-52
1.14	Steps for Quality Improvement: Getting Started	1-53
1.15	U. S. Air Force Approach to Total Quality Management (TQM)	1-56

- R & M 2000 Variability Reduction Process

Problem Set		1-72
Bibliography		1-74

Chapter 2   Conducting Simple Experimental Designs and Analysis

2.1   Introduction   2-1

2.2   Experimental Design Example (Full Factorial)   2-2

- Steps
- Analysis of the example data

2.3   Fractional Factorial Designs for 2-Level Factors   2-24

2.4   Analysis of a 2-Level Fractional Factorial Design   2-29

- Analysis steps

2.5   Two-Level Screening Example   2-36

2.6   Example of a Full Factorial Design for 3-Level Factors   2-39

2.7   Example of Identifying System Variability   2-42

2.8   Using Software for Analysis   2-46

2.9   Summary   2-53

Problem Set   2-54

Bibliography   2-57

Chapter 3   Design Types

3.1   Introduction   3-1

3.2   Full Factorial Designs   3-6

3.3   Fractional Factorials for 2-Level Designs   3-10

3.4   Foldover Designs   3-20

3.5   Plackett-Burman Designs   3-21

3.6   Non-Orthogonal 2-Level Designs   3-24

3.7   Fractional Factorials and Latin Squares for 3-Level Designs   3-28

3.8   Box-Behnken Designs   3-30

3.9   Box-Wilson (Central Composite) Designs   3-34

3.10   Comparison of 3-Level Designs   3-40

3.11   Taguchi Designs   3-42

3.12   Design Selection Scenarios   3-49

3.13   Nested Designs   3-52

3.14   Mixture Designs   3-53

3.15   Design Type Summary   3-56

Problem Set   3-57

Bibliography   3-62

Chapter 4   Statistical Techniques

4.1   Introduction   4-1

4.2   Analysis of Variance   4-1

4.3   Two-Factor ANOVA   4-12

4.4   Introduction to Simple Linear Regression   4-21

4.5   Strength of the Linear Relationship and ANOVA for
       Simple Linear Regression   4-26

4.6   Prediction Using Simple Linear Regression   4-30

4.7   Polynomial Regression   4-33

4.8   Multiple Regression   4-36

4.9   Example Demonstrating the Need to Use Coded Data
       When Examining Second-Order Models   4-43

4.10   Crossvalidation   4-48

4.11   Notes of Caution   4-50

4.12   ANOVA for Factors with More Than 2 Levels (Optional)   4-52

4.13   Two-Factor ANOVA for 1 Factor at 2 Levels and
       1 Factor at 3 Levels (Optional)   4-54

4.14   Determining Sample Size (Optional)   4-61

4.15   Comparison of Confidence Interval, t-Test, ANOVA, and
       Regression Approaches for Analyzing a Factor with
       2 Levels (Optional)   4-64

Problem Set   4-69

Bibliography   4-74

Chapter 5   Analysis of Experimental Data

5.1	Introduction	5-1
5.2	Analyzing a 2-Level Design	5-2
5.3	Importance of Modeling Interactions	5-13
5.4	Identifying Dispersion Factors (Factors Which Shift the Response Variability)	5-16
5.5	Variance Reduction Case Study	5-24
5.6	Residual Analysis for Replicated or Unreplicated Designs	5-28
5.7	Robust Design Analysis	5-32
5.8	Variance Reduction Using Multi-Variate Charts	5-37
5.9	Three-Level Modeling Example	5-41
5.10	Three-Level Design Analysis and Comparison	5-47
5.11	A Monte Carlo Simulation for Comparing Dispersion Factor Identification Methods (Optional)	5-65
	Problem Set	5-74
	Bibliography	5-80

Chapter 6   Taguchi Philosophy, Design, and Analysis

6.1	Introduction	6-1
6.2	Taguchi Philosophy and Methodology	6-3
	6.2.1 Loss Function	
	6.2.2 Quality Engineering	
	6.2.3 Quality by Design	
	6.2.4 Controllables and Noise	
6.3	Taguchi Strategy of Experimentation	6-11
	6.3.1 Saturated Fractional Factorial Experimental Designs	
	6.3.2 Designs That Assess Information About Noise	
	6.3.3 Signal-to-Noise Ratio	
6.4	Example of a Taguchi Experimental Design	6-13

6.5   Comparison of Classical and Taguchi Approaches   6-21

Problem Set   6-27

Bibliography   6-28

Chapter 7   Optimization and Advanced Response Surface Methods

7.1   Introduction and Overview   7-1

7.2   Response Surface Methodology   7-7

Problem Set   7-18

Bibliography   7-19

Chapter 8   Case Studies

See Case Study and Example Topics   xi

Appendixes

A   Table of Random Digits
B   Binomial Probability Distribution
C   Poisson Probability Distribution
D   Standardized Normal Distribution
E   Percentiles of the t Distribution
F   Percentiles of the F Distribution
G   Percentiles of the χ^2 Distribution
H   Studentized Ranges q_T
I   Power of the Analysis-of-Variance Test
J   Answers to Selected Problems
K   Time Line for Design of Experiment Techniques
L   List of Symbols and Abbreviations
M   Rules of Thumb
N   DOE KISS Student Version Software Manual

Glossary

Index

Case Study and Example Topics

Results of a First Run Fractional Factorial for the TRE 10X
  Photo Process on CMOS III Gate Policy (Honeywell)   8-2

Reduction of Measurement Device Variability Using
  Experimental Design Techniques (Kurt Manufacturing)   8-7

Corrosion Study   8-10

Linear Motor Design (Digital Equipment Corp.)   8-12

Factorial Experiment for Bottom Electrode Stress
  (Ramtron Corporation)   8-19

Experimental Design on an Injection Molded Plastic Part   8-23

Consider the Metric: Dealing with Mixtures and
  Temperature Gradients (Honeywell)   8-28

Minimizing Variability (Honeywell)   8-37

Computer Network Performance Analysis (USAFA)   8-55

Optimizing the Anodize Process and Paint Adhesion
  for Sheet Metal Parts (Boeing Commercial Airplanes)   8-78

Building Paper Helicopters (USAFA)   8-81

Optimizing Lunar Concrete (USAFA)   8-87

Design of Experiments Silane Doping in GaAs (DoD Contractor)   8-102

Bonding Gold Wire to Gallium Arsenide Wafers:
  Overcoming Preconceived Notions (Cray Research)   8-110

Constant Velocity Joints on a Front Wheel Drive
  Analysis of Multiple Responses from Designed Experiments
  (Quinas, Inc.)   8-114

A Strategic Defense Case Study (Logicon/RDA)   8-133

Identifying a Plasma Etch Process Window (Honeywell and
United Technologies Microelectronics Center)   8-144

Applying Experimental Design Techniques to Operating
Systems and Software   8-156

Reducing Variability in a Machining Process: An Industrial
Application of Experimental Design (Maxcor, USAFA)   8-160

The Effect of the CE Rivet H Parameter on Head Protrusion
(Boeing Military Airplanes)   8-168

Catapulting Statistics into Engineering Curricula
(Univ. of Alabama, USAF Academy, HQ USAF)   8-181

Environmental Engineering Operation of an Activated Sludge System
(USAF Academy, Dept. of Civil Engineering)   8-188

Minimizing the Number of Large Scale Computer Simulation Runs:
The Testing, Validation, and Implementation of the Interagency
Working Group (IWG) for Acquired Immune Deficiency
Syndrome (AIDS) Model (Logicon RDA, USAF Academy)   8-197

Karl Fisher Moisture Designed Experiment (Cell Technology)   8-204

The End Count Method - Power with Small Samples
(Digital Equipment Corp.)   8-212

A Designed Experiment to Reduce Circular Runout
(Ampex Recording Systems)   8-218

Ball Weld Experiment (Morton International)   8-229

Application of DOE to an Analog Filter Design
(Texas Instruments, Inc.)   8-235

A Robust Active Filter Design (Texas Instruments, Inc.)   8-250

Identifying Design Standards for Improved PWA Solderability
(Texas Instruments, Inc.)   8-262

Chemiluminescence: Can We Rule Out 3-Way Interactions?
(Abbott Laboratories)   8-272

Chapter 1

Foundations

1.1   Intended Audiences for this Text

This text is written for managers, engineers, scientists and technicians. It is intended to bridge the gap between texts that are too mathematical and those that attain simplicity by omitting important concepts. Mathematical and statistical theory are not emphasized; rather, the presentation is intended to provide conceptual understanding of designed experiments and related topics in statistical quality control. For years statisticians have provided us with textbooks written primarily for those well versed in statistical jargon and mathematics and aimed at understanding statistical theory. We recommend those texts for anyone seeking more depth into these subjects. This text, however, is geared to enhancing the art of statistical thinking.

The presentation of material is arranged to include a foundation for understanding experimental design (Chapter 1); an introduction to design and analysis (Chapter 2); a discussion of design types (Chapter 3); a review of basic statistical concepts and techniques (Chapter 4); an analysis of experimental data, variance reduction, and robust designs (Chapter 5); a presentation of Taguchi philosophy and analysis techniques (Chapter 6); an introduction to optimization and response surface methods (Chapter 7); and numerous case studies from industry (Chapter 8).

1.2   Quality and the Role of Statistical Techniques

The American Society for Quality Control has defined quality as "the totality of features and characteristics of a product or service that bear on its ability to satisfy stated or implied needs." A simple interpretation of this definition is that quality is a measure of how well the user's needs are met. The user's interpretation of quality focuses on whether the product functions as advertised with no variation in performance. There are several aspects of quality: the quality of product knowledge gained in the research and development phase, the quality of the product design, the quality of the process design, the quality of incoming parts, and the quality of the manufactured product. Research and development departments are typically tasked with discovering new technologies and determining their feasibility. It is here where knowledge must begin to be accumulated. Such knowledge should include identification of the critical parameters (factors). The most efficient way to identify these factors is through designed experiments. In the design phase of the product or process, it is important to test which factors affect the quality of the design and determine the settings of these factors that optimize the desired output. In this phase, experimental designs and response surface techniques are used to gain knowledge on how to produce a high quality, low cost product. The quality of incoming parts and the quality of the manufacturing process are typically achieved through the use of statistical process control techniques; however, on-line experimental designs have also produced excellent results in these areas.

This text will focus on quality improvements through the applications of experimental design and present both sides of the controversy between Taguchi methods and classical techniques. The authors are not committed to either side, but advocate an approach which finds the "best" method for the practitioner to solve the problem at hand. Included are new ideas on how to combine certain aspects of both approaches in order to produce a synergism leading to improved methods of experimental design. Emphasis is on quality improvements and not on selling one approach or another. Also included are some important contributions from Shainin [2]. All approaches are presented using a Keep It Simple Statistically (KISS) philosophy.

1.3   What is Experimental Design

Experimental design consists of purposeful changes of the inputs (factors) to a process (or activity) in order to observe the corresponding changes in the outputs (responses). The process (or activity) is defined as some combination of *machines, materials, methods, people, environment,* and *measurement* which, when used together, perform a service, produce a product, or complete a task. Thus, experimental design is a scientific approach which allows the researcher to gain knowledge in order to better understand a process and to determine how the inputs affect the response(s).

Graphically, a process appears as shown in Figure 1.1. Some examples of processes are shown in Figures 1.2 through 1.7. Obviously, there exist many different kinds of processes; the ones provided here are from various applications of designed experiments.

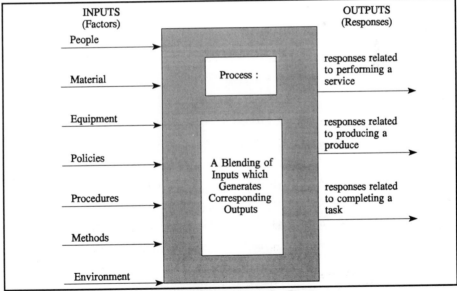

Figure 1.1 Illustration of a Process

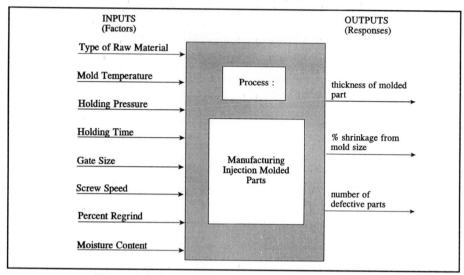

Figure 1.2 Manufacturing Injection Molded Parts

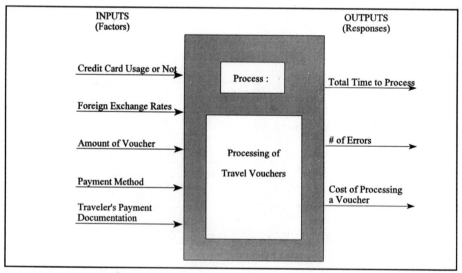

Figure 1.3 Processing of Travel Vouchers

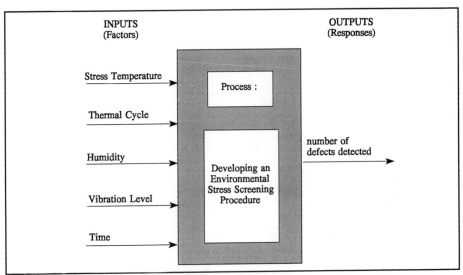

Figure 1.4 Developing an Environmental Stress Screen Test

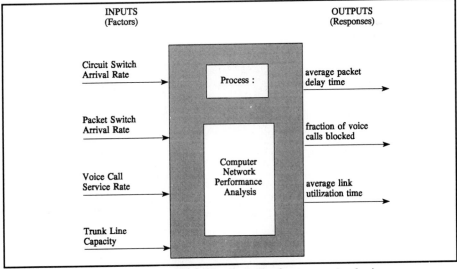

Figure 1.5 Computer Network Performance Analysis

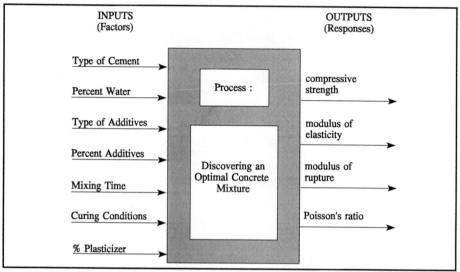

Figure 1.6 Discovering an Optimal Concrete Mixture

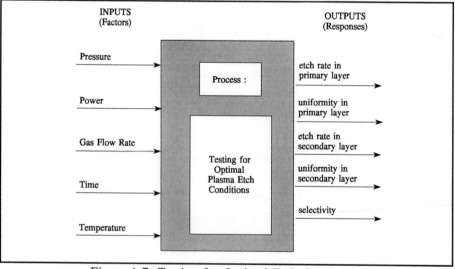

Figure 1.7 Testing for Optimal Etch Conditions

1.4 Why Use Experimental Design

Regardless of where one works (research, development, design, quality, test, reliability, manufacturing, packaging, etc.), the task for most scientists and engineers is to gain, document, and transfer product and/or process knowledge. True knowledge is more than theory or "gut feel." To really understand a process (or activity) one needs facts and data. Collecting data through the use of one-factor-at-a-time experimentation and/or a series of trial and error tests has resulted in very inefficient and ineffective attempts to understand and optimize product designs and processes. No longer can these archaic methods be tolerated if a company intends to survive in a highly competitive market.

Managers should be interested in experimental design because it will assist their people in gaining knowledge to: (1) improve performance characteristics; (2) reduce costs; and (3) shorten product development and production time. Improved performance characteristics result from the identification of the critical factors that optimize the mean response, minimize response variability, and provide robustness. This improved performance also leads to the reduction of scrap and rework, which greatly reduces costs. Understanding which factors are critical to improve performance will indicate which factors must be controlled and what their tolerances should be. Identifying the unimportant factors will allow one to widen tolerances and/or chose factor levels resulting in lower product manufacturing costs. Reduced time to market is accomplished by understanding what the customer really wants and using the most efficient and effective experimentation procedures to gain the knowledge required for product/process optimization.

For the engineer or scientist who needs to meet a manager's expectations in a timely manner, experimental designs are used as: (1) the most effective method for identifying the key input factors; (2) the most efficient way to gain an understanding of the relationship between the input factors and the response(s); (3) a method for building a mathematical model relating the response to the input factors, which is often referred to as process/product characterization; (4) a means of determining the settings of the input factors which optimize the response and minimize costs; and (5) a scientific method for setting tolerances.

Perhaps the easiest way to convey the importance of learning experimental design is to review some alternative methods for collecting data and discuss their weaknesses. The reader is presented three cases which sequentially illustrate the pitfalls associated with historical data, one-at-a-time designs, and full factorials. In addition, Case 4 reveals the need for improvement beyond traditional techniques in the pursuit of robust designs.

Case 1 (HISTORICAL DATA): A plant manager wishes to know if changes in temperature during a wire bonding process are related to the number of defects of the product. Since there was an abundance of historical or "happen-stance" data available, it was decided to use simple linear regression (see Chapter 4 for a detailed discussion) to correlate temperature and defect rate. The regression results indicate that a high positive correlation exists between these two variables, implying that as temperature increases the defect rate also increases (see the scatter diagram in Figure 1.8).

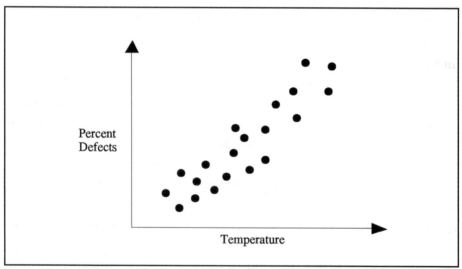

Figure 1.8 Scatter Diagram from Historical Data

Since the goal was to minimize defects, the engineer directed that temperature be maintained at some appropriate low level. What was not revealed from the data analysis was undocumented intervention in the production process by the operator. If the defect rate started to increase (for any reason), the operator (based on years of experience) would raise the temperature to minimize this increase. After the number of defects started to decline, he would reduce the temperature back to the standard operating setting. A low temperature had originally been specified to keep costs down; however, no experimentation had taken place to accurately examine the cause-and-effect relationship between temperature and defect rate. Analyzing this non-experimental data results in a confounding of temperature and operator manipulation which makes it impossible to separate the effects of the two inputs to the process. The apparent positive correlation of temperature and defect rate did not represent the true relationship, but rather a relationship induced by the operator. Historical data, although often valuable, can lead to this type of faulty conclusion because it probably does not provide sufficient information about how the data was collected and it frequently contains factors which are highly interrelated.

Case 2 (ONE-AT-A-TIME DESIGN): The plant manager is now convinced that historical data will not always provide a clear understanding of the process. In addition, he has been consulted on the problems associated with response variables which are attributes (defective/non-defective). Thus, bond strength has been selected as a better response. Next, the manager has decided to have the engineer set up an experiment to test temperature and pressure effects on bond strength. The experiment consists of the following: (1) select two levels of temperature, T_1 and T_2, and two levels of pressure, P_1 and P_2; and (2) holding pressure constant at level P_1, test the product at both T_1 and T_2 as shown in Table 1.1.

Pressure	Temperature	Response (y)
P_1	T_1	80
P_1	T_2	90

Table 1.1 One-At-A-Time Approach (part 1)

Since the objective is to maximize bond strength, the more desirable response occurred for T_2. Therefore, the best temperature setting appears to be T_2. The next step is to set temperature equal to T_2 and compare the bond strength at pressure P_1 and P_2. The results are displayed in Table 1.2.

Pressure	Temperature	Response (y)
P_1	T_2	90
P_2	T_2	100

Table 1.2 One-At-A-Time Approach (part 2)

At first glance one might say that the ideal settings of temperature and pressure are T_2 and P_2, resulting in a bond strength of 100 (see Figure 1.9a); however, this experimental method, referred to as a "one-at-a-time" design, did not allow for testing the interaction effect of temperature and pressure. Suppose that the untested combination P_2T_1 would produce a result as displayed in Figure 1.9b.

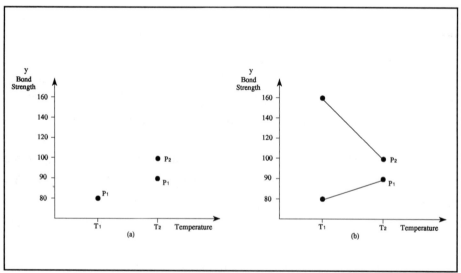

Figure 1.9 Graph of Temperature × Pressure Interaction

Figure 1.9b indicates that the relationship of y with temperature has a different slope depending on which pressure setting (P_1 or P_2) is used. Therefore, one can conclude that temperature and pressure interact. The interaction effect of temperature and pressure is interpreted as follows: as temperature changes from T_1 to T_2, the change in bond strength is different, depending on the pressure settings of P_1 and P_2 (i.e., the lines in the graph are non-parallel).

A more appropriate design would be to test all four combinations of temperature and pressure, referred to as a full factorial design (discussed further in Chapters 2 and 3).

Case 3 (FULL FACTORIAL DESIGN):

If one has only a few input factors to test, oftentimes a full factorial design (i.e., all possible combinations) is the best choice. However, assume that after understanding the pitfall discovered in Case 2 and brainstorming which other factors could potentially affect the bond strength, the manager requests further testing of temperature, pressure, and ten other input factors of interest. If all combinations are to be tested for each factor measured at two levels, there will be 2^{12} or 4096 experimental conditions in the full factorial design. This presents a new problem because of resource limitations in conducting the experiment. Fortunately, there is a solution to the problem. The material in this text includes fractional factorials and Plackett-Burman screening designs as efficient alternatives to full factorial designs. Obviously, you cannot get something for nothing. You'll see in Chapter 3 what you give up when conducting a fractional factorial of 32 experimental combinations versus the full factorial of 4096.

Case 4 (ROBUST DESIGN):

The manager and his engineering staff gave the 12 factors in Case 3 some more thought and they agree that 3 of the 12 factors cannot be controlled in the operating environment of the product. In other words, these three factors are considered uncontrollable or noise factors. What is needed is a design to determine the settings of the 9 controllable factor settings such that the variability of bond strength across the 3 uncontrollable factors is minimized. The solution to their problem is found in the study of robust designs which are described in Chapters 5 and 6. Basically, the goal is to find the settings of the 9 controllable factors that will make the process insensitive or robust to variation in the uncontrollable factors.

These four cases demonstrate that, unless experimenters are well informed on all the statistical tools available and their experiments are well designed, results can be misleading and large amounts of resources can be consumed unnecessarily.

An even more compelling reason for educating American industrial managers and engineers in the use of designed experiments is based on the need to compete with countries who have successfully adopted these techniques. In 1964, American industries were basking in a 6 billion dollar trade surplus, not realizing they were under attack by aggressive overseas companies. By 1984, the U.S. was looking at a 123 billion dollar trade deficit. Over this 20 year period, U.S. productivity increased a mere 35% as compared to about 60% for some European countries and an amazing 120% for Japan. The same country that had a reputation for making "junk" in the 50's and 60's was threatening the survival of U.S. industry in the 70's and 80's by producing high quality products at a lower price. How did this happen? It is becoming well known that much of this remarkable change in the quality of Japanese products is attributed to designing quality into the product through the use of designed experiments and monitoring quality during production by way of statistical quality control. Ironically, these techniques were developed mostly in the U.S. and Europe over 60 years ago, but U.S. companies failed to implement them on a continuous basis. In the 1950's, American statisticians, having been unsuccessful at convincing most of industry to adopt these methods, directed their efforts towards new theory instead of industrial implementation. U.S. industries became less and less interested in statistics primarily because there was little post World War II competition to warrant a quality improvement focus. Thus, industry in this country entered a stagnant mode with regard to quality. After World War II, the Japanese were very receptive to the ideas of two American statisticians, W. Edwards Deming and Joseph M. Juran. As a result, the Japanese soon became the quality experts and their companies prospered. The few American companies that did implement statistical techniques did not do so with the intensity seen in Japan. Japanese managers view the use of statistical methods as very important to their companies' success. Thus, everyone in the company, from management to on-line workers, are taught these techniques in training sessions and they all use them not just to monitor quality in production, but to steadily improve quality throughout the production process. This philosophy was initiated through Deming's 14 points which serve as a guide for management on how to improve quality and productivity.

The following 14 points of Deming are taken from Hogg and Ledolter [4] and can be further studied in Deming's book *Out of the Crisis* [3].

(1)   Create a constancy of purpose toward the improvement of product and service. Consistently aim to improve the design of your products. Innovation, money spent on research and education, and maintenance of equipment will pay off in the long run.

(2)   Adopt a new philosophy of rejecting defective products, poor workmanship, and inattentive service. Defective items are a terrible drain on a company; the total cost to produce and dispose of a defective item exceeds the cost to produce a good one, and defectives do not generate revenues.

(3)   Do not depend on mass inspection because it is usually too late, too costly, and ineffective. Realize that quality does not come from inspection, but from improvements on the process.

(4)   Do not award business on price tag alone, but consider quality as well. Price is only a meaningful criterion if it is set in relation to a measure of quality. The strategy of awarding work to the lowest bidder has the tendency to drive good vendors and good service out of business. Preference should be given to reliable suppliers that use modern methods of statistical quality control to assess the quality of their production.

(5)   Constantly improve the system of production and service. Involve workers in this process, but also use statistical experts who can separate special causes of poor quality from common ones.

(6)   Institute modern training methods. Instructions to employees must be clear and precise. Workers should be well trained.

(7)   Institute modern methods of supervision. Supervision should not be viewed as passive "surveillance," but as active participation aimed at helping the employee make a better product.

(8)   Drive out fear. Great economic loss is usually associated with fear when workers are afraid to ask a question or to take a position. A secure worker will report equipment out of order, will ask for clarifying instructions, and will point to conditions that impair quality and production.

(9)   Break down the barriers between functional areas. Teamwork among the different departments is needed.

(10)   Eliminate numerical goals for your work force. Eliminate targets and slogans. Setting the goals for other people without providing a plan on how to reach these goals is often counterproductive. It is far better to explain what management is doing to improve the system.

(11)   Eliminate work standards and numerical quotas. Work standards are usually without reference to produced quality. Work standards, piece work, and quotas are manifestations of the inability to understand and provide supervision. Quality must be built in.

(12)   Remove barriers that discourage the hourly worker from doing his or her job. Management should listen to hourly workers and try to understand their complaints, comments, and suggestions. Management should treat their workers as important participants

in the production process and not as opponents across a bargaining table.

(13) Institute a vigorous program of training and education. Education in simple, but powerful, statistical techniques should be required of all employees. Statistical quality control charts should be made routinely and they should be displayed in a place where everyone can see them. Such charts document the quality of a process over time. Employees who are aware of the current level of quality are more likely to investigate the reasons for poor quality and find ways of improving the process. Ultimately, such investigations result in better products.

(14) Create a structure in top management that will vigorously advocate these 13 points.

Other philosophies such as those of Juran, Crosby, and Taguchi may vary slightly from Deming, but all appear to center on four main aspects which lead to improved quality: **customer satisfaction, teamwork, statistical methods**, and **continuous improvement.**

There are many reasons why most of U.S. industry has been slow to implement the techniques and philosophies described above. The authors suggest the following 6 reasons:

(1) Failure of management to recognize the value of these techniques.

(2) Failure of statisticians to teach these techniques in an easy to understand fashion.

(3) Misperception that statistically designed experiments are costly, time-consuming, and impractical.

(4) Systems of reward which provide financial incentives primarily for short term planning and decision making.

(5) Misperception that quality is achieved by merely meeting specifications.

(6)   Until recent years, a lack of competition which imparted an air
of complacency within U. S. industry.

1.5   Variation and Its Impact on Quality

It should not be a surprise to the reader that, even in well controlled
laboratories or production processes, there will be variation in the final product.
The variation is caused by either uncontrolled factors or production noise. Too
much variation degrades the quality of the product and causes a loss to the
company.   Historically, the approach to this problem has been to set up
specification limits and perform inspections of finished products to ensure zero
defects out the back door. This approach will cause the company to classify the
product as either acceptable (within spec) or unacceptable (out of spec). Loss to
the company is based on being out of spec as shown in Figure 1.10. If large
numbers of the product are produced outside specifications, and if products are not
properly inspected before they are shipped, then (1) complaints will increase,
(2) extra resources will be expended to repair items under warranty, and (3)
eventually customers will become discouraged and seek a more reliable product.

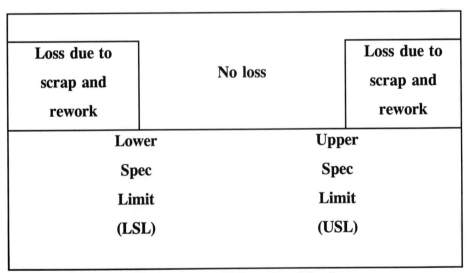

Figure 1.10 Old Philosophy of Quality

The optimal solution to this problem is not to rely on inspections at the end of the production line. According to Juran [6], these inspections are only 80% effective. Furthermore, this approach requires more manpower and results in rework and scrap costs that can become prohibitive. In addition, maintaining profits under these conditions will require sales prices to increase, resulting in a continued decrease in consumer satisfaction. It is also unreasonable to assume that a product just inside a specification limit results in no loss to the company. For example, consider building window glass and window frames. If the glass is at the lower limit and the frame at the upper limit (or vice versa) the quality of the product is reduced.

To develop a quality product at a reduced cost, quality must be designed into the process [13] and focus must be on the target (nominal) value instead of just being "in spec" as depicted in Figure 1.11. Deviations from the target result in loss to the producer in the form of inspection, scrap, rework, warranty costs, increased cycle time, design changes, tied up capital, increased inventories, etc. Losses to the purchaser are due to degraded performance and reliability, as well as increased maintenance costs. Eventually the purchaser (customer) will find a better producer (supplier) and, thus, the ultimate loss to the producer is reduced market share.

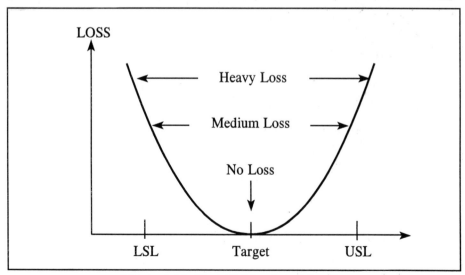

Figure 1.11 New Philosophy of Quality

To minimize the loss incurred due to deviation from the target, the variability of the product about the target must be reduced as shown in Figure 1.12. Product B has a much higher proportion of measured values close to the target than product A. Therefore, Product B will result in less overall loss. From whom would you rather buy, the producers of Product A or Product B? What about your customers? Do they see your product as one that looks like A or B?

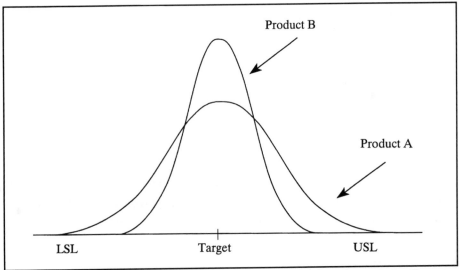

Figure 1.12 Higher Quality Implies Variance Reduction Around the Target

Whether you are in research, development, design, manufacturing, packaging or any other process related job, variance reduction is best achieved through a thorough understanding of the relationships between the output variables (responses or quality characteristics) and the input variables (factors or parameters). Two simple yet very important tools that should be used prior to collecting data are the process flow diagram and the cause-and-effect diagram [7]. If properly developed, a cause-and-effect diagram (see Figure 1.25) will contain a list of all possible sources (causes) of variation in the response (effect). These inputs should then be partitioned into those that should be held constant, those that cannot be held constant (noise) and those that represent the key factors to investigate. All identified sources of variation that can be easily removed at low cost should be dealt with prior to any experimentation in order to bring the process into an approximate state of control, (i.e., consistency). This will usually require a large number of inputs on the cause-and-effect diagram to be held constant. Holding these inputs constant will necessitate some type of documented operating procedure.

A designed experiment will provide the most efficient set of testing conditions of the key input factors in order to:

1)   Develop simple models that represent the relationship and sensitivity of the outputs with the inputs,

2)   Determine the critical input factors,

3)   Determine optimal factor settings.

A properly constructed process flow diagram and cause-and-effect diagram with a subsequent designed experiment will typically result in a large amount of variance reduction and thus improved quality. Ideally, these tools should be used in the early phases of product and process design. Companies that wait until manufacturing to solve product problems will soon be unable to compete.

1.6   Measurements of Quality

After a product has been properly designed and its manufacturing process has been brought into statistical control or stabilized (i.e., all special causes of variation have been detected and eliminated), we can assess the capability of the process through the use of metrics such as the C_p or C_{pk} indices. The C_p index (referred to as the process potential) was designed to compare the variability of some quality characteristic (y) with the distance between the lower specification limit (LSL) and the upper specification limit (USL). It assumes the average of the characteristic $(\overline{y})$ is equal to the desired target value (T), i.e., the process is centered between the specs. The formula for C_p is shown in Equation 1.1.

$$C_p = \frac{\text{specification width}}{\text{process width}} = \frac{\text{USL} - \text{LSL}}{6\hat{\sigma}}$$

$$\text{where } \hat{\sigma} = \sqrt{\frac{\sum_{1}^{n} (y_i - \bar{y})^2}{n-1}} \quad \text{and} \quad \bar{y} = \frac{\sum_{1}^{n} y_i}{n}$$

(Throughout this text, the symbols $\hat{\sigma}$ and s are both used to represent sample standard deviation; $\hat{\sigma}^2$ or s^2 are used to represent sample variance.)

Equation 1.1 Computing the C_p Index

According to acceptable standards in industry, C_p values less than 1.00 are unacceptable, values between 1.0 and 1.33 are marginally acceptable, and values greater than 1.33 are desired. Many quality oriented companies such as Ford Motor Company and Motorola are now requiring C_p values greater than or equal to 2.0. When C_p = 2.0 the process is defined by Motorola to have 6σ quality.

Since the C_p index does not take into account $\bar{y}$ deviations from the target, T, a more realistic estimate of the actual state of quality is the C_{pk} index computed as shown in Equation 1.2. Interpretation of C_{pk} values is similar to that described for C_p. Notice that once the process average is on the target, the only possible way to improve C_{pk} or C_p is through variance reduction. Companies that are successfully meeting their customer needs are able to estimate C_p and/or C_{pk} prior to manufacturing. Thus, if a design does not meet design for manufacturability criteria, it can be improved before being transferred to manufacturing.

$$C_{pk} = \text{minimum of} \left(\frac{\text{USL} - \bar{y}}{3\hat{\sigma}} \right) \text{ or } \left(\frac{\bar{y} - \text{LSL}}{3\hat{\sigma}} \right)$$

Equation 1.2 Computing the C_{pk} Index

A third measure of product quality is the number of defects per million (dpm). Obviously we need to minimize dpm in order to ensure high levels of quality. For attribute (pass/no pass) type quality characteristics, one would simply count the number of defects per number of opportunities and multiply by 10^6 to obtain an estimate of dpm. For quality characteristics that are quantitative, we can estimate dpm from our previous capability measurement, C_p. The relationship between C_p and dpm is shown in Table 1.3.

C_p	dpm
.50	133,600
.67	45,500
.80	16,400
.90	6,900
1.00	2,700
1.33	66
1.67	<1
2.00	.0018
3.00	<<1
4.00	<1 defect per billion

This information was obtained from an R&M 2000 Variability Reduction Process slide prepared by HQ USAF/LE-RD.

Table 1.3 Relationship Between C_p and dpm*

*The dpm shown for the given C_p assumes the process average is centered at the target which is the middle of the 2-sided spec limits. For shifted processes or 1-sided spec processes, C_{pk} is preferred over C_p. The disk in the back of this text contains software which will enable you to enter the mean, standard deviation, lower spec limit, and upper spec limit; and then the program will compute C_{pk} and dpm.

Taguchi has developed a loss function which associates a dollar value with the current state of process/product quality. This dollar amount can be used to identify key processes/products to be improved and to evaluate the amount of improvement. The quadratic loss function shown in Figure 1.13 will, in many cases, approximate the true loss due to deviations from the target. Actual losses result from a combination of inspection, scrap, rework, poor performance, lack of customer satisfaction, design changes, increased cycle time, tied-up capital, inventories, loss of market share, etc.

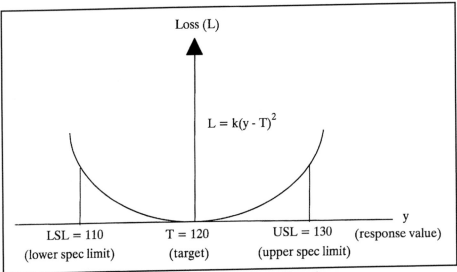

Figure 1.13 Taguchi Loss Function

The measured loss for a single product can be estimated using $L = k(y - T)^2$ where y is the response value, T is the target value, and k is the monetary constant. To determine k, we need to estimate the loss for any specific value of y. For example, if the estimated loss for y = 130 is \$100.00, then $k = 100 \div (130 - 120)^2 = 1.0$. The loss function $L = 1.0(y - T)^2$ can now be used to estimate the loss associated with other y values.

Another use of the quadratic loss function is the determination of average loss per product. Using expected loss to depict average loss, $\overline{L}$, it is shown in Chapter 6 that $\overline{L} = k(\sigma_y^2 + (\overline{y} - T)^2)$. If average loss is multiplied by the number of products per unit of time, the result will be total loss in dollars for the same time period.

Whether one uses the loss function or some other method to estimate loss, the Cost of Poor Quality (COPQ) in dollars should be determined for each project or activity. This information will help you prioritize your improvement efforts and motivate the team working the problem.

The route to improved C_p, C_{pk}, dpm, or average loss is the same. Improvements will occur as we reduce the variance, σ_y^2, and reduce the deviation of the response average from the designated target, i.e., reduce $|\overline{y} - T|$. The use of designed experiments can assist the practitioner in determining which inputs (factors) shift the average of the response, which shift the response variability, which shift both the average and the variability, and which have no effect on either the average or the variability of the response. This type of knowledge at the research, development, or design phase allows for a technology transfer which is manufacturable. In the manufacturing phase, the knowledge of input factor influence on the response can be used for achieving continuous process improvement.

1.7   Organizing the Experiment

The reason for any designed experiment is to provide data that can be used to answer well thought out questions. Failure to allocate sufficient time and thought in formulating these questions often results in wasting resources throughout the remainder of the experimental process. In many situations, up to 50% of the overall effort should go into the planning phase of the experiment. It is important that all of the key players (management, engineers, experts in the area, operators, and analysts) be involved in the planning phase. The goals should be specific, and statements of how the goals will be attained (objectives) must be complete.

One of the most effective approaches to the planning phase is the use of the "brainstorming" technique. This technique consists of an uninhibited method for creating ideas that pertain to the goals, objectives, and variables involved. Through the use of brainstorming and a team effort, there is a higher likelihood that the planning phase will be a success. If the planning phase fails, there is little chance of salvaging the experiment and further efforts could be a waste of resources. Figure 1.14 presents the guidelines for brainstorming that are given in Barker's text, *Quality by Experimental Design* [1]. While the brainstorming process appears simple, attention to a few additional details can help ensure success. Brainstorming groups should be small, i.e., 4 to 10 people. Someone should record the ideas and everyone should participate. Ideas are not to be challenged until after the brainstorming session is over. When the brainstorming is complete we should have the following: 1) a detailed process flow diagram depicting all the steps in the process; 2) a list of all the outputs (response variables) and possibly their customer driven specifications; 3) a completed cause and effect diagram containing every variable known to exist in the process; 4) every variable in step 3) labeled as either a constant (C) with standard operating procedures (SOPs) on where and how to hold them constant, a noise variable (N) which is uncontrolled, or an experimental variable (X). For the X variables, called factors, we should know the range over which we want to test them and whether we will test them at 2 or 3 levels.

At the completion of the brainstorming session, a form similar to the one shown in Figures 1.15 and 1.16 should be prepared and updated throughout the remainder of the experiment. The form is an expanded version from that of Barker [1]. The actual form you use may be a modification of this one; however, it is imperative that a similar document be used to facilitate an organized flow of events and provide a common basis for documentation and communication.

GUIDELINES FOR BRAINSTORMING

TEAM MAKEUP
Experts
"Semi" experts
Implementers
Analysts
Technical staff who will run the experiment
Operators

DISCUSSION RULES
Suspend judgement
Strive for quantity
Generate wild ideas
Build on the ideas of others

LEADER'S RULES FOR BRAINSTORMING
Be enthusiastic
Capture all the ideas
Make sure you have a good skills mix
Push for quantity
Strictly enforce the rules
Keep intensity high
Get participation from everybody

Figure 1.14 Guidelines for Brainstorming

STEPS FOR EXPERIMENTAL DESIGN

I. STATEMENT OF THE PROBLEM:

(The problem should be related to customer needs. If possible, this step should also include an estimate of your current level of quality by way of cycle time, C_{pk}, dpm, total loss, or Cost of Poor Quality (COPQ). This estimate will then be compared with improvements found after Step XII.)

II. OBJECTIVE OF THE EXPERIMENT:

(What do you want this experiment to do for you?)

III. START DATE: _____   **END DATE:** _____

IV. SELECT QUALITY CHARACTERISTICS (also known as responses, dependent variables, or output variables). These characteristics should be related to customer needs and expectations.

RESPONSE	TYPE	ANTICIPATED RANGE	How will you measure the response? Is the measurement method accurate and precise?
1.			
2.			
3.			

V. COMPLETE A LITERATURE REVIEW, PROCESS FLOW DIAGRAM, AND CAUSE/EFFECT DIAGRAM. FROM THE CAUSE/EFFECT DIAGRAM SELECT FACTORS (also known as parameters or input variables) which are anticipated to have an effect on the response. Write Standard Operating Procedures (SOPs) for all variables that are to be held constant.

FACTOR	TYPE	CONTROL-LABLE OR NOISE	RANGE OF INTEREST	LEVELS	ANTICIPATED INTERACTIONS WITH	HOW MEASURED
1.						
2.						
3.						
4.						
5.						

Figure 1.15 Part I of Steps for Experimental Design

STEPS FOR EXPERIMENTAL DESIGN (Continued)

VI. DETERMINE THE NUMBER OF RESOURCES TO BE USED IN THE EXPERIMENT.
(Consider the desired number, the cost per resource, time per experimental trial, the maximum allowable number of resources, and Rules of Thumb (Appendix M) recommendation.)

VII. WHICH DESIGN TYPES AND ANALYSIS STRATEGIES ARE APPROPRIATE?
(Discuss advantages and disadvantages of each. See Rules of Thumb (Appendix M) recommendation.)

VIII. SELECT THE BEST DESIGN TYPE AND ANALYSIS STRATEGY TO SUIT YOUR NEEDS.

IX. CAN ALL THE RUNS BE RANDOMIZED?
WHICH FACTORS ARE MOST DIFFICULT TO RANDOMIZE?

X. CONDUCT THE EXPERIMENT AND RECORD THE DATA. (Monitor both of these events for accuracy.)

XI. ANALYZE THE DATA, DRAW CONCLUSIONS, MAKE PREDICTIONS, AND DO CONFIRMATORY TESTS.

XII. ASSESS RESULTS, MAKE DECISIONS, AND DOCUMENT YOUR RESULTS. (Evaluate the new state of quality and compare with the quality level prior to the improvement effort. Estimate your return on investment. If necessary, conduct more experimentation.)

Figure 1.16 Part II of Steps for Experimental Design

It is important that any experimental effort be aimed at research, development, design, or production problems where there is the most to be gained. For example, assume a product consists of components A, B, C, D, and E. Failures of this product are the "company's" number one concern. Management directs the use of experimental design to solve the problem. You could attempt to design an experiment with parameters from all components; or you could use historical data such as that presented in the Pareto diagram of Figure 1.17 to identify the most critical component. Additional information on cost per defect type should also be considered.

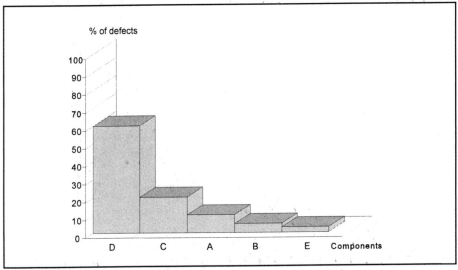

Figure 1.17 Pareto Diagram of the Contribution of
Each Component to the % of Defectives

Figure 1.17 reveals that component D contributes to 60% of the defects, component C contributes to 20% of the defects, etc. Since component D contributes the most to product defects, the best approach would be to build an experimental design to (1) determine the cause of defects within component D, and (2) improve the quality of component D, thereby improving the product quality. After completing the work on product D, the product quality is again monitored and defects categorized as to component contributions. If necessary, further experimentation can be conducted. This process of continuous improvement will lead to a high quality item. Other Pareto diagrams on estimated cost of correcting components or estimated return from an improvement effort might also prove valuable. In summary, we will want to attack that aspect of the product/process where the most gains can be made.

1.8   Selecting the Quality Characteristic(s)

Selecting the appropriate quality characteristic(s), also referred to as the response(s), is critical to good experimentation. For example, quantitative responses that are continuous will typically provide more information and require fewer resources than non-continuous or categorical type responses. If the response comes from a set of ordered categories (e.g., poor, fair, good, very good, excellent), we can assign the categories ordered integers (e.g., 1, 2, 3, 4, 5) and conduct the analysis in the same manner described in Chapters 2, 4, and 5. However, if binary responses (e.g., defective/non-defective) are used, we will need larger numbers of replications of each experimental condition in order to avoid ambiguous results. The authors have found that, on occasion, continuous variables highly correlated to the binary response can be used in the analysis, resulting in large savings of experimental resources. For example, the vibration of a device during processing can be highly related to whether the device will be defective or non-defective. Therefore, finding the optimal input values which minimize vibration will also minimize defects, but with fewer experimental resources. Some useful guidelines [11] in the selection of the proper response are as follows:

(1)   The quality characteristic should be related as closely as possible to the basic physical mechanism of the technology.

(2)   Use continuous responses if possible.

(3)   Use quality characteristics which can be measured precisely, accurately, and with stability. (The authors have found this can be a major stumbling block in effective experimentation with leading-edge technologies. Further brainstorming may be needed

to resolve this problem.)

(4) Ensure the quality characteristics cover the important aspects of the ideal function of the technology with respect to customer needs and expectations.

(5) Exercise a liberal blending of technical knowledge, information on customer requirements, statistical prowess, and common sense.

(6) For complex systems, select responses at the subsystem level and run experiments at this level before attempting overall system optimization.

1.9   Selecting the Design Matrix (Set of Test Conditions)

Although several issues should be addressed in choosing the best design matrix, one of the primary determining criteria is the number of factors to be investigated. Appendix M provides a set of "Rules of Thumb" on how to get started in choosing the proper matrix along with an ideal sample size per trial or run in the matrix. Basically, there are two concerns in design matrix selection:

(1)   a screening versus a modeling strategy, and

(2)   a linear (2 testing levels of each factor) versus a non-linear (3 testing levels of each factor) relationship of the output with the inputs (factors).

All the details for design matrix selection are provided in Chapter 3. At this point, however, the reader should become familiar with the basics. As our first example, consider a proposed experiment with two key factors. If each factor is to be tested at 2 levels, the full factorial design (set of all possible combinations) will consist of $2^2 = 4$ trials (runs). If each factor has a low (L) setting and a high (H) setting to be tested, then a tree diagram for two factors (A and B) will appear as shown in Figure 1.18.

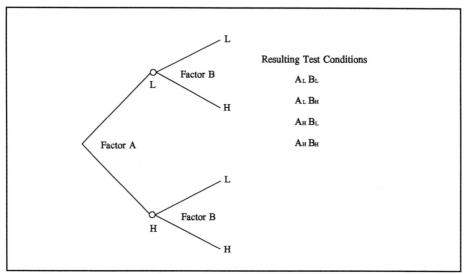

Figure 1.18 Tree Diagram for Testing 2-Factors, Each at 2 Levels

The set of 4 trials (runs) in Figure 1.18 can be displayed in a design matrix as shown in Table 1.4.

Run	A	B
1	L	L
2	L	H
3	H	L
4	H	H

Table 1.4 Design Matrix for Testing 2-Factors, Each at 2 Levels

Assume that factor A is Temperature where $A_L = 100°C$ and $A_H = 200°C$, and factor B is Time where $B_L = 5$ minutes and $B_H = 10$ minutes. A design matrix containing the actual factor settings is displayed in Table 1.5.

Run	Temperature (°C) A	Time (min) B	y_1 y_2 ..... y_9	$\bar{y}$	s
1	100	5			
2	100	10			
3	200	5			
4	200	10			

Table 1.5 2-Factor Design Matrix

The y_1 through y_9 columns represent the number of replicates or samples to be made and measured per run. If the process produced metal casted nuts and the output was nut thickness, then the columns y_1 through y_9 would contain the thickness values of 9 nuts per run.

Although our experiment would be run using the matrix in Table 1.5, the analysis requires that we resolve the problem of each factor having different units (i.e., °C for A and minutes for B) as well as different scales (i.e., 100 to 200 for A and 5 to 10 for B). The solution to this problem is to standardize units and scale for each factor. This standardization will enable us to later build powerful mathematical models. The standardization process is accomplished by coding the low test settings as a (-1) and the high test settings as a ($+1$). Table 1.6 displays the coded settings of A and B along with a column labeled A·B which will be used to evaluate the interaction of A and B. The A·B column is simply the product of columns A and B, row by row. The interaction of A and B, denoted here by A·B, is also sometimes written as AB, A*B or AxB. (Refer to the discussion about Figure 1.9 for a review of what is meant by an interaction.)

Run	A	B	A·B	y_1 y_2 ..... y_9	$\bar{y}$	s
1	-1	-1	1			
2	-1	1	-1			
3	1	-1	-1			
4	1	1	1			

Table 1.6 2-Factor Design Matrix with Coded Factors and Interaction Column

The A and B columns from the coded design matrix in Figure 1.6 can now be used as a generic set of test conditions for any 2 factor (2 level) experiment. Similarly, a 3 factor (2 level) generic coded design matrix would contain the $2^3 = 8$ runs and seven total columns shown in Table 1.7.

Run	A	B	AB	C	AC	BC	ABC	y_1 y_2 ... y_5	$\bar{y}$	s
1	−1	−1	1	−1	1	1	−1			
2	−1	−1	1	1	−1	−1	1			
3	−1	1	−1	−1	1	−1	1			
4	−1	1	−1	1	−1	1	−1			
5	1	−1	−1	−1	−1	1	1			
6	1	−1	−1	1	1	−1	−1			
7	1	1	1	−1	−1	−1	−1			
8	1	1	1	1	1	1	1			

Table 1.7 3 Factor Design Matrix with Coded Factors and Interaction Columns

The Rules of Thumb in Appendix M provide design choice guidance for most 2 level experiments. Since many beginners have found 2 level experiments to be very powerful, these design choices are a good place to start. As you develop a need for more complex experiments, the information in Chapter 3 will help you make the proper choice of design matrix.

1.10   Conducting the Experiment

The importance of exercising extreme discipline in the actual conduct of the experiment cannot be overstressed. During the brainstorming process, it is recommended that a wide open, creative environment be encouraged in which the free flow of wild ideas is welcomed. Conducting the experiment, however, requires a very disciplined approach. The authors find that groups who are not accustomed to functioning in a disciplined environment will lower the likelihood of success. Factor levels must be set as close as possible to the specific design settings. Samples must be prepared in exactly the same way each time. Operating procedures and data sheets need to be clearly prepared in advance. All parties involved in the conduct and analysis of the experiment must be well versed in the intimate details of how the experiment is conducted and not accept any deviations from the plan. You, as the champion of experimental design within your area, must communicate and enforce the need for discipline in the conduct of the experiment without alienating the other key team members. All too often, a lack of communication, leadership, or control dooms an otherwise brilliantly planned experimental scheme.

1.11   Causality versus Correlation

When testing the effects of input factors on the output, it is important to determine *causality*, not just correlational relationships. As pointed out in Section 1.4, a high correlation does not necessarily relate to the cause of the problem. To ascertain the causality of an input factor, it is important to ensure that changes in the output occurred because of changes in the input and not because of some other related factor. To accomplish this task, experiments must be designed such that input factors are uncorrelated and all *nuisance variables* (variables not controlled) have their effects averaged out over all experimental conditions. The designs presented in this text ensure that the input factors themselves are uncorrelated. With regard to nuisance variables, a procedure known as *randomization* is typically used to spread the effect of nuisance variables across the entire experimental region. Randomization consists of running the experimental conditions in a random sequence. This is especially important when the brainstorming process fails to identify all factors or when some factors cannot be measured or controlled in the experiment.

As an example demonstrating the need for randomization, consider the following: a new product is being tested using two different bonding agents. Assume that there are three batches of raw material being used to make these products and the batch factor was uncontrolled in the design. After testing bond strength, the data appear as shown in Table 1.8.

Measured Bond Strength for Bonding Agents A and B

A	Batch #	B	Batch #
2.4	3	2.7	3
2.5	3	3.3	1
2.1	3	3.1	1
3.1	1	3.7	2
2.7	3	3.5	2
2.9	3	3.6	1
3.0	2	2.8	2
Mean ($\overline{y}$)	2.670		3.243
STD Dev ($\hat{\sigma}$)	.359		.391

Table 1.8 Example Demonstrating the Need for Randomization

Initially, batch information was not considered and thus it appears that bonding agent B provides a higher mean bond strength. Therefore, it is concluded that product B is the better bonding agent. However, consider the consequences of batch #3 being defective. Since more observations using bonding agent A were conducted with batch #3 than any other batch, the results are now suspect. We cannot state unequivocally that bonding agent differences caused substantial differences in bond strength because the batch effect is confounded with bonding agents. To avoid this problem, we could have randomly assigned materials to bonding agents in an attempt to average out the effects of the defective batch over the two treatments. It is also possible to control the different batches beyond the randomization process by including them as an input factor. More than likely, the best design in this case will be one that provides information on which bonding agent will maximize strength and simultaneously provide insensitivity (robustness) to batch variations. Robust designs are discussed further in Chapters 5 and 6.

1.12   A Quality Improvement Example

Suppose you are responsible for vehicle maintenance in a large organization. As an initial attempt to implement continuous process improvement (CPI), you would probably form a team to assist you in defining your mission and developing a process flow diagram for vehicle repair. The diagram, shown in Figure 1.19, can be used to improve quality by identifying non-value added steps to be modified or eliminated. An obvious CPI objective would be to ensure that "the right repairs are done right the first time." Just imagine how many steps could be eliminated from the process flow diagram if you could educate, train, and motivate your people and suppliers to perform in this manner.

After reviewing the process flow diagram, assume you decide to implement a gas mileage improvement strategy. To benchmark your current state of vehicle quality, you would probably direct that gas mileage data be recorded on each vehicle. The best way to display this data is shown in Figure 1.20. As you can see, gas mileage does vary from week to week. Your concern is whether this variability is natural or due to specific causes. The use of control charts can assist you in answering this question.

To present a simple overview of control charts, we will assume that the 20 weeks of data in Figure 1.20 are from a vehicle whose performance is currently known to be stable. Computing the average gas mileage over the 20 weeks will give us a feel for the vehicle's expected gas mileage. This average is found as shown below:

$$\bar{y} = \frac{\sum_{1}^{20} y_i}{20} = \frac{[21+19+20+...+22]}{20} = \frac{400}{20} = 20$$

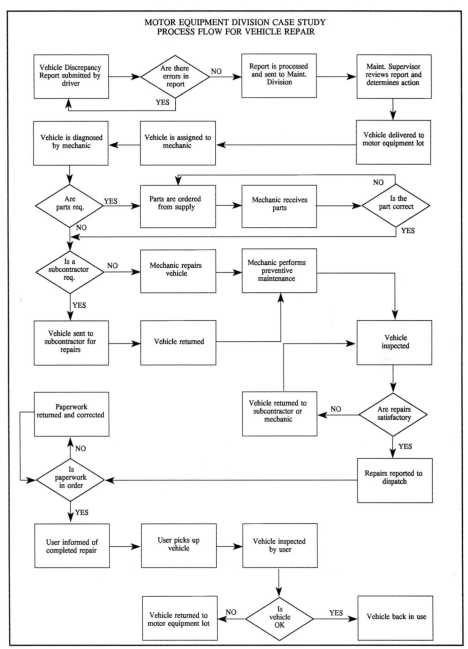

Figure 1.19 Process Flow Diagram for Vehicle Repair

Figure 1.20 Graph of Weekly Gas Mileage

To obtain a measure of the variability in vehicle gas mileage, we can calculate the sample variance, $\hat{\sigma}^2$, of the 20 mileage values. The equation for $\hat{\sigma}^2$ is:

$$\hat{\sigma}^2 = \frac{\sum (y - \bar{y})^2}{n-1}$$

which can be described as the sum of the squared deviations from the average, all divided by $(n-1)$. The deviations from the average and the variance calculations are displayed in Figure 1.21.

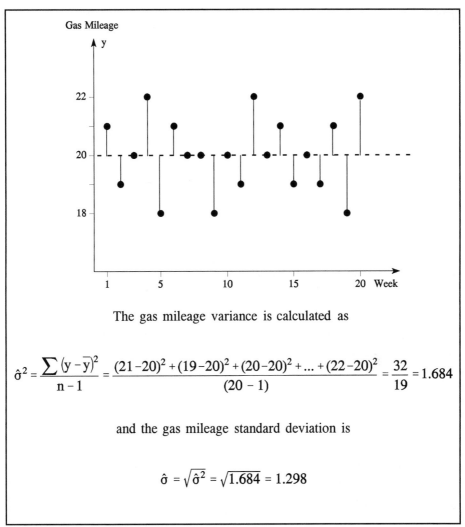

The gas mileage variance is calculated as

$$\hat{\sigma}^2 = \frac{\sum (y - \overline{y})^2}{n-1} = \frac{(21-20)^2 + (19-20)^2 + (20-20)^2 + ... + (22-20)^2}{(20-1)} = \frac{32}{19} = 1.684$$

and the gas mileage standard deviation is

$$\hat{\sigma} = \sqrt{\hat{\sigma}^2} = \sqrt{1.684} = 1.298$$

Figure 1.21 Graph of Deviations from the Mean Gas Mileage

A common technique which is a simple, yet powerful, way to get a visual feel for the way data is distributed is to draw a histogram. A histogram is a graph which displays how frequently a given outcome occurs. The histogram for our data is shown in Figure 1.22. If a smooth curved line were used to outline the histogram, it would appear to be symmetric and bell shaped, much the same as a normal curve. Assuming, then, that our gas mileage values are approximately normally distributed allows us to make use of the empirical rule shown in Figure 1.23.

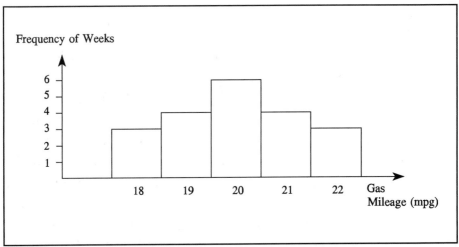

Figure 1.22 Histogram of Gas Mileage Data

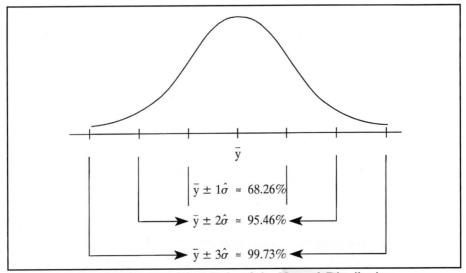

Figure 1.23 Empirical Rule of the Normal Distribution

The normal distribution of gas mileage values and the randomness of the data in Figure 1.20 supports our assumption that the vehicle's performance is stable. We can now estimate C_{pk} and use the empirical rule to assist us in developing control charts. If the lower specification for gas mileage is 17 then

$C_{pk} = \dfrac{20-17}{3(1.298)} = .77$ which converts to approximately 22,000 dpm (see

Table 1.3). To build the control chart you can compute $\bar{y} \pm 3\hat{\sigma}$ or $20 \pm 3(1.298)$ which results in a lower control limit (LCL) of approximately 16.1 and an upper control limit (UCL) of approximately 23.9 (see Figure 1.24). The empirical rule states that approximately 99.73% of all future gas mileage values for this vehicle should fall within these limits. Therefore, we can use these limits to evaluate the future performance of this specific vehicle. For instance, if week 21 produced a gas mileage of 23, chances are that nothing specific has caused this value; however, a value of 16 is outside the control limits (i.e., outside the natural variability) and indicates that there is a 99.73% chance this value was caused by some specific change in the vehicle. Other control chart rules exist for determining out of control conditions due to trends or increases in variability [7],[9].

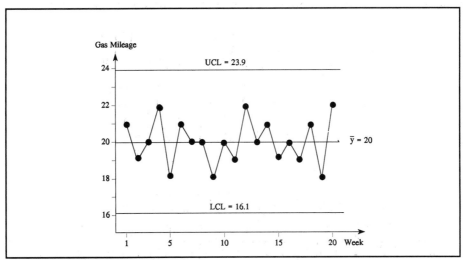

Figure 1.24 Control Chart Based on Gas Mileage Data

Once an out of control condition (e.g., 16 mpg on Week 21 in Figure 1.24) is determined, the obvious next step is to determine its cause. To obtain this information, a brainstorming session with the appropriate members should be conducted. The results of such a session might resemble the cause and effect diagram displayed in Figure 1.25. After completing the brainstorming, the group should try to reduce the information to those inputs most likely to affect the gas mileage. You now have a set of input factors to the process which can be investigated each time the mileage control chart detects a change.

Assume your investigation revealed an extremely low tire pressure which you determine to be the cause of the out of control mileage value. You would obviously correct the low tire pressure problem and possibly begin to chart the tire pressures on a weekly basis. In this way, you are now controlling a critical input to the process instead of waiting for a substantial change in the process output. Thus, you have now entered a problem prevention mode versus detecting poor quality through inspection of output values.

Control charting is an excellent process monitoring tool; however, engineers and scientists are not always sure which inputs are the important ones (vital few) and which ones are unimportant (trivial many). Furthermore, using control charting to gain an in-depth understanding of your process can be a long drawn out procedure. What is needed is a faster method for achieving "profound knowledge" of the gas mileage process. The statistical tool which provides this type of in-depth knowledge of a process in a timely manner is experimental design. The basic idea is to characterize the process by determining which inputs have a critical impact on the response. The design of experiments approach can be used effectively after control charts have been implemented; however, it is advantageous to characterize the process before production begins. This strategy allows for quality to be designed into the product from the beginning of the engineering design phase. Before the actual experiment is conducted, you should review the cause-and-effect diagram to remove unnecessary sources of variation. For example, it would be wise to develop training and/or a standard operating procedure to prevent operators from riding the brakes and making fast starts.

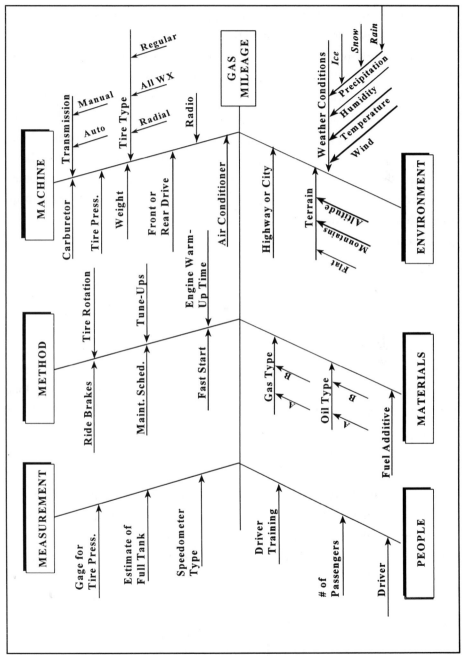

Figure 1.25 Cause-and-Effect Diagram for Gas Mileage

Assuming the brainstorming resulted in four input factors, the investigation of all possible combinations of four factors at two levels will require the 16 runs shown in Table 1.9. The (-1) values indicate a factor set at its low value and the $(+1)$ values represent a high factor setting. Notice that the design columns are all balanced vertically, i.e., there are an equal number of $(+1)$ and (-1) values in each column. That is, the sum of all the numbers in each column is zero. Furthermore, these columns are also horizontally balanced for each level of each column. For example, when A is at a (-1), the remaining columns have a balanced number of $(+1)$ and (-1) values. The same is true for all other columns evaluated at their $(+1)$ or (-1) settings. These balancing properties result in design orthogonality which allows us to estimate the effects of each factor independently of the others.

	1	2	3	4	5	6	7	8	9	10	11	12	13	14	15			
Run	A	B	AB	C	AC	BC	ABC	D	AD	BD	ABD	CD	ACD	BCD	ABCD	y_1 y_2 y_3	$\bar{y}$	s
1	-1	-1	1	-1	1	1	-1	-1	1	1	-1	1	-1	-1	1			
2	-1	-1	1	-1	1	1	-1	1	-1	-1	1	-1	1	1	-1			
3	-1	-1	1	1	-1	-1	1	-1	1	1	-1	-1	1	1	-1			
4	-1	-1	1	1	-1	-1	1	1	-1	-1	1	1	-1	-1	1			
5	-1	1	-1	-1	1	-1	1	-1	1	-1	1	1	-1	1	-1			
6	-1	1	-1	-1	1	-1	1	1	-1	1	-1	-1	1	-1	1			
7	-1	1	-1	1	-1	1	-1	-1	1	-1	1	-1	1	-1	1			
8	-1	1	-1	1	-1	1	-1	1	1	1	-1	1	-1	1	-1			
9	1	-1	-1	-1	-1	1	1	-1	-1	1	1	1	1	-1	-1			
10	1	-1	-1	-1	-1	1	1	1	1	-1	-1	-1	-1	1	1			
11	1	-1	-1	1	1	-1	-1	-1	-1	1	1	-1	-1	1	1			
12	1	-1	-1	1	1	-1	-1	1	1	-1	-1	1	1	-1	-1			
13	1	1	1	-1	-1	-1	-1	-1	-1	-1	-1	1	1	1	1			
14	1	1	1	-1	-1	-1	-1	1	1	1	1	-1	-1	-1	-1			
15	1	1	1	1	1	1	1	-1	-1	-1	-1	-1	-1	-1	-1			
16	1	1	1	1	1	1	1	1	1	1	1	1	1	1	1			

Table 1.9 Full Factorial Low (-1) and High $(+1)$ Settings
for a Four Factor Experiment

The design in Table 1.9 is referred to as a full factorial design. It is capable of estimating the effect of all possible factor combinations on the response. In this case, there are 15 possible "effects." There are 4 "linear" or "main" effects (A, B, C, D); 6 "two-way" interactions (AB, AC, AD, BC, BD, CD); 4 "three-way" interactions (ABC, ABD, ACD, BCD); and only 1 "four-way" interaction (ABCD). Seldom do experimenters find interactions beyond two ways to be significant; thus, time and resources can be conserved by the use of a fraction of the full factorial design. For the gas mileage experiment, it was decided to use a fractional factorial design with eight runs or rows. To build this half fraction design, you start with a 3-factor full factorial with 8 runs and 7 columns. Now to test 4 factors in 8 runs implies you will have to give something up. Since the ABC 3-way interaction is not likely to be important, you can sacrifice knowledge of a 3-way interaction in order to test a fourth factor. (Fractional factorial designs are discussed in depth in Chapter 3.) The resulting design matrix is shown in Table 1.10. Notice that this 8-run design also has the balancing properties required for orthogonality. The column for factor D was formed by using the elements of the ABC interaction column. The result is that the settings for factor D are "aliased" with (or identical to) the (-1) and $(+1)$ values for the ABC interaction. Since Table 1.10 is missing the columns to be used to evaluate interactions associated with factor D, you can build the AD, BD, CD, ABD, ACD, and BCD columns only to find them "aliased" with other columns in Table 1.10. See Table 1.11 for the complete set of aliasings in the 4 factor half fraction design. Terms that are aliased have the same $(+1)$ and (-1) sequence in their respective columns. The result of aliasing is that each aliased pair of terms cannot be evaluated separately. This loss of information is a consequence of saving resources by conducting fewer tests than the number required for a full factorial. For any fractional factorial design, the experimenter should assess the amount of aliasing prior to completing the experiment. As long as the amount of aliasing is acceptable, he/she should continue using that design; otherwise, a different design matrix should be chosen. More will be said regarding the consequences of aliasing once we have analyzed the data.

RUN	A 1	B 2	AB 3	C 4	AC 5	BC 6	D=ABC 7
1	−	−	+	−	+	+	−
2	−	−	+	+	−	−	+
3	−	+	−	−	+	−	+
4	−	+	−	+	−	+	−
5	+	−	−	−	−	+	+
6	+	−	−	+	+	−	−
7	+	+	+	−	−	−	−
8	+	+	+	+	+	+	+

Table 1.10 Half Fraction Design Settings for Four Factors
Each at 2 Levels (Low and High)

ALIAS PATTERN	
effect	**alias**
A	BCD
B	ACD
C	ABD
D	ABC
AB	CD
AC	BD
AD	BC
BC	AD
BD	AC
CD	AB
ABC	D
ABD	C
ACD	B
BCD	A

Table 1.11 Alias Pattern for the Half Fraction Design in Table 1.10

The four factors which were brainstormed to be used in the experiment and the low ($-$) and high ($+$) levels of each are shown in Table 1.12.

FACTOR	LEVELS
A: TIRE PRESSURE	28, 35
B: TIMING SETTING	LOW, HIGH
C: TYPE OF OIL	1, 2
D: TYPE OF GAS	1, 2

Table 1.12 Factors for the Gas Mileage Case Study

The completed experimental matrix with the response values (taken over three weeks for each run) are displayed in Table 1.13.

Run	A	B	AB	C	AC	BC	D	y_1	y_2	y_3	$\bar{y}$	s^*
1	$-$	$-$	$+$	$-$	$+$	$+$	$-$	22.27	21.12	21.37	21.59	.60
2	$-$	$-$	$+$	$+$	$-$	$-$	$+$	14.22	15.40	10.46	13.36	2.58
3	$-$	$+$	$-$	$-$	$+$	$-$	$+$	22.49	23.15	22.08	22.57	.54
4	$-$	$+$	$-$	$+$	$-$	$+$	$-$	9.96	13.80	11.92	11.89	1.92
5	$+$	$-$	$-$	$-$	$-$	$+$	$+$	17.35	18.60	17.97	17.98	.62
6	$+$	$-$	$-$	$+$	$+$	$-$	$-$	27.08	24.54	24.57	25.40	1.46
7	$+$	$+$	$+$	$-$	$-$	$-$	$-$	18.36	17.63	17.04	17.68	.66
8	$+$	$+$	$+$	$+$	$+$	$+$	$+$	22.78	26.97	27.14	25.63	2.47

Table 1.13 Complete Experimental Matrix with Response Values
for Gas Mileage Case Study

---

*s represents the response standard deviation for each run; the symbols s and $\hat{\sigma}$ are used interchangeably for sample standard deviation throughout the book.

The first step in analysis is to find the average $\bar{y}$ for each column set of (-) values and then for each column set of (+) values. See Table 1.14. The same set of calculations are then conducted using the s column (see Table 1.15).

	A	B	AB	C	AC	BC	D
$\bar{y}$ Avg @ (−)	17.35	19.58	19.46	19.95	15.23	19.75	19.14
$\bar{y}$ Avg @ (+)	21.67	19.44	19.56	19.07	23.80	19.27	19.88
Δ	4.32	−.14	.10	−.88	8.57	−.48	.74

Table 1.14 Analysis of Each Column y Effect (Δ) by Computing the Average $\bar{y}$ for Each Column (−) value and (+) value. $\Delta = [\bar{y} \text{ Avg @ } (+) - \bar{y} \text{ Avg @ } (-)]$

	A	B	AB	C	AC	BC	D
s Avg @ (−)	1.410	1.315	1.135	0.605	1.445	1.310	1.160
s Avg @ (+)	1.302	1.397	1.577	2.107	1.267	1.402	1.552
Δ	−.108	.082	.442	1.502	−.178	.092	.392

Table 1.15 Analysis of Each Column s Effect (Δ) by Computing the Average s for Each Column (−) value and (+) value. $\Delta = [s \text{ Avg @ } (+) - s \text{ Avg @ } (-)]$

A graphical analysis called a marginal means plot for each effect in the design matrix is obtained by plotting the average response at the (−) and (+) settings and then connecting these two dots with a straight line (see Figure 1.26). The large star (★) in Figure 1.26 was obtained by averaging the response values when factor A is set high (+). In other words, the average response for runs 5, 6, 7, and 8 is the 21.67 in the A column of Table 1.14. The other dots are found in a similar fashion. The steepness of the slope reflects the importance of each effect. From the graphical results in Figure 1.26, it is apparent that the most important effects are factor A and the AC or BD interaction. Since AC is aliased with BD (see Table 1.11), we must go back to the factors to determine which interaction is most likely responsible for the steep slope in Figure 1.26. Since it is highly unlikely to have a "tire pressure × type of oil" interaction (AC), it is decided that the steep slope in Figure 1.26 is due to the BD or "timing × type of gas" interaction. The BD interaction graph is shown in Figure 1.27 where the ★ was determined from the response average of the runs with B at (−) and D at (−);

i.e., the average of 22.27, 21.12, 21.37, 27.08, 24.54, and 24.57(from runs 1 and 6). The other points plotted in Figure 1.27 were found in a similar fashion.

Figure 1.26 Marginal Mean Graphs for Gas Mileage Case Study

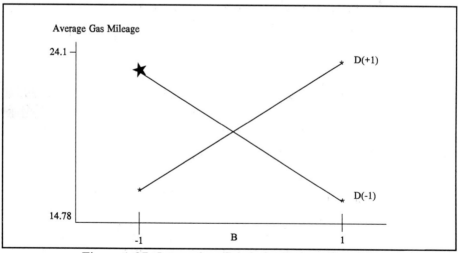

Figure 1.27 Interaction Graph for Factors B and D

Another graphical analysis can be used to determine if any of the factors shift the variability in the gas mileage. Figure 1.28 indicates that factor C has a big effect on variability.

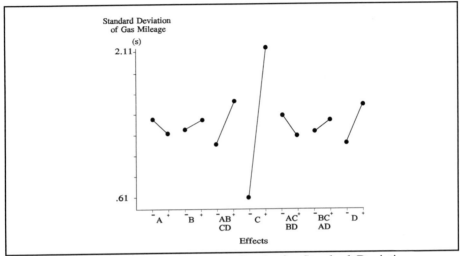

Figure 1.28 Marginal Average Graphs for Standard Deviation

To determine the best factor settings for maximizing gas mileage while minimizing variability, we must return to Figures 1.26, 1.27, and 1.28. Based on Figure 1.26, set factor A at (+). Figure 1.27 indicates that factors B and D should both be set at (+). Factor C is set at (−) to minimize variability (see Figure 1.28). Thus, the final factor setting decision is as follows:

Factor	(−) or (+) Setting	True Setting
A	+	35 psi
B	+	high timing
C	−	type 1 oil
D	+	type 2 gas

At this point, one should feel pretty good about solving a quality problem and thus satisfying the customer. Be careful, though, because we made the assumption that the customer wanted high gas mileage. When a survey was conducted in Madison, Wisconsin [5] as to what the customer (the driver and/or passenger) thought was important, the primary response was not mileage, but safety. The point is that designing experiments to solve problems which the customer may not think important will not always be a productive strategy. One should first survey the customer through the use of a procedure such as Quality Function Deployment (QFD). It is not enough to simply determine customer needs and expectations. To increase customer satisfaction, design processes and products to meet customer needs. To accomplish this task, crossfunctional teams from marketing, purchasing, engineering, and others must work closely together from the time the need is first conceived until the need is met. QFD provides a framework within which the crossfunctional teams can work.

To accomplish the QFD procedure you must answer the following questions: (Total Quality Management: A Guide for Implementation, DoD 5000.51.6):

(1) Who are your customers and what are their expectations?

(2) Are all the customer expectations equally important? If not, rank order them.

(3) Will meeting customer expectations enable you to obtain a competitive advantage?

(4) What are the engineering characteristics that are related to customer expectations?

(5) How does each engineering characteristic affect each customer expectation?

(6) How does each engineering change affect other characteristics?

(7) What should specifications be in order to satisfy minimum customer needs?

(8) What are the key process parameters (inputs, or factors) related to each characteristic?

For more information on QFD, see *Basic Statistics: Tools for Continuous Improvement* [7].

After assessing the customer needs and translating these needs into product/process design, then the designed experiments approach provides the type of profound knowledge for which we are striving.

1.13   Taguchi Approach to Designed Experiments

The objective of parameter (Robust) design is to determine factor settings that achieve desired response values (min., max., or specific value) while minimizing the variability due to uncontrolled or noise factors. To illustrate this concept in the gas mileage example, consider weather conditions which are outside the vehicle operator's control. It might be desirable to find the best settings of factors A, B, C, and D that maximize gas mileage while also making mileage insensitive (as much as possible) to weather conditions. In this case, we would make specific replicates at different weather conditions. In other words, the first replicate (first recorded column for the response) would all be obtained under rainy conditions (w_1), the second replicate obtained during snowy conditions (w_2), and the third replicate obtained under dry conditions (w_3). Thus, the test matrix would appear as shown in Table 1.16. The only difference between this strategy and that of previous sections is the specific condition of the replicated response. By minimizing the response variability over the noise factor levels we have found a condition for the process that makes it insensitive—or *robust*—to environmental conditions. Robustness can also be sought for process parameters which are difficult or expensive to control such as material types, different lots of material, other environmental conditions, etc. See Chapters 5 and 6.

Run	Factors & Interactions							Replicated Responses				
	A	B	AB	C	AC	BC	D	w_1	w_2	w_3	$\overline{w}$	s
1	-1	-1	1	-1	1	1	-1					
2	-1	-1	1	1	-1	-1	1					
3	-1	1	-1	-1	1	-1	1					
4	-1	1	-1	1	-1	1	-1					
5	1	-1	-1	-1	-1	1	1					
6	1	-1	-1	1	1	-1	-1					
7	1	1	1	-1	-1	-1	-1					
8	1	1	1	1	1	1	1					

Table 1.16 Replicated Responses Based on Weather Conditions

1.14   Steps for Quality Improvement: Getting Started

Across America, purchasers of products and services are demanding that their suppliers provide (1) **better** quality, (2) **faster** response times, and (3) **cheaper** prices. To meet these demands, suppliers must be able to use a host of statistical and non-statistical tools. Other texts that address some of the tools that are not included in this book are *Total Quality — A Textbook of Strategic Quality Leadership & Planning* (1991) by Lam, K.D., Watson, F. and Schmidt, S. R. [8] and *Basic Statistics: Tools for Continuous Improvement*, 3rd Edition (1993) by Kiemele, M. J. and Schmidt, S. R. [7]*. To improve quality, it is important to be familiar with all available tools; however, equally important is knowing when to use them. The following steps are suggested as an example of how to get started in the quality improvement cycle. We highly recommend that you use them as an outline for your company and that you make modifications as necessary.

---

*For more information on these texts, contact Air Academy Press at (719) 531-0777.

QUALITY IMPROVEMENT STEPS

WHAT	WHO	HOW
1. Commitment	Everyone, from the top to the on-line worker	Cultural Change Strategic Planning Awareness Training
2. Define products/ processes	Crossfunctional Teams	Process Flow Diagrams Brainstorming
3. Mission Statement	Crossfunctional Teams	Brainstorming Nominal Group Technique(NGT)
4. Identify customers and suppliers (Internal/ External)	Crossfunctional Teams	Brainstorming NGT
5. Identify customer expectations and what you expect from your suppliers	Crossfunctional Teams including Customer and Supplier Representatives	Surveys Brainstorming
6. Translate customer expectations into key quality characteristics	Crossfunctional Teams and the Customer	QFD Brainstorming NGT

7. Determine current state of quality for all products/processes; and use quality oriented metrics as a scorecard	Process Action Teams (PATs) and the Customer	Benchmarking
		Loss Function
		Process Capability
		Cycle Time
		Defective Parts per
		  Million (dpm)
		Surveys
		Cost of Poor
		  Quality (COPQ)

QUALITY IMPROVEMENT STEPS (cont.)

WHAT	WHO	HOW
8. Identify the best opportunity for improvement	PATs and the Customer	Loss Function Cost Analysis Pareto Charts
9. For the best opportunity, identify causes of degraded quality	PATs	Brainstorming Cause and Effect Diagrams Historical Data
10. Characterize product/ process, i.e., find relationship between outputs (effects) and inputs (factors)	PATs	Designed Experiments
11. Find optimal settings for input factors	PATs	Optimize models generated from Designed Experiments
12. Determine input tolerances	PATs	Use models in (11) as simulators for Designed Experiments to test proposed tolerances
13. Set up control charts on critical inputs and outputs	PATs	Statistical Process Control (SPC)
14. Measure Return on Investment	PATs	Estimate decrease in COPQ

15. Recycle the steps, implementing the philosophy of continuous process improvement (CPI)

1.15 U.S. Air Force Approach to Total Quality Management (TQM)

Total Quality Management is a phrase with which most industry people are quite familiar. In order to compete in the 21st century, a company must embrace improved quality as a new way of life. Numerous TQM success stories have been documented for all types of industry [7]. Several large companies such as Motorola, Boeing, IBM, Xerox, Ford Motor Company, Texas Instruments, and General Motors have instituted strong TQM type improvement strategies within their companies and for their suppliers. The U.S. Air Force has also implemented many aspects of TQM as shown in the following article on Reliability & Maintainability (R & M 2000).

R & M 2000 VARIABILITY REDUCTION PROCESS

Bruce A. Johnson, Captain, USAF
Office of the Special Assistant for Reliability and Maintainability
Headquarters, United States Air Force

THE PROCESS

Improving combat capability is a major Air Force objective. This is becoming increasingly difficult in the face of constrained manpower and fiscal resources. However, there is a solution. Substantial increases in combat capability are achievable through more reliable and maintainable weapon systems. Such systems are able to complete more missions with less spares, support equipment, facilities and maintenance personnel.

Weapon systems fail for many reasons. Some components, like tires, wear out. But most systems fail because of poor design, the use of defective parts and materials, or poor workmanship. The cause of these failures is *variability* in the design and manufacturing processes. The problem variability presents is that it exists in nearly all processes and it results in marginal or non-conforming products. The variability comes from the fact that conditions, under which these items are produced, change. Variability reflects the differences in raw material, machines, their operators and the manufacturing conditions. When process variation increases, the product's physical properties or functional performance can degrade, and the number of product defects increases. The significance variation has on a product's reliability and quality depends on the criticality of the manufacturing process and part characteristics.

There are two ways to reduce variability. Traditionally, the approach has been to tighten design tolerances and increase inspections. Costs climb as scrap and rework increase, and productivity drops. Inspections and tighter tolerances only treat the symptoms and do not resolve the actual problem.

The preferred method is to reduce the variability by improving the process. This can be done by eliminating the causes of variation through statistical techniques, and by developing more robust products which are insensitive to the causes of variation. The methods of reducing variability is aptly named the Variability Reduction Process (VRP).

VRP is a proven set of practices and technologies which yield more reliable and nearly defect free products at lower cost. It is a structured, disciplined design and manufacturing approach aimed at meeting customer expectations and improving the development and manufacturing process while minimizing acquisition time and cost (see Figure 1).

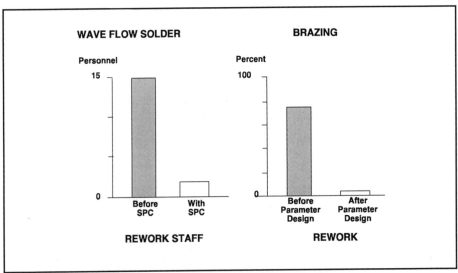

Figure 1 Process Improvements from VRP

The objectives of VRP are to design robust products which are insensitive to the causes of failure; to achieve capable manufacturing processes that produce nearly defect-free products; and to adopt the managerial attitude of continuously improving all processes. The basic tools are teamwork, statistical process control (SPC), loss function, design of experiments (DOE), parameter design and quality function deployment (QFD). VRP must span all of engineering, manufacturing and management, and include the suppliers (see Figure 2).

PURPOSE: MEET CUSTOMER EXPECTATIONS, IN MINIMUM TIME, AT LOWEST COST			
OBJECTIVES	Robust Designs	Capable Processes	Continuous Improvement
PRIMARY RESPONSIBILITY	Engineers	Manufacturing	Management
TEAMS	Interdisciplinary	Improvement	Multi-Functional
TOOLS/ TECHNOLOGIES	QFD/DOE/ Parameter Design	DOE/SPC	Loss Function

Figure 2 The Elements of VRP

CAPABLE MANUFACTURING PROCESSES

Capable manufacturing processes can only be achieved when the critical parameters are known and the causes of variation are eliminated or minimized. For most processes, SPC is highly effective (see Figure 3). It allows the operator to observe the process and distinguish between patterns of random and abnormal variation. It assists the operator in making timely decisions such as adjusting or shutting down the process before defects are produced. When combined with other statistical tools and problem solving techniques (see Figure 4), the worker can isolate and remove the causes of abnormal variations.

When the abnormal variations are removed from the process, the process is said to be under statistical control. In many processes, this will not be sufficient. The random variations alone can result in defective products, and their causes should be identified and removed until the process is capable of producing near defect-free products. However, causes of random variation are more difficult to identify, are usually systemic and normally require management action to remove.

PART	PROCESS				TARGET: 95 FT-LBS				
OPERATOR	MACHINE				SPEC LIMITS: ±20 FT-LBS				
DATE									
TIME									
Measurement	1	102	106	104	89	95	100	96	107	92
	2	94	97	91	94	105	95	85	103	100
	3	97	102	105	92	103	92	104	100	93
	4	106	107	98	96	92	93	90	105	90
Average, $\bar{x}$		100	103	99	93	99	95	94	104	94

Figure 3 SPC Control Chart

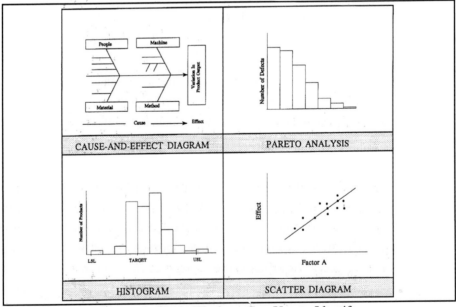

Figure 4 Tools Workers Can Use to Identify
the Causes of Variation

The difference between VRP and traditional methods of quality control is that improvements in quality are achieved through improvements in the manufacturing processes. No longer is better quality to be achieved through tightening specifications and more inspections. In the case of SPC, the manufacturing processes are improved by eliminating the causes of variability. Usually, the process can be centered around the design target and variation reduced well within the specifications (see Figure 5).

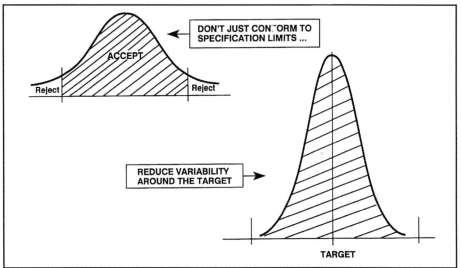

Figure 5   Design and Build to Target Values,
Not Specification Limits

When implemented correctly, the results can be impressive. For Parlex Nevada Inc., a circuit card manufacturer, SPC was used to cut scrap cost by 90 percent in one year, and it changed the company's losses into profits. Boeing used SPC to resolve a rivet flushness problem on the nose section of the 737 aircraft. The improvements saved a half-million dollars a year.

A more powerful method of resolving difficult or complex industrial problems is the statistical design of experiments (DOE). DOE methods have been around for 60 years and have been extensively used by the agricultural, pharmaceutical and chemical industries to advance their products. These techniques can greatly accelerate the rate of improving product designs and manufacturing processes. Such statistical experiments will aid the engineer in identifying the critical parameters for SPC, isolating the causes of variation, and improving the product's technical or operational characteristics.

DOE works by measuring the effects that different inputs have on a process. This is done by identifying a prospective set of input factors, varying the inputs over a series of experiments, collecting the data and analyzing the results. An input may be varied over a range of values such as can be done with an oven's curing temperature or conditionally, such as the decision to add or withhold a curing additive. The methods, whether they employ a full-factor, fractional-factor, orthogonal array or response surface technique, use a statistical approach that ensures accuracy and validity.

Well-planned experiments can have dramatic results. For example, a government-owned-company-operated (GOCO) munitions plant had a serious problem in producing the ADAM mine. Although SPC was in use and 12 of 13 processes were within their tolerances, 19 out of 25 lots were rejected. Aerojet Ordinance, the plant operator, decided to apply a Taguchi experiment to identify the critical parameters. They selected the 13 parameters used in the SPC program and tested parameters at three different levels. Only 27 experiments were conducted, firing 6 rounds each. The results were profound. Four parameters were found to be critical, and when set at their best levels, the process produced good lots without any rejects. The other nine parameters were less important and their tolerances can be relaxed. Results: production schedule met while achieving significant cost savings.

ROBUST DESIGNS

Having a capable manufacturing process is not enough. It may not be economical to remove or control some of the causes of variation. Therefore, it is necessary to develop robust manufacturing processes which are insensitive to the manufacturing conditions, materials, machines and operators. In most cases, the *greatest improvements come from robust designs.* These improvements are achieved through parameter design, a technique of selecting the optimum conditions (i.e., determining the ideal parameter settings) that minimize the variability without removing the causes of variation.

During parameter design, a set of parameters is identified to enhance the product, and a series of experiments is conducted to observe the effects of the parameters on the desired part characteristics. The result could identify new parameter settings that improve the product and increase yield. For example, the problem an engineer may want to solve is the variability of ceramic parts. The source of variation is the uneven temperatures in the kiln. Because modifying the kiln is too expensive, the engineer conducts several experiments to identify a way to minimize the effects of the uneven temperature. For the experiment, he selects as parameters the amounts of the ingredients, their textures, blending procedures and firing temperatures. For validity, the engineer should use DOE procedures for conducting the experiment. An orthogonal array may be used to minimize experimental time and cost. The results will enable the engineer to find new parameter settings that minimize the effects of uneven firing temperatures.

The problem with most design approaches is that parameter design is rarely done. Most engineers focus on the system design to develop the product, and immediately transition to tolerance design to establish the specification limits. Often, the result is an inferior product which is sensitive to variations in the manufacturing process. Parameter design should be done before tolerance design.

Parameter design can also be used to design and produce a more robust product that will perform better over a wider range of operating conditions and environments. It can be used to enhance a desired customer's need such as a smooth automobile ride, or to enhance an engineering requirement such as to lower the susceptibility of corrosion.

The success of parameter design and SPC hinges on the engineers' understanding of the customers' needs. Quality Function Deployment (QFD) is a systematic approach for developing and translating the customers' needs into the critical part characteristics and production requirements. The QFD requirements matrices are designed to minimize the chance of starting the design process with incomplete or erroneous requirements. They provide a methodology which assures an orderly translation of the customers' requirements throughout the product development process (see Figure 6)

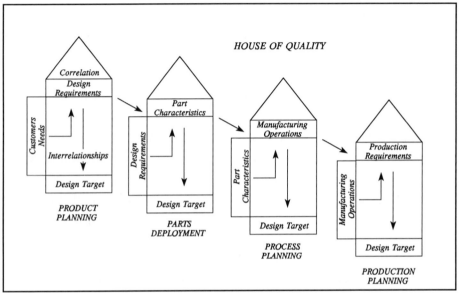

Figure 6 Quality Function Deployment

The basic approach used in QFD is conceptually similar to the practice employed by most companies. The difference lies in its structure. It compels the different disciplines and departments to communicate. QFD starts by defining the customers' requirements in the customers' terminology and translates these requirements into engineering requirements. These engineering requirements become the product characteristics which should be measurable and given target values. If properly executed, the product should fulfill the expectations of the customer.

The other matrices translate the engineering requirements into part characteristics, required manufacturing operations, and production requirements. Each matrix identifies the design targets, interrelationships and priorities. The end result should be a set of operating procedures which the factory can follow to consistently produce the critical part characteristics.

The design environment best suited to produce robust products is concurrent engineering (also referred to as simultaneous engineering). Concurrent engineering addresses all the customer, design and manufacturing issues up-front starting with concept exploration. The process employs good design practices, interdisciplinary teams and a structured requirements process, such as QFD, to concurrently develop the product and manufacturing processes. Its practice encourages communication between the design, product and production engineers. Concurrent engineering replaces the typical "sequential" approach to product design, which is more costly and time consuming. The effects concurrent engineering can have may be summed up by the following example. Using sequential engineering practices, the Allison Transmission Division estimated in 1982 it would cost $100 million in capital investment and $75,000 per unit to replace the transmission in the M-113 Armored Personnel Carrier. In 1987, using concurrent engineering, Allison's estimate dropped to $20 million for capital investment and $50,000 per unit (see Figure 7).

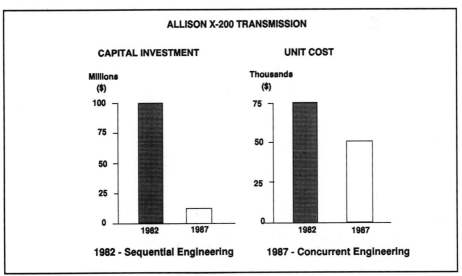

Figure 7 Concurrent Engineering

CONTINUOUS IMPROVEMENT

For VRP to succeed, management from the top down must adopt new attitudes about reliability and quality, and must become *directly involved* in continuously improving the design and manufacturing processes. They must implement programs to foster improvement, tear down the barriers that inhibit change, instill teamwork, establish goals for improvement, and provide education and training for successful implementation.

Management's primary objective should be to satisfy the customer and serve the customer's needs if a company is to stay in business and make a profit. Reliability and quality must come first - not profit. If done smartly, reliability and quality will reduce cost and increase profit. For example, Hewlett-Packard's Yokogawa plant implemented many of the VRP techniques and, after eight years, they achieved 240 percent increase in profit, 120 percent increase in productivity, 19 percent increase in market share, 79 percent decrease in failure rate, and 42 percent decrease in manufacturing costs.

Management must become *process-oriented* and stimulate efforts to improve the way employees do the job. Teamwork is the foundation for continuous improvement. An important part of team building is the assignment of people to multifunctional management teams, interdisciplinary design teams and process improvement teams. Everyone should be involved in process improvement.

Management should implement programs to foster continuous improvement, use education to change attitudes and provide training. Change will be gradual and will require a long-term outlook. In Japan, most of the small improvements come from the workers' suggestion system call *Kaizen Teian*. Kawasaki Heavy Industries Aircraft Works has one of the more impressive programs. In 1987, each employee submitted an average of 229 suggestions and 92 percent were adopted. Savings were estimated at $35 million. At a Texas Instruments plant, a five-year enhancement program reduced defects by over 90% (see Figure 8).

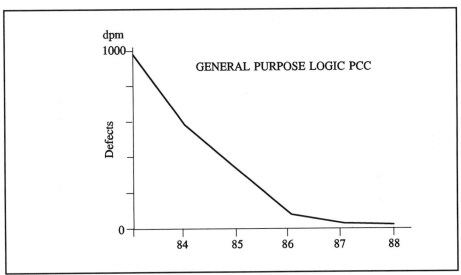

Figure 8 Continuous Improvement

Management must take responsibility for process improvement while giving the workers the responsibility of maintaining the process. Without the ability to maintain the existing process, there can be no improvement. This means management must give the worker ownership of his processes and allow the worker to improve or stop the process as necessary. In many progressive companies, workers are involved in the development of their own operating procedures, and in some cases, they write their procedures.

Management must change the accounting procedures. The notion that loss only occurs when the product is outside the specification limits is obsolete. Loss includes not only the cost of scrap and rework, but also the cost of warranties, excess inventory and capital investment, customer dissatisfaction, and eventual loss of market share. The traditional go/no-go approach to quality should be replaced with a powerful monetary loss function to better account for loss (see Figure 9). A quadratic loss function allows management to better assess the true cost of production processes and the benefits derived from process improvements (see Figure 10). Most important, the loss function supports continuous process improvement because minimizing the loss equates to reducing the variability around the target.

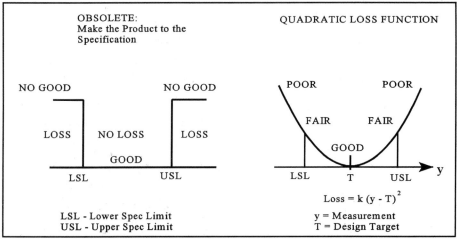

Figure 9 Quality Loss Functions

Figure 10 Table for Evaluating the Benefits from Process Improvement

IMPLEMENTATION

The adoption of VRP begins with the conviction that change is necessary and beneficial. Implementation is an evolutionary process. VRP may start out in a single product line at one factory with a single group of suppliers. But it must be grown with the long-range goal that it will encompass the entire enterprise. The Air Force strategy for VRP is to encourage defense contractors and suppliers to (1) foster top-level commitment to VRP, (2) involve all levels, departments and vendors, (3) apply the VRP in a systematic approach, and (4) create a culture of continuous process improvement. Within the Air Force, the Vice Chief of Staff has directed all commands involved in weapon system acquisition and support to implement VRP by 1993. The Air Force acquisition regulations have been rewritten to incorporate VRP in the acquisition process. VRP will be an essential part of Air Force's Total Quality Management program (see Figure 11).

I N C R E A S I N G	C O M P L E X I T Y ▼	**TQM TOOLS**	**R&M 2000 VRP**	
			Capable Processes	Robust Designs
		Team Work	X	X
		SPC	X	
		Loss Function	X	X
		Design of Experiments	X	X
		Parameter Design		X
		House of Quality (QFD)		X

Figure 11 TQM / VRP Relationship

SUMMARY

The Variability Reduction Process makes two seemingly contradictory goals compatible: to produce highly reliable and maintainable weapon systems while reducing development time and costs. The method is to design robust systems, produce them with capable manufacturing processes, and achieve continuous improvement.

VRP provides a win-win situation. The Air Force obtains more combat capability with the available dollars. Industry is able to satisfy their customers, improve productivity and lower costs.

References for the R & M 2000 Article

Statistical Process Control:

E. L. Grant and R. S. Leavenworth, Statistical Process Control, 5th ed., McGraw Hill, New York, 1979.

J. S. Oakland, Statistical Process Control, Wiley, New York, 1986.

D. J. Wheeler and D. S. Chambers, Understanding Statistical Process Control, Statistical Process Controls, Inc., Knoxville, TN, 1986.

Design of Experiment/Parameter Design:

G. C. P. Box, W. G. Hunter, and J. S. Hunter, Statistics for Experimenters, Wiley, New York, 1978.

G. C. P. Box, S. Bisgaard, and C. Fung, "An Explanation and Critique of Taguchi's Contribution to Quality Engineering," Quality and Reliability Engineering International, Vol. 4, pp. 123-131, 1988.

C. Daniel, Applications of Statistics to Industrial Experimentation, Wiley, New York, 1976.

R. V. Hogg and J. Ledolter, Engineering Statistics, Macmillan Publishing Company, New York, 1987.

S. R. Schmidt and R. G. Launsby, Understanding Industrial Designed Experiments, Air Academy Press, Colorado Springs, 1993.

G. Taguchi and Y. Wu, Introduction to Off-Line Quality Control, Central Japan Quality Control Association, Nagoya, Japan, 1980.

G. Taguchi, System of Experimental Design - Engineering Methods to Optimize Quality and Minimize Costs, UNIPUB/Kraus International Publications, White Plains, NY, 1987.

Quality Function Deployment:

J. R. Hauser and D. Clausing, "The House of Quality," <u>Harvard Business Review</u>, May-June, 1988.

B. King, <u>Better Designs in Half the Time: Implementing QFD Quality Function Deployment in America</u>, GOAL/QPC, Methuen, MA, 1987.

L. P. Sullivan, "Quality Function Deployment," <u>Quality Progress</u>, June, 1986, pp. 39-50.

Concurrent (Simultaneous) Engineering:

R. I. Winner, J. P. Pennell, H. E. Bertrand, and M. Slusarczuk, <u>The Role of Concurrent Engineering in Weapon System Acquisition</u>, (IDA Report R-338), Institute for Defense Analyses, Alexandria, VA, 1988.

Continuous Improvement:

M. Imai, <u>Kaizen</u>, Random House Business Division, New York, 1986.

W. W. Scherkenbach, <u>The Deming Route to Quality and Productivity - Road Maps and Roadblocks</u>, Mercury Press/Fairchild Publications, Rockville, MD, 1986.

Please feel free to duplicate and distribute this article. If you have any questions or comments, please contact Capt. Bruce Johnson, HQ USAF/LE-RD, (202) 697-9388 or av 227-9388.

Problem Set

1. What is experimental design?

2. Discuss how the use of experimental design can result in higher quality products at lower costs.

3. List the disadvantages of one factor at a time experiments.

4. List the disadvantages of full factorial (all possible combinations) experiments.

5. What is variability in the response and how can it be reduced?

6. Describe what is meant by a robust design.

7. Assuming your process is in control, use the following information to calculate C_p and C_{pk}. Interpret these values and discuss the proper use of each.

    Upper Spec = 2100 $\bar{y}$ = 2040
    Target   = 2000 $\hat{\sigma}$ = 20
    Lower Spec = 1900

8. What is meant by randomization and why is it important?

9. How important is the brainstorming session? How much time should be devoted to brainstorming?

10. Discuss the difference in the following two philosophies:
    (1) Manufacture the product to meet specifications.
    (2) Manufacture the product to the designed target value.

    What are the advantages/disadvantages of each philosophy?

11. Your travel time to work tends to vary by several minutes. As a result of this variation problem, you have been late on several occasions. Your boss is concerned and has asked that you brainstorm the factors that contribute to travel time variation. Use the fishbone diagram below to organize the brainstorming process.

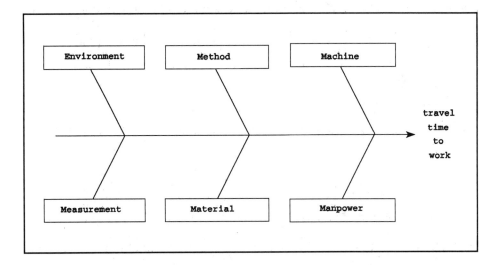

12. Processed material thickness is a critical dimension with specifications of [0.49, 0.51] and a designed target value of 0.5. Your internal customer has indicated that any time the thickness deviates from the design target by ± 0.004, they are required to complete an $8.00 extra processing step.

(1) Compute the monetary constant, k.

(2) Compute the average loss if your process has a $\bar{y}$ = 0.499 and s = .008.

(3) Find the total loss for one year if n = 100,000 per year.

(4) If reducing s to .004 is estimated to cost $700,000, will this improvement be a good investment? (Assume $\bar{y}$ remains .499.)

Bibliography

1. Barker, Thomas B., (1985) *Quality by Experimental Design*, Marcel Dekker, Inc.

2. Bhote, Keki R., (1988) *World Class Quality*, American Management Association, New York.

3. Deming, W.E., (1982) *Out of the Crisis*, MIT Center for Advanced Engineering Study, Cambridge, MA.

4. Hogg, Robert V. and Johannes Ledolter, (1987) *Engineering Statistics*, Macmillan Publishing Company, New York.

5. Hunter,W., J.O'Neill, and C. Wallen, (1988) *Doing More with Less in the Public Sector: A progress report from Madison, Wisconsin*, Center for Quality and Productivity Improvement, University of Wisconsin.

6. Juran, J.M., (1988) *Juran's Quality Control Handbook*, 4th Edition, McGraw-Hill, Inc., New York.

7. Kiemele, M.J., S.R. Schmidt, and R.J. Berdine, (1997) *Basic Statistics: Tools for Continuous Improvement,* 4th Edition, Air Academy Press, Colorado Springs, CO.

8. Lam, K.D., Frank D. Watson, and S.R. Schmidt, (1991) *Total Quality – A Textbook of Strategic Quality, Leadership and Planning*, Air Academy Press, Colorado Springs, CO.

9. Montgomery, D., (1991) *Introduction to Statistical Quality Control*, 2nd Edition, John Wiley and Sons, Inc.

10. Ott, Ellis R., (1975) *Process Quality Control*, McGraw-Hill Inc., New York.

11. Phadke, M.S., (1989) *Quality Engineering Using Robust Design*, Prentice-Hall, New Jersey.

12. Prince, George M., (1972) *The Practice of Creativity*, Collier Books.

13. Taguchi, G., (1986) *Introduction to Quality Engineering*, Asian Productivity Organization.

Chapter 2

Conducting Simple Experimental Designs and Analysis

2.1   Introduction

Before beginning an experimental design or any data collection process, it is important that the following preparatory steps be taken:

(1)   Management must establish an environment suitable for action. Any statistical method will ultimately fail unless management is receptive to "facts and data" and has prepared an environment where employees are motivated toward quality improvement, willing to work in teams, and unafraid to be candid with their inputs [1]. If you believe that your management is lacking in these areas, it is highly recommended that you educate them.

(2)   Employees must be well trained in their jobs. Whether an employee is a supervisor, operator, engineer, scientist, or analyst, the experiment is likely to be conducted improperly unless everyone knows his/her job.

(3)   Establish a team consisting of people from different backgrounds. An ideal team would have engineers, supervisors, analysts, scientists, operators, and possibly someone from maintenance, materials, and marketing. Many well designed experiments have

been unsuccessful because someone did not provide the necessary support. Support is easier to obtain from someone if you make him/her part of the team.

(4)   Review the literature of other work in your area of interest and develop a plan for documentation of your results.

(5)   Define the process (use a process flow diagram) and ensure standard operating procedures are being adhered to.

(6)   Complete a thorough cause-and-effect diagram analysis. Remove all extraneous sources of variation by establishing procedures to ensure consistency.

2.2   Experimental Design Example (Full Factorial)

Perhaps the simplest way to discuss design and analysis is to take an example and walk through the 12-step process described in Chapter 1. Consider the following nickel plating process.

Step I: Statement of the Problem

In the manufacture of high-technology disks for computer disk drives, several layers of materials are typically plated by electrolytic processes. The base material used in some of these products is aluminum. A pictorial cross-section of the product is shown below:

⊗⊗⊗⊗⊗⊗⊗⊗	Chromium Layer
/////////////////////////////	Nickel Layer
\\\\\\\\\\\\\\\\\\\\\\\\\\\\\\	Aluminum Base Material

The customer of the nickel plated layer is unsatisfied with product performance. A process action team has identified the problem as inability to achieve the target plating thickness resulting in unsatisfactory process capability.

Step II: Objective of the Experiment

The objective of the experiment is to determine a way to increase process capability. To achieve this objective, it is necessary to characterize the process (i.e., determine which are the important factors and how they are related to plating thickness). Figure 2.1 depicts the types of factors we will try to identify.

EXPERIMENTAL DESIGN WILL HELP YOU
IDENTIFY THE FOLLOWING TYPES OF FACTORS

i)   Factor A shifts the average.

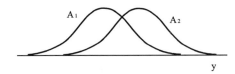

ii)   Factor B shifts the standard deviation.

iii)   Factor C shifts both the average and the standard deviation.

iv)   Factor D has no affect on the average or the standard deviation.

Figure 2.1 Objectives of an Experimental Design

The subscripts "1" and "2" in Figure 2.1 relate to the settings or levels for the various factors. As shown in example (i), changing factor A from its high setting (level 2) to its low setting (level 1) shifts the average of the response, but not the variability. Factor B, however, shifts the variability, but not the average. Factor C shifts both the response average and variability, while Factor D appears to be a benign factor, i.e., it does not shift either the response average or variability. (Note: Because of the introductory nature of this chapter, our first example will only discuss identifying factors that shift the response average. Variability shifting factors are discussed in Section 2.4 and Chapter 5.)

Step III: **Start Date -**   May 7
            End Date -   June 1

Step IV: Select the Response

The response or quality characteristic to be used in the experiment should be measurable and related to the customer's needs and expectations.   Plating thickness variability has been determined to be related to customer dissatisfaction. The thickness measurement process has been studied through a gage capability study [1] and found to be precise, accurate, and stable.

Step V: Select the Factors

Although numerous factors can be tested in a plating process (e.g., bath temperature, time, ph, phosphorous, percent nickel, bath volume, hydrogen ion concentration, etc.), this first example will test only time and bath temperature. The other factors from the cause and effect diagram are held constant at predetermined settings.

Bath temperature and plating time are both quantitative factors whose low and high experimental levels are determined to be (16°C, 32°C) and (4 sec, 12 sec), respectively.   These levels were determined from prior experience, engineering knowledge of the process and results of a literature review.   In addition, engineering knowledge and experience suggests a strong possibility that these two factors interact.

Step VI: Determine the Number of Resources for Experimentation

The process is available for experimentation during the week of 14 - 18 May. Each experimental test will take approximately 50 minutes (including setup time). The estimated cost per test is $400. Due to cost and time constraints, it is determined to allocate 20 resources (products) for experimentation and another 5 for confirmatory tests.

Step VII: Which Experimental Designs and Analysis Strategies Are Appropriate

A wide variety of design types and analysis strategies are available. The advantages/disadvantages of each design type are presented in Chapter 3 and the advantages/disadvantages of each analysis strategy are presented in Chapter 5. With only two factors selected for testing in this example, some of the design type possibilities are:

(i)   2 factor (2 levels) full factorial
(ii)   2 factor (3 levels) full factorial
(iii) central composite design (3 or 5 levels)

Numerous data analysis approaches are available. You have a choice ranging from simple graphs to more complex statistical techniques such as Analysis of Variance (ANOVA) or regression modeling.

Step VIII: Select the Best Design Type and Analysis Strategy to Suit Your Needs

Assuming that quadratic effects are not felt to be important, the best design type for 2 factors (2 levels) that will provide information on both linear effects and their interaction is the 2-factor (2-levels) full factorial. The number of test combinations can be enumerated through the use of the tree diagram displayed in Figure 2.2. Notice that we are now using a ($+1$) to indicate the high factor setting and a (-1) to indicate the low factor setting. This coded notation will allow us to perform more powerful analysis than the previously used 1 and 2 for low and high settings.

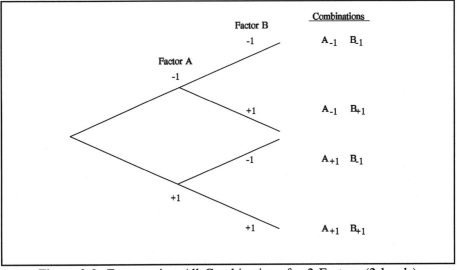

Figure 2.2 Enumerating All Combinations for 2 Factors (2 levels)
in a Tree Diagram

For our 2 factor experiment we will let factor A be bath temperature where $(-1) = 16°C$ and $(+1) = 32°C$. Similarly, factor B will represent plating time where $(-1) = 4$ seconds and $(+1) = 12$ seconds.

The combinations produced in the tree diagram can be represented in a design (or test) matrix as shown in Table 2.1. The actual setting design matrix (Table 2.1a) is used to run the experiment and the coded setting matrix (Table 2.1b) is used for analysis purposes.

The analysis strategy selected for this introductory example will be a graphical approach with a prediction equation relating plating thickness to the factors time and temperature.

a)	**Actual Settings**		b)	**Coded Settings**	
Run	**A (Temp)**	**B (Time)**	**Run**	**A (Temp)**	**B (Time)**
1	16°C	4 sec	1	−1	−1
2	16°C	12 sec	2	−1	+1
3	32°C	4 sec	3	+1	−1
4	32°C	12 sec	4	+1	+1

Table 2.1   a) Design Matrix for Actual Settings and
b) Design Matrix for Coded Settings

Step IX: Randomization of the Runs in the Design Matrix

The run order should be conducted randomly to compensate for the effects of potential uncontrolled factors which were left out of the experiment. If a factor is difficult to randomize, such as bath temperature, one can randomize runs within each temperature level (this is sometimes referred to as a blocked design because the randomization process applies to each block separately).

Step X: Conducting the Experiment and Recording the Data

This is one of the most critical steps in experimentation. It is imperative that everyone involved in the process understand the discipline required in conducting the experimental runs. It is recommended that a very specific test procedure be laid out for the designed experiment. This allows you to do two things: 1) it forces you to think through the details of the experiment, and 2) it helps you and your teammates communicate on exactly how the experiment is to be conducted.   Even after this has been done, things can still go wrong in conducting the experiment; therefore, it is also recommended that you be present for the setup and running of the experiment.

Based on the 20 allocated resources for this experiment, we have decided to collect thickness measurements on a sample of size 5 (i.e., 5 tested products) for each of the 4 design combinations. Each sample of size five can be either repetitions or replications. With replications every experimental trial represents a total reconfiguration of the test setup. To do five replications would typically result in randomly testing one product for each of the four combinations and then repeating this procedure four more times. Repetitions are conducted by testing the entire sample of the same design combination (or run) before adjusting factor levels and conducting tests on another combination. From a statistical standpoint, replications are better (they provide for a better estimate of experimental error), but they also tend to be more expensive. It is also possible for an experimental run to contain both replications and repetitions. See Appendix M for a Rule of Thumb on the desired sample size per combination.

Continuing with our example, the full factorial design is conducted as specified with 5 measures of each experimental run. The design and data collected are shown in Table 2.2.

Run	Temp	Time	Thickness Readings	Average
1	16°C	4 sec	116.1, 116.9, 112.6, 118.7, 114.9	115.8
2	16°C	12 sec	116.5, 115.5, 119.2, 114.7, 118.3	116.8
3	32°C	4 sec	106.7, 107.5, 105.9, 107.1, 106.5	106.7
4	32°C	12 sec	123.2, 125.1, 124.5, 124.0, 124.7	124.3

Table 2.2 Example Design Matrix with Response Data

Step XI: Analyze Data, Draw Conclusions, Make Predictions, and Do Confirmatory Tests

Our choice of analysis strategy is a graphical approach with a prediction equation relating plating thickness to time and temperature. The steps for this analysis approach are:

(1) Conduct "eye-ball" analysis.

(2) Calculate the response average at each level for each design column.

(3) Calculate the effects and half effects for each column.

(4) Plot the averages from Step 2.

(5) Generate a Pareto chart of the absolute value of each half effect from Step 3.

(6) Determine the "importance" of factors/interactions.

(7) Generate a prediction equation using "important" factors and interactions.

(8) Based upon the objective, select the best settings for the factors.

(9) Using the prediction equation generated in Step 7, predict the response. This value will be used as a target for the confirmation runs.

Analysis of the Example Data

Each of the nine steps for analysis will be individually discussed for the plating example.

Step (1) "Eye-ball" analysis − First, review the response values, looking for anything that appears unusual. Investigate all unusual response values and correct any discovered errors. Once we are certain the data is accurate, we can continue.

Suppose we wish to maximize the nickel thickness. Using a "pick the winner" technique, each experimental run is evaluated in light of the objective. The run which most closely meets our objective is declared the winner. Run #4 (temperature = 32°C, time = 12 seconds) is the best of the four possible combinations with an average thickness of 124.3 microinches. The obvious advantage of this approach is in its simplicity. Disadvantages lie in its inability to indicate which factors have the greatest effect, in possible suboptimization, and in its inability to predict untested combinations (this will be very important with fractional factorial designs discussed in Chapter 3). Another disadvantage of simply picking a design run winner occurs when trying to achieve a target thickness and none of the run averages equals the target value.

Steps (2) and (3) Determine the effect and half effect for each column by first calculating the response average at each level for each column. This technique will be demonstrated using the temperature column. For the design in Table 2.2, temperature is at the low setting (16°C) for runs #1 and #2. There are ten thickness readings (116.1, 116.9, 112.6, ... , 114.7, 118.3) when temperature is at 16°C. The average of these ten values is 116.3 microinches. When temperature is at 32°C, the average reading is 115.5 microinches. The difference (average at the high minus the average at the low) between these two averages is −0.8 microinches. This difference (or Δ) is defined as the effect of the temperature factor. On the average, we estimate that changing temperature from 16°C to 32°C decreases the plating thickness by 0.8 microinches. Analysis of the time column gives us an average of 111.3 microinches when time is 4 seconds and 120.6 microinches when time is 12 seconds. The estimated effect for time is 120.6 − 111.3 = 9.3 microinches. Comparing the effect of temperature with the effect of time indicates that the time factor appears to be approximately twelve times more important than temperature (over the tested levels) in effecting plating thickness.

The simple analysis approach used above to estimate the effects of time and temperature is possible because of the nature of the design matrix. All full factorial designs (as well as many other families of designs) are orthogonal designs. These types of designs are simple, but powerful in that each column in the design can be used to independently estimate the effect of the assigned factor. Had our design not been orthogonal, this simple analysis would not have been possible. Table 2.3 is a summary of the analysis up to this point.

A   B

Run	Temp	Time	Thickness Readings	Avg
1	— 16°C	4 sec —	116.1, 116.9, 112.6, 118.7, 114.9	115.8
2	— 16°C	12 sec +	116.5, 115.5, 119.2, 114.7, 118.3	116.8
3	+ 32°C	4 sec —	106.7, 107.5, 105.9, 107.1, 106.5	106.7
4	+ 32°C	12 sec +	123.2, 125.1, 124.5, 124.0, 124.7	124.3
	116.3	111.3	Average Thickness at Low Setting	
	115.5	120.6	Average Thickness at High Setting	
	−0.8	9.3	Effect (Δ)	

Table 2.3 Analysis Summary for Factor Effects

$$\bar{\bar{y}} = 115.93 \quad \text{Grand Mean}$$

In addition to providing independent estimates of the effects of factors, full factorial designs (and other types of designs — see Chapter 3) can be used to provide independent estimates of the factor interaction effects. Interactions refer to dependencies between a factor's effect on the response and the levels of another factor.

Let's leave our example briefly to discuss the concept of interactions and how orthogonal designs can be used to analyze interactions. We first look at an example where factors do not interact. Consider two factors, dwell time and cure temperature, where dwell time is being investigated at two levels and cure temperature is being investigated at three levels. The one response being considered is tensile strength. Using a full factorial design, the runs are shown in Table 2.4.

A plot of the data in Table 2.4 with strength on the y-axis, cure temperature on the x-axis, and dwell times displayed as two lines is shown in Figure 2.3.

Run	Dwell Time	Cure Temp	Tensile Strength (using ASTM D3916)
1	60 min	250°F	100
2	60 min	300°F	150
3	60 min	350°F	200
4	100 min	250°F	125
5	100 min	300°F	175
6	100 min	350°F	225

Table 2.4 Design Matrix and Response Values for
an Example with No Interaction Effects

To understand how an interaction graph is generated, verify for yourself that the ★ on Figure 2.3 represents the average response when cure temperature is 250°F and dwell time is 100 minutes, i.e., the average response for run number 4. The other dots in Figure 2.3 are obtained in a similar manner.

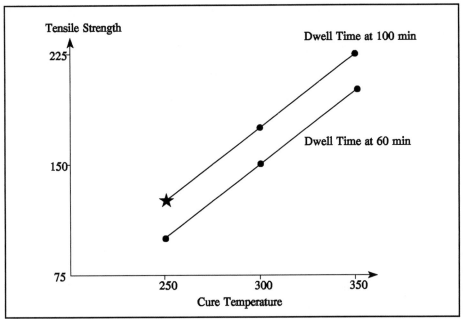

Figure 2.3 Graph of 2 Factors that Do Not Interact

Note that the lines in Figure 2.3 are parallel. With this type of plot, parallel or near parallel lines are indicators of little or no interaction. Figure 2.3 allows us to quantify the impact of increasing dwell time from 60 minutes to 100 minutes independent of the setting for cure temperature. The effect is an increase of 25 units regardless of the level for cure temperature.

Next, the scenario is modified to look at an example of factors which do interact. In this case, suppose the same full factorial design is conducted for cure temperature and dwell time, but now the response values are somewhat different (see Table 2.5 and Figure 2.4).

Run	Dwell Time	Cure Temp	Tensile Strength
1	60 min	250°F	200
2	60 min	300°F	200
3	60 min	350°F	200
4	100 min	250°F	100
5	100 min	300°F	200
6	100 min	350°F	300

Table 2.5   Design Matrix and Response Values of 2 Factors That Do Interact

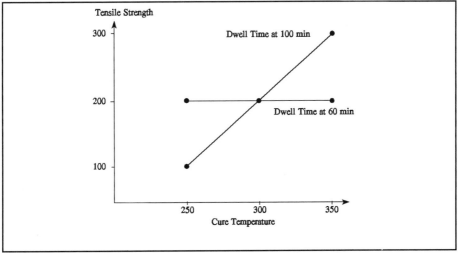

Figure 2.4   Graph of a 2-Factor Interaction

No longer are the lines parallel. In this scenario we can no longer discuss the impact of dwell time without first specifying the level for cure temperature. There now appears to be a dependency between cure temperature's effect on tensile strength and the levels of dwell time.

Returning to our plating example with two factors at two levels, we will plot the interaction graph and then mathematically calculate the effect of the interaction. The interaction plot for temperature and time in the plating example is shown in Figure 2.5. This interaction graph indicates that there may be an interaction between temperature and time. To mathematically evaluate the effect of the time and temperature interaction, we must convert the design matrix from the actual factor settings to the coded values (for 2-level designs, the low level is always a -1 setting and the high level is always a $+1$ setting). The coded matrix for temperature and time is shown in Table 2.6.

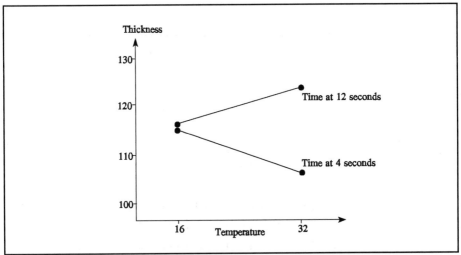

Figure 2.5 Interaction Graph for Temperature and Time
in the Plating Example

Run	A(Temp)	B(Time)
1	-1	-1
2	-1	$+1$
3	$+1$	-1
4	$+1$	$+1$

Table 2.6 Coded Design Matrix for the Plating Example

In order to quantify the temperature and time interaction effect, a coded column is generated to represent the interaction. An interaction column AB is generated by simply multiplying together the appropriate column values of temperature (A) and time (B) for each row. For row #1, the value for AB is +1. This value is obtained by multiplying together −1 (for temperature) and −1 (for time). Continuing this operation for all four rows we obtain the columns shown in Table 2.7.

Run	A(Temp)	B(Time)	AB
1	−1	−1	+1
2	−1	+1	−1
3	+1	−1	−1
4	+1	+1	+1

Table 2.7 Coded Design Matrix with AB Interaction Column

We can now use the AB* column to mathematically determine the effect of the interaction. The actual calculations are performed in the same manner as for the factor columns. For the AB interaction column, the response average associated with −1 values is $\frac{116.8 + 106.7}{2} = 111.75$. The response average associated with +1 values is $\frac{115.8 + 124.3}{2} = 120.05$. Thus, the interaction effect is $120.05 - 111.75 = 8.3$.

The next part of the analysis is to calculate the half effects ($\Delta/2$) for each column. The half effects (as the name implies) are simply one half of the effects

---

*Throughout this book you will see interaction terms (such as the one representing the A and B interaction) denoted as AB, A•B, A×B or A*B. They are all equivalent.

(Δ's); these values will be used later to build a prediction equation. For our example, the values are:

	A (Temp)	B (Time)	AB
$\Delta/2$	-0.4	4.65	4.15

Step (4) in our 9 Step Data Analysis Procedure is to plot the averages from steps (2) and (3). The graphical analysis of the averages will reveal the relative importance of all of the effects (see Figure 2.6).

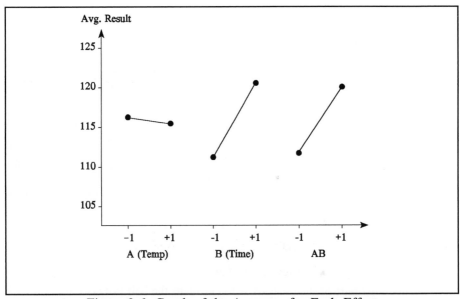

Figure 2.6 Graph of the Averages for Each Effect
(Also referred to as a graph of the Marginal Means or Effects Plot)

Step **(5)** directs us to generate a Pareto chart of the absolute value of each half effect. The Pareto chart for our example is shown below.

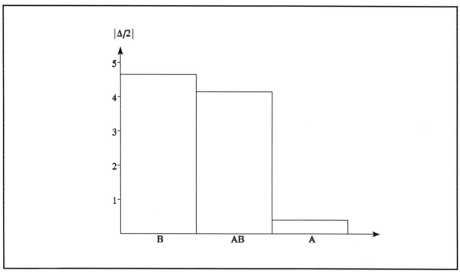

Figure 2.7 Pareto Chart for $|\Delta/2|$ Values or Half Effects

In **Step (6)** we must determine which factors and interactions are important and which are trivial. The three approaches available to assist in making this decision are:

    (1) Graphical

    (2) Analysis of Variance (ANOVA)

    (3) Regression Modeling

In Chapters 4 and 5, ANOVA and regression approaches are discussed. At this point we present only the graphical approach. There are two reasons for determining which factors/interactions should be included in our prediction equation: (1) If too many are included, we tend to overpredict; whereas, if too few are included, we tend to underpredict. Neither of these situations is good. (2) Optimal factor levels will be based on the $\Delta/2$ for factor effects that appear in the model as well as on the cost, convenience, etc., for benign factors whose effects are not strong enough to be in the model. Our objective is to develop a prediction equation which includes only those factor effects and interactions which are important in describing the response. The calculated effect for important factors/interactions should be non-zero; however, even benign factor effects can possibly be non-zero due to sampling error. Using the shape of the Pareto chart, the size of the half effects, and prior knowledge, we should be able to make a fairly good decision concerning which effects should be in the model and which should not. If in doubt, we will normally place the term in the model. A more statistical approach to significance will be presented in Chapters 4 and 5.

The shape of the Pareto chart can reveal a separation between important and benign effects. For example, the Pareto chart in Figure 2.7 indicates that effects B and AB are important and effect A is potentially benign. The size of the half effect can also be used in helping us make a decision with regard to which factors/interactions should be included in the prediction equation. For example, the calculated half effect for time in our example is 4.65 microinches. Leaving this term out of the model when it is truly important would result in an inaccurate prediction of $\pm$ 4.65 microinches. If 4.65 microinches is technically significant, time should be included in our prediction equation. On the other hand, leaving the temperature effect out of the prediction equation would only result in a $\pm$. 4 microinch inaccuracy.

Using our technical knowledge of a process can also be useful in determining what needs to be included in the prediction equation. Suppose it is known beforehand that a particular factor has a positive effect upon the response. In this case the factor will usually be included in our prediction equation regardless of the size of the calculated effect.

If prior technical knowledge suggests that both time and temperature are important factors for the plating example, then the Pareto chart for our process suggests the Time $\times$ Temperature interaction (A$\times$B) is important as well.

Step (7) The general form of the prediction equation is shown below.

$$* \quad \hat{y} = \overline{\overline{y}} + \left(\frac{\Delta_A}{2}\right) A + \left(\frac{\Delta_B}{2}\right) B + \left(\frac{\Delta_{AB}}{2}\right) AB + \dots \quad \text{(other terms for larger}$$

designs), where

$\hat{y}$	$=$	the predicted response
$\overline{\overline{y}}$	$=$	the average of all response values from the experimental data, or the grand mean
$\dfrac{\Delta_A}{2}$	$=$	half effect for factor A
$\dfrac{\Delta_B}{2}$	$=$	half effect for factor B
$\dfrac{\Delta_{AB}}{2}$	$=$	half effect for the AB interaction

* NOTE

(1) We only include the important effects in the prediction equation.

(2) This prediction equation applies only to orthogonally - coded factor settings for 2-level designs. See Chapter 4 techniques for other designs.

(3) The origin of this equation for only 1 factor is the slope and intercept model, i.e.,

$$\hat{y} = \text{intercept} + \text{slope}(A) = \overline{\overline{y}} + \frac{\Delta_A}{2} A.$$

Equation 2.1 Prediction Equation for a Coded 2-Level Orthogonal Design

Although the use of coded factor settings in the prediction equation may appear at first glance to be an inconvenience, models built on the true factor settings will frequently provide equations whose coefficients are not independent.

Therefore, only equations for coded settings will provide a clear understanding of the relationship between the response and the inputs.

The prediction equation for our example is:

$$\hat{y} = 115.93 - \left(\frac{.8}{2}\right)A + \left(\frac{9.3}{2}\right)B + \left(\frac{8.3}{2}\right)AB = 115.93 - (.4)A + (4.65)B + (4.15)AB$$

Notice that the A effect is included in the model even though it was thought to be unimportant. The hierarchy rule for modeling dictates that if a higher order term (quadratic or interaction) is included in the model, then all linear effects represented in the higher order term will be included in the model regardless of their significance. In our example, the inclusion of the AB term as an important effect dictates (using the hierarchy rule) that both the linear A and B terms be included in the model. The prediction equation represents a Taylor Series approximation of the true physical process model relating the observed response to the factors tested. An important observation is that the prediction equation can be obtained quickly and is based on actual process data.

Step (8) Select the best settings for the factors. If our objective is to minimize the response, then our optimum predicted thickness occurs for $A = (+1)$ and $B = (-1)$ (i.e., temperature $= 32°C$ and time $= 4$ seconds). The predicted minimum thickness for $A = (+1)$ and $B = (-1)$ is:

$$\hat{y} = 115.93 - .4(+1) + 4.65(-1) + 4.15(+1)(-1) = 106.73$$

Analyzing the plot of averages at each level (see Figure 2.6) reveals the same optimal settings as we found with the prediction equation; however, the graphical approach does not provide an accurate prediction value.

If the objective is to maximize thickness, then the best settings appear to be temperature at 32°C and time at 12 seconds (i.e., both A and B at the high setting). The predicted maximum thickness is:

$$\hat{y} = 115.93 - .4(+1) + 4.65(+1) + 4.15(+1)(+1) = 124.33$$

The two previous analysis approaches agree with the conclusions reached by the "eye-ball" analysis approach. This will always be the case with full factorial designs when maximizing or minimizing the response. Why then do we need to create a prediction equation? There are three important reasons: 1) we will not always conduct a full factorial design and the prediction equation can be used to estimate the untested combinations when a fractional factorial design is used, 2) the equation is useful in predicting a target value, and 3) the coded prediction equation can be used to complete sensitivity analysis and tolerance design.

Assume that our objective is to achieve a target thickness of 120 microinches. Since we have one prediction equation with 2 variables, suppose we decide to set factor A at $32°C$ ($+1$). The prediction equation can then be used as follows:

$$\hat{y} \quad = \quad 115.93 - .4(A) + 4.65(B) + 4.15(A)(B)$$
$$120 \quad = \quad 115.93 - .4(+1) + 4.65(B) + 4.15(+1)(B)$$
$$4.47 \quad = \quad 8.80(B) \quad \Rightarrow \quad B = .51$$

In order for the coded value $B = .51$ to be meaningful, it must be translated to actual units for the factor time.[*] The following picture indicates what is happening.

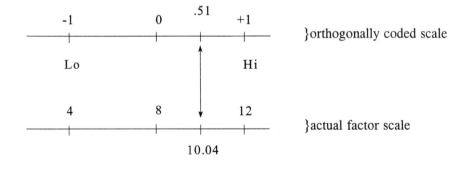

---

[*]Normally, this transformation from coded to actual settings (and vice versa) is computed by a software package such as DOE KISS, RS/Discover, DOE Wisdom, JMP, etc.

To obtain the actual setting corresponding to a coded value of .51, the computations must be conducted:

$$\text{Actual} = \left[\frac{\text{Hi} + \text{Lo}}{2} \right] + \left[\frac{\text{Hi} - \text{Lo}}{2} \right] (\text{Coded})$$

$$= \left[\frac{12 + 4}{2} \right] + \left[\frac{12 - 4}{2} \right] (\text{Coded})$$

$$= 8 + 4(.51) = 10.04$$

Thus, setting time at 10.04 seconds and temperature at 32°C should give us a thickness of 120 microinches.

Step (9) involves obtaining a predicted best value and conducting a confirmation test. Suppose (as discussed earlier) the target thickness is 120 microinches. From the prediction equation, setting time at 10.04 seconds and temperature at 32°C should result in a thickness of 120 microinches.

Confirmation tests should now be conducted to validate our model (prediction equation). We recommend a confirmatory test minimum sample size between 4 and 20, depending on what is affordable. The confirmation process will consist of repeated trials with time at 10.04 seconds and temperature at 32°C. If the response average of the confirmation test is close to the predicted value, one confirms the previous assumptions and the accuracy of the model. If, on the other hand, the confirmatory response average is not close to the predicted value, one will conclude that one or more of the assumptions made are not valid and/or the model is inaccurate.

For example, suppose that five verification tests produce the following results: 120.5, 122.7, 118.4, 125.3, 117.0. Pictorially, the data would appear as follows:

Since the actual values are distributed nicely about 120, it appears reasonable to conclude that our confirmation test verified our assumptions and our model. If, on the other hand, our results had been 130.6, 125.9, 127.3, 133.3, 119.1:

or 110.5, 115.6, 120.6, 112.3, 111.0:

then our conclusions are questionable. The previous discussion is based on a simple graphical approach for determining if the experimental results have been confirmed. Chapter 5 provides a more in-depth discussion of confirmation.

Step XII   Assess Results, Make Decisions, and If Necessary, Conduct More Experimentation

Let's assume, for example, that the confirmation results closely match the predicted value with relatively small variability. We would then set up and run our plating process at the predicted best settings and conduct a capability study to evaluate our new state of quality. The new C_{pk} would be compared with the old process C_{pk} described in Step I. In Section 2.4, you will also learn how to model the standard deviation, s. Using the $\hat{s}$ prediction model you will also be able to optimize the standard deviation. If you are a researcher, developer, or designer, you can use both the $\hat{y}$ and $\hat{s}$ models to estimate C_{pk} values before technology transfer. Obviously, if the estimated C_{pk} is unsatisfactory, it is advisable to redesign the product to insure it is manufacturable.

The previously discussed plating example was intended to provide an overview of the steps in setting up a designed experiment, as well as a discussion of simple data analysis. The design analyzed was a full factorial design. Full factorial designs are good in that they provide estimates of all factor effects, as well as the effects of all possible interactions. A disadvantage is that full factorial designs will require a large number of runs when more than a few factors are tested. For example, it is not unusual to want to experiment with ten factors at two levels. A full factorial design would require 2^{10} or 1024 experiments. This number is excessive for even the largest budget. Fortunately, alternatives to full factorial designs exist. One very popular alternative is the fractional factorial design. As the name implies, we run only a portion or fraction of the full factorial. The good news with fractional factorials is that we can dramatically reduce the total number of runs required to estimate the effects of factors. The bad news is that some (or all) information about interactions is lost. See Appendix M for Rules of Thumb on when to use full and fractional factorials.

2.3   Fractional Factorial Designs for 2-Level Factors

Let's discuss how fractional factorial designs can be generated. Suppose we have two factors (A and B) to be tested at two levels. The coded design matrix is shown in Table 2.8.

Run	A	B
1	-1	-1
2	-1	$+1$
3	$+1$	-1
4	$+1$	$+1$

Table 2.8 Coded Design Matrix for 2 Factors at 2 Levels

We could then generate the AB interaction column as demonstrated previously in the plating example. Our coded design matrix would now appear as shown in Table 2.9.

Run	A	B	AB
1	−1	−1	+1
2	−1	+1	−1
3	+1	−1	−1
4	+1	+1	+1

Table 2.9 Coded Design Matrix with the Interaction Column

Now suppose a third factor C exists which we want to include in our experiment. The two choices are: 1) full factorial with eight runs, or 2) a fractional factorial of only four runs. The mechanics of creating a complex fractional factorial are somewhat burdensome (see Chapter 3 for the details). In this case we simply take the AB column and assign the settings for Factor C to it. Our design for 3 factors in 4 runs is shown in Table 2.10.

Run	A	B	C = (AB)
1	−1	−1	+1
2	−1	+1	−1
3	+1	−1	−1
4	+1	+1	+1

Table 2.10 Fractional Factorial Design of 3 Factors (at 2 Levels) in 4 Runs

The tree diagram for three factors at two levels is presented in Figure 2.8.

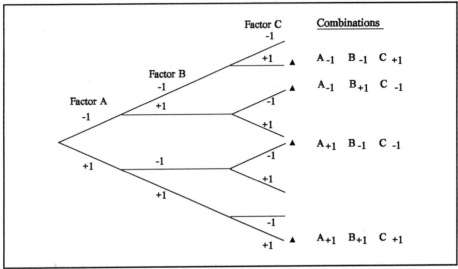

Figure 2.8 Tree Diagram for a 3-Factor Fractional Factorial in 4 Runs

Note that the branches labeled with triangles (▲) are the four runs specified in the fractional factorial which is a subset of the full factorial. It is, however, a special subset in that it is an orthogonal subset. If we had chosen the remaining four runs in the above tree diagram, it would have been the complementary fractional factorial which is another orthogonal subset of the full factorial.

Returning to the coded design matrix of Table 2.10, let's generate all possible interaction columns for 3 factors at 2 levels: AB, AC, BC, ABC (see Table 2.11).

Run	A	B	C	AB	AC	BC	ABC
1	-1	-1	$+1$	$+1$	-1	-1	$+1$
2	-1	$+1$	-1	-1	$+1$	-1	$+1$
3	$+1$	-1	-1	-1	-1	$+1$	$+1$
4	$+1$	$+1$	$+1$	$+1$	$+1$	$+1$	$+1$

Table 2.11 Coded Matrix for All Terms in a 3-Factor (2-Level) 4-Run
Half Fractional Factorial Design

Notice that column A is identical to column BC. B is identical to AC, while C is identical to AB. Thus, A is said to be aliased with BC, B with AC, and C with AB. The impact of aliasing is encountered when we evaluate the effects for each column. Since A is aliased with BC, analysis for these columns will provide the same effect values. The practical implication of A aliased with BC is that we are unable to determine whether the effect calculated for column A is due to A, due to BC, or due to some combination of the two. Fortunately for industrial experimenters, the effects of interactions are usually small relative to the effects of factors, thus providing justification for the use of fractional factorial designs. In addition, not all fractional factorial designs result in the loss of all interaction information. Chapter 3 provides a detailed discussion of fractional factorial designs. See Appendix M for Rules of Thumb on the use of full and fractional factorials.

Consider another example of testing 4 factors in a half fraction of 8 runs or combinations. The full factorial for an 8-run matrix is based on only 3 factors (see Table 2.12).

Run	A	B	C	AB	AC	BC	ABC
1	−1	−1	−1	+1	+1	+1	−1
2	−1	−1	+1	+1	−1	−1	+1
3	−1	+1	−1	−1	+1	−1	+1
4	−1	+1	+1	−1	−1	+1	−1
5	+1	−1	−1	−1	−1	+1	+1
6	+1	−1	+1	−1	+1	−1	−1
7	+1	+1	−1	+1	−1	−1	−1
8	+1	+1	+1	+1	+1	+1	+1

Table 2.12 3-Factor (2-Level) Full Factorial Coded Matrix

Now consider which column in Table 2.12 you would be willing to give up to estimate a fourth factor D without going to the 16 run full factorial for 4 factors. The least likely column in Table 2.12 to produce an important effect is the 3 way interaction ABC; therefore, it is the best choice for aliasing with the settings for factor D. The half fraction (8-run matrix versus 16) for 4 factors at 2 levels is displayed in Table 2.13. Note that aliasing factor D with the ABC interaction also created other effect aliasings, i.e., A = BCD, B = ACD, C = ABD, AB = CD, AC = BD and BC = AD. Table 2.13 illustrates the two names for each column.

Run	BCD A	ACD B	ABD C	CD AB	BD AC	AD BC	D ABC
1	− 1	− 1	− 1	+ 1	+ 1	+ 1	− 1
2	− 1	− 1	+ 1	+ 1	− 1	− 1	+ 1
3	− 1	+ 1	− 1	− 1	+ 1	− 1	+ 1
4	− 1	+ 1	+ 1	− 1	− 1	+ 1	− 1
5	+ 1	− 1	− 1	− 1	− 1	+ 1	+ 1
6	+ 1	− 1	+ 1	− 1	+ 1	− 1	− 1
7	+ 1	+ 1	− 1	+ 1	− 1	− 1	− 1
8	+ 1	+ 1	+ 1	+ 1	+ 1	+ 1	+ 1

Table 2.13 A Half Fraction Coded Matrix for 4 Factors at 2 Levels

Thus, if any 2-factor interaction column has a significant effect, it could be due to one or both of the aliased effects for that column. Which effect is actually placed in the prediction equation is based on prior knowledge as to which interaction is most likely to be important. It is important to understand that the term "aliased effects" does not imply that their physical effects upon the response are the same. Rather, aliased effects have identically coded columns and, thus, evaluation of that column will not separate the two effects. In practice, an experimenter typically chooses a fraction of the full factorial design and evaluates the aliasing patterns before running any tests. If the aliasing patterns are unacceptable, then more runs or fewer factors must be selected. Those experimenters not familiar with the generation of fractional factorial designs could simply refer to the tabled designs in Section 3.3.

2.4   Analysis of a 2-Level Fractional Factorial Design

Suppose you are a process engineer for an aircraft company and are responsible for the painting process. You wish to conduct a designed experiment in order to determine which of 4 factors are influential on paint thickness and so that you can hit a target thickness of .5 mm. The factors being tested are shown in Table 2.14.

Factor	Low	High
A - Thinner Amount	10	20
B - Fluid Pressure	15	30
C - Temperature	68	78
D - Vendor	1	2

Table 2.14 Factors and Settings for Aircraft Painting Example

An 8-run fractional factorial design is conducted with 4 replicates of each run. The design and results are displayed in Table 2.15.

Target = 0.5

Run	Thinner Amount	Fluid Pressure	Temp	Vendor	Paint Thickness (y)	Avg (ȳ)	Std Dev (s)
1	10	15	68	1	.638, .489, .477, .541	.536	.073
2	10	15	78	2	.496, .509, .513, .495	.503	.009
3	10	30	68	2	.562, .493, .529, .531	.529	.028
4	10	30	78	1	.564, .632, .507, .757	.615	.108
5	20	15	68	2	.659, .623, .636, .584	.626	.031
6	20	15	78	1	.549, .457, .604, .386	.499	.096
7	20	30	68	1	.842 .910, .657, .953	.840	.131
8	20	30	78	2	.910, .898, .913, .875	.899	.017
					Overall average	.631	.0616

Table 2.15 Experimental Data for Aircraft Painting Process

Analysis Steps

Step (1) Runs #2 and #6 come closest to meeting our target value of .5 inches.

Steps (2) and (3) The averages, effects, and half effects for the applicable factors and interactions are shown in Table 2.16.

Factor/Interaction	Avg −	Avg +	effect (Δ)	Δ/2
A - Thinner amount	.546	.716	.17	.085
B - Fluid pressure	.541	.721	.18	.090
C - Temperature	.633	.629	−.004	−.002
D - Vendor	.623	.639	.016	.008
*AB(CD)	.567	.695	.128	.064
*AC(BD)	.646	.616	−.030	−.015
*BC(AD)	.593	.669	.076	.038

Table 2.16 Table of Effects

---

* Since we are evaluating 4 factors in an 8-run design, the 2-factor interactions are aliased with each other, i.e., AB(CD) implies AB is aliased with CD.

Step (4) A plot of the averages at the high and low for each effect is shown in Figure 2.9.

Step (5) Generate a Pareto chart of the absolute value of each half effect as shown in Figure 2.10.

Step (6) Determine the important factors/interactions. Based upon the Pareto chart, in Figure 2.10, and engineering knowledge, we decide that the factors A and B and the interaction AB(CD) are the important effects. Since AB and CD are aliased, we are unable to determine whether the associated effect is due to AB or CD. It is, however, unusual for an interaction to be important without one or both of the constituent factors being important. Therefore, the important interaction is probably AB which is plotted in Figure 2.11.

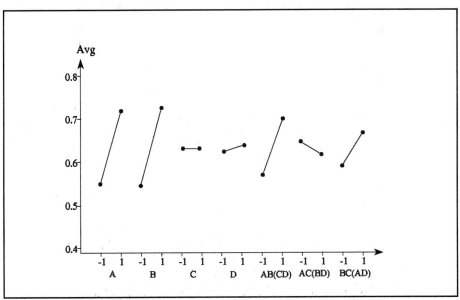

Figure 2.9 Graph of the Averages for Each Effect

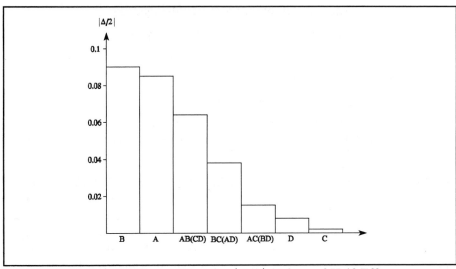

Figure 2.10 Pareto Chart for $|\Delta/2|$ Values of Half Effects

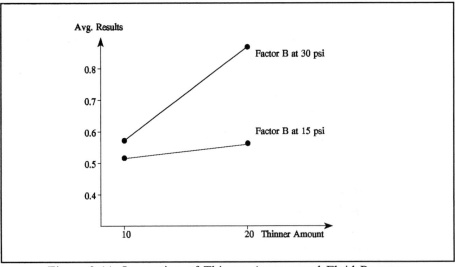

Figure 2.11 Interaction of Thinner Amount and Fluid Pressure

A more quantitative method for determining effect significance in $\hat{y}$ models is discussed in Chapters 4 and 5.

Step (7) Generate the following prediction equation:

$$\hat{y} = .631 + .085(A) + .09(B) + .064(A)(B)$$

To model the standard deviation for any orthogonal design, use the same mechanics discussed for modeling y. For this example the average s at the low and high for each column is shown in Table 2.17.

Factor/Interaction	s Avg. -	s Avg. +	s effect (Δ)	Δ/2
A - Thinner Amount	.0545	.0687	.0142	.0071
B - Fluid Pressure	.0522	.0710	.0188	.0094
C - Temperature	.0657	.0575	-.0082	-.0041
D - Vendor	.1020	.0212	-.0808	-.0404
AB(CD)	.0657	.0575	-.0082	-.0041
AC(BD)	.0697	.0535	-.0162	-.0081
BC(AD)	.0660	.0572	-.0088	-.0044

Table 2.17 Average s for Highs and Lows

The Pareto chart for the s half effects (Δ/2) is shown in Figure 2.12.

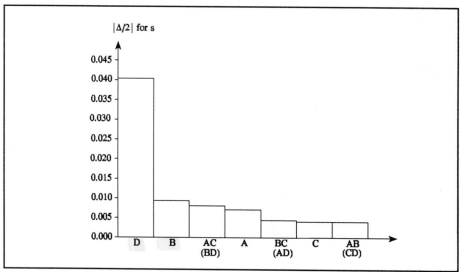

Figure 2.12 Pareto Chart for s Half Effects, |Δ/2|

Using the results of the Pareto chart in Figure 2.12 and Rule of Thumb (6) in Appendix M, the only term which appears to have a significant effect on s is factor D. Therefore, the general ŝ model, which is

$$\hat{s} = \bar{s} + \left(\frac{\Delta_A}{2}\right)A + \left(\frac{\Delta_B}{2}\right)B + \left(\frac{\Delta_{AB}}{2}\right)AB + \left(\frac{\Delta_C}{2}\right)C + \left(\frac{\Delta_{AC}}{2}\right)AC + \left(\frac{\Delta_{BC}}{2}\right)BC + \left(\frac{\Delta_D}{2}\right)D,$$

can be reduced to ŝ = .0616 - .0404D.

Step (8) Using both the ŷ and ŝ equations we can optimize the process to a target average thickness of .5 with minimum standard deviation. Starting with the ŝ equation, ŝ = .0616 - .0404D, we will need to set D = +1 (coded), which is vendor #2, to obtain a minimum ŝ of .0212. To achieve a target average thickness of .5, we set the ŷ equation equal to .5 as shown below:

$$\hat{y} = .631 + .085(A) + .09(B) + .064 \ (A)(B)$$

$$.5 = .631 + .085(A) + .09(B) + .064 \ (A)(B)$$

Since there is 1 equation with 2 variables (factors), we set 1 variable based on cost, throughput, convenience, etc., and solve for the other. Assume we set thinner amount at the low (-1) then the equation becomes

$$.5 = .631 + .085(-1) + .09(B) + .064(-1)B$$

which results in B = -1.77. This coded value of B is outside the experimental region; however, it may still be worth trying. Whether or not the ŷ will extrapolate outside the experimental region will be checked during the confirmation tests. The actual setting for B = -1.77 can be found using the following generic transformation equation from actual to coded values.

$$Actual = \left[\frac{Hi + Lo}{2}\right] + \left[\frac{Hi - Lo}{2}\right] Coded$$

For factor B, the transformation equation is:

$$Actual = \left[\frac{30 + 15}{2}\right] + \left[\frac{30 - 15}{2}\right] Coded$$

$$Actual = 22.5 + 7.5 \ Coded$$

Thus, a coded setting of B $= -1.77$ is equivalent to an actual setting of

$$Actual = 22.5 + 7.5(-1.77) = 9.225$$

Therefore, setting A $= -1$ coded (or 10 actual), B $= -1.77$ coded (or 9.22 actual), C $=$ any setting within the region 68 to 78, and D $= 1$ coded (or vendor 2 actual) should produce a $\hat{y} = .5$ and $\hat{s} = .0212$.

If we know the customer specs we can also estimate C_{pk} at this point, i.e.,

$$\hat{C}_{pk} = \frac{min(US - \hat{y}, \ \hat{y} - LS)}{3\hat{s}}$$

Assume your customer specs are LS $= .49$ and US $= .51$, then

$$\hat{C}_{pk} = \frac{min(.51 - .5, \ .5 - .49)}{3(.0212)} = \frac{.01}{.0636} = .157$$

This estimate of C_{pk} equates to more than 133,600 defects per million using Table 1.3. (Note: most companies would benefit substantially if C_{pk} could be estimated early in the product or process design phase. Low C_{pk} estimates like this one indicate downstream manufacturing problems unless improvements are made.)

Step (9) We are now ready to confirm the conclusions in Step (8). Usually 4 to 20 minimum parts or tests at the optimum settings will allow us to determine if we have confirmed the $\hat{y}$ ($\hat{s}$) model. Oftentimes, the number of parts or tests conducted during confirmation is constrained by cost or time. Obviously, one should try to meet or exceed the minimums described above, if possible.

Graphically, if the results are close to the predicted $\hat{y}$, the experimental results are verified. A more quantitative method for determining verification is to compute $\hat{y} \pm 3\hat{s}$ which, for our example, is $.5 \pm 3(.0212) = [.4364, .5636]$. Assuming a distribution close to normal, 99.73% of all confirmatory part values should fall within these limits. If the models cannot be confirmed, we will have to investigate what caused the lack of confirmation. See Chapter 5 for a detailed discussion of this topic.

2.5   Two-Level Screening Example

Suppose we have 11 factors to evaluate. A full factorial is $2^{11} = 2048$ runs which is just too many to consider. Even a half or quarter fraction of the full factorial has too many runs. When we have large numbers of factors (greater than 5) it is often advisable to split the experimental strategy into two parts: screening and modeling. To screen large numbers of factors simply means to separate the vital few from the trivial many. Appendix M, Rule of Thumb (4), indicates that one suggested approach is a Taguchi L_{12} design.

In Chapter 3 you can read how this design is built and how it is related to the Plackett-Burman design and a Hadamard matrix. For the purposes of this chapter, we will use the L_{12} for the experiment and discuss how to analyze it using graphical techniques.

The L_{12} design matrix shown in Table 2.18 contains 11 columns which are tested over 12 combinations (runs). Thus, up to 11 factors can be tested in this screening experiment. The suggested 4 replicates indicate the minimum number to adequately detect a reasonable shift in standard deviation. Using this L_{12} design to test our 11 factors, we choose the low and high settings for each factor and randomly (if possible) make the 12 runs. Assume the data from our screening experiment is shown in Table 2.19.

Factors															
Run	x_1	x_2	x_3	x_4	x_5	x_6	x_7	x_8	x_9	x_{10}	x_{11}	y_1	y_2	y_3	y_4
1	$-$	-1	-1	-1	-1	-1	-1	-1	-1	-1	-1				
2	-1	-1	-1	-1	-1	$+1$	$+1$	$+1$	$+1$	$+1$	$+1$				
3	-1	-1	$+1$	$+1$	$+1$	-1	-1	-1	$+1$	$+1$	$+1$				
4	-1	$+1$	-1	$+1$	$+1$	-1	$+1$	$+1$	-1	-1	$+1$				
5	-1	$+1$	$+1$	-1	$+1$	$+1$	-1	$+1$	-1	$+1$	-1				
6	-1	$+1$	$+1$	$+1$	-1	$+1$	$+1$	-1	$+1$	-1	-1				
7	$+1$	-1	$+1$	$+1$	-1	-1	$+1$	$+1$	-1	$+1$	-1				
8	$+1$	-1	$+1$	-1	$+1$	$+1$	$+1$	-1	-1	-1	$+1$				
9	$+1$	-1	-1	$+1$	$+1$	$+1$	-1	$+1$	$+1$	-1	-1				
10	$+1$	$+1$	$+1$	-1	-1	-1	-1	$+1$	$+1$	-1	$+1$				
11	$+1$	$+1$	-1	$+1$	-1	$+1$	-1	-1	-1	$+1$	$+1$				
12	$+1$	$+1$	-1	-1	$+1$	-1	$+1$	-1	$+1$	$+1$	-1				

Table 2.18   Taguchi L_{12} Design Matrix

Run	y_1	y_2	y_3	y_4	$\bar{y}$	s
1	532.412	460.059	646.676	500.971	535.094	80.11
2	647.412	435.147	624.000	493.706	550.066	102.18
3	1270.882	1356.265	1198.647	1248.000	1268.449	65.85
4	895.471	966.794	837.853	1043.265	935.846	88.94
5	1572.118	1699.882	1693.059	1518.882	1620.985	88.87
6	1584.706	1395.088	1334.059	1388.677	1425.632	109.53
7	1092.835	1100.765	1049.576	1072.729	1078.976	22.88
8	1205.729	1235.765	1224.047	1253.153	1229.674	19.94
9	694.459	681.506	748.741	643.518	692.056	43.54
10	1423.235	1461.318	1411.247	1416.706	1428.126	22.66
11	806.859	849.635	779.353	773.212	802.265	34.80
12	950.706	966.518	923.259	957.388	949.468	18.64

Table 2.19   Data From a Taguchi L_{12} Screening Design

The result of finding the average response and the average standard deviation at the low (-1) and the high ($+1$) settings is shown in Table 2.20.

	x_1	x_2	x_3	x_4	x_5	x_6	x_7	x_8	x_9	x_{10}	x_{11}
Avg $\bar{Y}$ @ -1	1056.012	892.386	744.132	1052.235	970.026	1032.660	1057.829	1035.097	1033.807	1041.071	1050.368
Avg $\bar{Y}$ @ $+1$	1030.094	1193.720	1341.974	1033.871	1116.080	1053.446	1028.277	1051.009	1052.299	1045.035	1035.738
Δ	-25.918	301.334	597.842	-18.364	146.054	20.786	-29.552	15.912	18.492	3.964	-14.630

Avg S @ -1	89.41	55.75	61.37	55.57	62.03	49.85	56.14	54.81	56.09	60.79	60.76
Avg S @ $+1$	27.08	60.74	55.12	60.92	54.46	66.64	60.35	61.68	60.40	55.70	55.73
Δ	-62.33	4.99	-6.25	5.35	-7.57	16.79	4.21	6.87	4.31	-5.09	-5.03

Table 2.20 Analysis of Effects (Δ) for Taguchi L_{12} Design

To determine the critical factors that shift the average and the standard deviation we can look at a Pareto diagram of the average and standard deviation effects (Δs). See Figures 2.13 and 2.14.

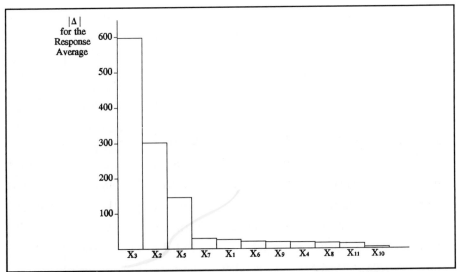

Figure 2.13 Pareto Diagram for Response Average Effects

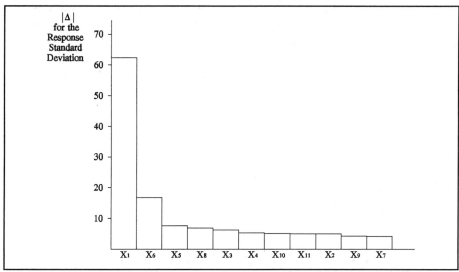

Figure 2.14 Pareto Diagram for Response Standard Deviation Effects

Close examination of Figures 2.13 and 2.14 indicates that factor X_1 is a possible variance reducing factor, and that factors X_3, X_2, and X_5 stand out as the key average shifting factors. From Table 2.20 we can see that factor X_1 at the high setting has roughly one third the variation as X_1 at the low setting. Therefore, we can reduce variation by setting X_1 at the high setting. Factors X_2, X_3 and X_5 might be considered for a subsequent modeling experiment where the remaining factors are held constant at the low cost or high throughput settings, etc.

2.6 Example of a Full Factorial Design for 3-Level Factors

Suppose you are a chemical process engineer in charge of rubber compounding formulations for a particular butadiene based formulation. Your objective is to maximize the durometer (hardness) reading. The two factors you choose to experiment with are cure temperature and amount of filler in the formulation. Each factor is to be tested at 3 levels in a full factorial design with all other non-experimental factors held constant. The design and data are shown in Table 2.21.

Run	Cure Temperature (°F)	Filler (lbs)	Durometer Reading	Average
1	250	50	20, 21, 20	20.3
2	250	60	23, 22, 20	21.7
3	250	70	24, 24, 25	24.3
4	275	50	29, 30, 31	30.0
5	275	60	31, 32, 30	31.0
6	275	70	34, 33, 34	33.7
7	300	50	27, 26, 25	26.0
8	300	60	27, 27, 29	27.7
9	300	70	28, 24, 26	26.0

Table 2.21 Design and Response Data for a 2-Factor (3-Level) Full Factorial

Step (1) The pick-the-winner approach indicates run #6 provides the maximum durometer reading. The average for this run is 33.7.

Step (2) directs us to calculate the average at each level for each column in the design. The average values for our factors are:

	Cure Temperature	Filler
Avg Low	22.1	25.4
Avg Middle	31.6	26.8
Avg High	26.6	28.0

Step (3) calculates the effects and half effects for each column. The steps to simple analysis (p. 2-8) break down at this point for 3-level designs. Regression analysis or ANOVA must be used for complete analysis of the data.

Some graphical analysis, however, can be used with 3-level designs. Plots of averages for each factor (shown in Figure 2.15), as well as interaction plots (see Figure 2.16), can be generated.

Figure 2.15 Plot of Averages

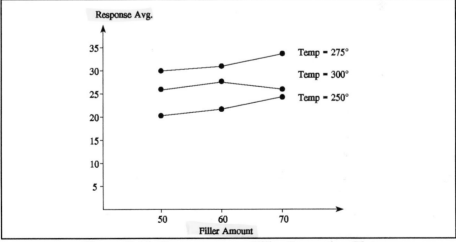

Figure 2.16 Temperature × Filler Interaction Plot

Since our objective is to maximize hardness, it appears from Figures 2.15 and 2.16 that cure temperature = 275°F and filler amount = 70 lbs provide the best settings. The next step would be the confirmation tests.

2.7   Example of Identifying System Variability

A diagnostic laboratory uses a chemical dispensing process for sample testing. The process consists of a probe which is inserted into a chemical container. A suction pump is then used to draw the chemical substance into the probe. The probe is then mechanically positioned on top a sample testing tray. Finally, the chemical substance is dispensed onto the tray. (See Figure 2.17.)

Figure 2.17.   Chemical Dispense System Schematic

A critical performance measure of this process is the dispense volume. Based on customer needs, the ideal dispense volume is 100±5 microliters.

The purpose of this experiment is to take the current system and determine which variables or factors (operator, probe lot, power supply, electric motors, chemical lots, pump, specifications on chemical temperature, and a specification on probe height at dispense) are the major causes of the dispense volume variability.

A process input/output diagram is shown below.

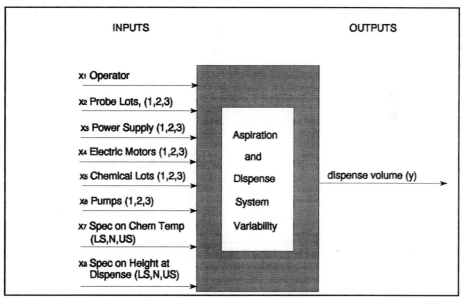

Figure 2.18 Process Input/Output Diagram

Using the Rules of Thumb inside the front cover, a Taguchi L_{18} screening design was selected with four repiclates per run. The L_{18} matrix and data appear in Table 2.22.

Run	x_1	x_2	x_3	x_4	x_5	x_6	x_7	x_8	y_1	y_2	y_3	y_4	$\overline{y}$	s
1	-1	-1	-1	-1	-1	-1	-1	-1	103.1	102.9	103.2	103.0	103.00	0.129
2	-1	-1	0	0	0	0	0	0	99.1	98.5	99.0	98.9	98.87	0.263
3	-1	-1	1	1	1	1	1	1	103.9	104.5	104.1	104.0	104.00	0.263
4	-1	0	-1	-1	0	0	1	1	101.2	101.3	100.9	100.8	101.00	0.238
5	-1	0	0	0	1	1	-1	-1	103.2	103.4	103.5	103.2	103.30	0.150
6	-1	0	1	1	-1	-1	0	0	98.1	98.2	98.3	98.1	98.17	0.096
7	-1	1	-1	0	-1	1	0	1	96.7	96.8	96.9	96.7	96.77	0.096
8	-1	1	0	1	0	-1	1	-1	103.1	102.9	103.0	103.3	103.10	0.171
9	-1	1	1	-1	1	0	-1	0	103.1	103.1	103.2	103.2	103.10	0.058
10	1	-1	-1	1	1	0	0	-1	98.0	97.5	98.6	98.7	98.20	0.560
11	1	-1	0	-1	-1	1	1	0	104.1	104.5	103.5	103.2	103.80	0.585
12	1	-1	1	0	0	-1	-1	1	99.1	99.6	98.1	98.3	98.77	0.699
13	1	0	-1	0	1	-1	1	0	100.0	99.7	98.3	101.1	99.80	1.153
14	1	0	0	1	-1	0	-1	1	100.1	100.4	100.7	100.0	100.30	0.316
15	1	0	1	-1	0	1	0	-1	102.8	102.1	101.6	102.0	102.10	0.499
16	1	1	-1	1	0	1	-1	0	103.9	103.9	102.4	102.8	103.20	0.768
17	1	1	0	-1	1	-1	0	1	103.2	101.4	103.1	103.8	102.90	1.031
18	1	1	1	0	-1	0	1	-1	100.0	101.3	102.1	100.7	101.00	0.892

$$\overline{\overline{y}} = 101.20 \qquad \overline{s} = 0.433$$

Table 2.22 Taguchi L_{18} Matrix

The average and standard deviation values for each column at the -1, 0, $+1$ settings are shown in Table 2.23.

	x_1	x_2	x_3	x_4	x_5	x_6	x_7	x_8
Avg $\overline{y}$ @ -1	101.3	101.1	100.3	102.7	100.5	101.0	102.0	101.8
Avg $\overline{y}$ @ 0	*****	100.8	102.0	99.8	101.2	100.4	99.5	101.2
Avg $\overline{y}$ @ $+1$	101.1	101.7	101.2	101.2	101.9	102.2	102.1	100.6
Avg s @ -1	0.163	0.416	0.491	0.423	0.352	0.546	0.353	0.400
Avg s @ 0	*****	0.409	0.419	0.542	0.440	0.388	0.424	0.487
Avg s @ $+1$	0.723	0.503	0.418	0.362	0.536	0.393	0.550	0.440

Table 2.23 Analysis of L_{18} Effects for the Taguchi L_{18} Example

A graphical plot for ȳ and s is shown in Figure 2.19.

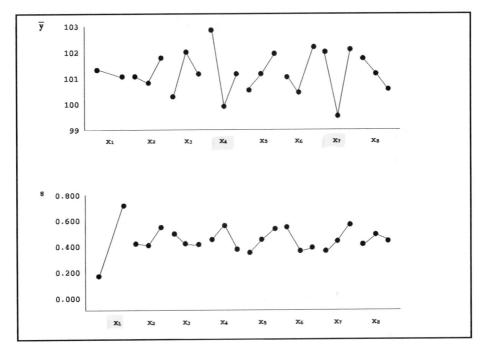

Figure 2.19 Graphical Plots for ȳ and s

The final results of our analysis indicate that the system is currently performing slightly higher than the desired average, i.e., our grand average was 101.2 versus a desired value of 100. The average (ȳ) plots in Figure 2.19 indicate that the most variation for the average performance is coming from x_4 (Electric Motors) and x_7 (Spec on Chemical Temp). However, the range in average fluctuations (99.50 to 102.6) does not appear to be a serious problem with a specification of 100 ± 5. The graph of s effects does indicate a serious problem with operators. Operator 2 (the +1 setting) has a standard deviation of .723 which is more than four times the standard deviation of .163 for operator 1 (the −1 setting). The calculated worst case C_{pk} for each operator is shown on the following page:

$$\text{UL} \quad Y_{x_i} \varphi 1$$

$$\text{Operator 1: } C_{pk} = \frac{(105 - 101.3)}{3(.163)} = \frac{3.7}{.489} = 7.57$$

$$\text{SD} \quad Y_{x_i} \varphi 2$$

$$\text{Operator 2: } C_{pk} = \frac{(105 - 101.1)}{3(.723)} = \frac{3.9}{2.169} = 1.80$$

$$\text{SD}$$

Since the difference in performance is substantial, a meeting has been planned where the operators can review the data and their standard operating procedures in order to get a more consistent C_{pk}. Also on the agenda for the meeting will be a discussion on how to best move the system average performance from 101.2 back down to the desired 100.

2.8 Using Software for Analysis

The designs evaluated so far have been fairly easy to analyze "by hand." However, "by hand" analysis is tedious, prone to mistakes and in some cases infeasible. Therefore, software is needed to assist in analysis. Most experimental design software will contain the desired design in a stored file and all that is required is to call it up and enter the variable lows and highs plus the number of replicates.

Consider the 3-level modeling design in Table 2.24. This is a 2 factor Central Composite Design (CCD) with one output and 4 replicates. See the flow chart on the inside cover of this text to understand when one should choose this design.

	Factors			Replicates			
Row #	A	B		Y1	Y2	Y3	Y4
1	160	3		43.75	43.00	42.50	43.25
2	160	5		67.75	67.50	67.00	66.75
3	180	3		90.00	91.00	90.00	90.00
4	180	5		95.00	96.00	94.25	98.00
5	170	4		93.00	92.50	97.50	91.50
6	170	4		92.50	92.00	92.50	92.00
7	160	4		73.00	74.00	73.00	74.50
8	180	4		109.00	110.50	110.50	109.00
9	170	3		68.00	67.25	68.00	68.00
10	170	5		82.00	82.75	81.25	80.50

Table 2.24 2 Factor CCD with 1 Output and 4 Replicates

The statistical analysis will appear as shown in Table 2.25. To interpret the analysis you should refer to the Rules of Thumb in Appendix M. For example, the most important pieces of the analysis in Table 2.25 are numbered. Each of these numbered entries has a Rule of Thumb to assist you. For items ① and ② see Rule of Thumb 5. For item ③ see Rule of Thumb 8. For items ④ and ⑤ see Rule of Thumb 7. The unnumbered entries in Table 2.25 are discussed in subsequent chapters.

Y-hat Model

Factor	Name	① Coeff	② P(2 Tail)	③ Tol	Active
Const		92.754	**0.0000**		
A	A	18.635	**0.0000**	1	X
B	B	7.2500	**0.0000**	1	X
AB		-4.64063	**0.0000**	1	X
AA		-0.88393	**0.0307**	0.972	X
BB		-17.853	**0.0000**	0.972	X

Rsq	0.9963	④
Adj Rsq	0.9957	⑤
Std Error	1.1980	
F	1819.8128	
Sig F	0.0000	

Source	SS	df	MS
Regression	13058.4	5	2611.7
Error	48.8	34	1.4
Total	13107.2	39	

Table 2.25 Statistical Analysis of a 2 Factor CCD

The optimization process can be performed graphically or by using a software optimization routine. Two commonly used graphs are the surface and contour plots in Figures 2.20 and 2.21. Assume your customer driven target is Y = 110. After studying Figures 2.20 and 2.21, we should conclude that Factor A will need to be set somewhere between 178 and 180 when Factor B is set equal to 4. A software optimization routine would more precisely obtain A = 179.7 and B = 4.0.

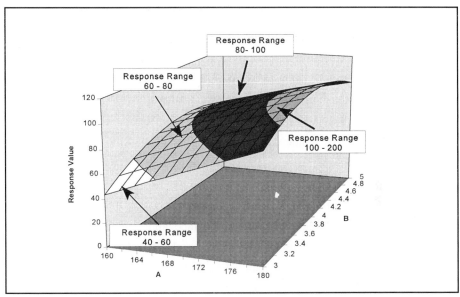

Figure 2.20 Surface Plot of Y with A and B from 2 Factor CCD

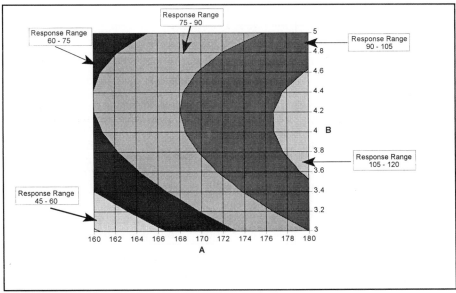

Figure 2.21 Contour Plot from 2 Factor CCD

A similar analysis could be performed for the standard deviation, ŝ. Likewise, if we had multiple outputs, they could all be modeled and optimized simultaneously with tradeoffs if necessary.

Another reason for using software for analysis is missing data. When some runs or parts of runs have missing data, statistical software will ignore these empty cells in the spreadsheet and continue to model the process. Missing data will generate some concerns which can be investigated using Rule of Thumb 8 (see Appendix M).

Consider what would have happened had we encountered missing data for experimental run (row) 9. The data will appear as shown in Table 2.26. The analysis is shown in Table 2.27 and the surface and contour plots appear in Figures 2.22 and 2.23. Careful study of these graphs reveals little difference from the previous analysis using all the data. As a Rule of Thumb, you can miss up to 10% of the experimental runs (rows) in an experiment and still obtain fairly accurate results.

	Factors			Replicates			
Row #	A	B		Y1	Y2	Y3	Y4
1	160	3		43.75	43.00	42.50	43.25
2	160	5		67.75	67.50	67.00	66.75
3	180	3		90.00	91.00	90.00	90.00
4	180	5		95.00	96.00	94.25	98.00
5	170	4		93.00	92.50	97.50	91.50
6	170	4		92.50	92.00	92.50	92.00
7	160	4		73.00	74.00	73.00	74.50
8	180	4		109.00	110.50	110.50	109.00
9	170	3					
10	170	5		82.00	82.75	81.25	80.50

Table 2.26 2 Factor CCD with Missing Data

When we let the software optimize the process for a target Y = 110, we obtained exactly the same A and B settings of 179.7 and 4. For more in-depth discussions on statistical analysis, see Chapter 5.

Y-hat Model					
Factor	**Name**	**Coeff**	**P(2 Tail)**	**Tol**	**Active**
Const		92.706	**0.0000**		
A	A	18.635	**0.0000**	1	X
B	B	7.30625	**0.0000**	0.909	X
AB		-4.64063	**0.0000**	1	X
AA		-0.78750	0.1190	0.833	X
BB		-17.925	**0.0000**	0.857	X

Rsq	0.9961
Adj Rsq	0.9954
Std Error	1.2670
F	1523.1112
Sig F	0.0000

Source	**SS**	**df**	**MS**
Regression	12224.5	5	2444.9
Error	48.2	30	1.6
Total	12272.6	35	

Table 2.27 2 Factor CCD Analysis with Missing Data

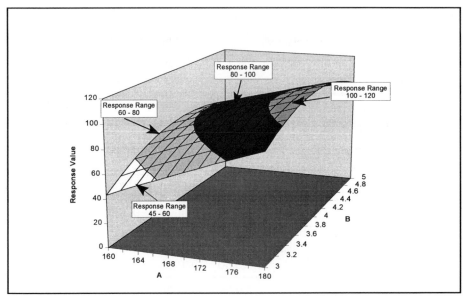

Figure 2.22 Surface Plot from Missing Data Analysis

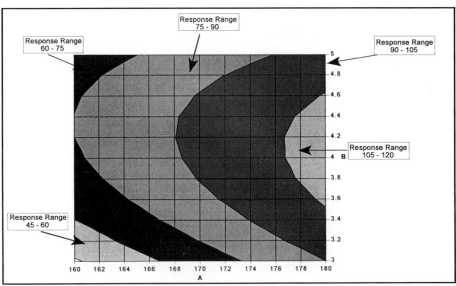

Figure 2.23 Contour Plot from Missing Data Analysis

2.9   Summary

This chapter was designed to show you how to set up and analyze experiments with minimum use of mathematics and statistics. Subsequent chapters will expand on design types and statistical analysis techniques; however, a clear understanding of Chapters 1 and 2 is enough to get you started. See Appendix M for Rules of Thumb on sample size, design selection, and statistical analysis. These Rules of Thumb can be a great aid in planning, organizing, designing, conducting, and analyzing the experiment.

Problem Set

1. A company that manufactures can-forming equipment wants to set up an experiment to help understand the factors influencing surface finish on a particular steel subassembly. Three factors were chosen to be varied in the experiments. The factors were:

Factor Name	Low Level	High Level
A - RPM	588	1182
B - Feed Rate	.004	.008
C - Tool Radius	1/64	1/32

    a) Generate a tree diagram which represents the possible combinations of a full factorial at two levels.
    b) Generate a design matrix representing the possible combinations.
    c) Suppose changing the tool radius requires an 8-hour teardown procedure. Discuss strategies for limiting the number of times tool radius must be changed.
    d) Suppose the part being machined is 4130 preheated steel. The part is a 1" diameter rod which is 4" long. Should the part be measured at one point or several points? Discuss various approaches for making multiple measures on the part.

2. The company in the previous example decides to test an 8-run full factorial design with the 3 factors at 2 levels. A brainstorming session conducted with the operator, supervisor, and engineer resulted in the finished part being measured at 4 places. (Two measures were taken on each end 180° from each other.) The design and resultant response values were:

Run	RPM	Feed Rate	Tool Radius	Surface Finish
1	588	.004	1/64	50, 50, 55, 50
2	588	.004	1/32	145, 150, 100, 110
3	588	.008	1/64	160, 160, 155, 160
4	588	.008	1/32	180, 200, 190, 195
5	1182	.004	1/64	60, 60, 60, 55
6	1182	.004	1/32	25, 35, 35, 30
7	1182	.008	1/64	160, 160, 160, 160
8	1182	.008	1/32	80, 70, 70, 80

a)   Calculate the averages, effects, and half effects for all factors and interactions. Plot the average values for each factor.

b)   Generate a Pareto chart of the absolute value of each half effect. Which effects appear to be most important?

3.   For the previous problem,

a)   Generate an interaction plot for any interactions with large effects.

b)   Suppose the objective in Problem 2 is to minimize the response. What are the best settings for the factors to achieve this objective?

4.   For a process you work with, assemble a cross functional team and begin brainstorming the setup of a designed experiment. Be sure to use the check sheet from Chapter 1 as a guide in assisting you through the proper steps.

5. a) Using Factsim on the disk in the back of the book, select 3 factors at 2 levels, and set up a designed experiment (full factorial).

    b) Conduct the experiment using four repeated measures per run.

    c) Calculate the average response at each level for all factors and interactions.

    d) Generate a Pareto chart of the absolute value for each half effect. Decide which factors and interactions appear to be most important in influencing the response.

    e) Build the prediction equation.

    f) Determine one set of factor settings which can achieve a target of .5 inches.

    g) Using Factsim, validate your set of optimal conditions.

Bibliography

1. Kiemele, Mark J., Stephen R. Schmidt, and Ronald J. Berdine, (1997) Basic Statistics: Tools for Continuous Improvement, 4th Edition, Air Academy Press, Colorado Springs, CO.

2. Ott, Ellis R., (1975) Process Quality Control, McGraw-Hill, Inc., New York.

Chapter 3

Design Types

3.1   Introduction*

As introduced in Chapters 1 and 2, a design matrix displays the levels of each factor for all of the run combinations in the experiment. For example, Table 3.1 contains 3 factors and 4 runs where the first run is made with factors A, B, and C set at values of 20, 800, and 50, respectively.

Run	Factors		
	A	B	C
1	20	800	50
2	20	600	30
3	10	800	30
4	10	600	50

Table 3.1 Actual Settings for 3 Factors in a 4-Run Design Matrix

---

*This chapter will show the reader the mechanics required for developing good experimental designs. To avoid the complexity contained in this chapter please refer to the design selection Rules of Thumb in Appendix M and a software package such as DOE KISS, RS/Discover, DOE Wisdom, Statgraphics, SAS, ECHIP, JMP, etc.

Although the actual settings are used to run the experiment, you will need to standardize the different factor units and scale before analysis. The standardization is performed by coding the low actual settings as -1 and the high actual settings as +1. To convert from coded to actual and vice versa we used the following formula in Chapters 1 and 2:

$$\text{Actual} = \left[\frac{\text{Hi} + \text{Lo}}{2} \right] + \left[\frac{\text{Hi} - \text{Lo}}{2} \right] \text{Coded}$$

A more formal approach is presented in Equation 3.1.

$$x_j = 2 \frac{f_j - \overline{f_j}}{d_j} \quad \text{or} \quad f_j = \overline{f_j} + \frac{d_j}{2} x_j$$

where x_j is the coded setting for factor j

   f_j is the actual setting for factor j

   $\overline{f_j}$ is the average of all the actual settings for factor j

   d_j is the distance between the largest and smallest actual settings
      of factor j

Equation 3.1 Converting Actual Factor Settings to Coded Values

Converting the actual values in Table 3.1 to coded values results in the design matrix shown in Table 3.2a. For convenience, researchers will sometimes abbreviate the "+1" and "−1" with "+" and "−" as shown in Table 3.2b.

(a)	Factors			(b)	Factors		
Run	A	B	C	Run	A	B	C
1	+1	+1	+1	1	+	+	+
2	+1	−1	−1	2	+	−	−
3	−1	+1	−1	3	−	+	−
4	−1	−1	+1	4	−	−	+

Table 3.2 Coded Values for 3 Factors in a 4-Run Design Matrix

(Note: Column A shows the sequence + + − − which is the opposite (or negative) of that shown in Chapter 2 where we used the sequence − − + +; either choice can be used with the same end result.)

Each column of factor values is referred to as a vector in the design matrix. A design of n runs is considered to be **balanced vertically** when the coded values, x_{ij} , sum to zero, i.e., $\sum\limits_{i=1}^{n} x_{ij} = 0$ for each factor (or column) j. Balanced designs are desirable because they simplify the calculations during analysis, and under certain conditions they lend themselves to orthogonal designs. A 2-level design matrix is said to be **orthogonal** if it is balanced vertically and if the dot product of all possible column pairs is zero, i.e., $\sum\limits_{i=1}^{n} x_{ij}x_{ik} = 0$ (for all n combinations of columns j and k, where j ≠ k). In other words, multiplying the + 1 and -1 values in each row for any 2 columns should sum to zero. The dot product test for orthogonality is similar to testing horizontal balancing described in Chapter 1. The reader should verify that the design in Table 3.2 is balanced and orthogonal.

Designs for 2-level factors that exhibit only vertical balance will not be orthogonal (see Table 3.3). In this case, factors B and C have the same set of coded values and their columns are referred to as **identical, aliased,** or

perfectly confounded. Notice that the B and C dot product in Table 3.3 equals 4, not 0. This aliasing will prevent the independent assessment of the effects of factors B and C.

The reason why orthogonality is so important is that it allows the desired effects to be evaluated independently. A high degree of non-orthogonality results in a dependency of effects which is particularly undesirable.[*] For example, if factors B and C are tested in a non-orthogonal or dependent fashion and only B is truly important, C may appear important during analysis due to its test dependency with B. Even worse, if B and C are both truly important, but one in a positive direction and the other in a negative direction, their dependency may cause both to appear unimportant.

	Factors		
Run	A	B	C
1	+1	+1	+1
2	+1	−1	−1
3	−1	+1	+1
4	−1	−1	−1

Table 3.3 Non-Orthogonal Matrix with Aliased Columns B and C

Design matrices can be non-orthogonal even without containing aliased columns. For example, the design in Table 3.4 is a vertically balanced, non-orthogonal design without any two vectors (or columns) identical. In Table 3.4, each possible dot product is not equal to zero and the factors are not

---

[*]Most experimental design software packages will automatically generate orthogonal or nearly orthogonal designs. Thus, the reader need not be overly concerned about non-orthogonality problems when using these packages.

aliased; instead, all factors are partially correlated (confounded) with each other. In other words, aliased design columns are perfectly confounded (in a positive or negative sense); whereas partially confounded design columns are neither independent nor aliased. (See Chapter 4 for information on how to use regression to determine if any two columns are aliased, partially correlated, or independent.)

The designs in Tables 3.1 through 3.4 all represent factors measured at only two levels. For that reason, they are typically referred to as **2-level designs** which are particularly useful for factor screening and for estimating first order (linear) models with or without interactions. The inclusion of a center point (or 0 coded values for all factors) in a 2-level design will create a third level to allow for curvature evaluation. There are several different ways to test factors at 3-levels, most of which are discussed in detail later in this chapter. For now, an example of an orthogonal 3-level design is shown in Table 3.5. The reader should be able to verify that all 4 columns in Table 3.5 are balanced vertically and horizontally.

	Factors			
Run	A	B	C	D
1	+1	−1	−1	−1
2	−1	+1	−1	−1
3	−1	−1	+1	−1
4	−1	−1	−1	+1
5	+1	+1	+1	+1
6	−1	+1	+1	+1
7	+1	−1	+1	+1
8	+1	+1	−1	+1
9	+1	+1	+1	−1
10	−1	−1	−1	−1

Table 3.4 Non-Orthogonal Matrix without Aliased Columns

	Factors			
Run	A	B	C	D
1	−	−	−	−
2	−	0	0	0
3	−	+	+	+
4	0	−	0	+
5	0	0	+	−
6	0	+	−	0
7	+	−	+	0
8	+	0	−	+
9	+	+	0	−

Table 3.5 Example of an Orthogonal Design for 4 Factors, Each at 3 Levels
(Interactions cannot be evaluated independently in this design.)

3.2   Full Factorial Designs

Given that the experimenter desires to examine 3 factors (A, B, and C) each at two levels, the design used to estimate all possible effects would be a **full factorial design**. The set of all possible effects for this example is described in Table 3.6.

Main Effects	2-Way Interactions	3-Way Interactions
A	AB	ABC
B	AC	
C	BC	

Table 3.6 List of All Possible Effects for a Full Factorial Design
for 3 Factors, Each at 2 Levels

The number of combinations or runs (n) for a full factorial design of k factors will be $n = 2^k$. The number of orthogonal columns representing all possible effects will be $n - 1$. In this example, a total of 7 effects can be analyzed in 8 runs. The full factorial design matrix for 3 factors (2 levels) is designated a 2^3 design matrix and is displayed in Table 3.7. Notice the pattern used to generate the A, B, and C columns. (Any similar scheme can be used to obtain orthogonality, i.e., vertical and horizontal balancing of these columns.) Remember that the experimenter will use the actual setting values for columns A, B, and C when running the experiment. The remaining columns are obtained by multiplying the associated coded settings in each of the factor columns and they are used only in the analysis phase of the experiment. Graphically, the runs in Table 3.7 have the geometric interpretation shown in Figure 3.1.

Run	A	B	C	AB	AC	BC	ABC
1	-1	-1	-1	$+1$	$+1$	$+1$	-1
2	-1	-1	$+1$	$+1$	-1	-1	$+1$
3	-1	$+1$	-1	-1	$+1$	-1	$+1$
4	-1	$+1$	$+1$	-1	-1	$+1$	-1
5	$+1$	-1	-1	-1	-1	$+1$	$+1$
6	$+1$	-1	$+1$	-1	$+1$	-1	-1
7	$+1$	$+1$	-1	$+1$	-1	-1	-1
8	$+1$	$+1$	$+1$	$+1$	$+1$	$+1$	$+1$

Table 3.7 Coded Design Matrix for a Full Factorial of 3 Factors
Each at 2 Levels (2^3 Design)

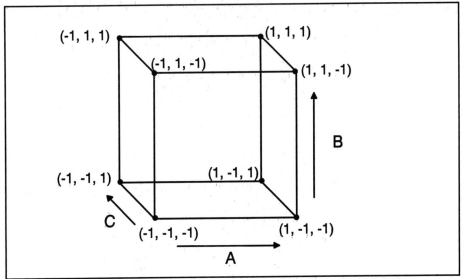

Figure 3.1 Geometric Representation of All Runs in Table 3.7

For 3 factors, each with 3 levels, a full factorial design consists of $3^3 =$ 27 runs. The reader should verify the vertical and horizontal balancing of the orthogonal 3-factor (3-level) full factorial design in Table 3.8. The full factorial design in Table 3.8 is graphically displayed in Figure 3.2.

The advantages of full factorial designs are: (i) orthogonality, (ii) no aliasing concerns, and (iii) all main factors and all interactions can be evaluated. The obvious disadvantage is the cost, time, and resources needed to make all the runs required by a full factorial, especially when the number of factors is large. A full factorial for 7 factors, each at 2 levels, is 2^7 or 128 runs. Three-level factor full factorials are even more cost, time, and resource demanding. For example, a full factorial for 7 factors, each at 3 levels, is 3^7 or 2187 runs. See Appendix M for Rules of Thumb on when to use full factorial designs.

	Factors				Factors				Factors		
Run	A	B	C	Run	A	B	C	Run	A	B	C
1	+	+	+	10	0	+	+	19	−	+	+
2	+	+	0	11	0	+	0	20	−	+	0
3	+	+	−	12	0	+	−	21	−	+	−
4	+	0	+	13	0	0	+	22	−	0	+
5	+	0	0	14	0	0	0	23	−	0	0
6	+	0	−	15	0	0	−	24	−	0	−
7	+	−	+	16	0	−	+	25	−	−	+
8	+	−	0	17	0	−	0	26	−	−	0
9	+	−	−	18	0	−	−	27	−	−	−

Table 3.8 Full Factorial for 3 Factors, Each at 3 Levels
(To save space, interaction and quadratic columns are not displayed.)

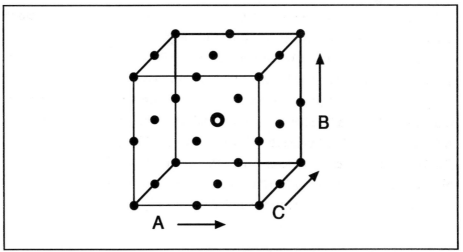

Figure 3.2 Graphical Representation of All Runs in Table 3.8

3.3   Fractional Factorials for 2-Level Designs

Often a researcher does not have the time, funds, or resources to complete a full factorial design. Obviously, it doesn't take too many factors, k, before 2^k or 3^k runs are infeasible. Additionally, it is not often that you would expect to find 3-way (and higher) interactions to be significant. A type of orthogonal design which allows you to estimate all linear effects and desired interactions while requiring fewer runs is a **fractional factorial**. Given any number of factors, k, we can build a 2-level fractional factorial design with $n = 2^{k-q}$ runs by completing the following:

>   (1)   Write the full factorial design for k - q factors (q is some positive integer < k where q = 1 indicates a half fraction, q = 2 a quarter fraction, etc).

>   (2)   Alias the extra q factors with higher order interactions which can be assumed to be insignificant. When q > 1, try to use interactions with the same number of letters.

Consider the 2^{4-1} design in Table 3.9 as a simple example. The full factorial for k = 4 requires $2^4 = 16$ runs. If we can assume an insignificant ABC interaction, then a fractional factorial can be accomplished in $2^4/2$ or $2^{4-1} = 8$ runs. This design is referred to as a half fraction. The ABC column is used to generate the coded settings for factor D, i.e., we purposely aliased factor D coded settings with the coded values for ABC. Since each letter in the design matrix represents a column of (+) and (−) elements, aliasing D with ABC implies D = ABC. This does not imply that the D and ABC effects are the same, only that the evaluation of the D and ABC column effects cannot be separated. If the column effect is significant, it is either due to D, ABC, or some combination of the two. However, the only justification for attempting this design is the assumption that the ABC interaction is very unlikely to be important. If this assumption is not well founded this design choice should be abandoned. Assuming we continue with this design choice, the complete aliasing (confounding) pattern development is discussed next.

	Factors						
Run	A	B	C	AB	AC	BC	D=ABC
1	-1	-1	-1	$+1$	$+1$	$+1$	-1
2	-1	-1	$+1$	$+1$	-1	-1	$+1$
3	-1	$+1$	-1	-1	$+1$	-1	$+1$
4	-1	$+1$	$+1$	-1	-1	$+1$	-1
5	$+1$	-1	-1	-1	-1	$+1$	$+1$
6	$+1$	-1	$+1$	-1	$+1$	-1	-1
7	$+1$	$+1$	-1	$+1$	-1	-1	-1
8	$+1$	$+1$	$+1$	$+1$	$+1$	$+1$	$+1$

Table 3.9 A 2^{4-1} Fractional Factorial Design Matrix

It should be obvious that if any column is multiplied by itself, the result is an identity column, I, which contains all $+1$ values. Thus, DD or $D^2 = I$. Multiplying both sides of the **half fraction design generator** $D = ABC$ by D, we conclude $D^2 = ABCD = I$. The term ABCD is called **a defining word**, and the relation $I = ABCD$ is referred to as **the defining relation** which defines all of the aliased effects contained in our fractional factorial design. Since there is only one generator $(D = ABC)$ represented in the defining relation for this design, the **aliasing pattern** is easily constructed by multiplying both sides of the defining relation $(I = ABCD)$ by the desired effect. For example, $AB(I) = AB(ABCD) = A^2B^2CD = IICD = CD$; therefore, $AB = CD$. The right side then becomes the alias of the effect on the left side (see Table 3.10). Aliasing patterns should be evaluated before experimentation is begun.* If the design produces certain

---

*Aliasing patterns can usually be obtained from a software package. The intent of this discussion is to show the reader the actual mechanics required.

undesirable aliased effects, then one should (i) select different generators, (ii) make more runs, and/or (iii) hold one or more of the factors constant.

effect	alias
A	BCD
B	ACD
C	ABD
D	ABC
AB	CD
AC	BD
AD	BC
BC	AD
BD	AC
CD	AB
ABC	D
ABD	C
ACD	B
BCD	A
ABCD	I

Table 3.10 Aliasing Pattern for a 2^{4-1} Fractional Factorial Design
with Generator D = ABC

In the event you run a 2^{4-1} design and later desire to add more runs to overcome the alias problems shown in Table 3.10, you need only run the other half fraction using the negative of the generator ABC = D, i.e., build another 8 runs with factor settings for A, B, and C the same as in the first 8 runs and this time let D = −ABC. Then augment the second set of 8 runs with the previous 8 runs to make a 2^4 full factorial. For the augmented 16-run design, the D factor will be de-aliased from the ABC interaction; however, the ABCD interaction will now be aliased with any effect produced by differences in the two blocks or sets of 8 runs.   If this column has a significant effect, the experimenter should investigate the cause, i.e., changes in materials, manpower, environment, etc., or possible ABCD interactions.   More details on overcoming aliasing problems are discussed in the foldover design section.

A more complex example of a fractional factorial would be to have 7 factors each at 2 levels.   Not wanting to do n $= 2^7 = 128$ runs, you choose to fractionalize the design so that n=16, i.e., you want a 2^{7-3} design.   The design consists of a full factorial for (7 − 3 = 4) factors with 3 higher order interactions of the original 4 factors (A, B, C, D) chosen as generators for factors E, F, and G, respectively.

Run	A	B	C	D	E=ABC	F=BCD	G=ABD
				Factors			
1	+	+	+	+	+	+	+
2	+	+	+	−	+	−	−
3	+	+	−	+	−	−	+
4	+	+	−	−	−	+	−
5	+	−	+	+	−	−	−
6	+	−	+	−	−	+	+
7	+	−	−	+	+	+	−
8	+	−	−	−	+	−	+
9	−	+	+	+	−	+	−
10	−	+	+	−	−	−	+
11	−	+	−	+	+	−	−
12	−	+	−	−	+	+	+
13	−	−	+	+	+	−	+
14	−	−	+	−	+	+	−
15	−	−	−	+	−	+	+
16	−	−	−	−	−	−	−

Table 3.11 Coded Matrix for a 2^{7-3} Design
(Interactions other than generators are not shown for convenience;
evaluation of the matrix would include all 15 orthogonal columns.)

For the 1/8 fraction design shown in Table 3.11, we have chosen the generators E = ABC, F = BCD, and G = ABD which result in the following 3 defining words: I = ABCE = BCDF = ABDG. Multiplying any 2 defining words at a time results in I^2 = I = ADEF = CDEG = ACFG and multiplying all 3 at one time yields I^3 = I = BEFG. Therefore, the complete defining relation consists of:

$$I = ABCE = BCDF = ABDG = ADEF = CDEG = ACFG = BEFG$$

This defining relation will allow the computation of the aliasing pattern for the entire design. For example, if you want to determine all effects aliased with factor A, just multiply the defining relation by A which then appears as:

$$A = BCE = ABCDF = BDG = DEF = ACDEG = CFG = ABEFG$$

Similarly, the aliasing pattern for EF is:

$$EF = ABCF = BCDE = ABDEFG = AD = CDFG = ACEG = BG$$

In general, there exists 2^q different fractions of a 2^{k-q} design. For our 2^{7-3} example, there exists $2^q = 2^3 = 8$ different 1/8 fractional designs, represented by "+" or "−" in the previous set of generators, i.e.,

$$E = \pm ABC, \quad F = \pm BCD, \text{ and } G = \pm ABD.$$

To de-alias any particular factor from its 3-way interaction generator, simply add a different 1/8 fraction to the first one. For example, to de-alias factor E from ABC, augment two 1/8 fractions, one with E = ABC and the other with E = −ABC.

A term commonly used by experimenters to communicate information about a design is **resolution** (R). Formally, the resolution of a 2-level fractional factorial design is the length (in number of letters) of the shortest word in the defining relation. For our 2^{7-3} example, you can see that the length of every defining word in the defining relation is 4. Hence, this 2^{7-3} fractional factorial has a resolution of IV or R_{IV} (i.e., the 2^{7-3} design is a resolution four design). The meaning of different resolutions is tied to the degree of aliasing in the fractional factorial design matrix. Specific definitions follow:

R_{II}:   A design which contains main effects aliased with other main effects. This is not a desired experimental approach.

R_{III}:   A design which does not alias main effects with one another, but does alias mains with 2-way interactions. R_{III} designs are typically used when screening a large number of factors to find only the most important factors for future experimentation.

R_{IV}:   A design which does not alias mains with 2-way interactions, but does alias 2-way interactions with other 2-way interactions. R_{IV} designs are used for building prediction equations when resource limitations do not permit R_V designs. Prior knowledge must be used to determine which aliased effect goes into the prediction equation.

R_V: A design that does not alias the main effects with each other or with 2-way interactions. In addition, 2-way interactions are not aliased with one another. Thus, this design provides for unconfounded main and 2-way interaction effects. R_V designs are excellent for building prediction equations that will typically not have serious interaction concerns.

When n is approximately equal to k and 2-way interactions are assumed insignificant, or if you are screening a large number of factors, a R_{III} design is used. R_{III} designs are sometimes referred to as **saturated** designs because all or most of the orthogonal columns are assigned factors. If 2-way interactions are of interest, **unsaturated** designs of R_V or greater are suggested.

NOTE: A question often asked is, "what is the resolution of a full factorial design?" Since a full factorial does not contain aliasing, it will not have a defining relation. Therefore, we either do not use the term resolution with a full factorial or simply state its resolution as "full factorial."

For a few fractional factorial examples, consider the following:

i) 2^{5-2}

Run	A	B	C	AB	AC	D = BC	E = ABC
1	−	−	−	+	+	+	−
2	−	−	+	+	−	−	+
3	−	+	−	−	+	−	+
4	−	+	+	−	−	+	−
5	+	−	−	−	−	+	+
6	+	−	+	−	+	−	−
7	+	+	−	+	−	−	−
8	+	+	+	+	+	+	+

I = BCD = ABCE = ADE
Design is of Resolution III.

ii) 2^{6-3}

Run	A	B	C	AB	D = AC	E = BC	F = ABC
1	−	−	−	+	+	+	−
2	−	−	+	+	−	−	+
3	−	+	−	−	+	−	+
4	−	+	+	−	−	+	−
5	+	−	−	−	−	+	+
6	+	−	+	−	+	−	−
7	+	+	−	+	−	−	−
8	+	+	+	+	+	+	+

I = ACD = BCE = ABCF = ABDE = BDF = AEF = CDEF
Design is of Resolution III.

iii) 2^{5-1}

RUN	1 A	2 B	3 AB	4 C	5 AC	6 BC	7 ABC DE	8 D	9 AD	10 BD	11 ABD CE	12 CD	13 ACD BE	14 BCD AE	15 ABCD E	$y_1 \dots y_3$	$\bar{y}$	s
1	-1	-1	1	-1	1	1	-1	-1	1	1	-1	1	-1	-1	1			
2	-1	-1	1	-1	1	1	-1	1	-1	-1	1	-1	1	1	-1			
3	-1	-1	1	1	-1	-1	1	-1	1	1	-1	-1	1	1	-1			
4	-1	-1	1	1	-1	-1	1	1	-1	-1	1	1	-1	-1	1			
5	-1	1	-1	-1	1	-1	1	-1	1	-1	1	1	-1	1	-1			
6	-1	1	-1	-1	1	-1	1	1	-1	1	-1	-1	1	-1	1			
7	-1	1	-1	1	-1	1	-1	-1	1	-1	1	-1	1	-1	1			
8	-1	1	-1	1	-1	1	-1	1	-1	1	-1	1	-1	1	-1			
9	1	-1	-1	-1	-1	1	1	-1	-1	1	1	1	1	-1	-1			
10	1	-1	-1	-1	-1	1	1	1	1	-1	-1	-1	-1	1	1			
11	1	-1	-1	1	1	-1	-1	-1	-1	1	1	-1	-1	1	1			
12	1	-1	-1	1	1	-1	-1	1	1	-1	-1	1	1	-1	-1			
13	1	1	1	-1	-1	-1	-1	-1	-1	-1	-1	1	1	1	1			
14	1	1	1	-1	-1	-1	-1	1	1	1	1	-1	-1	-1	-1			
15	1	1	1	1	1	1	1	-1	-1	-1	-1	-1	-1	-1	-1			
16	1	1	1	1	1	1	1	1	1	1	1	1	1	1	1			

I = ABCDE

Design is of Resolution V.

This design is particularly useful because all 2-way interactions can be evaluated without serious confounding problems. Notice that all of the 2-way interactions are confounded with 3-way interactions. Since 3-way interactions are typically insignificant, columns 7, 11, 13, and 14 will most likely represent the 2-way interaction effects of Factor E.

iv) 2^{6-2}

Run	A	B	C	D	AB AC AD BC BD CD ABC ABD E=ACD F=BCD ABCD
1	−	−	−	−	
2	−	−	−	+	
3	−	−	+	−	
4	−	−	+	+	
5	−	+	−	−	
6	−	+	−	+	
7	−	+	+	−	
8	−	+	+	+	Note: The reader should be able to develop
9	+	−	−	−	these columns by using appropriate column
10	+	−	−	+	value multiplication.
11	+	−	+	−	
12	+	−	+	+	
13	+	+	−	−	
14	+	+	−	+	
15	+	+	+	−	
16	+	+	+	+	

$$I = ACDE = BCDF = ABEF$$

Design is of Resolution IV.

Note: A Rule of Thumb for selecting interaction generators of a 2^{k-q} when $q > 1$ is to choose interactions (when possible) that have the same number of letters, i.e., avoid using both a 3-way and a 4-way interaction as generators in the same design. The reader should verify that for the 2^{6-2} design above, if $E = ABCD$, the resolution would have been III instead of IV.

v) 2^{7-3} (see Table 3.11)

vi) 2^{8-4}

Run	A B C D AB AC AD BC BD CD E=ABC F=ABD G=ACD H=BCD ABCD
1	
2	
3	
4	
5	
6	
7	Note: By now the reader should be able to
8	develop the coded values for all
9	of these columns.
10	
11	
12	
13	
14	
15	
16	

I = ABCE = ABDF = ACDG = BCDH = CDEF = BDEG = ADEH =
BCFG = ACFH = ABGH = AEFG = BEFH = DFGH = CEGH =
ABCDEFGH

This design is of Resolution IV.

In the very recent past, classical statistics texts and short courses typically spent a large amount of time on the concept of alias patterns, fractionalization, and resolution. With the advent of powerful software, computers can quickly determine what takes the student many hours. Since this is the case, we shall leave you with the previous set of examples and forge on into additional useful design concepts. (Also see Appendix M for Rules of Thumb on when to use fractional factorial designs.)

3.4   Foldover Designs

It may happen that an experimenter has run a R_{III} design to save runs and feels that interactions have confused the analysis. In this case, some interaction information can be gained by making more runs and adding or augmenting them to the original set of runs.

A **foldover design** of resolution IV consists of a R_{III} design augmented with a second R_{III} design whose factor settings "+1" or "−1" are just the reverse of the first R_{III} design. The new design will have twice as many runs as the original R_{III}. An additional blocking variable will be naturally added by placing a column consisting of "−1" values for the first R_{III} design and "+1" values for the second R_{III} runs. As an example, consider the 2^{3-1} design of resolution III in Table 3.12. The defining relation for the design in Table 3.12 is I = ABC.

	Factors		
Run	A	B	C = A × B
1	−	−	+
2	−	+	−
3	+	−	−
4	+	+	+

Table 3.12   Coded Matrix for a 2^{3-1} R_{III} Design

Therefore, the 2^{3-1} design is of resolution III. By folding over this R_{III} design, we obtain the 8 runs displayed in Table 3.13. For both sets of 4 runs the blocking variable, D = −ABC; therefore, the defining relation is I = −ABCD. Thus, the resulting 8-run design resolution is R_{IV}. The blocking variable D is used only to evaluate differences between the two blocks of 4 runs. If factor D appears significant during the analysis, the experimenter should investigate the cause of this difference. For more information on foldover designs, see references [5] and [8].

	Factors			
Run	A	B	C	D = new factor
1	−1	−1	+1	−1
2	−1	+1	−1	−1
3	+1	−1	−1	−1
4	+1	+1	+1	−1
5	+1	+1	−1	+1
6	+1	−1	+1	+1
7	−1	+1	+1	+1
8	−1	−1	−1	+1

Table 3.13 Foldover Design Matrix

3.5 Plackett-Burman Designs

These R_{III} designs were developed by Plackett and Burman in 1946 [10] for screening a large number of factors, i.e., reducing a large number of factors to a smaller set of important factors for subsequent experimentation. The disadvantage of using only fractional factorial designs for screening large numbers of factors is that the number of runs is always 2 to some power, i.e., $n = 2^{k-q} =$ 4, 8, 16, 32, 64, 128 ... . The **Plackett-Burman** designs are based on Hadamard matrices which have more flexibility. The number of runs in a **Hadamard matrix** is a multiple of 4, i.e., n = 4, 8, 12, 16, 20 ... . Each column contains elements which are 1 or −1 and all (n−1) columns are balanced and pairwise orthogonal (see [7]). Plackett and Burman use the Hadamard matrices for their designs in order to study up to (n−1) factors in n runs. When the number of runs in a Plackett-Burman design is $n = 2^{k-q}$, the design is referred to as a geometric Plackett-Burman design which is essentially the same as the 2-level fractional factorial discussed previously. If we built a fractional factorial design, multiplied the even ordered interactions by (−1) and then permuted the rows and columns, we would get a geometric Plackett-Burman design. All other 2-level Plackett-Burman designs, such as n = 12, 20, 24, 28, 36, 40, etc., are non-geometric and have a different confounding structure of the main effects and 2-way interactions, i.e., the 2-way interactions are only partially confounded with main effects and other 2-ways.

For geometric Plackett-Burman designs of R_{III}, it can be shown that each 2-way interaction will be positively or negatively aliased with a main effect (see Table 3.14 and verify D = -AB). However, the non-geometric Plackett-Burman designs contain main effects which are only partially confounded (intercorrelated) with 2-way interactions (see Table 3.15). In fact, each 2-way interaction is partially confounded with each of the other main effects, i.e., the BC interaction of Table 3.15 is intercorrelated with the A, D, E, F, G, H, I, J, and K main effects. The simplest way to determine the amount of intercorrelation between any two columns is to find the correlation coefficient which is discussed in Chapter 4.

				Factors			
Run	A	B	C	D	E	F	G
1	+	–	–	+	–	+	+
2	+	+	–	–	+	–	+
3	+	+	+	–	–	+	–
4	–	+	+	+	–	–	+
5	+	–	+	+	+	–	–
6	–	+	–	+	+	+	–
7	–	–	+	–	+	+	+
8	–	–	–	–	–	–	–

Table 3.14 Plackett-Burman 8-Run Design

To build a Plackett-Burman design, take a single generating vector for a particular size n design and generate the remaining vectors. For example, if n = 8 combinations, the unique generating vector consists of the following (n − 1) values (+ + + − + − −). The first (n − 1) rows of the design matrix can be constructed as follows: (1) use the generating vector as column A, (2) build column B by making the last value in A the first value in B and then slide the rest of the A values below that value, (3) vector C is made by using the last value in B as the first value in C with the rest of B below it, and (4) continue until all (n − 1) columns are complete. Finally, add an nth row of all "−" values as shown in Tables 3.14 and 3.15. Another way of viewing the construction of a Plackett-Burman design is to do a sequence of downward circular shifts of the

Run						Factors					
	A	B	C	D	E	F	G	H	I	J	K
1	+	−	+	−	−	−	+	+	+	−	+
2	+	+	−	+	−	−	−	+	+	+	−
3	−	+	+	−	+	−	−	−	+	+	+
4	+	−	+	+	−	+	−	−	−	+	+
5	+	+	−	+	+	−	+	−	−	−	+
6	+	+	+	−	+	+	−	+	−	−	−
7	−	+	+	+	−	+	+	−	+	−	−
8	−	−	+	+	+	−	+	+	−	+	−
9	−	−	−	+	+	+	−	+	+	−	+
10	+	−	−	−	+	+	+	−	+	+	−
11	−	+	−	−	−	+	+	+	−	+	+
12	−	−	−	−	−	−	−	−	−	−	−

Table 3.15 Plackett-Burman 12-Run Design

generating vector, disregarding the last row which always consists of all "−" values. Generating columns for various Plackett-Burman designs are as follows:

n = 8   (+ + + − + − −)

n = 12 (+ + − + + + − − − + −)

n = 16 (+ + + + − + − + + − − + − − −)

n = 20 (+ + − − + + + + + − + − + − − − − + + −)

n = 24 (+ + + + + + − + − + + − − + + − − + − + − − − −)

The obvious advantage of Plackett-Burman designs is the limited number of runs to evaluate large numbers of factors. Since interactions are typically not evaluated for Plackett-Burman designs, the effects for all the factors are placed in a Pareto chart and the obviously important factors can then be selected for more in-depth testing. (See DOE PRO, from Air Academy Associates, to evaluate the partially confounded 2-way interactions.)

The disadvantages of Plackett-Burman designs are tied to the assumptions required to evaluate up to (n − 1) factors in n runs. The assumptions are that interactions (if they exist) are not strong enough to mask main effects and quadratic effects (if they exist) are closely related to the factor linear effects. For most processes and most types of factors evaluated over reasonable ranges, these assumptions are often fairly accurate.

As a final note, any R_{III} Plackett-Burman design can be folded over to achieve a design of resolution IV. For more information on Plackett-Burman designs, see the article by Plackett and Burman [10]. (Also see Appendix M for Rules of Thumb on the use of screening designs.)

3.6   Non-Orthogonal 2-Level Designs

There are also non-orthogonal, 2-level designs worthy of discussion: the random balanced, foldover Koshal, and D-optimal designs. The **random balanced** design is typically used as a supersaturated or saturated design and is formed by randomly generating the first n/2 runs for k factors. In order to balance the design, you fold it over, generating an equal number of "+" and "−" values in each column. In this case, n can be determined independently of k, which can be advantageous. The primary disadvantage is that the confounding structure for main effects and interactions is totally random and outside of the experimenter's control. However, a design can be generated and intercorrelations calculated (see Chapter 4) for all possible pairs of effects. Then, if any paired correlations are too high, they can be avoided or the design can be regenerated randomly until satisfactory levels of intercorrelation are achieved. The random balanced design is not highly recommended.

A balanced non-orthogonal design which is closely related to the one-at-a-time design is a **Foldover Koshal**. Basically, the Foldover Koshal is a one-at-a-time design for 2 levels which is folded over to achieve balance. Except for $n=8$, the design is far from orthogonal, which is why simulation studies indicate it is only about 60 to 70% effective in discriminating significant main and 2-way interaction effects when the number of factors k is less than 20. Intercorrelations will remain low enough, provided k is less than or equal to 20. Due to increased main effect correlations as k increases, the effectiveness deteriorates for larger k. There are three reasons for including this design in this book: (1) some consultants are using it to screen large numbers of variables even though better alternatives exist, such as Plackett-Burman and fractional factorials; (2) if a one-at-a-time or Koshal design as shown in Table 3.16 has already been run, then folding over the design results in the Foldover Koshal of Table 3.17, providing the

Run	Factors			
	A	B	C	D
1	−	−	−	−
2	+	−	−	−
3	−	+	−	−
4	−	−	+	−
5	−	−	−	+

Table 3.16 One-at-a-Time Design (First Half of a Foldover Koshal)

means for possibly salvaging a previously bad design; and (3) if you are conducting a failure analysis study and desire to swap components between a good

(+) and bad (−) system for the purpose of isolating a bad component. Shainin [2] has popularized a component swapping technique based on the Foldover Koshal matrix.

			Factors	
Run	A	B	C	D
1	−	−	−	−
2	+	−	−	−
3	−	+	−	−
4	−	−	+	−
5	−	−	−	+
6	+	+	+	+
7	−	+	+	+
8	+	−	+	+
9	+	+	−	+
10	+	+	+	−

Table 3.17 A Foldover Koshal Design of 10 Runs

The last and most important non-orthogonal design to be discussed is the **D-optimal** design. If we were to call our design matrix X, then premultiplying X with its transpose, X′, yields the matrix X′X. D-optimality refers to a design that maximizes the determinant of X′X which is called the D value, i.e., $D = |X'X|$. Optimal values of D vary with n (the number of runs), and for some n values, D-optimal designs will be orthogonal. When n is a value associated with a non-orthogonal design, the columns of " + " and " − " values must be generated to maximize $|X'X|$. Sophisticated software packages such as RS/Discover, SAS, E-Chip, and Catalyst will do this for you, i.e., generate a D-optimal or close to D-optimal design for any desired number of runs, n. It should be noted that less than orthogonal designs ought to be avoided when possible. The lack of orthogonality will confound effects and complicate the analysis. However, in situations where it is too costly or impossible to generate a fractional design which is orthogonal, the D-optimal alternative is very helpful.

Consider the following three examples where the D-optimal design is useful. As Example 1, assume we are interested in 4 factors at 2 levels, along with all 2-way interactions. Suppose, due to cost and time constraints, we need the number of runs to be 13 instead of 16. First of all, the odd number of runs will prevent the design from being balanced or orthogonal. The D-optimal design will generate 13 runs where the corresponding $|X'X|$ is maximized, thus, minimizing the amount of non-orthogonality.

For Example 2, consider a design with 5 factors: A, B, C, D, and E. Factor A is qualitative with 3 levels, B is qualitative with 4 levels, C is qualitative with 2 levels, and D and E are quantitative with 3 levels. Assume the C × D and C × E interactions are also of interest and the number of runs must be less than 25. In this case, standard designs will not be of much assistance. However, the D-optimal alternative will provide a design with $|X'X|$ close, if not equal, to its maximum for some specific n < 25.

Example 3 consists of 4 quantitative factors, each at 3 levels where the (+, +, +, +) and (−, −, −, −) combinations are infeasible. Using the D-optimal option of the appropriate software package, you can generate a near orthogonal design excluding the infeasible points. Thus, the advantages of D-optimal designs include (i) ability to handle different types of factors with different numbers of levels, (ii) exclusion of infeasible points, and (iii) desired number of runs. Disadvantages include non-orthogonality and specific software needs. (See Chapters 5 and 8 for some more D-optimal design matrices contained in this text.)

D-optimality is just one of several related design optimality criteria which can be used to select the best design points from a list of candidate points. The other most common criteria are A-, G-, and V-optimality. All four types employ slightly different variance minimizing criteria. If we let b̲ be the column vector of regression coefficients (see multiple regression in Chapter 4), it can be shown that:

1) $\underline{b} = (X'X)^{-1} X'Y$

2) $Var\ (\underline{b}) = (X'X)^{-1} \sigma^2$

3) $Var\ (\hat{Y}\ (X)) = X'(X'X)^{-1}X\ \sigma^2$, where $\hat{Y}\ (X)$ is the predicted response at some point X.

A-optimality seeks to minimize the average variance of the elements of $\underline{b}$. G-optimality seeks to minimize the maximum prediction variance, max $[d = X'(X'X)^{-1}\ X\sigma^2]$, for a given set of design points. V-optimality minimizes the average value of d over a specified set of design points [6].

3.7 Fractional Factorials and Latin Squares for 3-Level Designs

Fractional factorials in designs of 3 levels, 3^{k-q} designs, are a bit more complicated. They are designs which include a mixture of corner points and mid-level points. Because of this combination of experimental points, these designs can be used to test all linear effects, all second-order (quadratic) effects, and limited linear and quadratic interactions. The 3-level fractional factorial and full factorial designs are the only designs discussed in this book which can evaluate quadratic interactions. An example of a quadratic interaction term is AB^2 (see more on quadratic interactions in Chapter 4). The drawbacks of the 3^{k-q} design are the number of runs required to estimate all desired effects, the limited number of interactions which can be evaluated, and the complexity of the confounding/aliasing patterns. Even if quadratic interactions and simple 2-way interactions are not estimated, n must still be large to estimate second-order terms for all factors. The aliasing patterns for 3^{k-q} designs are difficult to obtain and will not be discussed in this book. Several software packages will do this for you, or you can see reference [9] or [1] for more details.

One method of formulating a 3^{k-1} fractional design is through the use of **Latin Squares**. For example, given the Latin Square design in Table 3.18, a list of all combinations of the 3 factors results in a balanced orthogonal 3^{k-1} design shown in Table 3.19. When k = 4, a Latin cube is constructed to build a 3^{4-1} design and Latin hypercubes are the basis for designs with k > 4. As with the use of regular Latin Square designs, interactions are not typically available for evaluation.

	Factor B			
	−1	0	+1	
Factor C −1	−1	0	+1	
0	0	+1	−1	Factor A (coded values
+1	+1	−1	0	are inside the square)

Table 3.18 Latin Square Used to Generate a Latin Square Design

	Factors		
Run	A	B	C
1	−	−	−
2	0	0	−
3	+	+	−
4	0	−	0
5	+	0	0
6	−	+	0
7	+	−	+
8	−	0	+
9	0	+	+

Table 3.19 Coded Matrix for Latin Square Design for 3 Factors
Each at 3 Levels

The 3-level fractional factorial designs are especially useful for qualitative factors (method, material type, lot, vendor, operator, etc.) which have few (if any) anticipated significant interactions. These designs are also used for 3-level screening. The Taguchi L_{18} (page 3-47) is the most commonly used 3-level screening design, and it is easily applied to both qualitative and quantitative factors.

If all the factors are quantitative (time, temperature, pressure, flow, speed, concentration, etc.) and you want to evaluate factor interactions, the 3-level fractional factorial design will be very inefficient. A better choice would be the Box-Behnken or the Central Composite Design discussed in the next two sections. Also, see the comparisons of 3-level design types in Section 3.10.

3.8   Box-Behnken Designs

An efficient and frequently used 3-level design for modeling quantitative factors (time, temperature, pressure, flow, speed, concentration, etc.) with 3 levels is the **Box-Behnken** design. Through the use of embedded 2^2 designs while holding certain factors at their centerpoint, the Box-Behnken designs are much more efficient than 3^k full factorials, and some types are easily blocked. In their 1960 article, Box and Behnken [3] provide tabled Box-Behnken designs for k up to 16 (excluding k = 8). When k = 4, the design, in shorthand notation, appears as shown in Table 3.20a. This design is divided into 3 orthogonal blocks of 9 runs where each ± 1 refers to alternating columns of $(- - + +)$ and $(- + - +)$. The completed 27 run Box-Behnken for k = 4 is shown in Table 3.20b.

Factors			
A	B	C	D
± 1	± 1	0	0
0	0	± 1	± 1
0	0	0	0
± 1	0	0	± 1
0	± 1	± 1	0
0	0	0	0
± 1	0	± 1	0
0	± 1	0	± 1
0	0	0	0

Table 3.20a Shorthand Notation for a Box-Behnken Design for 4 Factors

Run	A	B	C	D	
1	−	−	0	0	
2	−	+	0	0	
3	+	−	0	0	
4	+	+	0	0	
5	0	0	−	−	
6	0	0	−	+	
7	0	0	+	−	
8	0	0	+	+	
9	0	0	0	0	
10	−	0	0	−	
11	−	0	0	+	Note: Interaction and quadratic columns
12	+	0	0	−	can be developed by the appropriate
13	+	0	0	+	column value multiplications.
14	0	−	−	0	
15	0	−	+	0	The orthogonal blocks are separated by
16	0	+	−	0	dashed lines.
17	0	+	+	0	
18	0	0	0	0	
19	−	0	−	0	
20	−	0	+	0	
21	+	0	−	0	
22	+	0	+	0	
23	0	−	0	−	
24	0	−	0	+	
25	0	+	0	−	
26	0	+	0	+	
27	0	0	0	0	

Table 3.20b Complete Set of 27 Runs for a Box-Behnken Design
Based on 4 Factors

The Box-Behnken design does not contain any corner points in the design space, which may or may not be of concern. In some experiments where corner points are infeasible, the Box-Behnken may be an attractive alternative. If only 3 factors were used, the design would appear graphically as shown in Figure 3.3. The Box-Behnken designs are nearly orthogonal, resolution V designs that allow for estimating linear effects, quadratic effects, and all linear 2-way interactions. The primary disadvantage of Box-Behnken designs is that the number of runs is large enough to estimate all factor second-order effects and all linear 2-way interactions, whether you want to or not. When the number of factors is greater than 4, the Box-Behnken design will be less efficient with regard to the number of runs than the central composite design (see Sections 3.9 and 3.10). The slight non-orthogonality is not a disadvantage when the analysis is conducted by way of least squares regression (see Chapter 4).

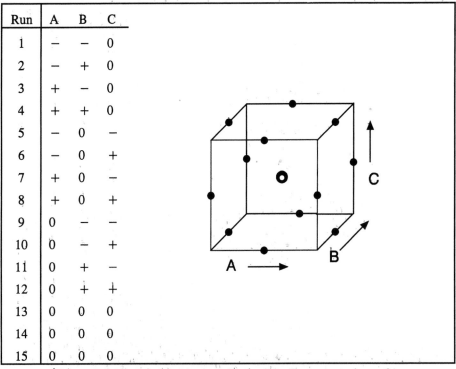

Run	A	B	C
1	−	−	0
2	−	+	0
3	+	−	0
4	+	+	0
5	−	0	−
6	−	0	+
7	+	0	−
8	+	0	+
9	0	−	−
10	0	−	+
11	0	+	−
12	0	+	+
13	0	0	0
14	0	0	0
15	0	0	0

Figure 3.3 Design Matrix and Geometric Representation of a
Box-Behnken Design for 3 Factors

A shorthand notation summary of Box-Behnken designs for k = 3, 4, and 5 is displayed in Table 3.21.

						Box-Behnken Designs		
Number of factors (k)		Design Matrix				Number of Points	Blocking and Centerpoint Information	
3		± 1	± 1	0		4		No orthogonal blocking with 3 replicated center points
		± 1	0	± 1		4		
		0	± 1	± 1		4		
		0	0	0		3		
						n = 15		
4	± 1	± 1	0	0		4		3 blocks of 9 runs each and 1 center point per block
	0	0	± 1	± 1		4		
	0	0	0	0		1		
	± 1	0	0	± 1		4		
	0	± 1	± 1	0		4		
	0	0	0	0		1		
	± 1	0	± 1	0		4		
	0	± 1	0	± 1		4		
	0	0	0	0		1		
						n = 27		
5	± 1	± 1	0	0	0	4		2 blocks of 23 runs each and 3 replicated center points per block
	0	0	± 1	± 1	0	4		
	0	± 1	0	0	± 1	4		
	± 1	0	± 1	0	0	4		
	0	0	0	± 1	± 1	4		
	0	0	0	0	0	3		
	0	± 1	± 1	0	0	4		
	± 1	0	0	± 1	0	4		
	0	0	± 1	0	± 1	4		
	± 1	0	0	0	± 1	4		
	0	± 1	0	± 1	0	4		
	0	0	0	0	0	3		
						n = 46		

Table 3.21 Box-Behnken Designs for k = 3, 4, and 5

3.9   Box-Wilson (Central Composite) Designs

A very flexible and efficient second-order modeling design for quantitative factors is the Box-Wilson or **Central Composite Design** (CCD). For 3 factors ($k = 3$), CCD examples appear as shown in Tables 3.22a and b.

a) CCD $k = 3$, using 2^3					b) CCD $k = 3$, using 2^{3-1}				
	FACTOR					FACTOR			
Run	A	B	C		Run	A	B	C	
1	−	−	−		1	−	−	−	
2	−	−	+		2	−	+	+	F
3	−	+	−		3	+	−	+	
4	−	+	+	F	4	+	+	−	
5	+	−	−		5	0	0	0	
6	+	−	+		6	0	0	0	C
7	+	+	−		7	0	0	0	
8	+	+	+		8	α	0	0	
9	0	0	0		9	$-\alpha$	0	0	
10	0	0	0		10	0	α	0	
11	0	0	0	C	11	0	$-\alpha$	0	A
12	0	0	0		12	0	0	α	
13	0	0	0		13	0	0	$-\alpha$	
14	0	0	0						
15	α	0	0						
16	$-\alpha$	0	0		F =	factorial portion			
17	0	α	0	A	C =	centerpoint portion			
18	0	$-\alpha$	0		A =	axial portion			
19	0	0	α						
20	0	0	$-\alpha$						

Table 3.22 Central Composite Designs for 3 Factors
a) Full Factorial 2-Level Portion   b) Fractional Factorial 2-Level Portion

Graphically, the design from Table 3.22a will appear as indicated in Figure 3.4.

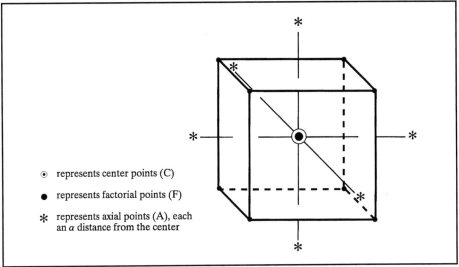

Figure 3.4 Graphical Representation of a k=3 CCD

If 2-way interactions are potentially important, you can use a R_{IV} or R_V design in the factorial or F portion of the design; whereas, if they are not, you can save runs by using a R_{III}. The number of axial points will typically be 2k unless some factors don't lend themselves to quadratic effects. The value for α is usually set at $(n_F)^{1/4}$ where n_F is the number of runs in the factorial (2 level) portion of the design. These values are chosen to produce **rotatability**, which simply implies that the predicted response is capable of being estimated with equal variance regardless of the direction from the center of the design space. Rotatability is a desired design property; however, it is not necessary for successful designs. You should notice that, in order to satisfy rotatability, α will take on values greater than 1 which means that ±1 no longer represents the factor minimum and maximum. An easy way to resolve the transformation to or from the design notation of 0, ±1, $\pm\alpha$ is as follows:

(1) Set up a number line as shown below.

design values:	$-\alpha$	-1	0	1	α

(2) Place the real factor value min and max (use 100 and 300 as an example) under $-\alpha$ and α, respectively. Place $\dfrac{\text{max} + \text{min}}{2}$ under 0. The values under -1 and 1 will be computed as the value $200 \pm$ some Δ.

design values:	$-\alpha$	-1	0	1	α
real values:	100	(200 - Δ)	200	(200 + Δ)	300

(3) To find the value for Δ, we need to know the α value (assume it is 1.5) and then form the following ratio:

$$\frac{300 - 200}{1.5} = \frac{\Delta}{1} \; ; \text{ therefore, } \quad \Delta = 66.\overline{66},$$

implying $200 + \Delta = 266.\overline{66}$ and $200 - \Delta = 133.\overline{33}$

The final result appears as follows:

	$-\alpha$				α
design values:	(-1.5)	-1	0	1	(1.5)
real values:	100	133.33	200	266.67	300

which indicates the real factor values for all 5 possible design levels. Sometimes it may be infeasible to run 5 levels of a factor, yet efficient second-order modeling is desired. In this case you can set $\alpha = 1$ for a Central Composite Face (CCF) design. This design will not have rotatability and will suffer some loss of orthogonality for the quadratic terms; however, it has been used successfully in numerous case studies.

The number of center points, n_c, will vary, but basically, enough are needed to get a good estimate of pure experimental error and to maintain orthogonality. According to John [7], to retain orthogonality $n_c = 4(\sqrt{n_F} + 1) - 2k$. Since John's formula calls for an unusually large number of center points, we have made a slight modification to be $n_c = 4\sqrt{n_F + 1} - 2k$, which will save runs at the center and still produce a nearly orthogonal design. For more information, see Box and Draper [4] or Myers [9]. Depending on which resolution you use in the 2^{k-q} (factorial) portion of a CCD, it can be up to 47% more efficient in experimental runs than the Box-Behnken design [4]. In addition, the CCD is rotatable for $\alpha = (n_F)^{1/4}$ whereas the Box-Behnken is only rotatable for specific values of k.

The CCD has several advantages:

(i)   It is the most effective and efficient second-order modeling design for factors which are quantitative.

(ii)   It has flexibility in resolution selection, i.e., you can build the factorial portion for any resolution you might require.

(iii)   Aliasing for R_{III} or R_{IV} designs is computed the same way as for 2-level fractional factorials.

(iv)   CCD designs can be run sequentially to possibly save resources, i.e., run factorial and center point portions first. Build a linear model from the 2-level portion and predict the center point results. If the linear model is not confirmed at the center points, then add the axial points to complete a quadratic model.

The primary disadvantage is that the CCD is best suited for quantitative factors. If only one qualitative factor exists, a CCD can still be modified to handle it, but when you have several qualitative factors you would need to use a fractional factorial or D-optimal design. The slight non-orthogonality of the CCD is not a disadvantage when the analysis is conducted by way of least squares regression (see Chapter 4).

Consider another example where the CCD might be your choice of design type. Assume you have 5 factors (all quantitative) and you wish to build a second-order model with all important 2-way linear interactions. Since interactions are a big concern, you require a R_{IV} or R_V design. Starting with the factorial (or 2 level) portion, you could:

(i)   Let $n_F = 2^5 = 32$ runs, i.e., use a full factorial which would allow for estimating all interactions, or

(ii)   Let $n_F = 2^{5-1} = 16$ runs where factor E = ABCD and I = ABCDE. This design is R_V and a big savings over the full factorial, or

(iii) Let $n_F = 2^{5-2} = 8$ runs where D = AB, E = ABC, and I = ABD = ABCE = CDE. Since this design is less than R_{IV}, it appears that the best 2-level design for the factorial portion is a 2^{5-1}.

Next, you need to determine the number of center points:

$$n_c = 4\sqrt{n_F} + 1 - 2k \text{ which results in } n_c = 4\sqrt{16} + 1 - 2(5) \doteq 7.$$

The number of axial points is $n_a = 2k = 10$ with a recommended α value of $\alpha = (n_F)^{1/4} = (16)^{1/4} = 2.0$. The complete CCD for this example is shown in Table 3.23.

Run	A	B	C	D	E =ABCD	
1	−	−	−	−	+	
2	−	−	−	+	−	
3	−	−	+	−	−	
4	−	−	+	+	+	
5	−	+	−	−	−	
6	−	+	−	+	+	
7	−	+	+	−	+	
8	−	+	+	+	−	
9	+	−	−	−	−	
10	+	−	−	+	+	
11	+	−	+	−	+	
12	+	−	+	+	−	
13	+	+	−	−	+	Note: All quadratic and
14	+	+	−	+	−	2-way linear interaction
15	+	+	+	−	−	columns can be developed by
16	+	+	+	+	+	appropriate column value
17	0	0	0	0	0	multiplications.
18	0	0	0	0	0	
19	0	0	0	0	0	
20	0	0	0	0	0	
21	0	0	0	0	0	
22	0	0	0	0	0	
23	0	0	0	0	0	
24	α	0	0	0	0	
25	$-\alpha$	0	0	0	0	
26	0	α	0	0	0	
27	0	$-\alpha$	0	0	0	
28	0	0	α	0	0	
29	0	0	$-\alpha$	0	0	
30	0	0	0	α	0	
31	0	0	0	$-\alpha$	0	
32	0	0	0	0	α	
33	0	0	0	0	$-\alpha$	

Table 3.23 CCD for k = 5 with R_V

3.10   Comparison of 3-Level Designs

The information in Table 3.24 is provided as a comparison of the number of runs required for the 3^{k-q}, Box-Behnken, CCD, D-optimal (D-opt), and the full factorial designs, given 3-level design scenarios with various numbers of quantitative factors and 2-way interactions of interest. Numbers in parentheses indicate the number of replicated center points included in the design. The 3^{k-q} designs are similar to the Taguchi 3-level designs discussed in Section 3.11.

Scenario	3^{k-q}	Box-Behnken	CCD**	D-opt***	Full Factorial
k = 3 with no interactions	9(0)	15(3)	13(3)*	7(0)	27(1)
k = 4 with 3 2-way linear interactions	27(0)	27(3)	20(4)*	12(0)	81(1)
k = 5 with 1 2-way linear interaction	27(0)	46(6)	20(2)*	12(0)	243(1)
k = 5 with all 2-way linear interactions	81(0)	46(6)	33(7)*	21(0)	243(1)
k = 7 with 7 2-way linear interactions	81(0)	62(6)	33(3)*	22(0)	2187(1)

$*n_c = 4\sqrt{n_F + 1} - 2k$

**The total number of runs for the CCD includes 2k axial points. In many cases all the factors will not necessarily have a significant second-order term. Sequentially adding axial points in the CCD until there is no more evidence of another quadratic term will typically lead to less than 2k axial points.

***This is a minimum number of required runs; however, most software packages will add 2 to these numbers.

Table 3.24 Comparison of Design Types for Quantitative Factor Scenarios

As you can see, the CCD is far more efficient than the 3^{k-q}, Box-Behnken, and Full Factorial in all but one of the situations presented. Keep in mind, though, the objective for each of the designs as shown in Table 3.25.

Design	Objective
3^{k-q}	• Used for qualitative or quantitative factors • Estimate all linear and quadratic effects and, when higher order resolutions are desired, you can also estimate linear and quadratic 2-way interactions. • Typically the 3^{k-q} is a good choice if you have qualitative factors with few or no interactions. This is not a good choice for factors that are quantitative.
Box-Behnken	• Used only for quantitative factors; however, for some k you can have one qualitative blocking factor. • Estimate all linear, quadratic, and 2-way linear interactions plus experimental error.
CCD	• Used primarily for quantitative factors; however, if you only have 1 qualitative factor, the CCD can still be useful. • Estimate all linear effects, selected quadratics, and selected 2-way linear interactions plus experimental error. • This is typically your best choice for quantitative factors, each at 3 levels.
D-optimal	• Used for quantitative, qualitative, or a mix of both. • Can estimate any desired effect with reduced n. • This design will typically not be orthogonal, but it does minimize the non-orthogonality of the design.
Full Factorial	• Used for qualitative or quantitative factors. • Estimate all linear and quadratic effects plus all possible simple and higher order interactions.

Table 3.25 Summary of Objectives for 3-Level Design Types

3.11 Taguchi Designs

The next set of designs to be discussed are those of Taguchi [11]. These designs are typically orthogonal with respect to main effects, but contain either aliased or confounded main effects with 2-way interactions. They are essentially R_{III} type screening designs. The designs used by Taguchi are not new, but simply a form of Plackett-Burman, Fractional Factorial, or Latin Square designs. The fact that they are tabled provides for easy use. The most frequently used Taguchi designs are shown in Table 3.26. Tables 3.27 through 3.33 provide more information on each of these design types. Notice that the 2-level Taguchi designs (L_4, L_8, L_{16}) are the same as the 2^{k-q} designs except that all even-way interactions have been multiplied by (-1). The L_{12} is a Plackett-Burman or Hadamard matrix with permuted rows and columns. Note that, while Plackett-Burman designs have all (-1)'s in the last row, Taguchi designs have all (-1)'s in the first row.

The most important contribution of Taguchi is not the orthogonal arrays. They have been around for some time. Instead, it is Taguchi's loss function and robust designs which are significant contributions. The robust designs investigate and detect factors which affect the response value and the variability or dispersion of the response. The result is a design which determines factor settings to optimize the response and minimize response variability. A more detailed discussion on Taguchi methods, including how to set up robust designs, appears in Chapters 5 and 6.

Design	# of Levels	# of Factors for the Full Factorial	# of Factors to Maintain Resolution V	# of Factors for Screening
L_4	2	2	2	3
L_8	2	3	3	7
L_9	3	2	–	4
L_{12}	2	–	–	11
L_{16}	2	4	5	15
L_{18}	mixed	–	–	8
L_{27}	3	3	–	13

Table 3.26 Most Frequently Used Taguchi Designs

	L₄ Design		
Run	1	2	3
1	−1	−1	−1
2	−1	+1	+1
3	+1	−1	+1
4	+1	+1	−1
Original factors and interactions used to generate the matrix	a	b	−ab
Strategy	# Factors	Which Columns?	
Full Factorial	2	1, 2	
Screening	3	1, 2, 3	

Table 3.27 L₄ Design [11] for 2-Level Factors

	L₈ Design						
Run	1	2	3	4	5	6	7
1	−1	−1	−1	−1	−1	−1	−1
2	−1	−1	−1	+1	+1	+1	+1
3	−1	+1	+1	−1	−1	+1	+1
4	−1	+1	+1	+1	+1	−1	−1
5	+1	−1	+1	−1	+1	−1	+1
6	+1	−1	+1	+1	−1	+1	−1
7	+1	+1	−1	−1	+1	+1	−1
8	+1	+1	−1	+1	−1	−1	+1
Original factors and interactions used to generate the matrix	a	b	−ab	c	−ac	−bc	abc
Strategy	# Factors			Which Columns?			
Full Factorial	3			1, 2, 4			
Screening	4-7			1, 2, 4, 7, ...			

Table 3.28 L₈ Design [11] for 2-Level Factors

Note: If you use the L_8 for screening 7 factors, each 2-way interaction will be aliased with a main effect and other 2-way interactions. For this reason, many experimenters prefer the L_{12} to screen 6 to 11 factors. See Appendix M for more information on screening recommendations.

Run	L_9 Design			
	1	2	3	4
1	-1	-1	-1	-1
2	-1	0	0	0
3	-1	$+1$	$+1$	$+1$
4	0	-1	0	$+1$
5	0	0	$+1$	-1
6	0	$+1$	-1	0
7	$+1$	-1	$+1$	0
8	$+1$	0	-1	$+1$
9	$+1$	$+1$	0	-1
Strategy	# Factors		Which Columns?	
Full Factorial	2		1, 2	
Screening	3-4		1, 2, 3, ...	

Table 3.29 L_9 Design [11] for 3-Level Factors

Due to confounding problems with interactions, the L_9 is not often recommended. As an alternative design see the L_{18}.

					L_{12} **Design**						
Run	1	2	3	4	5	6	7	8	9	10	11
1	-1	-1	-1	-1	-1	-1	-1	-1	-1	-1	-1
2	-1	-1	-1	-1	-1	+1	+1	+1	+1	+1	+1
3	-1	-1	+1	+1	+1	-1	-1	-1	+1	+1	+1
4	-1	+1	-1	+1	+1	-1	+1	+1	-1	-1	+1
5	-1	+1	+1	-1	+1	+1	-1	+1	-1	+1	-1
6	-1	+1	+1	+1	-1	+1	+1	-1	+1	-1	-1
7	+1	-1	+1	+1	-1	-1	+1	+1	-1	+1	-1
8	+1	-1	+1	-1	+1	+1	+1	-1	-1	-1	+1
9	+1	-1	-1	+1	+1	+1	-1	+1	+1	-1	-1
10	+1	+1	+1	-1	-1	-1	-1	+1	+1	-1	+1
11	+1	+1	-1	+1	-1	+1	-1	-1	-1	+1	+1
12	+1	+1	-1	-1	+1	-1	+1	-1	+1	+1	-1

Strategy	# Factors	Which Columns?
Screening	6-11	any

Table 3.30 L_{12} Design [11] for 2-Level Factors
(This is the same matrix as the Plackett-Burman 12-run design;
however, the columns and rows have been permuted.)

Note 1: If you use less than 11 factors in this design, choose the first k columns
you need and let the remaining columns represent experimental noise.

Note 2: See DOE PRO from Air Academy Associates to evaluate the partially
confounded 2-way interactions.

L₁₆ Design

Run	1	2	3	4	5	6	7	8	9	10	11	12	13	14	15
1	−	−	−	−	−	−	−	−	−	−	−	−	−	−	−
2	−	−	−	−	−	−	−	+	+	+	+	+	+	+	+
3	−	−	−	+	+	+	+	−	−	−	−	+	+	+	+
4	−	−	−	+	+	+	+	+	+	+	+	−	−	−	−
5	−	+	+	−	−	+	+	−	−	+	+	−	−	+	+
6	−	+	+	−	−	+	+	+	+	−	−	+	+	−	−
7	−	+	+	+	+	−	−	−	−	+	+	+	+	−	−
8	−	+	+	+	+	−	−	+	+	−	−	−	−	+	+
9	+	−	+	−	+	−	+	−	+	−	+	−	+	−	+
10	+	−	+	−	+	−	+	+	−	+	−	+	−	+	−
11	+	−	+	+	−	+	−	−	+	−	+	+	−	+	−
12	+	−	+	+	−	+	−	+	−	+	−	−	+	−	+
13	+	+	−	−	+	+	−	−	+	+	−	−	+	+	−
14	+	+	−	−	+	+	−	+	−	−	+	+	−	−	+
15	+	+	−	+	−	−	+	−	+	+	−	+	−	−	+
16	+	+	−	+	−	−	+	+	−	−	+	−	+	+	−
	a	b	−ab	c	−ac	−bc	abc	d	−ad	−bd	abd	−cd	acd	bcd	−abcd

Strategy	# Factors	Which Column?
Full Factorial	4	1, 2, 4, 8
Resolution V	5	1, 2, 4, 8, 15
Screening	6-15	1, 2, 4, 8, 14, 7, 11, 13, 15, 12, 10, ...

Table 3.31 L₁₆ Design [11] for 2-Level Factors

Note: For screening up to 8 factors use columns 1, 2, 4, 7, 8, 11, 13, and 14 to reduce the aliasing problem. Beyond 8 factors you will have 2 ways aliased with main factors and thus you might prefer the L₁₂ if k ≤ 11. See Appendix M.

				L$_{18}$ Design				
Run	1	2	3	4	5	6	7	8
1	-1	-1	-1	-1	-1	-1	-1	-1
2	-1	-1	0	0	0	0	0	0
3	-1	-1	+1	+1	+1	+1	+1	+1
4	-1	0	-1	-1	0	0	+1	+1
5	-1	0	0	0	+1	+1	-1	-1
6	-1	0	+1	+1	-1	-1	0	0
7	-1	+1	-1	0	-1	+1	0	+1
8	-1	+1	0	+1	0	-1	+1	-1
9	-1	+1	+1	-1	+1	0	-1	0
10	+1	-1	-1	+1	+1	0	0	-1
11	+1	-1	0	-1	-1	+1	+1	0
12	+1	-1	+1	0	0	-1	-1	+1
13	+1	0	-1	0	+1	-1	+1	0
14	+1	0	0	+1	-1	0	-1	+1
15	+1	0	+1	-1	0	+1	0	-1
16	+1	+1	-1	+1	0	+1	-1	0
17	+1	+1	0	-1	+1	-1	0	+1
18	+1	+1	+1	0	-1	0	+1	-1

Strategy	# Factors	Which Column?
Screening	8 (max)	2-level factor in column 1
		3-level factors in columns 2-8

Table 3.32 L$_{18}$ Design [11] for (1) 2-Level Factor and (7) 3-Level Factors

Note: Using DOE PRO from Air Academy Associates, it is possible to evaluate 2-way interactions and to build a quadratic model for prediction purposes.

L$_{27}$ Design

Run	1	2	3	4	5	6	7	8	9	10	11	12	13
1	−	−	−	−	−	−	−	−	−	−	−	−	−
2	−	−	−	−	0	0	0	0	0	0	0	0	0
3	−	−	−	−	+	+	+	+	+	+	+	+	+
4	−	0	0	0	−	−	−	0	0	0	+	+	+
5	−	0	0	0	0	0	0	+	+	+	−	−	−
6	−	0	0	0	+	+	+	−	−	−	0	0	0
7	−	+	+	+	−	−	−	+	+	+	0	0	0
8	−	+	+	+	0	0	0	−	−	−	+	+	+
9	−	+	+	+	+	+	+	0	0	0	−	−	−
10	0	−	0	+	−	0	+	−	0	+	−	0	+
11	0	−	0	+	0	+	−	0	+	−	0	+	−
12	0	−	0	+	+	−	0	+	−	0	+	−	0
13	0	0	+	−	−	0	+	0	+	−	+	−	0
14	0	0	+	−	0	+	−	+	−	0	−	0	+
15	0	0	+	−	+	−	0	−	0	+	0	+	−
16	0	+	−	0	−	0	+	+	−	0	0	+	−
17	0	+	−	0	0	+	−	−	0	+	+	−	0
18	0	+	−	0	+	−	0	0	+	−	−	0	+
19	+	−	+	0	−	+	0	−	+	0	−	+	0
20	+	−	+	0	0	−	+	0	−	+	0	−	+
21	+	−	+	0	+	0	−	+	0	−	+	0	−
22	+	0	−	+	−	+	0	0	−	+	+	0	−
23	+	0	−	+	0	−	+	+	0	−	−	+	0
24	+	0	−	+	+	0	−	−	+	0	0	−	+
25	+	+	0	−	−	+	0	+	0	−	0	−	+
26	+	+	0	−	0	−	+	−	+	0	+	0	−
27	+	+	0	−	+	0	−	0	−	+	−	+	0

Strategy	# Factors	Which Columns?
Full Factorial	3	1, 2, 5
Screening	8-13	1, 2, 5, ...

Table 3.33 L$_{27}$ Design [11] for 3-Level Factors

3.12   Design Selection Scenarios

The following scenarios have been written to give you a feel for the options available when selecting a design. Each time you run an experiment a similar design selection process should take place. (Also see Appendix M for Rules of Thumb on design selection).

Scenario 1: Suppose you have 11 quantitative factors and you wish to i) separate the important factors from the unimportant factors (screening phase), and ii) build a model relating the important factors to the response (modeling phase).

Screening Phase Options:

(A)   Use a Plackett-Burman 12-run design and Pareto the effects to select the important factors.

(B)   Use a 2^{11-7} fractional factorial design with 16 runs to separate the important factors and also assess a few potentially strong interactions.

(C)   Use an L_{27} design to screen important factors from a linear and a quadratic perspective.

(D)   Use a D-optimal design built for studying all linear effects, desired quadratics, and desired interactions.

Modeling Phase Options: (Assume 4 quantitative factors were screened out in the previous phase. In addition, quadratics and 2-way linear interactions are possible.)

(A)   Use an L_{27} design which provides most of the interaction information plus all linear and all quadratic information.

(B)   Use a Box-Behnken design for 4 factors with $n = 27$ (see Table 3.21). This design will be R_V, allowing you to estimate all linear, all quadratic, and all 2-way linear interactions.

(C)   Use a CCD design with either

        (i)   a factorial portion of $n_F = 2^{4-1} = 8$ runs (R_{IV})

$$n_c = 4\sqrt{8+1} - 2(4) = 4 \text{ runs}$$

$$n_a = 2(k) = 8 \text{ runs}$$

$$\alpha = (n_F)^{1/4} = (8)^{1/4} = 1.68 \text{ or use } \alpha = 1 \text{ for a CCF}$$

$$n(\text{total}) = 20 \text{ runs, or}$$

        (ii)   a factorial portion of $n_F = 2^4 = 16$ runs (full factorial)

$$n_c = 4\sqrt{16+1} - 2(4) = 9 \text{ runs}$$

$$n_a = 2(k) = 8 \text{ runs}$$

$$\alpha = (n_F)^{1/4} = (16)^{1/4} = 2.0 \text{ or use } \alpha = 1 \text{ for a CCF}$$

$$n(\text{total}) = 33 \text{ runs.}$$

(D)   Use a D-optimal design where $n_{min} = 4 + 4 + 6 + 1 = 15$ is derived from $\underline{4}$ linear terms, $\underline{4}$ quadratics, $\binom{4}{2} = \dfrac{4!}{2! \, 2!} = \underline{6}$ 2-way linear

interactions, and $\underline{1}$ constant. The actual design(s) to address scenario 1 will depend on several key issues such as:

        (i)   time and dollars available

        (ii)   time and cost per run

        (iii)   amount of prior knowledge of the process

        (iv)   software support for generating and analyzing designs

        (v)   experience level of the experimenter

In most cases, there is no perfect choice of designs. All of the choices have advantages and disadvantages − some choices are better than others. You eventually have to choose which strategy best suits your needs.

Scenario 2: Suppose you have 1 qualitative factor with 2 levels and 5 quantitative factors with potential quadratics and 2-way interactions. Your objective is to model the response with respect to the input factors. Consider the following approaches:

(A) (i) Screen the factors with an L_{18} to identify important factors based on linear and quadratic effects.

  (ii) Model important factors with a CCD. If the qualitative factor is found important from step (i), then you could run 2 modeling experiments for each level of the qualitative factor. Another approach would be to leave the qualitative factor in the CCD and run half the center points at (-1) and the other half at ($+1$). The axial points can all be run at either the ($-$) or ($+$) setting for the qualitative factor. Some orthogonality is lost in the approach, but it is usually not enough to be overly concerned.

(B) (i) Screen the factors with a L_8 design.

  (ii) Model the important factors with a CCD described in (A) (ii) above or use a D-optimal design.

(C) (i) Model the 6 factors using a D-optimal design where $n_{min} = \underline{1}$ (for the qualitative factor) + $\underline{5}$ (for the quantitative linear effects) + $\underline{5}$ (for the quantitative quadratic effects) + $\underline{15}$ (for all 2-way linear interactions) + $\underline{1}$ (for estimating the constant). Therefore, $n_{min} = 27$.

As in Scenario 1, we cannot recommend any one particular approach. Obviously, the choice depends on many concerns. The important point in this section is to realize that any experimental design can be approached in different ways. It is always best to first brainstorm all the possibilities and then select the choice that best meets your objectives while satisfying budgetary considerations.

3.13   Nested Designs

A special type of design is required when the settings of one factor are dependent on the settings of another. For example, suppose you have 2 cleaning solutions (1, 2) which have different effective temperature ranges, i.e., solution 1 operates over 70 to 90 degrees while solution 2 operates over 100 to 120 degrees. A tree depicting the test conditions along with time and method is shown in Figure 3.5.

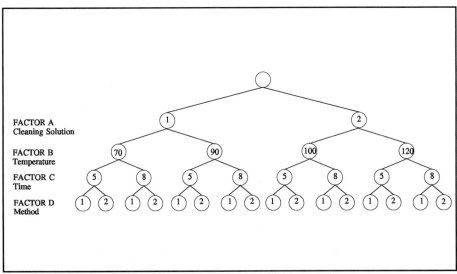

Figure 3.5 Design Tree Depicting Combinations for
4 Factors with B Nested in A

Obviously, factor B is not a 2-level factor because its levels are different for each level of A. In this case, factor B is said to be "nested in A" or "conditioned (dependent) on A" and thus a normal 2-level design does not apply. Since nested design analysis can become quite complex, it is recommended that non-statisticians use the "Keep It Simple Statistically" (K.I.S.S.) approach which is described in Figure 3.6. Notice that the design has been split in half, where separate simple experiments can be run for each half. The splitting process has now created two designed experiments and the nested problem has disappeared. Which experiment is run first will depend on cost, convenience or some other criteria. It is even possible that only one of the halves needs to be tested.

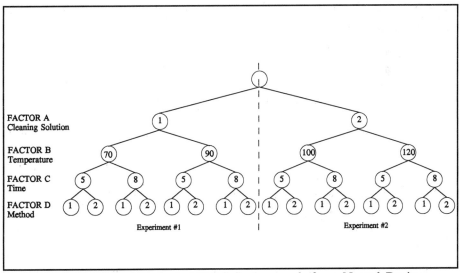

Figure 3.6   Recommended K.I.S.S. Approach for a Nested Design

Some other typically encountered nested scenarios include lots nested in vendor, or parts nested in lot, etc. For these types of nested problems it might be more reasonable to investigate robustness to lots, vendors or parts and thus avoid the nesting problem altogether by using a robust design (see Chapters 5 and 6). For more information on nested designs see [8].

3.14 Mixture Designs

Another special design type occurs when the sum of the actual factor settings must equal the same fixed total for every run. For example, suppose we mix gases A, B, and C, where A + B + C = TOTAL. Similar scenarios might involve "amount of materials" or "amount of chemicals" for A, B, and C. Assume the factors A, B, and C have lows and highs of (20, 40), (30, 40) and (25, 50), respectively. A standard 2^3 (8-run) design appears as follows:

A	B	C	TOTAL (A+B+C)
20	30	25	75
20	30	50	100
20	40	25	85
20	40	50	110
40	30	25	95
40	30	50	120
40	40	25	105
40	40	50	130

Although the design appears to be orthogonal, the total amount of the substance differs for each run and it is possible that the differing totals could bias the results. It should also be noted that TOTAL is partially confounded with A, B, and C. The classical approach to this type of problem is by way of a methodology called mixture experiments [6]. Since mixture experiments are extremely complex, a K.I.S.S. recommendation is to use ratios. For example, if we choose factor C to be the ratio denominator, then two ratios A/C and B/C are needed to describe these 3 factors. If we choose lows and highs for each ratio, then a standard orthogonal design can be used. For example, the two ratios A/C and B/C can be tested in the 4 runs shown below.

Run	A/C	B/C
1	−	−
2	−	+
3	+	−
4	+	+

To determine the amount of A, B, and C for each run, we set up 3 equations in 3 unknowns and solve. If A/C at the low is .2, B/C at the low is .4 and the desired TOTAL for each run is 100, then run 1 (A/C_ and B/C_) will require amounts of A, B, and C as derived below.

---

(1)   A/C = .2 →   A = .2C

(2)   B/C = .4 →   B = .4C

(3)   A + B + C = 100

Replacing A and B in (3) with .2C and .4C, respectively, results in

   .2C + .4C + C = 100   or   C = 100/1.6 = 62.5

Then A = .2C implies   A = .2(62.5) = 12.5

and B   = .4C implies B = .4(62.5) = 25

Thus, run 1 requires the amounts of A, B, and C to be 12.5, 25 and 62.5, respectively. The other actual run amounts for A, B, and C would be calculated in a similar fashion.

The use of ratios has two primary advantages over the more sophisticated mixture design approach found in Cornell [6]:   (1) you can use standard experimental designs previously discussed, and (2) you can use mixture type factors together with process type factors. For example, in the last example you might have desired to vary temperature, time, pressure or any other process type factor as D. The design matrix with additional process factors is then easy to construct. To add a factor D, the matrix would appear as follows:

Run	A/C	B/C	D
1	−	−	−
2	−	−	+
3	−	+	−
4	−	+	+
5	+	−	−
6	+	−	+
7	+	+	−
8	+	+	+

For more information on the ratio approach, see the case study on *"Consider the Metric: Dealing with Mixtures and Temperature Gradients"* in Chapter 8.

3.15 Design Type Summary

Choosing the right experimental design can appear to be a difficult process. The problem most experimenters have is wanting to select the "perfect" design for their specific problem. It is important to remember that the reason we typically experiment is that our knowledge level is insufficient to solve our problem. If this is the case, then we will also lack the knowledge to choose the "perfect" design type. The most important thing to remember is to spend sufficient time on the first two boxes of the flow chart inside the front cover of the text. (These boxes are similar to steps I through VI in the section titled *Steps for Experimental Design* on pages 1-26 and 1-27.) Without proper planning even the "perfect" design will most likely fail to produce the knowledge we need. Our experience shows that over 90% of the time the best design type can be selected by following the KISS flow chart inside the front cover of the text. For more 2-level design options, see the inside back cover of the text.

We suggest using a lot of critical thinking in choosing your design. Brainstorm at least 3 options and then discuss the advantages/disadvantages of each. This approach may lead to a quick consensus of the best design choice.

Problem Set

1. Describe the similarities among the following design types: (i) a Taguchi L_8 design, (ii) a fractional factorial 2^{7-4}, and (iii) a fractional factorial 2^{4-1}.

2. What are the similarities/differences between fractional factorial, geometric Plackett-Burman, non-geometric Plackett-Burman, and Taguchi designs?

3. Define the following terms:
   - Orthogonal designs
   - Resolution of a design
   - Aliased
   - Confounded
   - Generator
   - Defining word
   - Defining relationship

4. Discuss when you would use the following designs:
   - 3-level fractional factorial (Taguchi)
   - Box-Behnken
   - Box-Wilson (CCD)
   - D-optimal

5. How can you use sequential experimentation to increase resolution from III to IV?

6. Given a 2^{7-2} design where F = ABCD and G = BCDE, determine the defining relationship.

7. For the example on page 3-13, let E = ABCD. How does this affect the defining relationship?

8. Why should you use balanced and orthogonal designs?

9. What is a saturated design?

10. Given the generating vector or column for a n=16 Plackett-Burman design (see page 3-24), write the entire design matrix. Compare this design with a 16-run fractional factorial and a L_{16}.

11. What 3-level designs can you use for qualitative factors? Quantitative factors? Mixtures of both?

12. If Taguchi designs are not new, what has been his major contribution to experimental design?

13. A chemical engineer with a major oil refinery wanted to try a number of different additives to boost the octane level of the unleaded fuel produced at the refinery. For the sake of secrecy his additives were uniquely coded: A, B, C, D, E. All additives were varied from their lowest to their highest. The engineer knew from prior experience that the following additives would interact with each other: AB, AE, BD, DE, CE, and AD. To make matters worse, his boss told him that if his experiment did not produce positive results, he could forget about any further raises. The engineer ran the following screening design:
(Note: D = AC, E = BC)

Runs	A	B	C	D = AC	E = BC
1	+1	+1	+1	+1	+1
2	+1	+1	−1	−1	−1
3	+1	−1	+1	+1	−1
4	+1	−1	−1	−1	+1
5	−1	+1	+1	−1	+1
6	−1	+1	−1	+1	−1
7	−1	−1	+1	−1	−1
8	−1	−1	−1	+1	+1

(a) What is the defining relation?

(b) Are any important interactions aliased with main effects? If you answer "yes", list the alias problems you found.

(c) Generate a foldover of the design generated for this problem.

(d) What is the defining relation for the foldover design?

(e) Would a foldover design salvage this experiment?

(f) Should the engineer forget about his future raises?

14. Given the following factor evaluation scenarios:

(a) 3 quantitative factors (3 levels for each factor)

(b) 3 quantitative factors (3 levels for each factor)
    1 qualitative factor (3 levels for each factor)

(c) 3 quantitative factors (3 levels for each factor)
    2 qualitative factors (3 levels for each factor)
    2 qualitative factors (4 levels for each factor)

For each scenario, determine the minimum number of runs for a D-optimal design (Hint: you will have to count up the total degrees of freedom of the effects to be estimated).

15. A scrubber is used to remove particles from a wafer surface in semiconductor processing. An engineer would like to determine the most important variables that optimize the scrubber's ability to remove particles. There are 8 factors that can be manipulated. The engineer does not want to spend too much time doing this experiment. Given the following data:

Factors	Levels	Units
A: Pre-Rinse Speed (PRS)	500 — 1500	RPM
B: Pre-Rinse Time (PRT)	10 — 25	Sec
C: Scrub Speed (SRS)	500 — 1500	RPM
D: Scrub Time (SRT)	10 — 50	Sec
E: Post-Rinse Speed (PSRS)	500 — 1500	RPM
F: Post-Rinse Time (PSRT)	10 — 25	Sec
G: Dry Speed (DRS)	4000 — 5000	RPM
H: Dry Time (DRT)	25 — 50	Sec

Note: The interactions of interest are AB, CD, EF, and GH

(a) Generate the most effective screening experiment.

(b) What is the defining relationship?

(c) Are any of your important interactions aliased with any main effects?

(d) What is the resolution of your design?

16. A researcher in immunology wanted to determine what effect various concentrations of nitrogen dioxide, as well as the effects of environmental humidity and temperature, have on the ability of the lungs to fight off airborne infections such as pneumonia. Her factors were as follows:

Factors	Absolute Low and High
A: Nitrogen Dioxide Concentration	.5% to 2%
B: Environmental Humidity	10% to 80%
C: Environmental Temperature	50°F to 90°F

She decides to create a Central Composite Design with a factorial portion consisting of 8 runs (i.e., a full factorial).

(a) What should the α distance be for this experiment if she wants to make the design rotatable?

(b) What are the actual and coded values for each factor?

(c) How many center points should this design have in order to be orthogonal?

(d) How many total runs are needed for this experiment?

17. In the following design matrix, determine the degree of confounding between factors (i.e., factor intercorrelations).

Runs	A	B	C
1	+1	+1	+1
2	+1	+1	-1
3	-1	+1	+1
4	-1	-1	-1
5	+1	-1	+1
6	+1	-1	-1
7	-1	+1	+1
8	-1	-1	-1

Hint: To find the correlation between A and B, regress A on B and look at R^2.

18. If an engineer was running an 8-factor screening experiment, how many possible 2-factor interactions are there?

Bibliography

1. Barker, Thomas B., (1985) *Quality By Experimental Design*, Marcel Dekker Inc.

2. Bhote, Keki R., (1988) *World Class Quality*, AMA Membership Publications Division.

3. Box, George E.P., and D.W. Behnken, (1960) "Some New Three Level Designs for the Study of Quantitative Variables," *Technometrics*, Vol. 2, pp. 455-475.

4. Box, George E.P. and Norman R. Draper, (1987) *Empirical Model-Building and Response Surfaces*, John Wiley and Sons, Inc., New York.

5. Box, George E.P., William G. Hunter, and J. Stuart Hunter, (1978) *Statistics for Experimenters*, John Wiley and Sons, Inc., New York.

6. Cornell, John A., (1990) *Experiments with Mixtures*, 2nd Edition, John Wiley and Sons, Inc., New York.

7. DOE PRO, Digital Computations, Inc., and Air Academy Associates, Colorado Springs, CO, 1997.

8. John, Peter W.M., (1971) *Statistical Design and Analysis of Experiments*, Macmillan Publishing Company, New York.

9. Montgomery, Douglas C., (1991) *Design and Analysis of Experiments*, 3rd Edition, John Wiley and Sons, Inc.

10. Myers, Raymond H. and Montgomery, D.C., (1995) *Response Surface Methodology*, John Wiley and Sons, Inc., New York.

11. Plackett, R.L. and J.P. Burman, (1946) "The Design of Optimum Multifactorial Experiments," *Biometricka*, Vol. 33, pp. 305-325.

12. Taguchi, G. and S. Konishi, (1987) *Taguchi Methods: Orthogonal Arrays and Linear Graphs*, American Supplier Institute, Inc.

Chapter 4

Statistical Techniques

4.1   Introduction

In order to fully comprehend subsequent discussions on analysis of experimental designs, the reader should have a working knowledge of inferential statistical techniques. This chapter covers the broad topics of Analysis of Variance (ANOVA), Simple Linear Regression and Multiple Linear Regression in enough detail to provide the required level of expertise for subsequent use. In no way does this chapter exhaust these three topics. For those who might not be comfortable with this material, the authors recommend using software such as DOE KISS to generate analysis results and Appendix M, Rules of Thumb, to determine significant effects.

4.2   Analysis of Variance

In Chapter 2 we discussed a non-statistical approach for determining which factor effects and/or interactions should go into the prediction equation and which ones should be left out. In this section we provide a statistical approach to assist in making that decision.

As an example, suppose a company wants to investigate ways to improve a process measured by the output variable y. It is decided to test the process at two different temperatures (70° and 90°) to determine if there is a temperature effect. In order to infer that temperature did or did not cause a difference in response, random assignments of experimental units to each temperature are made. The purpose of randomization is to spread all possible uncontrolled variable effects evenly over the two treatments. Thus, if a significant difference in means is found, it can be concluded that temperature did in fact cause that difference, and not some uncontrolled factor.

To conduct the randomization procedure, you could simply flip a coin where a head implies 70° and a tail is 90°. Say the first coin flip resulted in a head; then test number 1 would be made with the temperature at 70°. The process continues until a predetermined number of replicated tests are made for each temperature. Random numbers, such as those found in Appendix A, are also helpful in conducting randomized tests.

Sample data from one input factor (temperature) at 2 levels (70° and 90°) is presented in Table 4.1. This set of data will be used to demonstrate how to conduct ANOVA.

	Factor	Replicated Response Values						
Run	Temperature (A)	y_1	y_2	y_3	y_4	$\bar{y}_r$	S^2_r	df_r^*
1	70° (−)	2.2	2.8	3.2	3.6	2.95	.357	3
2	90° (+)	4.6	5.0	5.4	5.8	5.20	.267	3
Average @ (−)	2.95							
Average @ (+)	5.20		$\bar{\bar{y}}$ = grand mean = 4.075					
Δ	2.25							

Table 4.1 Sample Data for 1-Factor ANOVA

The average response ($\bar{y}_r$) and variance (S^2_r) for each run, (r = 1, 2), in the design matrix are found by using Equations 4.1 and 4.2.

---

*The df$_r$ column, which refers to degrees of freedom, is (n_r − 1) for each run r.

$$\bar{y}_r = \frac{\sum\limits_{i=1}^{n_r} y_i}{n_r} \qquad \text{where } n_r = \text{number of response values for run r}$$

Equation 4.1 Average Response for each Run in the Design Matrix

$$S_r^2 = \frac{\sum\limits_{i=1}^{n_r} (y_i - \bar{y}_r)^2}{(n_r - 1)}$$

Equation 4.2 Sample Variance for each Run in the Design Matrix

Using run number 1 (r = 1) as an example,

$$\bar{y}_1 = \frac{2.2 + 2.8 + 3.2 + 3.6}{4} = 2.95$$

$$S_1^2 = \frac{(2.2 - 2.95)^2 + (2.8 - 2.95)^2 + (3.2 - 2.95)^2 + (3.6 - 2.95)^2}{3} = .357.$$

The Δ in Table 4.1 represents the response average for temperature at the $(+)$ minus the response average when temperature is at the $(-)$ (i.e., $\Delta = \bar{y}_{(+)} - \bar{y}_{(-)}$). It is obvious from the data that $\bar{y}_{(+)} > \bar{y}_{(-)}$, but is this difference large enough, based on sample variations, to conclude that the population (or true) means, $\mu_{(+)}$ and $\mu_{(-)}$, are different? If we conclude a significant difference in $\mu_{(+)}$ and $\mu_{(-)}$, then we are concluding temperature is an important factor and it would be included in a prediction equation (i.e., $\hat{y} = \bar{\bar{y}} + (\Delta/2)A$, where A is in coded values of -1 through $+1$). However, if $\mu_{(+)}$ and $\mu_{(-)}$ are not found to be significantly different, then our one factor experiment would have a simple prediction equation, $\hat{y} = \bar{\bar{y}}$. To complete the statistical test requires some knowledge of hypothesis tests.

A statistical hypothesis is a statement or claim about some unrealized true state of nature. The actual hypotheses to be tested consist of two complementary statements about the true state of nature. For our example, the hypotheses appear as follows:

H_0:   Mean response at low temperature is equal to the mean response at high temperature.

H_1:   Mean response at low temperature differs from the mean response at high temperature.

These two complementary statements are defined as the null hypothesis (H_0) and the alternative hypothesis (H_1). If we conclude H_0, temperature does not go into the prediction equation; whereas, concluding H_1 implies temperature does go into the prediction equation. Since the true state of nature is seldom (if ever) known with 100% certainty, the two statements serve as strawmen to be tested based on sample data results. Probability and statistics are combined with sample data results to infer something about the entire population (the true state of nature) with a measurable amount of uncertainty.

An analogy of hypothesis testing can be drawn from our legal system where the accused on trial is presumed to be innocent unless the prosecution presents overwhelming evidence to convict him. In this example, the hypotheses to be tested are stated as:

H_0: Defendant is Innocent

H_1: Defendant is Guilty

Regardless of the jury's conclusion, they are never really sure of the true state of nature. Concluding "H_0: Defendant is Innocent" does not mean that the defendant is in fact innocent. An H_0 conclusion simply means that the evidence was not overwhelming enough to justify a conviction. On the other hand, concluding H_1 does not prove guilt; rather, it implies that the evidence is so overwhelming that the jury can have a high level of confidence in a guilty verdict.

Since verdicts are arrived at with less than 100% certainty, either conclusion has some probability of error. Consider Table 4.2.

		True State of Nature	
		H_0	H_1
Conclusion Drawn	H_0	Conclusion is correct	Conclusion results in a Type II error
	H_1	Conclusion results in a Type I error	Conclusion is correct

Table 4.2 Type I or II Error Occurs if Conclusion is Not Correct

The probability of committing a Type I error is defined as α where $(0 \le \alpha \le 1)^*$ and the probability of committing a Type II error is β where $(0 \le \beta \le 1)$.

In the courtroom example, α, the probability of convicting an innocent person, is of critical concern. To minimize the risk of such an erroneous conclusion, our court system requires overwhelming evidence to conclude H_1. Although minimizing α has its advantages, it should be obvious that requiring overwhelming evidence to conclude H_1 will in turn increase β, the probability of a Type II error. To resolve this dilemma, statistical hypothesis tests are designed such that

(1)   The most critical decision error is a Type I error.

(2) α is set at a minimum level, usually .10, .05, .01, or .001, depending on the degree of criticality of such an error.

(3)   Based on (1) and (2) above, the hypothesis statement to be tested with $(1 - \alpha)100\%$ confidence is placed in H_1.

---

*This is not the same α as used for the axial points in the Central Composite Designs of Chapter 3.

(4)   The nature of most statistical hypothesis tests requires that the equality condition be placed in H_0.

(5)   To minimize β while holding α constant requires increased sample sizes. (See Appendix M for Rules of Thumb on Sample Sizes.)

In our example, α represents the risk of placing temperature in the prediction equation when it is really an unimportant factor. When optimizing any prediction equation, the optimal settings for a factor represented in the equation will depend on the size and sign of its coefficient ($\Delta/2$ for 2-level designs). Therefore, placing an unimportant factor in the prediction equation (committing a Type I error) might limit our ability to also minimize cost. To minimize the probability of this type of error, α is usually set at .05 or .01 for industrial experiments.

The term "analysis of variance" indicates we will be testing the hypotheses

$$H_0: \quad \mu_{(+)} = \mu_{(-)}$$
$$H_1: \quad \mu_{(+)} \neq \mu_{(-)}$$

with an analysis based on estimates of variance. Since variation exists within each run of replicated response values and also between each low and high average response, we can obtain two estimates of the overall population variability, σ^2 (also referred to as experimental variance).

The variance from each run (S_r^2) could be used to estimate σ^2; however, a better estimate will be obtained by pooling all the S_r^2 values. This pooled estimate of run variances is referred to as the mean square error (MSE) and is calculated as shown in equation 4.3. Note that it is essentially a weighted (by sample size) average of the S_r^2 values.

$$MSE = \frac{\sum (df_r)(S_r^2)}{\sum (df_r)} = \frac{\sum (n_r - 1)(S_r^2)}{\sum (n_r - 1)}$$

where each sum, $\sum$, is over all design matrix runs.

Equation 4.3 Estimating σ^2 with MSE

Assuming H_0 is true implies that $\bar{y}_{(+)}$ and $\bar{y}_{(-)}$ are both estimates of μ and they both vary about μ with variance $\sigma_{\bar{y}}^2 = \sigma^2 / n'$ where n' is the number of data values used to determine each $\bar{y}$. In terms of the population variance, $\sigma^2 = (n')(\sigma_{\bar{y}}^2)$. Thus, another estimate for σ^2 called mean square between (MSB) can be found using Equation 4.4.

$MSB = (n')(S_{\bar{y}}^2)$ which for 2-level designs results in the simpler form of

$MSB = \dfrac{N}{4}(\Delta^2)$ where N is the total number of response values obtained in the entire experimental matrix.

Equation 4.4 Estimating σ^2 with MSB

Since MSE and MSB are independent estimates of σ^2, their observed ratio, $F_0 = MSB/MSE$, behaves according to what is called an F distribution. Furthermore, if H_0 is true, then the F_0 value is usually close to 1.0. If H_0 is not true, there should not be any discernable change in MSE; however, MSB should increase due to an expected increased difference in $\bar{y}_{(+)}$ and $\bar{y}_{(-)}$, causing $S_{\bar{y}}^2$ to increase. Therefore, values of $F_0 = MSB/MSE$ larger than 1.0 are evaluated using the F distribution to assess their probability (p) of being that large due to chance alone or because H_1 is really true. Assuming H_0 is true, we interpret low probabilities (i.e., less than .05) to mean that the ratio is large not due to chance, but rather because H_0 is really not true. This leads to the conclusion that there is a significant factor effect. The test is formalized by the following steps:

Step (1)   Define the hypotheses. H_0: $\mu_{(+)} = \mu_{(-)}$
                                       H_1: $\mu_{(+)} \neq \mu_{(-)}$

Step (2)   Select an appropriate α.

Step (3)   Compute MSB, MSE, and $F_0 = MSB/MSE$. The degrees of freedom for MSB, df_B, is the number of levels minus 1, i.e., $df_B = 1$ for 2-level designs. The degrees of freedom for MSE,

$df_E = \sum (n_r - 1)$. These degrees of freedom are used to determine a critical F value, F_C, which is compared with F_0.

Step (4)   In an F table (see Appendix F), look up $F(1 - \alpha, df_B, df_E) = F_C$.

Step (5)   Compare the ratio $F_0 = MSB/MSE$ to F_C from Step (4).

If $F_0 \leq F_C$, fail to reject H_0 because the evidence is not overwhelming.

If $F_0 > F_C$, reject H_0 in favor of H_1, with at least $(1 - \alpha)$ 100% confidence of being correct.

(See Appendix M for Rules of Thumb on significant F values from a designed experiment.)

For our example, the five steps appear as follows:

Step (1) $H_0: \mu_{(+)} = \mu_{(-)}$
$H_1: \mu_{(+)} \neq \mu_{(-)}$

Step (2)   Select $\alpha = .05$

Step (3) $MSB = \dfrac{N}{4}(\Delta^2) = \dfrac{8}{4}(2.25)^2 = 10.125$

$MSE = \dfrac{\sum (n_r - 1)(S_r^2)}{\sum (n_r - 1)} = \dfrac{(4-1)(.357) + (4-1)(.267)}{(4-1) + (4-1)} = .312$

$F_0 = \dfrac{10.125}{.312} = 32.452$

Step (4) $F_C = F(.95, 1, 6) = 5.99$ (from Appendix F)

Step (5)   Since $F_0 > F_C$, reject H_0 and conclude H_1 with at least 95% confidence. Also notice that $F(.99, 1, 6) = 13.75$ implies we could be as high as 99% confident in concluding H_1. Thus, the prediction equation generated from our experiment will be

$$\hat{y} = \overline{\overline{y}} + \left(\frac{\Delta}{2}\right)A = 4.075 + 1.125A, \quad \text{where} \quad A \quad \text{is} \quad \text{the} \quad \text{coded}$$

temperature value.

For a second 1-factor design example, assume you are working on a new composite material and you desire to test the effect of two different processing temperature settings on the strength of the material. Based on scientific knowledge, the autoclave temperature settings to be tested are 200°F and 300°F. Samples of 9 composite materials are taken at each setting. The sample results are shown in Table 4.3.

		Strength Measurements										
Run	Factor Temp(A)	y_1	y_2	y_3	y_4	y_5	y_6	y_7	y_8	y_9	$\overline{y}_r$	S^2_r
1	200°(−)	2.8	3.6	6.1	4.2	5.2	4.0	6.3	5.5	4.5	4.6889	1.3761
2	300°(+)	7.0	4.1	5.7	6.4	7.3	4.7	6.6	5.9	5.1	5.8667	1.1575

Avg (−)	4.6889	
Ave (+)	5.8667	$\overline{\overline{y}}$ = Grand Average = 5.2778
Δ	1.1778	

Table 4.3 Composite Material Data (Low (−) = 200°, High (+) = 300°)

The hypotheses to be tested are:

$$H_0: \mu_{(+)} = \mu_{(-)}$$
$$H_1: \mu_{(+)} \neq \mu_{(-)}$$

These hypotheses are portrayed graphically in Figure 4.1.

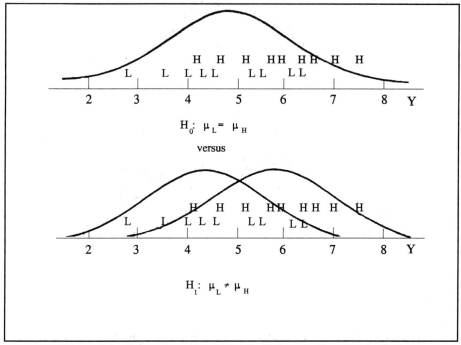

Figure 4.1 Illustration of a 1 Factor (2-Level) Hypothesis Test
(where Low and High setting responses are identified with
an L and H, respectively)

The graphics in Figure 4.1 seem to indicate that H_1 is very likely; however, without conducting the statistical test, it is difficult to determine if the evidence is overwhelming. The actual hypothesis test using ANOVA produces the following results.

Step (1) H_0: $\mu_{(+)} = \mu_{(-)}$

                 H_1: $\mu_{(+)} \neq \mu_{(-)}$

Step (2) $\alpha = .05$

Step (3) $\qquad$ $MSB = \dfrac{N}{4}(\Delta^2) = \dfrac{18}{4}(1.1778)^2 = 6.2425$

$\qquad$ $MSE = \dfrac{\sum (n_r - 1)(S_r^2)}{\sum (n_r - 1)} = \dfrac{8(1.3761) + 8(1.1575)}{8 + 8} = 1.2668$

$\qquad$ $F_0 = \dfrac{MSB}{MSE} = \dfrac{6.2425}{1.2668} = 4.9278$

Step (4) $\qquad$ $F_C = F(.95, 1, 16) = ???$ Since Appendix F does not include the values for $\nu_2 = 16$, we will use the most conservative estimate ($\nu_2 = 15$) to find
$\qquad$ $F_C = F(.95, 1, 15) = 4.54$.

Step (5) $\qquad$ Since $F_0 > F_C$, we reject H_o and set

$\qquad$ $\hat{y} = 5.2778 + \dfrac{1.1778}{2}$ (Temp), where Temp can only take on coded

$\qquad$ values.

When factors in the experiment have only 2 levels, we can also complete the hypotheses test using the t distribution. A t-test for this same problem would be:

Step (1) $\quad$ H_0: $\qquad$ $\mu_{(+)} = \mu_{(-)}$

$\qquad\quad$ H_1: $\qquad$ $\mu_{(+)} \neq \mu_{(-)}$

Step (2) $\quad$ $\alpha = .05$

Step (3) $\quad$ $t_0 = \dfrac{\bar{y}_{(+)} - \bar{y}_{(-)}}{S_P\sqrt{\dfrac{1}{n_{(+)}} + \dfrac{1}{n_{(-)}}}}$ $\qquad$ where $\; S_P = \sqrt{MSE}$

$\qquad\quad$ $t_0 = \dfrac{5.8667 - 4.6889}{\sqrt{1.2668}\sqrt{\dfrac{1}{9} + \dfrac{1}{9}}} = 2.2199$

Step (4) $t_C = t \left[1 - \frac{\alpha}{2}, \sum (n_r - 1) \right]$

             $= t \,(.975,\ 16) = 2.120$

Step (5)   Since $|t_0| > t_C$, we reject H_0 just as we did using the F test. Note that for 2-level designs, these tests are equivalent. The resulting prediction equation is $\hat{y} = 5.2778 + \dfrac{1.1778}{2}$ (Temp), where Temp can only take on coded values.

Note: A quick and easy way to accomplish the same procedure as an F or t test can be found in Appendix M. The End Count method described in Appendix M is almost as accurate (statistically) as the F or t test but with much less complexity. Also see the Case Study on End Count in Chapter 8.

4.3   Two-Factor ANOVA

Assume that in the first example problem of Section 4.2 there exists another variable of interest: pressure. If we were to use a one-factor-at-a-time experimental approach, we would collect data on two temperatures as shown in Section 4.2 and then collect more data on a desired number of pressure levels. In each case, we would hold one factor at some constant level while varying the levels of the other factor. As previously discussed in Chapter 1, this approach is grossly inefficient and it is possible to determine that neither temperature nor pressure has an effect when in fact the contrary may be true. The missing effect when conducting one-at-a-time designs is the interaction of two effects when analyzed together.   To estimate whether or not a **main effect** and/or an **interaction effect** exists in our example, a **factorial design** must be used. Assuming that we are interested in only two levels of pressure, 100 and 200 psi, Table 4.4   displays the different combinations possible for temperature and pressure in a factorial design.

2 Factor Combinations of Temperature and Pressure

Run	Temp	Pressure
1	70	100
2	70	200
3	90	100
4	90	200

Table 4.4 Combinations for 2 Factors, Temperature and Pressure

If the average observed values in runs 1, 2, 3, and 4 are 10, 30, 30, and 50, respectively, they would appear graphically as shown in Figure 4.2.

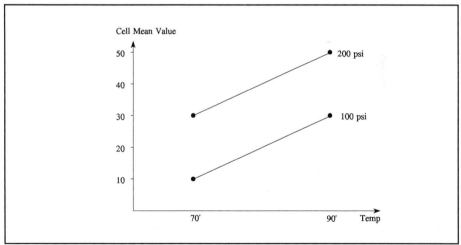

Figure 4.2 Graphical Display of 2 Factors that Do Not Interact

The graph in Figure 4.2 indicates a temperature effect (average response over pressure levels increases as temperature changes from 70° to 90°), a pressure effect (average response for 200 psi is consistently above that for 100 psi), and there appears to be no 2-way interaction because the change in response from 70° to 90° for 100 psi is equal to the change at 200 psi (i.e., the lines are approximately parallel). An example of run average values which indicate a potential 2-way interaction would be 20, 50, 40, and 30 for runs 1, 2, 3, and 4, respectively. Graphically, these values appear as shown in Figure 4.3. The non-parallel nature of these lines (whether they cross or not) is an indicator of an interaction. The non-horizontal nature of each pressure line over the two temperatures indicates a positive temperature effect when pressure is 100 psi and a negative temperature effect when pressure is 200 psi. Notice that if you average the response over levels of pressure, there will appear to be no temperature effect; thus, measuring the effect of the interaction term is of utmost importance for understanding the process.

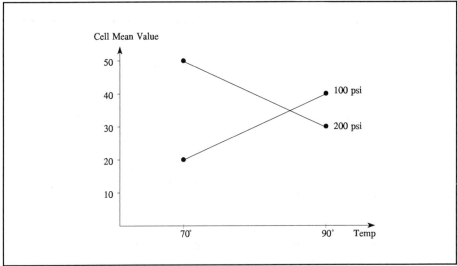

Figure 4.3 Graphic of a 2-Factor Interaction

For an example of a 2-factor (2-level) ANOVA, see the data presented in Table 4.5.

	Temp	Press	Response			
Run	A	B	y_1	y_2	$\bar{y}_r$	S_r^2
1	70	100	2.2	2.8	2.5	.18
2	70	200	3.2	3.6	3.4	.08
3	90	100	5.0	4.6	4.8	.08
4	90	200	5.8	5.4	5.6	.08

Table 4.5 Data for a 2-Factor (2-Level) ANOVA Example

Replacing the low factor settings with a (-1) and the high factor settings with a $(+1)$, the above design matrix becomes the coded matrix shown in Table 4.6.

				Response				
Run	A	B	A × B	y_1	y_2	$\bar{y}_r$	S_r^2	df_r
1	−1	−1	+1	2.2	2.8	2.5	.18	1
2	−1	+1	−1	3.2	3.6	3.4	.08	1
3	+1	−1	−1	5.0	4.6	4.8	.08	1
4	+1	+1	+1	5.8	5.4	5.6	.08	1

Table 4.6 Coded Matrix and Response Values for 2-Factor (2-Level) Example

Notice that the A × B column is generated by multiplying the $(+1)$ and (-1) values in columns A and B for each row. This new column represents the 2-way linear interaction between A and B. To analyze the significance of the A, B, and A × B effects, you need only complete a table similar to the one shown in Table 4.7.

	A	B	A × B
(-1) Level Response Avg	2.95	3.65	4.10
$(+1)$ Level Response Avg	5.20	4.50	4.05
$(+1)$ Avg $-$ (-1) Avg $= \Delta$	2.25	.85	$-.05$
MSB_{effect}	10.125	1.445	.005
F_{effect}	96.428	13.762	.0476

Note: MSE $= .105$ $\qquad$ $\overline{\overline{y}} = 4.075$

Table 4.7   Analysis Table for 2-Factor (2-Level) ANOVA Example

To complete Table 4.7, you will need the following hypotheses tests for each column in the matrix:

Test I: Testing for a significant A effect.

Step (1) H_0: $\mu_{A(+)} = \mu_{A(-)}$
$\qquad\qquad$ H_1: $\mu_{A(+)} \neq \mu_{A(-)}$
Step (2) $\alpha = .05$

Step (3) $MSB_A = \dfrac{N}{4}(\Delta_A)^2 = \dfrac{8}{4}(2.25)^2 = 10.125$

$\qquad\qquad$ $MSE = \dfrac{\sum (n_r - 1)(S_r^2)}{\sum (n_r - 1)} = \dfrac{\sum (df_r)(S_r^2)}{\sum (df_r)}$

$\qquad\qquad\qquad$ $= \dfrac{1(.18) + 1(.08) + 1(.08) + 1(.08)}{1 + 1 + 1 + 1} = \dfrac{.42}{4} = .105$

The observed F_0 for the A effect is

$\qquad\qquad$ $F_0 = \dfrac{MSB_A}{MSE} = \dfrac{10.125}{.105} = 96.428$

Step (4) $F_C = F(.95, 1, \sum(n_r - 1)) = F(.95, 1, 4) = 7.71$

Step (5) Since $F_0 > F_C$, conclude H_1, i.e., factor A is significant and it belongs in the prediction equation. (See Appendix M for Rules of Thumb on significant F values from a designed experiment.)

Test II: Testing for a significant B effect.

Step (1) H_0: $\mu_{B(+)} = \mu_{B(-)}$

H_1: $\mu_{B(+)} \neq \mu_{B(-)}$

Step (2) $\alpha = .05$

Step (3) $MSB_B = \dfrac{8}{4}(.85)^2 = 1.445$

$MSE = .105$ (same as before)

$F_0 = \dfrac{1.445}{.105} = 13.762$

Step (4) $F_C = F(.95, 1, 4) = 7.71$ (same as before)

Step (5) Since $F_0 > F_C$, conclude H_1, i.e., factor B is significant and it belongs in the prediction equation.

Test III: Testing for a significant A × B interaction effect.

Step (1) H_0: $\mu_{AB(+)} = \mu_{AB(-)}$

H_1: $\mu_{AB(+)} \neq \mu_{AB(-)}$

Step (2) $\alpha = .05$

Step (3) $MSB_{AB} = \dfrac{8}{4}(-.05)^2 = .005$

$MSE = .105$ (same as before)

$F_0 = \dfrac{.005}{.105} = .0476$

Step (4) $F_C = F(.95, 1, 4) = 7.71$ (same as before)

Step (5) Since $F_0 < F_C$, conclude H_0, i.e., the A × B interaction effect is not significant and it does not belong in the prediction equation.

Now that all the statistical tests are complete, you should review the statistical conclusions to see if they make sense from a practical and physical perspective.   Assuming the statistical conclusions match the non-statistical conclusions, the resulting prediction equation will be as follows:

$$\hat{y} = 4.075 + 1.125A + .425B.$$

Using the ANOVA approach just demonstrated, it now becomes a fairly easy task to perform ANOVA for any number of factors provided they all have 2 levels. For example, consider the 3-factor ANOVA in Table 4.8.

Run	A	B	AB	C	AC	BC	ABC	$y_1\ y_2\ y_3$	$\bar{y}_r$	S^2_r	df_r
1	−	−	+	−	+	+	−	5 7 6	6.00	1.00	2
2	−	−	+	+	−	−	+	20 21 23	21.33	2.33	2
3	−	+	−	−	+	−	+	6 4 5	5.00	1.00	2
4	−	+	−	+	−	+	−	23 19 18	20.00	7.00	2
5	+	−	−	−	−	+	+	21 21 20	20.67	.33	2
6	+	−	−	+	+	−	−	4 5 6	5.00	1.00	2
7	+	+	+	−	−	−	−	24 21 19	21.33	6.33	2
8	+	+	+	+	+	+	+	5 8 7	6.67	2.33	2

	A	B	AB	C	AC	BC	ABC	
Avg (−)	13.08	13.25	12.67	13.25	20.83	13.16	13.08	$\bar{\bar{y}}=13.25$
Avg (+)	13.42	13.25	13.83	13.25	5.67	13.33	13.42	
Δ=Avg(+) −Avg(−)	.34	0.0	1.16	0.0	-15.16	.17	.34	$F_C =$ F(.95,1,16) = 4.49
MSB_{effect}	.6936	0.0	8.0736	0.0	1378.95	.1734	.6936	
F_{effect}	.260	0.0	3.029	0.0	517.43	.065	.260	

$$MSE = \frac{2(1) + 2(2.33) + \ldots + 2(2.33)}{2 + 2 + \ldots + 2} = \frac{42.64}{16} = 2.665$$

Table 4.8 Eight Run Matrix and ANOVA Results for a 3-Factor Design

Notice that most of the effects are not significant. In fact, the only significant terms for the prediction equation are the grand mean and the A × C interaction. Using the rule of hierarchy, our prediction equation becomes:

$$\hat{y} = 13.25 + .17(A) + 0(C) - 7.585(A \times C) = 13.25 + .17(A) - 7.585(A \times C).$$

In order to use this equation to predict the response for various factor settings, you must ensure the levels for A and C are in the coded form, i.e., somewhere between -1 and $+1$.

If your objective is to minimize or maximize the response, you can determine the optimum settings by using the prediction equation. For example, consider minimizing the average predicted response, $\hat{y}$. Since the coefficient for A is positive, setting A at (-1) will move the $\hat{y}$ in the minimum direction. The negative coefficient of A × C and the previously determined negative setting for A will dictate that factor C be set to (-1). The resulting predicted average response will be:

$$\hat{y} = 13.25 + .17(-1) - 7.585(-1)(-1) = 5.495.$$

If your objective is to hit a target value for $\hat{y}$, say 16, then the factor settings are determined as follows:

(1) Set $\hat{y} = 16$ as shown below.
$$16 = 13.25 + .17A - 7.585 (A \times C)$$

(2) Since we have one equation with two variables, select a coded setting for either A or C based on economics. For example, suppose A is twice as expensive to set at $(+1)$ than at (-1). Therefore, you would want to set A at (-1) to minimize cost.

(3) Solve the equation for C.
$$16 = 13.25 + .17(-1) - 7.585(-1)(C)$$

$$C = \frac{2.92}{7.585} = .385 \text{ which is a coded value.}$$

(4)   The actual values for A and C are easily decoded using the following equation from Chapter 3.

$$f_j = \overline{f_j} + \frac{d_j}{2} x_j$$

where f_j is the actual value for factor j
$\overline{f_j}$ is the actual average for factor j
d_j is the range for factor j actual values (i.e., $f_{max} - f_{min}$)
x_j is the coded value for factor j

Equation 4.5 Transformation Equation for
Coded to Actual Factor Values

As an example, assume the actual low and high values for factor C are 100 and 400. To convert the coded value of C = .385 to an actual factor value, see the following calculations:

$$f = 250 + \left(\frac{300}{2} \right) .385 = 307.75.$$

Graphically, the linear interpolation appears as shown in Figure 4.4.

Figure 4.4 Linear Interpolation from Coded to Actual Factor Value

4.4   Introduction to Simple Linear Regression

Consider the scenario of measuring the same response (y) of a certain process at four levels of temperature: 70°, 80°, 90°, and 100°. Suppose our objective is to develop a model which will allow us to estimate the response at levels other than those mentioned above and be able to place prediction intervals about these estimates. We also need a measure of model effectiveness. The technique to be discussed next, **Simple Linear Regression**, will assist us in accomplishing these objectives. Assume we have collected three response values or observations from each of the four temperature settings. The data is shown in Table 4.9.

	Temperature		
70°	80°	90°	100°
2.3	2.5	3.0	3.3
2.6	2.9	3.1	3.5
2.1	2.4	2.8	3.0

Table 4.9 Three Response Values at Each of Four Temperatures

A graph of the data is given in Figure 4.5. The line represents an "eyeball fit" or free-hand regression line. The closeness of all the observations to the line is an indicator of how well the regression line will predict values of y for any given temperature.

The objective in placing the line is to minimize the distance that observations are from the line. Using the slope-intercept formula for an expected y value, $[E(y) = \beta_0 + \beta_1 x]$, we can estimate the intercept "β_0" graphically as .1 by looking at Figure 4.5. The slope, β_1, is estimated by measuring the change in y, (Δy), for some specific change in x, (Δx), i.e.,

$$\frac{\Delta y}{\Delta x} = \frac{2.7 - 2.3}{80 - 70} = 0.04.$$

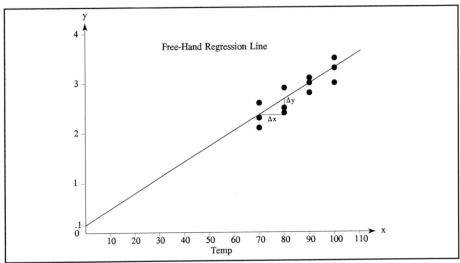

Figure 4.5 Free-Hand Regression Line for Table 4.9 Data

Thus, our free hand regression line takes the form of $\hat{y} = .1 + .04x$, where x is now in actual values of temperature. Since all observations do not lie exactly on the line, there is obviously some error in our straight line estimate. To incorporate this *error* in the formula for predicting y given any x value, we use $y = \beta_0 + \beta_1 x + \epsilon$, where ϵ represents the error term which is assumed to be distributed normally about 0; the ϵ's have equal variability for all levels of x; and the error values are assumed to be independent.

The relationship or model $y = \beta_0 + \beta_1 x + \epsilon$ is applicable for the population data (i.e., the set of *all* possible x and y values). The true regression line imbedded in this model is represented by $E(y) = \beta_0 + \beta_1 x$. Unfortunately, β_0 and β_1 are population parameters which are not known. Thus, the true regression line is not known. Data obtained from experimentation are almost always sample data (a subset of the population data); therefore, we use $\hat{y} = b_0 + b_1 x$ to approximate the true regression line; i.e., $\hat{y}$, b_0, and b_1 estimate $E(y)$, β_0, and β_1, respectively. In addition, $e_i = y_i - \hat{y}_i$ is called the ith residual which estimates the ith error value ϵ_i. These terms are displayed graphically in Figure 4.6. The subscript " i " as used here is associated with the " i^{th} " observation.

Figure 4.6 Partitioning of $(y_i - \bar{y})$

From Figure 4.6, it can also be seen that each observation's deviation from $\bar{y}$ can be partitioned such that

$$(y_i - \bar{y}) = (y_i - \hat{y}_i) + (\hat{y}_i - \bar{y}).$$

Squaring both sides and then summing both sides over all observations (i.e., over all i) yields, after some algebra, the following equation:

$$\sum_{i=1}^{n} (y_i - \bar{y})^2 = \sum_{i=1}^{n} (y_i - \hat{y}_i)^2 + \sum_{i=1}^{n} (\hat{y}_i - \bar{y})^2$$

where n is the number of paired data points and

$$SST = \sum_{i=1}^{n} (y_i - \bar{y})^2$$

is the numerator of what we have already seen to be the variance of y. SST is called the sum of squares total (or total sum of squares). The two terms partitioning SST are called the sum of squares due to error, SSE, and the sum of squares due to regression, SSR. SSE and SSR are calculated as follows:

$$SSE = \sum_{i=1}^{n} (y_i - \hat{y}_i)^2 \quad \text{and} \quad SSR = \sum_{i=1}^{n} (\hat{y}_i - \bar{y})^2.$$

Thus, in abbreviated notation, SST = SSE + SSR

If we don't use x to predict y, then the best prediction for some future y is $\bar{y}$. The variance of observations about $\bar{y}$ is

$$S_y^2 = \frac{\sum_{i=1}^{n} (y_i - \bar{y})^2}{n - 1} = \frac{SST}{n - 1}.$$

When information on x is used to predict y, then $\hat{y} = b_0 + b_1 x$ is the prediction equation and the variance of observations about $\hat{y}$ is given by

$$\frac{\sum_{i=1}^{n} (y_i - \hat{y}_i)^2}{n - 2} = \frac{SSE}{n - 2} = MSE.$$

MSE is referred to as the mean square error or the error variance. The objective of Simple Linear Regression modeling is to develop a line such that MSE is much smaller than S_y^2. The size of the error variance determines how good a fit the regression line will be and it plays a key role in the prediction intervals to be developed later in this chapter.

One mathematical method for finding b_0 and b_1 is the **method of least squares**. The name is derived from the property of minimizing the sum of squared deviations from the line, i.e., minimizing $\sum e_i^2$. The appropriate formulas are presented in Equation 4.6.

$$b_1 = \frac{\sum_{i=1}^{n} x_i y_i - \dfrac{\sum_{i=1}^{n} x_i \sum_{i=1}^{n} y_i}{n}}{\sum_{i=1}^{n} x_i^2 - n\bar{x}^2} \qquad \text{and} \qquad b_0 = \bar{y} - b_1 \bar{x}$$

Equation 4.6 Least Squares Estimates for β_1 and β_0
in Simple Linear Regression

For the example data, the calculations are shown in Table 4.10. As you can see, the estimates from the free-hand regression, namely $b_1 = .04$ and $b_0 = .1$, are fairly close to those of least squares for this example. However, when the data are more spread out and more variables are added, the free hand regression becomes an impossible task.

n	x	y	x^2	xy	y^2
1	70	2.3	4900	161	5.29
2	70	2.6	4900	182	6.76
3	70	2.1	4900	147	4.41
4	80	2.5	6400	200	6.25
5	80	2.9	6400	232	8.41
6	80	2.4	6400	192	5.76
7	90	3.0	8100	270	9.00
8	90	3.1	8100	279	9.61
9	90	2.8	8100	252	7.84
10	100	3.3	10000	330	10.89
11	100	3.5	10000	350	12.25
12	100	3.0	10000	300	9.00
	1020	33.5	88200	2895	95.47

$$\bar{x} = 85 \qquad \bar{y} = 2.79$$

$$b_1 = \frac{\sum xy - \dfrac{\sum x \sum y}{n}}{\sum x^2 - n\bar{x}^2} = \frac{2895 - \dfrac{1020(33.5)}{12}}{88200 - 12(85)^2} = \frac{47.5}{1500} = 0.032$$

$$b_0 = \bar{y} - b_1\bar{x} = 2.79 - 0.032(85) = 0.07$$

$$\hat{y} = .07 + .032x \qquad \text{The } \hat{y} \text{ model indicates that a predicted average value of y is a function of x.}$$

Table 4.10 Simple Linear Regression Calculations

4.5 Strength of the Linear Relationship and ANOVA for Simple Linear Regression

A measure of strength of the linear relationship of y with temperature (x) is the correlation coefficient, R, given by

$$R = \frac{\sum xy - \dfrac{\sum x \sum y}{n}}{\sqrt{\left(\sum x^2 - n\overline{x}^2\right)\left(\sum y^2 - n\overline{y}^2\right)}}$$

Equation 4.7 Correlation Coefficient

For our example,

$$R = \frac{2895 - \dfrac{1020(33.5)}{12}}{\sqrt{[88200 - 12(85)^2][95.47 - 12(2.792)^2]}} = \frac{47.5}{53.763} = .883.$$

The value of R is limited to the interval $[-1, 1]$ where -1 indicates perfect negative correlation, 1 indicates a perfect positive correlation, and 0 implies no linear relationship between y and x. For simple linear regression, the sign of R will be the same as that for b_1.

The proportion of variability in y which is explained by y's relationship with x is measured by R^2; to understand the meaning of R^2, recall that SST = SSR + SSE. If we divide both sides of this equation by SST, we obtain

$$1 = \frac{SSR}{SST} + \frac{SSE}{SST}.$$

Since SST represents the sum of squares total, the ratio SSE/SST represents the proportion of total variability remaining about the regression line and SSR/SST is the proportion of the total variability that has been explained (or removed) by the regression line. Using algebra, it can be shown that

$$\frac{SSR}{SST} = \frac{\sum_{i=1}^{n} (\hat{y}_i - \bar{y})^2}{\sum_{i=1}^{n} (y_i - \bar{y})^2}$$

is equivalent to R^2 and thus the proportion of noise in the model is

$$\frac{SSE}{SST} = 1 - R^2.$$

If there is no linear relationship between x and y, then SSE = SST and thus $R^2 = 0$ (see Figure 4.7a). If all the observations fall on the y (or regression) line, then SSE = 0 which implies that $R^2 = 1$ (see Figure 4.7b).

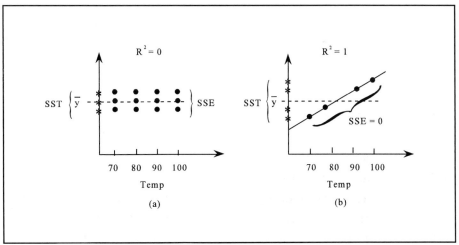

Figure 4.7 Extremes in Strength of Linear Relationship

In the example data, $R^2 = .780$ indicates that about 78.0% of the variability in y is explained through y's linear relationship with x and about 22%

of the data is noise or unaccounted variation. The strength of the linear relationship between the two variables is directly related to the amount of variability in y that can be accounted for by the model. In this example, the strength of the linear relationship between the two variables is considered to be fairly strong.

Recall that the variance of y without regard to x is

$$S_y^2 = \frac{SST}{n-1} = \frac{\sum y^2 - n\bar{y}^2}{n-1} = \frac{95.47 - 12(2.792)^2}{11} = .1752.$$

We can now solve for SST in the above relationship by computing

$$SST = S_y^2 \cdot (n-1) = .1752(11) = 1.927.$$

Since we know SST = 1.927 and $R^2 = .78$, we can solve for SSR in the relationship $R^2 = SSR/SST$ to get

$$SSR = R^2 \cdot SST = .78(1.927) = 1.503.$$

The last of the three sums of squares, SSE, can be obtained from the previous two sums of squares by way of the relationship SSE = SST − SSR. For example,

$$SSE = SST - SSR = 1.927 - 1.503 = .424.$$

It is customary to formalize the partitioning of the total sums of squares in what is called an Analysis of Variance (ANOVA) table. Table 4.11 depicts the format of a Simple Linear Regression ANOVA table.

Source	SS	df	MS	F_0
Regression	$SSR = R^2 \cdot SST$	1	SSR/1	MSR/MSE
Error	$SSE = (1 - R^2) \cdot SST$	n − 2	SSE/(n − 2)	
Total	$SST = s_y^2 \cdot (n-1)$	n − 1		

Table 4.11 Format for Simple Linear Regression Analysis of Variance (ANOVA) Table

The degrees of freedom (df) for the error term in a regression model are equal to n minus the number of parameters estimated.   In Simple Linear Regression, b_0 and b_1 are estimates (obtained from sample data) of the two population parameters β_0 and β_1, respectively.   Therefore, $df_E = n - 2$.   The degrees of freedom for regression (df_{Reg}) in Simple Linear Regression are the number of parameters estimated   minus   1,   which   is $2 - 1 = 1$.   The F_0 in Table 4.11 is compared with $F_C = F(1 - \alpha; df_{Reg}, df_E)$ from Appendix F.   If $F_0 \leq F_C$, the model (with all the terms included for this test) is insignificant, i.e., the best model is $\hat{y} = \bar{y}$.   If $F_0 > F_C$, the model represented in the test is significant.

Since R^2 is computed from sample data, it is only an estimate of $\mathbb{R}^2$, the unknown true strength of the population linear relationship between y and x.   In order to test the value of $\mathbb{R}^2$ for significance (different from zero), consider the following hypotheses:

$$H_0: \quad \mathbb{R}^2 = 0$$
$$H_1: \quad \mathbb{R}^2 \neq 0$$

For Simple Linear Regression the hypotheses for $\mathbb{R}^2$ are equivalent to testing the slope parameter, i.e.,

$$H_0: \quad \beta_1 = 0$$
$$H_1: \quad \beta_1 \neq 0$$

In either case, concluding H_0 implies that the sample data did not provide overwhelming evidence to indicate a significant linear relationship between y and x.   Concluding H_1 indicates the presence of a significant linear relationship with $(1 - P)100\%$ confidence.   The symbol P refers to the minimum value that α (the risk of a Type I error) can have in order to still conclude H_1.   The MSE is an estimate of the variance of y at any level of x.   Table 4.12 displays the example data in the format of Table 4.11.   Note that the F-test indicates the model, $\hat{y} = b_0 + b_1 x$, is significant in representing y's relationship with x.

Source	SS	df	MS	F_0
Regression	1.503	1	1.503	35.448*
Error	.424	10	.0424	
Total	1.927	11		

*   From Appendix F, F(.999; 1, 10) = 21.0 implies
    P < .001. Thus, we can conclude H_1: $R^2 \neq 0$ or
    H_1: $\beta_1 \neq 0$ with at least 99.9% confidence.

Table 4.12 Simple Linear Regression ANOVA Table for Example Data

4.6   Prediction Using Simple Linear Regression

To obtain a point estimate for y, given any x value (x_h), we simply insert the value for x_h in $\hat{y} = b_o + b_1 x_h$. (Note: whereas the subscript "i" is associated with "observed" values, the subscript "h" is used to denote *any* possible or hypothetical level of x.) Point estimates, however, do not include any reference to precision. In order to indicate how precise our estimate is, a **prediction interval** is formed in which the researcher is $(1 - \alpha)100\%$ confident that the true value of y will occur. Since the data used to generate b_0 and b_1 represents just one sample, it is understandable that another sample would produce different values of b_0 and b_1. This implies that there exists sample variability in the slope and intercept of the regression line, in addition to observation variability about the regression line itself. The slope and intercept variability will cause larger deviations in the predicted response, ($\hat{y}$), as the desired level of x, (x_h), moves further away from the center of the domain of x, ($\bar{x}$). For these reasons, the prediction limits for the response at any given value of x (labeled x_h) will appear as shown:

$$\hat{y}\big|_{x_h} \pm t\left(\frac{\alpha}{2}, n - 2\right)\sqrt{MSE\left(1 + \frac{1}{n} + \frac{(x_h - \bar{x})^2}{\sum x^2 - n\bar{x}^2}\right)}$$

While MSE represents the variability of observations about the regression line, the quantity

$$MSE \left[\frac{1}{n} + \frac{(x_h - \bar{x})^2}{\sum x^2 - n\bar{x}^2} \right]$$

represents the variability in the average response due to the variability in slope, b_1, and intercept, b_0.

For our temperature example, the prediction intervals (y_L , y_U) based on 95% confidence are displayed in Table 4.13 for all x values previously used to develop the model. For example, if the process temperature is set at 80°, we would be 95% confident that any future y value will occur between 2.15 and 3.11.

x	$\hat{y}$	y_L	y_U	Δ
70	2.31	1.80	2.82	1.02
* 80	2.63	2.15	3.11	.96
90	2.95	2.47	3.43	.96
100	3.27	2.76	3.78	1.02

* Using $x_h = 80$, $t(.975, 10) = 2.228$, MSE = .0424, $\bar{x} = 85$ and $(\Sigma x^2 - n\bar{x}^2) = 1500$

$\hat{y} | x_h = .07 + .032(80) = 2.63$

$$y_L = 2.63 - 2.228 \sqrt{.0424 \left(1 + \frac{1}{12} + \frac{(80 - 85)^2}{1500}\right)} = 2.15$$

$$y_U = 2.63 + 2.228 \sqrt{.0424 \left(1 + \frac{1}{12} + \frac{(80 - 85)^2}{1500}\right)} = 3.11$$

Table 4.13 Prediction Intervals for y Based on 95% Confidence

Predicting values of y for $x_h < x_{min}$ or $x_h > x_{max}$ is called **extrapolation**, which assumes the same relationship between y and x outside the region of the sampled x as it is within the sampled region. Notice that the 95% prediction interval would be wider outside the interval [70, 100] than it will be inside. This lack of accurate predictability and uncertainty of the relationship between y and x outside the region of sampled x is why extrapolation is not always recommended. If we need to predict outside the sampled region for x, it is best to expand the sampled region. If extrapolation is used, confirmation data should be collected as soon as possible to validate the prediction.

When prediction intervals are too wide for a particular application, there are ways to reduce these intervals. The researcher can:

(1) decrease the confidence level $(1 - \alpha)$ which will decrease
   $t(1 - \alpha/2, n - 2)$

(2) increase the sample size which will decrease t and $1/n$

(3) increase $[\sum x^2 - n\bar{x}^2]$

It is not recommended that one alter the confidence level to reduce interval width. In fact, the subject of experimental design discussed in this text addresses only items (2) and (3) as the major areas of interest. For non-experimental data, the primary method of reducing prediction interval widths is increasing sample size. However, in conducting industrial experiments, the sample size is desired to be a minimum, thus putting even more emphasis on the remaining option, i.e., maximizing $[\sum x^2 - n\bar{x}^2]$. This is equivalent to maximizing the variance of x, which is accomplished by placing n/2 observations at each of x_{min} and x_{max}. The problem with sampling at only two levels of x is that we are restricted to linear estimates. For this reason, center points can be added to 2-level designs to allow for testing the linear assumption and, if necessary, axial points are added to complete a Central Composite Design (CCD) as described in Chapter 3. The idea of maximizing $[\sum x^2 - n\bar{x}^2]$ has another effect in that

$$S^2(b_1) = \frac{MSE}{\sum x^2 - n\bar{x}^2}$$

can be minimized. Therefore, the variance of the slope is minimized if $[\sum x^2 - n\bar{x}^2]$ is maximized, creating more power in the F test for H_0: $R^2 = 0$ or H_0: $\beta_1 = 0$. This implies that if we minimize $S^2(b_1)$, we are more likely to find a real linear relationship if it in fact exists. Once multiple regression is covered, the reader will also see that maximizing $[\sum x^2 - n\bar{x}^2]$ is related to the D-optimal experimental designs discussed in Chapter 3.

4.7 Polynomial Regression

Suppose that the previous temperature data included 3 more observations taken at $x = 50°$, which when added to Table 4.9 produces the example data shown in Table 4.14. Graphically, the data are presented in Figure 4.8. This graph indicates the danger in **extrapolation** because the function takes on a different shape outside the original sampled space of x.

Temperature				
50°	70°	80°	90°	100°
3.3	2.3	2.5	3.0	3.3
2.8	2.6	2.9	3.1	3.5
2.9	2.1	2.4	2.8	3.0

Table 4.14 Example Data for Polynomial Regression

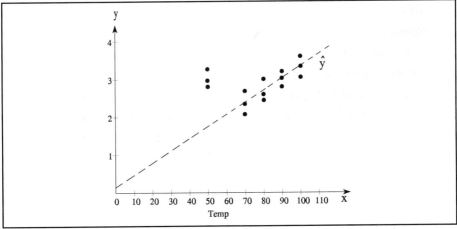

Figure 4.8 Graph of Example Data for Polynomial Regression

As you can see, the linear dashed line is no longer a good fit. In fact, if we try simple linear regression on the new data set, the result is $b_1 = 0$ which implies $R^2 = 0$. However, the more appropriate thing to do is to try a quadratic term. One modeling approach would be $\hat{y} = b_0 + b_{11}x_1^2$, where the subscript "11" is not read "eleven" but rather "one, one" to indicate that b_{11} is the coefficient of x_1x_1 or x_1^2. Table 4.15 shows the computations required to estimate b_0 and b_{11}.

Computations for $\hat{y} = b_0 + b_{11}x_1^2$				
(x_1^2)	y	$(x_1^2)^2$	y^2	$(x_1^2)y$
2500	3.3			
2500	2.8			
2500	2.9			
4900	2.3			
4900	2.6			
4900	2.1			
6400	2.5			
6400	2.9			
6400	2.4			
8100	3.0			
8100	3.1			
8100	2.8			
10000	3.3			
10000	3.5			
10000	3.0			
95700	42.5	710490000	122.61	276810
$\hat{y} = 2.472 + 0.000056(x_1^2)$			$R^2 = .146$	

Table 4.15 Simple Linear Regression Using a Quadratic Term

The low R^2 indicates that something is wrong because we would expect a better fit. What happened? Creating x^2 as a new variable and using it in the simple linear regression model above resulted in fitting the points with a parabola whose vertex (lowest point) is on the y-axis. Graphically, you can see that the vertex should be far from $x = 0$.

To properly conduct polynomial regression we should include both x_1 and x_1^2 in the model. That is, the correct model should be $\hat{y} = b_0 + b_1 x_1 + b_{11} x_1^2$. This second-order polynomial regression model has more than just the linear term and thus requires the use of multiple regression analysis techniques. The computer output shown in Table 4.16 indicates that this model is a very good fit, i.e., the probability of an incorrect model is very small (P = .001).

Summary of Computer Output for Polynomial Regression

Model: $\hat{y} = 7.96048 - .15371 x_1 + .00108 x_1^2$ $R^2 = .673$ P = .001

Parameter	Estimate	P-value
b_0	7.96048	.0001
b_1	−0.15371	.001
b_{11}	0.00108	.001

Table 4.16 Polynomial Regression on Example Data

The graph of $\hat{y} = 7.96048 - .15371 x_1 + .00108 x_1^2$ in Figure 4.9 shows that the minimum of y does indeed occur when $x_1 = 71.162$.

Note: Differential calculus can be used to find the x values which produce a minimum for y. That is, setting dy/dx =0 and solving for x, we have

$$\frac{dy}{dx} = -.15371 + .00216 x_1 = 0$$

and

$$x_1 = \frac{.15371}{.00216} = 71.162$$

Thus, y is a minimum at $x_1 = 71.162$.

Figure 4.9 Graph of Polynomial Regression

4.8   Multiple Regression[**]

If a variable for pressure was also included in the original data discussed in section 4.7, the multivariable model would be referred to as a **Multiple Regression Model**. A new set of example data shown in Table 4.17 will be used to discuss multiple regression analysis.

[**]This material is slightly more complex than the previous sections. Although some hand calculation techniques are presented, the reader should realize that evaluating 3-level designs will require the use of multiple regression and the interpretation of software output. Readers having difficulty with this section should skim the rest of the chapter and move on to Chapter 5.

Multiple Regression Data		
y	x_1 = Temp	x_2 = Pressure
3.1	50	100
3.3	50	100
2.6	50	200
2.4	50	200
2.5	70	100
2.6	70	100
3.0	70	200
3.3	70	200
2.4	80	100
2.3	80	100
3.2	80	200
3.5	80	200
2.8	90	100
2.6	90	100
3.1	90	200
3.0	90	200
3.2	100	100
3.4	100	100
2.5	100	200
2.4	100	200

Table 4.17 Example Data for Multiple Linear Regression

The initial model to be tested is $y = \beta_0 + \beta_1 x_1 + \beta_2 x_2 + \epsilon$, which can be represented in an equivalent matrix notation as shown below.

$$\underline{Y} = X \underline{\beta} + \underline{\epsilon} \qquad \text{where}$$

$$\underline{Y} = \begin{bmatrix} y_1 \\ y_2 \\ y_3 \\ . \\ . \\ . \\ y_n \end{bmatrix} \qquad X = \begin{bmatrix} 1 & x_{11} & x_{12} \\ 1 & x_{21} & x_{22} \\ 1 & x_{31} & x_{32} \\ . & . & . \\ . & . & . \\ . & . & . \\ 1 & x_{n1} & x_{n2} \end{bmatrix} \qquad \underline{\beta} = \begin{bmatrix} \beta_0 \\ \beta_1 \\ \beta_2 \end{bmatrix} \qquad \underline{\epsilon} = \begin{bmatrix} \epsilon_1 \\ \epsilon_2 \\ \epsilon_3 \\ . \\ . \\ . \\ \epsilon_n \end{bmatrix}$$

As in simple linear regression, $\underline{b}$ is used to estimate β and $\underline{e}$ is an estimate of $\underline{\epsilon}$. Suppose n is the number of observations and p is the number of parameters to be estimated. In general, $\underline{Y}$ is a (n × 1) vector of the response values; X is a (n × p) matrix whose first column is all 1's (to estimate β_0) and whose other columns are the levels of the k = p − 1 independent variables; $\underline{b}$ is a (p × 1) vector of p regression coefficients; and $\underline{e}$ is a (n × 1) vector of random error values (one for each of the n observations). For the example data, $\underline{Y}$, X, $\underline{b}$, and $\underline{e}$ appear as shown below.

$$
\underline{Y} = \begin{bmatrix} 3.1 \\ 3.3 \\ 2.6 \\ 2.4 \\ 2.5 \\ \cdot \\ \cdot \\ \cdot \\ 2.4 \end{bmatrix}
\qquad
X = \begin{bmatrix} 1 & 50 & 100 \\ 1 & 50 & 100 \\ 1 & 50 & 200 \\ 1 & 50 & 200 \\ 1 & 70 & 100 \\ \cdot & \cdot & \cdot \\ \cdot & \cdot & \cdot \\ \cdot & \cdot & \cdot \\ 1 & 100 & 200 \end{bmatrix}
$$

$$
\underline{b} = \begin{bmatrix} b_0 \\ b_1 \\ b_2 \end{bmatrix}
\qquad \text{and} \qquad
\underline{e} = \begin{bmatrix} e_1 \\ e_2 \\ e_3 \\ \cdot \\ \cdot \\ \cdot \\ e_n \end{bmatrix}
$$

Values of $\underline{b}$ and $\underline{e}$ are determined from the least squares multiple regression which has the objective of minimizing [$\underline{e}' \underline{e}$]. Note: Let $\underline{e}'$ denote the transpose of $\underline{e}$ and z^{-1} the inverse of z. To find $\underline{b}$ such that $\hat{\underline{Y}} = X\underline{b}$, use $\underline{b} = (X'X)^{-1}X'\underline{Y}$, and then after computing $\hat{\underline{Y}} = X\underline{b}$ compute $\underline{e} = \underline{Y} - \hat{\underline{Y}}$. The variance-covariance matrix of the regression coefficients is the (p × p) matrix given by

$$\text{cov}(\underline{b}) \;=\; \hat{\sigma}^2 \cdot (X'\,X)^{-1} \;=\; \begin{bmatrix} \sigma^2_{b_0} & \sigma_{b_0 b_1} & \sigma_{b_0 b_2} \\ & \sigma^2_{b_1} & \sigma_{b_1 b_2} \\ & & \sigma^2_{b_2} \end{bmatrix}$$

where $\sigma^2_{b_0}$ is the variance of b_0 and $\sigma_{b_1 b_2}$ denotes the covariance of b_1 and b_2. The estimate for σ^2, $\hat{\sigma}^2$, is based on the variance of the residuals which have n − p degrees of freedom. It is given by

$$\text{MSE} \;=\; \frac{\displaystyle\sum_{i=1}^{n} (y_i - \hat{y}_i)^2}{n - p}.$$

MSE = SSE/(n − p) is referred to as the regression error variance. For our multiple linear regression example data in Table 4.17, the computer analysis is displayed in Table 4.18.

Dependent Variable: Y N: 20 Multiple R: .108 Squared Multiple R: .012						
Adjusted Squared Multiple R: .000				Standard Error of Estimate: 0.410		
Variable	Coefficient	Std Error	Std Coeff	Tolerance	T	P(2 Tail)**
Constant	2.695	0.506	0.000	1.0000000	5.322	0.000
x_1	0.001	0.005	0.026	1.0000000	0.108	0.915
x_2	0.001	0.002	0.105	1.0000000	0.437	0.668
Analysis of Variance						
Source	Sum-of-Squares	DF	Mean-Square		F-Ratio	P
Regression	0.034	2	0.017		0.101	0.904
Residual*	2.854	17	0.168			

Table 4.18 Computer Output for the Model $\hat{y} = b_0 + b_1 x_1 + b_2 x_2$

---

* The *MYSTAT Software* which accompanies [1] refers to the error as Residual.
**See Appendix M for Rules of Thumb on significant P(2 Tail) values.

Since $R^2 = .012$ with $P = .904$, the model is inadequate for predicting the response. In order to calculate the model's lack of fit, partition SSE into pure error and lack of fit error. That is, $SSE = SSE_{pure} + SSE_{LF}$. The pure error, SSE_{pure}, is acquired from repeated observations of y at the same levels of x. Since the example data has 2 observations of y for each of the 10 different combinations of x_1 and x_2 (i.e., $n_j = 2$ for $j = 1, 2, ..., 10$), the pure error* is estimated by

$$SSE_{pure} = \sum_{j=1}^{10} (n_j - 1) \; s_j^2 = 0.19$$

$MSE_{pure} = SSE_{pure}/df_{pure} = .19/10 = .019$, where $\quad df_{pure} = \sum_{j=1}^{10} (n_j - 1)$. The sum of squares for lack of fit, SSE_{LF}, is calculated using $SSE_{LF} = SSE - SSE_{pure}$. For our example, $SSE_{LF} = 2.854 - .19 = 2.664$ and the degrees of freedom for testing lack of fit are $df_{LF} = df_E - df_{pure}$ where $df_E = n - p$ and p is the number of parameters estimated. Therefore, the example data has a mean square for lack of fit of

$$MSE_{LF} = \frac{SSE_{LF}}{df_{LF}} = \frac{2.664}{17 - 10} = 0.3806.$$

The F test for lack of fit is $F_0 = MSE_{LF}/MSE_{pure} = .3806/.019 = 20.032$. Comparing F_0 to $F(1 - \alpha, df_{LF}, df_{pure}) = F(.95, 7, 10) = 3.14$, you can see that the lack of fit is significant. Thus, it is necessary to investigate a better model using interactions and quadratic effects. If you test a full second-order model, it will appear as

$$\hat{y} = b_0 + b_1 x_1 + b_2 x_2 + b_{12} x_1 x_2 + b_{11} x_1^2.$$

Notice that a quadratic effect for x_2 is not estimated because it has only 2 levels. The computer output is shown in Table 4.19.

---

*If the design does not contain any replicated points at designated levels of x, replications can be made at the center point (in our example, this would be 75° and 150 psi) and the SSE_{pure} would then be calculated for this one combination of independent variables. If using replicated center points, then $df_{pure} = n_c - 1$, where n_c is the number of center points.

Computer Output for $\hat{y} = b_0 + b_1x_1 + b_2x_2 + b_{12}x_1x_2 + b_{11}x_1^2$

Dependent Variable: Y   N: 20   Multiple R: .115   Squared Multiple R: .013

Adjusted Squared Multiple R: .000   Standard Error of Estimate: .0436

Variable	Coeff	STD Error	STD Coeff	Tolerance	T	P(2 Tail)
Constant	2.973	2.371	0.000	1.0000000	1.254	0.229
x_1	-0.004	0.057	-0.203	.0100250	-0.079	0.938
x_2	0.000	0.009	-0.054	.0463950	-0.046	0.964
$x_1 x_2$	0.000	0.000	0.194	.0327289	0.137	0.893
$x_{11} x_2$	0.000	0.000	0.124	.0110192	0.051	0.960

Analysis of Variance

Source	Sum-of-Squares	DF	Mean-Square	F-Ratio	P
Regression	0.038	4	0.010	0.05	0.995
Residual	2.850	15	0.190		

Table 4.19 Computer Output for Full Second-Order Regression Model

Testing this model for lack of fit indicates that $SSE_{LF} = SSE - SSE_{pure} = 2.850 - .19 = 2.66$ (about the same as before). The F test for lack of fit is $F_0 = \dfrac{2.66/(15-10)}{0.19/10} = 28.0$ which is still significant at the .05 level. What's wrong? The problem is that second-order models can possibly have quadratic interaction terms. Therefore, the new model to be tested will include a quadratic interaction of x_1^2 with variable x_2:

$$\hat{y} = b_0 + b_1x_1 + b_2x_2 + b_{12}x_1x_2 + b_{(11)2}x_1^2x_2 + b_{11}x_1^2$$

The results are shown in Table 4.20. It should be of concern that adding the $x_1^2x_2$ term increased R^2 to .901 which is significant, but the coefficient for $x_1^2x_2$ is 0. The reason the coefficient is 0 is due to the use of only 3 decimal place accuracy. Formatting the output to 4 or more places will produce non-zero coefficients. There is, however, another more insidious problem occurring in this analysis.

Computer Output for $\hat{y} = b_0 + b_1x_1 + b_2x_2 + b_{12}x_1x_2 + b_{(11)2}x_1^2x_2 + b_{11}x_1^2$

Dependent Variable: Y   N: 20   Multiple R: .949   Squared Multiple R: .901

Adjusted Squared Multiple R: .866   Standard Error of Estimate: .0143

Variable	Coeff	Std Error	Std Coeff	Tolerance	T	P(2 Tail)
Constant	23.815	2.015	0.000	1.0000000	11.818	0.000
x_1	-0.596	0.056	-26.984	.0011019	-10.656	0.000
x_2	-0.139	0.013	-18.337	.0025125	-10.934	0.000
$x_1\,x_2$	0.004	0.000	49.539	.0003606	11.192	0.000
$x_1\,x_1$	0.004	0.000	27.053	.0006840	10.683	0.000
$x_1\,x_1\,x_2$	0.000	0.000	-36.030	.0006840	-11.210	0.000

Analysis of Variance

Source	Sum-of-Squares	DF	Mean-Square	F-Ratio	P
Regression	2.602	5	0.520	25.504	0.000
Residual	0.286	14	0.020		

Table 4.20 Computer Output for Quadratic Interaction Model

The tolerance column represents $1 - R_j^2$, where R_j^2 is the squared correlation coefficient obtained by regressing predictor variable j on all other predictor variables. Tolerance values indicate the proportion of orthogonality for each term and thus values other than 1.0 indicate that a phenomenon, referred to as **multicollinearity**, exists. Tolerance values close to zero imply strong multicollinearity conditions which can grossly affect the accuracy of the estimated coefficients. Since the vectors of actual values for x_1, x_2, and $x_1^2x_2$ are intercorrelated, the predictor variable matrix is not orthogonal and, therefore, the estimates of the β_i's can be very ambiguous. In fact, sometimes a variable with a known positive effect on the response will have a negative coefficient due to high multicollinearity. In addition, multicollinearity increases the off-diagonal values of $(X'X)^{-1}$ causing inflated variance of the coefficients which is estimated by $MSE(X'X)^{-1}$. This can result in a significant F-ratio for the model without significant regression weights. When including terms for a second order model, multicollinearity is difficult to avoid unless a transformation of the predictor values is used. Commonly used transformations of non-experimental data are "centering of the data" or "standardized data." To center each predictor column, transform each x_{ij} into $(x_{ij} - \bar{x}_j)$. To generate standardized data, each x_{ij} is transformed to $Z_{ij} = (x_{ij} - \bar{x}_j)/\sigma_j$. When possible, orthogonal coding should be used as discussed in Chapter 3. A final note is that the multiple linear regression prediction limits for a future observation at some independent level $\underline{x}_h' = (x_{1h}, x_{2h}, \ldots, x_{kh})$ can be obtained by using

$$\hat{y}\big|_{\underline{x}_h} \pm t\left(1 - \frac{\alpha}{2}, n - p\right) \sqrt{MSE(1 + \underline{x}_h'(X'X)^{-1}\underline{x}_h)}$$

4.9   Example Demonstrating the Need to Use Coded Data When Examining Second-Order Models

Table 4.21 contains data from a 3-factor Box-Behnken design where the run order has been randomized. Building a second-order model on the actual data results in the multiple regression output displayed in Table 4.22.

Run	Actual Factors			Coded Factors			$\bar{y}$
	A	B	C	x_1	x_2	x_3	
1	75	60	71	0	−	+	1332
2	70	60	68	−	−	0	1370
3	75	60	65	0	−	−	1386
4	80	60	68	+	−	0	1377
5	75	65	68	0	0	0	1387
6	80	65	71	+	0	+	1365
7	80	65	65	+	0	−	1392
8	75	65	68	0	0	0	1411
9	70	65	71	−	0	+	1372
10	70	65	65	−	0	−	1417
11	75	65	68	0	0	0	1397
12	75	70	71	0	+	+	1394
13	80	70	68	+	+	0	1405
14	75	70	65	0	+	−	1410
15	70	70	68	−	+	0	1399

Table 4.21   Data from a 3-Factor Box-Behnken Design

Dependent Variable: Y   N: 15   Multiple R: .951   Squared Multiple R: .904

Adj Squared Multiple R: .731   Std Error of Estimate: 11.397

Variable	Coeff	Std Error	Std Coeff	Tolerance	T	P(2 Tail)
Constant	-934.713	4599.512	0.000	1.0000000	-0.203	0.847
A	-6.475	46.409	-1.113	0.0003015	-0.140	0.894
B	4.375	43.719	0.752	0.0003398	0.100	0.924
C	74.602	97.241	7.696	0.0001908	0.767	0.478
A^2	-0.092	0.237	-2.365	0.0005126	-0.386	0.715
B^2	-0.332	0.237	-7.416	0.0006823	-1.398	0.221
C^2	-1.060	0.659	-14.876	0.0002245	-1.609	0.169
A·B	-0.010	0.228	-0.171	0.0012674	-0.044	0.967
A·C	0.300	0.380	4.207	0.0006763	0.790	0.465
B·C	0.633	0.380	8.539	0.0007318	1.667	0.156

Analysis of Variance

Source	Sum-of-Squares	DF	Mean-Square	F-Ratio	P
Regression	6116.183	9	679.576	5.232	0.042
Residual	649.417	5	129.883		

Table 4.22 Computer Output Using Actual Data in Table 4.21

The high R^2 value of .904 with no significant coefficients, i.e., no P(2 Tail) values less than .10, is initially puzzling. However, the low tolerance values in Table 4.22 indicate high multicollinearity which is due to modeling the actual factor levels. High multicollinearity inflates the standard error estimates which results in large P(2 Tail) values. This can result in a significant model without any significant effects. The following simple example illustrates the problem which can result from using non-coded data in quadratic models.

Actual Data		vs	**Coded Data**	
A	A^2		x	x^2
1	1		−1	1
1	1		−1	1
2	4		0	0
2	4		0	0
3	9		1	1
3	9		1	1
$R^2 = 0.979$			$R^2 = 0.0$	

Notice that a model with actual values for A equal to 1, 2, and 3 will have 97.9% collinearity of the A and A^2 terms. That means the analysis cannot separate the effects of A and A^2, thus leading to confusion. The coded data for A is shown by the x column. Note that the x and x^2 terms are uncorrelated, indicating that they can be evaluated independently during the analysis.

Using the coded data in Table 4.21 to build a second-order model, you will obtain output as shown in Table 4.23. The Analysis of Variance table has not changed. MSE is the same and the overall R^2 is also the same as in Table 4.22. However, the high tolerance values in Table 4.23 indicate there is little or no multicollinearity. Because of the reduced multicollinearity, the P(2 Tail) values are useful in finding which effects are significant, and variable selection is much easier. The tolerance column indicates that there is some correlation among the variables, although not as much as before. Specifically, the quadratic terms are slightly correlated with each other. Since only two terms are significant in Table 4.23, it will be beneficial to re-run the regression analysis with only x_2 and x_3. Table 4.24 contains the analysis summary of the regression with only x_2 and x_3 in the model. The coefficients for x_2 and x_3 have not changed but the constant term, MSE, and R^2 have changed.

Dependent Variable: Y	N: 15	Multiple R: .951		Squared Multiple R: .904		

Adj Squared Multiple R: .731   Std Error of Estimate: 11.397

Variable	Coeff	Std Error	Std Coeff	Tolerance	T	P(2 Tail)
Constant	1398.333	6.580	0.000		212.517	0.000
x_1	−2.375	4.029	−0.082	1.0000	−0.589	0.581
x_2	17.875	4.029	0.615	1.0000	4.436	0.007
x_3	−17.750	4.029	−0.610	1.0000	−4.405	0.007
x_1^2	−2.292	5.931	−0.054	0.9890	−0.386	0.715
x_2^2	−8.292	5.931	−0.195	0.9890	−1.398	0.221
x_3^2	−9.542	5.931	−0.224	0.9890	−1.609	0.169
$x_1 \cdot x_2$	−0.250	5.698	−0.006	1.0000	−0.044	0.967
$x_1 \cdot x_3$	4.500	5.698	0.109	1.0000	0.790	0.465
$x_2 \cdot x_3$	9.500	5.698	0.231	1.0000	1.667	0.156

Analysis of Variance

Source	Sum-of-Squares	DF	Mean-Square	F-Ratio	P
Regression	6116.183	9	679.576	5.232	0.042
Residual	649.417	5	129.883		

Table 4.23 Computer Output Using Coded Data in Table 4.21

Dependent Variable: Y	N: 15	Multiple R: .866		Squared Multiple R: .750		

Adj Squared Multiple R: .709   Std Error of Estimate: 11.864

Variable	Coeff	Std Error	Std Coeff	Tolerance	T	P(2 Tail)
Constant	1387.600	3.063	0.000	.0000	452.990	0.000
x_2	17.875	4.194	0.615	1.0000	4.262	0.001
x_3	−17.750	4.194	−0.610	1.0000	-4.232	0.001

Analysis of Variance

Source	Sum-of-Squares	DF	Mean-Square	F-Ratio	P
Regression	5076.625	2	2538.312	18.034	0.000
Residual	1688.975	12	140.748		

Table 4.24 Computer Output Using Coded Data in Table 4.21

The final prediction equation for coded inputs is:

$$\hat{Y} = 1387.600 + 17.875(x_2) - 17.750(x_3).$$

Assume you desire to predict Y for $x_2 = 1$ and $x_3 = 0.5$ (x_1 can be any value since it is not in the model), then the prediction interval is given by the following formula.

$$\hat{Y}\big|_{\underline{x}_h} \pm t\left(1 - \frac{\alpha}{2}, n - p\right)\hat{\sigma}\sqrt{1 + \underline{x}'_h (X'X)^{-1}\underline{x}_h}$$

where $\underline{x}_h = (1, x_{2h}, x_{3h}) = (1, 1, 0.5)$, $n = 15$, $p = 3$ and the design matrix X is given by:

$$X = \begin{bmatrix} 1 & x_{12} & x_{13} \\ 1 & x_{22} & x_{23} \\ . & . & . \\ . & . & . \\ 1 & x_{15,2} & x_{15,3} \end{bmatrix}$$

The X'X and (X'X)⁻¹ matrices will be:

$$X'X = \begin{bmatrix} 15 & 0 & 0 \\ 0 & 8 & 0 \\ 0 & 0 & 8 \end{bmatrix} \qquad (X'X)^{-1} = \begin{bmatrix} \dfrac{1}{15} & 0 & 0 \\ 0 & \dfrac{1}{8} & 0 \\ 0 & 0 & \dfrac{1}{8} \end{bmatrix}$$

Therefore, a 95% prediction interval at $x_2 = 1$ and $x_3 = 0.5$ is

$$1396.6 \pm (2.179)(11.864) \sqrt{1 + (1, 1, 0.5) \begin{bmatrix} \dfrac{1}{15} & 0 & 0 \\ 0 & \dfrac{1}{8} & 0 \\ 0 & 0 & \dfrac{1}{8} \end{bmatrix} \begin{bmatrix} 1 \\ 1 \\ 0.5 \end{bmatrix}}$$

$= 1396.6 \pm (2.179)(11.864) \sqrt{1.2229}$

$=- 1396.6 \pm 28.588$

$= [1368.012, \ 1425.188]$

4.10   Crossvalidation

When conducting simple linear or multiple regression, researchers typically use R^2 as a measure of the strength of the model. It should be noted that R^2 is dependent on the model, the sample, and the sample size. It is possible to continue to add variables up to $k = n - 1$ to produce an R^2 which is monotonically increasing up to 1.0. This results in an overfit condition which inaccurately predicts the strength of the model. To compensate for models with large numbers of terms compared to number of observations, some software packages such as MYSTAT will calculate an adjusted R^2 value. The formula for adjusted R^2 is

$$\text{Adj } R^2 = 1 - \left[\frac{n-1}{n-p} \right] (1 - R^2)$$

where n = total number of observations, and

p = total number of terms in the model (including the constant).

A more accurate measure of model strength is to use part of the sample data to build the model and the other part for crossvalidation. Yet another strategy is to build the model on sample data and obtain a crossvalidated R^2 from confirmation experiments. To get the crossvalidated estimate of R^2, calculate $e_i = y_i - \hat{y}_i$, where the $\hat{y}_i$ are the estimates based on the model obtained from the original data, and the y_i are the observed y values in the crossvalidation sample data. Next, compute

$$R^2 = 1 - \frac{\sum e_i^2}{SST_2}$$

where SST_2 is the total sum of squares for the data in the crossvalidation sample. It is also possible to correlate the y_i from the original data with the $\hat{y}_i$ model based on the second data set.

Since splitting a sample into two parts requires a large amount of data, a more efficient measure for estimating model strength is the Relative Press (Rel PRESS) value. This measure of the strength of a model is calculated through the following steps:

(1) Compute the first of a series of regression model coefficients by withholding (or setting aside) the first observation (i).

(2) Use the model obtained in (1) to predict the observation (i) not used in building the model, namely $\hat{y}_{i, -i}$.

(3) Calculate the press residual $(e_{i, -i}) = (y_i - \hat{y}_{i, -i})$ for observation (i) not in the model.

(4) Repeat steps (1) through (3) n times, deleting a different observation each time.

(5) Calculate the Prediction Sum of Squares, referred to as

$$PRESS = \sum_1^n (e_{i, -1})^2.$$

(6) Calculate relative Press as follows:

$$\text{Rel PRESS} = 1 - \frac{PRESS}{SST},$$

where SST is the Total Sum of Squares for all n actual values of the response y. For a more efficient approach to calculating PRESS and more information on Rel PRESS see [2].

One way to reduce overfit is to ensure that for any regression model, n is sufficiently large compared to the number of parameters estimated. A good Rule of Thumb for the final prediction model is to have n be at least ten times more than the number of parameters in the model.

4.11   Notes of Caution

When using simple linear or multiple regression on "happenstance or non-experimental data," one should be cautious of high correlations, for they may be significant only because a spurious condition unknown to the analyst exists. This problem can be averted by designing proper experiments instead of analyzing non-experimental data. Furthermore, there are times when a researcher has properly designed an experiment, but obtains unanticipated results due to improper levels for independent variables. For example, consider the relationship between y and x in Figure 4.10. If the levels of x are a and d, you will find a significant linear relationship between y and x. However, if the range on x is b to c, this relationship may not be discovered.

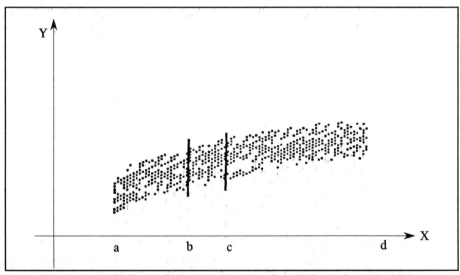

Figure 4.10   Effect of Varying Levels of an Independent Variable

A less likely, but certainly possible, scenario appears in Figure 4.11. Due to the quadratic relationship, if the sampled levels of x are a and b for a simple linear regression, a non-significant relationship will be found even though a significant quadratic relationship exists. To find a significant linear relationship, the sampled region of x must be [a,c] or [c,b]. Obviously, using a quadratic (or second-order) model over [a,b], one could detect a significant quadratic relationship. Thus, experimenters should carefully plan not only which variables to put in the model, but also which levels are necessary to detect anticipated relationships.

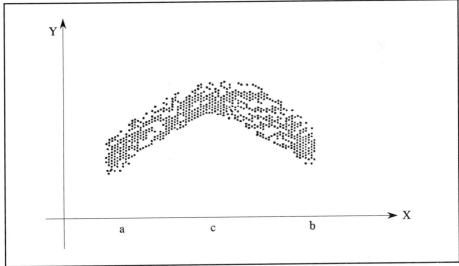

Figure 4.11   Quadratic Relationship on the Interval [a,b]

Finally, it will be briefly stated that outliers in the data can significantly alter regression results. Before any regression is conducted or any other data analysis takes place, one should explore the data distribution for each variable, identify outliers, ascertain their accuracy, and take the appropriate action. One can also use sophisticated techniques to identify outliers in a multi-variable setting through the use of residual analysis [2]. Obviously, you don't want to just throw out data to make your model look good. However, leaving outliers in a data set will result in a distorted model that may not predict future events. Sometimes it is wise to build one model with all the data and another model without the outliers and then test these models with confirmation data. The key to any good modeling practice is whether or not the model confirms.

4.12   ANOVA for Factors with More Than 2 Levels (Optional)

Consider the following one-way ANOVA example where factor A has 5 levels (L = 5).

Factor A

	A_1	A_2	A_3	A_4	A_5
$\bar{y}_j$	8	6	7	12	10
s_j^2	4.888	7.333	12.222	9.333	10.667
n_j	10	10	10	10	10

(1) H_0: $\mu_1 = \mu_2 = \mu_3 = \mu_4 = \mu_5$

$\quad\;\;\;$ H_1: not all μ_j are equal

(2) $\alpha = .01$

(3) $\displaystyle \bar{\bar{y}} = \frac{\sum\limits_{j=1}^{L} n_j \bar{y}_j}{\sum\limits_{j=1}^{L} n_j} = \frac{10\,(8 + 6 + 7 + 12 + 10)}{50} = 8.6$

$\displaystyle \text{MSB} = \frac{\sum\limits_{j=1}^{L} n_j (\bar{y}_j - \bar{\bar{y}})^2}{L - 1} = \frac{10(-.6)^2 + 10(-2.6)^2 + 10(-1.6)^2 + 10(3.4)^2 + 10(1.4)^2}{5 - 1} = 58$

$\displaystyle \text{MSE} = \frac{\sum\limits_{j=1}^{L} (n_j - 1)S_j^2}{\sum\limits_{j=1}^{L} (n_j - 1)} = \frac{9(4.888) + 9(7.333) + 9(12.222) + 9(9.333) + 9(10.667)}{9 + 9 + 9 + 9 + 9}$

$\displaystyle \quad\;\; = \frac{400}{45} = 8.889$

$\displaystyle F_0 = \frac{\text{MSB}}{\text{MSE}} = \frac{58}{8.889} = 6.52$

(4) $F(.99, 4, 45) = 3.83$

(5)   Since $F_0 > 3.83$, reject H_0 and conclude H_1 with 99% confidence.

In this case the conclusion is H_1: μ_j are not all equal. The next question is, "Which levels are different and which are not?" Using a regular t test or F test on each possible pair of treatments, i.e., H_0: $\mu_j = \mu_k$ for all $j < k$, would result in an increase in the overall α known as the experiment-wise α, α_{EW}. The α_{EW} is also referred to as the probability of committing at least one Type I error after conducting all of the individual pairwise tests. A procedure which allows for testing all pairwise levels at a specified α_{EW} is the Tukey Post Hoc comparison test. Use of the test is valid only when the factor overall F_0 is significant. The test is fairly simple and requires the use of the studentized ranges provided in Appendix H. The Tukey test will be illustrated using the previous example.

Since factor A is significant, it is desired to determine which levels differ and which do not. The procedure is as follows:

(a)   Use the Tukey comparison test and compute the critical Tukey distance $d_T = q_T \sqrt{\dfrac{MSE}{n_j}}$ where q_T is found in Appendix H and all n_j are assumed equal.

(b)   In Appendix H, p is the number of factor levels. Thus, for our example, $p = 5$. Since $df_E = 45$ and given $\alpha = .05$, then $q_T = 3.171$. Therefore,

$$d_T = 3.171 \sqrt{\frac{8.889}{10}} = 2.99.$$

(c)   Compute all factor level mean pairwise differences and conclude those   differences greater than $d_T = 2.99$ as significant at the chosen α level (see the chart shown below).

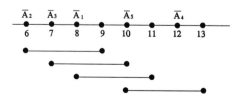

Conclude μ_4 is significantly different than μ_2, μ_3, and μ_1. Also conclude μ_5 is significantly different than μ_2 and μ_3; fail to reject H_0 for all other pairwise comparisons. What does this mean to the experimenter? If your objective is to maximize the response, then you are 95% confident that material 4 produces a larger response than materials 2, 3, and 1. In addition, material 5 produces a larger response than 2 or 3. Therefore, the best material type is either 4 or 5. Unless 5 is substantially cheaper than 4, the most conservative decision is to select type 4 because you failed to reject μ_5 different from μ_1. However, there exists probabilities of error, α (rejecting H_0 incorrectly) and β (failing to reject H_0 when you should). Small sample sizes contribute to large β values. Our sample size of ten for each factor level may not provide enough overwhelming information to reject H_0: $\mu_5 = \mu_1$ when we should.

4.13 Two-Factor ANOVA for 1 Factor at 2 Levels and 1 Factor at 3 Levels (Optional)*

Assume you are interested in testing factor A (3 lots of incoming material) and factor B (2 different processing temperatures). The qualitative variable (lot) complicates the modeling process. Thus, ANOVA is an easier approach to identify i) significant lot-to-lot differences, ii) significant temperature differences, and iii) a significant interaction effect.

The experimental matrix for testing all possible combinations (the full factorial) is shown on the next page. A sample of size 9 was chosen to comply as closely as possible with Rule of Thumb #1 in Appendix M.

---

*This section provided with permission from Dr. Mark Kiemele

Run	A (Lot)	B (Temp)	y_1	y_2	y_3	y_4	y_5	y_6	y_7	y_8	y_9	$\bar{y}$	s	s^2
	FACTORS					RESPONSE VALUES								
1	− (1)	− (Lo)	2.1	2.5	2.3	2.6	2.3	2.4	2.0	2.3	2.4	2.322	.186	.034
2	− (1)	+ (Hi)	3.9	4.0	4.1	4.0	4.0	4.1	3.9	4.0	4.1	4.011	.078	.006
3	0 (2)	− (Lo)	3.1	2.9	2.7	3.3	2.9	3.0	3.2	3.1	2.9	3.011	.183	.034
4	0 (2)	+ (Hi)	3.8	4.1	3.9	4.0	4.1	4.0	4.1	3.9	4.0	3.989	.105	.011
5	+ (3)	− (Lo)	2.4	2.7	2.1	2.2	2.3	2.4	2.4	2.6	2.3	2.378	.186	.034
6	+ (3)	+ (Hi)	4.0	3.9	4.1	3.9	4.0	4.0	4.1	3.9	4.0	3.989	.078	.006

Avg @ −	3.166	2.570
Avg @ 0	3.500	N/A
Avg @ +	3.183	3.996

$\bar{\bar{Y}} = 3.283$

The computational formulas for completing the ANOVA follow:

$$(1) \quad MSB_{(A)} = \frac{SSB_{(A)}}{df_{(A)}} = \frac{\sum_{j=1}^{L_A} n_{Aj}(\bar{Y}_{Aj} - \bar{\bar{Y}})^2}{L_A - 1}$$

where

$MSB_{(A)}$ = mean square between for A

$SSB_{(A)}$ = sum of squares between for A

$df_{(A)}$ = degrees of freedom for A

L_A = number of levels of A

n_{Aj} = number of data values for A at level j

$\bar{Y}_{Aj}$ = average response for A at level j

$\bar{\bar{Y}}$ = grand average of all the data

(2)
$$MSB_{(B)} = \frac{SSB_{(B)}}{df_{(B)}} = \frac{\sum_{i=1}^{L_B} n_{Bi}(\overline{Y}_{Bi} - \overline{\overline{Y}})^2}{L_B - 1}$$

where
$$\begin{cases}
MSB_{(B)} & = \text{ mean square between for B} \\[6pt]
SSB_{(B)} & = \text{ sum of squares between for B} \\[6pt]
df_{(B)} & = \text{ degrees of freedom for B} \\[6pt]
L_B & = \text{ number of levels of B} \\[6pt]
n_{Bi} & = \text{ number of data values for B at level i} \\[6pt]
\overline{Y}_{Bi} & = \text{ average response for B at level i} \\[6pt]
\overline{\overline{Y}} & = \text{ grand average of all the data}
\end{cases}$$

(3)
$$MSE = \frac{SSE}{df_{(E)}} = \frac{\sum_{k=1}^{L_{AB}} (n_r - 1) S_r^2}{\sum_{k=1}^{L_{AB}} (n_r - 1)}$$

where
$$\begin{cases}
MSE & = \text{ mean square error} \\[6pt]
SSE & = \text{ sum of squares error} \\[6pt]
df_{(E)} & = \text{ degrees of freedom for error} \\[6pt]
L_{AB} & = \text{ number of combinations (runs)} \\[6pt]
n_r & = \text{ number of data values in combination (run) r} \\[6pt]
S_r^2 & = \text{ variance of the data in combination (run) r}
\end{cases}$$

(4)
$$MSB_{(AxB)} = \frac{SSB_{(AxB)}}{df_{(AxB)}} = \frac{SS_{(TOT)} - SSB_{(A)} - SSB_{(B)} - SSE}{df_{(TOT)} - df_{(A)} - df_{(B)} - df_{(E)}}$$

where

$MSB_{(AxB)}$ = mean square between for the interaction of AxB

$SSB_{(AxB)}$ = sum of squares between for AxB

$SS_{(TOT)}$ = sum of squares total which is $S^2(N-1)$ where S^2 is the variance of all N response values

$df_{(TOT)}$ = degrees of freedom total which is N-1

$df_{(AxB)}$ = degrees of freedom for AxB which is found by subtraction

(5) $F_{(A)} = \dfrac{MSB_{(A)}}{MSE}$ $F_{(A)}$ is the F test value for A

$F_{(B)} = \dfrac{MSB_{(B)}}{MSE}$ $F_{(B)}$ is the F test value for B

$F_{(AxB)} = \dfrac{MSB_{(AxB)}}{MSE}$ $F_{(AxB)}$ is the F test value for AxB

Using formulas (1) through (5), the computations for our example are as follows:

$$MSB_{(A)} = \frac{18(3.166 - 3.283)^2 + 18(3.500 - 3.283)^2 + 18(3.183 - 3.283)^2}{3 - 1}$$

(1)

$$= \frac{1.274}{2} = .637$$

(2) $MSB_{(B)} = \dfrac{27(2.570 - 3.283)^2 + 27(3.996 - 3.283)^2}{2 - 1} = \dfrac{27.452}{1} = 27.452$

(3) $MSE = \dfrac{8(.034) + 8(.006) + 8(.034) + 8(.011) + 8(.034) + 8(.006)}{8 + 8 + 8 + 8 + 8 + 8}$

$= \dfrac{.998}{48} = .0208$

(4) $SS_{(TOT)} = (.587)53 = 31.111$

$MSB_{(AxB)} = \dfrac{31.111 - 1.274 - 27.452 - .998}{53 - 2 - 1 - 48} = \dfrac{1.387}{2} = .694$

(5) $F_{(A)} = \dfrac{.637}{.0208} = 30.625$

$F_{(B)} = \dfrac{27.452}{.0208} = 1319.808$

$F_{(AxB)} = \dfrac{.694}{.0208} = 33.365$

Using a statistical software package such as MYSTAT to generate the ANOVA table below, we see that our manual calculations agree quite well with the computer output. The differences are due to roundoff error.

ANOVA TABLE FROM MYSTAT

SOURCE	SS	df	MS	F	P
A	1.270	2	.635	30.278	.000
B	27.449	1	27.449	1308.830	.000
AxB	1.369	2	.685	32.645	.000
ERROR	1.007	48	.021		
TOTAL	31.095	53			

As you can see from the size of the F and P values, all the terms (A, B and AxB) are significant.   Thus, there is a significant lot effect, temperature effect and interaction effect.   A plot of the interaction is shown next.

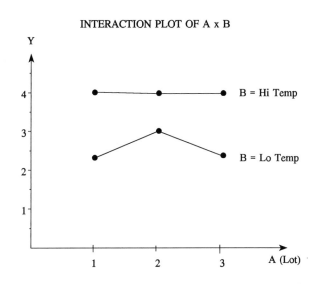

INTERACTION PLOT OF A x B

The interaction plot indicates that the variability across lots is possibly reduced when the temperature is high.

The next step will be to examine the high and low temperature data separately.

HIGH TEMPERATURE DATA

	A_1	A_2	A_3
$\bar{y}$	4.011	3.989	3.989
s^2	.006	.011	.006
n	9	9	9

Analysis of variance for only the high temperature data results in the MYSTAT computer output shown as follows:

ANOVA FOR HIGH TEMPERATURE DATA ONLY[*]

SOURCE	SS	df	MS	F	P
A(Lots)	.003	2	.001	.190	.828
ERROR	.187	24	.008		
TOTAL	.190	26			

Based on the ANOVA table above for the high temperature data only, there is no significant difference in lots. The low temperature data is next.

LOW TEMPERATURE DATA

	A_1	A_2	A_3
$\bar{y}$	2.322	3.011	2.378
s^2	.034	.034	.034
n	9	9	9

Analysis of variance for only the low temperature data results in the MYSTAT computer output shown below:

ANOVA FOR LOW TEMPERATURE DATA ONLY[*]

SOURCE	SS	df	MS	F	P
A(Lots)	2.636	2	1.318	38.580	.000
ERROR	.820	24	.034		
TOTAL	3.456	26			

---

[*]Both of these tables can be computed manually using the method described in section 4.12.

Based on the ANOVA table for the low temperature data only, the lots have a significantly different effect.

Therefore, the separate analyses of the low and high temperature data indicate that setting the temperature high will produce uniform results over the lots, i.e., high temperature produces a robust process for lot-to-lot variability.

4.14 Determining Sample Size (Optional)

Often the question is asked, "How large should my sample be?" The answer to this question depends on the following:

(1)   The desired level for α.

(2)   The desired level for $1 - \beta$ (β is the probability of concluding H_0 given that H_1 is true; therefore, $1 - \beta$ is the probability of concluding H_1 correctly. The term $1-\beta$ is also referred to as the power of the test).

(3)   Estimated amount of experimental error, $\hat{\sigma}^2$.

(4)   A specific alternative hypothesis deemed critical to detect with $(1 - \beta)100\%$ confidence.

(5)   Whether or not we're estimating $\hat{y}$ models or $\hat{s}$ models or both, items (1) through (4) are easily used for sample size determination. When modeling variation, $\hat{s}$, we should as a Rule of Thumb have 16 degrees of freedom for any average standard deviation used in building the model. (Chapter 5 provides more details. Also see Appendix M for simple Rules of Thumb.)

The appropriate equation relating terms (1) through (4) is:

$$\Phi^2 = \frac{n' \left[\dfrac{\sum\limits_{1}^{k} (\mu_i - \mu)^2}{k} \right]}{\hat{\sigma}^2}$$

where $\hat{\sigma}^2$, α, n', and degrees of freedom are used to enter the power charts in Appendix I. A step-by-step procedure for determining the approximate n', as well as an example, follows:

(1)   Determine k, the number of factor levels you intend to investigate.

(2)   Determine the alternative hypothesis of interest, either with specific values of μ_i or with the anticipated range, R, of the μ_i values and their suspected variability (low, medium, or high) over that range.

Low variability implies that one μ_i is at the maximum, one μ_i is at the minimum and the other $(k - 2)$ μ_i values occur at a point in between.

Medium variability implies that the μ_i are approximately normally distributed, i.e., $\sum\limits_{1}^{k} \dfrac{(\mu_i - \mu)^2}{k} \doteq \dfrac{R^2}{16}$.

High variability implies that half of the μ_i are at the maximum and the other half are at the minimum.

(3)   Estimate $\hat{\sigma}^2$ via
      (a) historical data,
      (b) pilot study, or
      (c) $\dfrac{R_i^2}{16}$ where R_i is the anticipated range of raw data for some factor level i. This method assumes $Y_{ij} \sim N(\mu_i, \sigma_i^2)$ for each level i.

(4)   Select an estimated value of a factor level sample size, n'.

(5)   Compute Φ.

(6)   Enter the appropriate chart from Appendix I where v_1 is the degrees of freedom for the factor effect and v_2 is the degrees of freedom for error.

(7)   If $(1 - \beta)$ is larger than desired, lower n'; otherwise increase n'.

(8)   Iterate steps (4) - (7) until the tabular value of $(1 - \beta)$.

This procedure is applied in the following example:

Assume $\hat{\sigma}^2 = 20$, $k = 4$, $\sigma = 0.5$ and the desired power is 0.8 for detecting $\mu_1 = 16$, $\mu_2 = 21$, $\mu_3 = 23$, $\mu_4 = 20$

$$\Phi^2 = \frac{n'\left[\dfrac{(16-20)^2 + (21-20)^2 + (23-20)^2 + (20-20)^2}{4}\right]}{20} = .325n'$$

$$\Phi = .57\sqrt{n'}$$

Using the table with $v_1 = 3$, you could iterate n' as follows:

n'	v_2	Φ	$1 - \beta$	Comments
5	16	1.28	.45	Too small, increase n'
7	24	1.51	.64	Too small, increase n'
9	32	1.72	.78	Slightly less than .8
10	36	1.80	.83	Use $n = 10$ for each level

4.15 Comparison of Confidence Interval, t-Test, ANOVA, and Regression Approaches for Analyzing a Factor with 2 Levels (Optional)

Given some factor A measured at 2 levels, A_1 and A_2, this section will compare four approaches which could be used to test the following hypotheses:

H_0: $\mu_1 = \mu_2$ versus H_1: $\mu_1 \neq \mu_2$.

The following sample data will be used to make the comparison.

	SAMPLE DATA		
	A_1	A_2	
	4	14	
	10	8	
	2	16	
	12	6	
Mean ($\bar{A}_j$)	7	11	Grand Mean, $\bar{A} = 9$
Variance (S_j^2)	22.667	22.667	
Sample size (n_j)	4	4	

Table 4.25 Sample Data

For the sake of this comparison, α will be set to .05.

(1) Confidence Interval Approach

$(1 - \alpha)100\%$ confidence interval for $\mu_2 - \mu_1$ is:

$$\left(\bar{A}_2 - \bar{A}_1 \right) \pm t\left(1 - \frac{\alpha}{2}, \; n_1 + n_2 - 2 \right) \sqrt{ MSE \left(\frac{1}{n_1} + \frac{1}{n_2} \right) }$$

where $MSE = \dfrac{\sum (n_j - 1) s_j^2}{\sum (n_j - 1)}$. The sample data results are:

$$\left[(11 - 7) \pm t\,(.975,\ 6) \sqrt{22.667 \left(\frac{1}{4} + \frac{1}{4}\right)} \right]$$

$$= [(4) \pm (2.447)(3.366)]$$

$$= [-4.236,\ 12.236]$$

Since the 95% confidence interval for $\mu_2 - \mu_1$ contains zero, your conclusion will be to fail to reject H_0.

(2)   t-Test Approach

The 5 step method for the t-test is:

(1) H_0: $\mu_1 = \mu_2$
     H_1: $\mu_1 \neq \mu_2$

(2) $\alpha = .05$

$$(3)\ \ t_0 = \frac{\overline{A}_2 - \overline{A}_1}{\sqrt{MSE\left(\frac{1}{n_1} + \frac{1}{n_2}\right)}} = \frac{4}{\sqrt{22.667\left(\frac{1}{4} + \frac{1}{4}\right)}} = 1.188$$

$$(4)\ \ t_{CRITICAL} = t\left[1 - \frac{\alpha}{2},\ n_1 + n_2 - 2\right] = t\,(.975,\ 6) = 2.447$$

(5) Since $[|t_0| = 1.188]$ is less than $[t_{CRITICAL} = 2.447]$ your conclusion will be to fail to reject H_0.

(3)   Analysis of Variance (ANOVA) Approach

The 5 step method for ANOVA is:

(1) H_0: $\mu_1 = \mu_2$
     H_1: $\mu_1 \neq \mu_2$

(2) $\alpha = .05$

(3) $F_0 = \dfrac{MSB}{MSE} = \dfrac{32}{22.667} = 1.412$

	ANOVA TABLE			
SOURCE	SS	df	MS	F
BETWEEN	32	1	32	1.412
ERROR	136	6	22.667	
TOTAL	168	7		

Table 4.26 ANOVA for Sample Data

(4) $F_{CRITICAL} = F(.95, 1, 6) = 5.99$

(5) Since $[F_0 = 1.412]$ is less than $[F_{CRITICAL} = 5.99]$, your conclusion will be to fail to reject H_0.

Before completing a regression approach to this problem, it is worth reviewing the following:

- All previous conclusions were the same.

- $t_0^2 = F_0$

- $\left[t \left(1 - \dfrac{\alpha}{2}, 6 \right) \right]^2 = F(1 - \alpha, 1, 6)$

(4)   Regression Approach

To solve a two group problem by way of regression, you must form a group membership column. Although there are several ways to code this column, the most common would be using effect coding as shown on the following page.

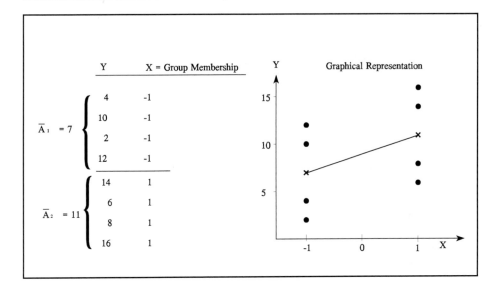

The regression equation is built based on the slope-intercept equation, i.e.,

$$\hat{y} = \text{intercept} + \text{slope } X = b_0 + b_1 X$$

$$= 9 + \frac{11 - 7}{2} X$$

$$= 9 + 2X$$

Using the formula for R^2 in Section 4.5, R^2 is calculated to be .1904.

The 5 step hypotheses test would be:

(1a) H_0: $R^2 = 0$ (1b) H_0: $\beta_1 = 0$

 H_1: $R^2 \neq 0$ H_1: $\beta_1 \neq 0$

In either case, H_0 implies no group differences, i.e., $\mu_1 = \mu_2$.

(2) $\alpha = .05$

(3a) To test H_0: $R^2 = 0$

$$F_0 = \frac{\dfrac{R^2}{1}}{\dfrac{(1 - R^2)}{(N - 2)}} = \frac{.1904}{\dfrac{(1 - .1904)}{(8 - 2)}} = 1.412$$

(4a) $F_{CRITICAL} = F(.95, 1, 6) = 5.99$

(5a) Since $F_0 < F_{CRITICAL}$, you fail to reject H_0.

(3b) To test H_0: $\beta_1 = 0$

$$t_0 = \frac{b_1}{\sqrt{\dfrac{MSE}{\left(\sum x^2 - n\bar{x}^2\right)}}} = \frac{2}{\sqrt{\dfrac{22.667}{8}}} = 1.188$$

(4b) $t_{CRITICAL} = t(.975, 6) = 2.447$

(5b) Since $|t_0| < t_{CRITICAL}$, you fail to reject H_0.

The conclusion is that all 4 methods produced identical results.[*] In addition, the t_0 values were all the same and equal to $\sqrt{F_0}$. Likewise, the $t_{CRITICAL}$ values were all the same and equal to $\sqrt{F_{CRITICAL}}$.

---

[*]Note: A similar conclusion would be made using the end count technique described on page 8-215 and in Appendix M.

Problem Set

1. Given 2 different soft bake methods for a photoresist process, determine if the mean critical dimension in microns for the feature of interest is the same for each process. (Conduct this test using Confidence Intervals, t-test, and ANOVA.)

Setup #	Convection Bake	Hot Plate Bake
1	4.1	3.8
2	3.9	3.5
3	4.3	4.1
4	4.1	4.1
5	4.1	3.8

    All data is in microns, and the exposure was constant in all cases.

2. A precision bearing manufacturer wants to determine a minimum sample size to check the accuracy of two groups of bearings manufactured 2 months apart. The manufacturer wants to insure that he can detect a specific H_1: $\mu_1 - \mu_2 > 10$ microns with 90% confidence given: $\sigma^2 = 81$ and $\alpha = 0.05$.

    What is the minimum sample size?

3. (a) Given the following etch rate data as a function of DC bias voltage, determine which levels of DC bias voltage are significantly different.

Level
A1 = -220 volts
A2 = -240 volts
A3 = -260 volts
A4 = -280 volts
A5 = -300 volts

Etch Rate Data: Angstroms/Min.

A1	A2	A3	A4	A5
1280	1285	1301	1345	1401
1277	1296	1310	1350	1390
1275	1291	1298	1340	1395
1282	1287	1316	1356	1402
1272	1294	1300	1351	1391
1289	1281	1285	1360	1405
1271	1285	1305	1347	1406

$\alpha = 0.01$

(b) If the objective is to maximize the etch rate, which level(s) should be selected?

4. A manufacturer of a proprietary plating additive wants to understand the reaction kinetics of the additive. The manufacturer assumes that the following reaction represents his process:

Concentration of Additive (C_A) > Rate of Reaction $(-r_A)$.

Given this Data:

n	1	2	3	4	5
C_A	5.6	9.2	15.6	22.5	26.2
$-r_A$	1.5	3.2	7.3	13.9	22

(a) Calculate the slope and intercept of the linear model between C_A and $-r_A$ where $-r_A$ is the dependent variable.

(b) Find the correlation coefficient, R, and also the proportion of variability explained by the regression.

(c) Graph C_A versus $-r_A$. Is a linear relationship a good assumption?

(d) The manufacturer now thinks the relationship is:

$-r_A = k \, C_A^N$, where k = rate constant and N = order of reaction.

Taking the natural log of this equation yields:
$\ln(-r_A) = \ln(k) + N \cdot \ln(C_A)$

Which is a linear model? Take the natural log of the raw data and repeat (a) and (b) to find the order of the reaction, N, and the rate constant, k.

5. A graduate student was asked to determine if a relationship existed between sulfuric acid concentration (wt. %) and temperature (°C), and their effect on a 1st order rate constant (k) with respect to the hydrolysis of cellulose.

(a) Using the data shown below, generate a ANOVA table.

(b) Is there an interaction between wt. % H_2SO_4 and temperature?

(c) Is there a significant effect by temperature or wt. % H_2SO_4?

T(°C)	wt. % H_2SO_4	k_1	k_2
170	0.5	0.0196	0.0201
190	0.5	0.0376	0.0355
210	0.5	0.8710	0.9000
170	1.0	0.0808	0.0799
190	1.0	0.1660	0.1500
210	1.0	0.7420	0.7210
170	2.0	0.0736	0.0691
190	2.0	0.1630	0.1700
210	2.0	0.4480	0.4350

6.   Given the following data:

Run	A	B	−AB	C	−AC	D	E	y_1	y_2	$\bar{y}$	s
1	−1	−1	−1	−1	−1	−1	−1	66	62	64.0	2.83
2	−1	−1	−1	+1	+1	+1	+1	68	63	65.5	3.54
3	−1	+1	+1	−1	−1	+1	+1	88	80	84.0	5.66
4	−1	+1	+1	+1	+1	−1	−1	63	65	64.0	1.41
5	+1	−1	+1	−1	+1	−1	+1	73	71	72.0	1.41
6	+1	−1	+1	+1	−1	+1	−1	37	42	39.5	3.54
7	+1	+1	−1	−1	+1	+1	−1	38	39	38.5	0.71
8	+1	+1	−1	+1	−1	−1	+1	57	48	52.5	6.36

a)   Complete the following table:

	A	B	−AB	C	−AC	D	E
Avg (−)							
Avg (+)							
Δ							
Δ/2							
MSB							
F							

b)   Complete a standard ANOVA table.

c)   Determine the prediction equation: $\hat{Y}$ =

d)   What are the optimal settings to minimize $\hat{Y}$?

e)   Using the optimal settings from (d), what is the predicted response for Y?

7.   If only one variable is being tested at 2 levels, is there any difference between using ANOVA or a t-test?

8. What is the effect of using pairwise t-tests if multiple μ_i are found not to be equal? Why is it bad to increase α_{EW}?

9. What does R^2 measure in regression modeling?

10. $$\left[t \left(1 - \frac{\alpha}{2}, n \right) \right]^2 = F(\underline{}, \underline{}, \underline{})$$

11. What is an example of a blocking variable? How can an experimenter deal with a blocking variable?

12. What is a prediction interval and how would you interpret it?

Bibliography

1. Kiemele, M.J., S.R. Schmidt, and R.J. Berdine, (1997) *Basic Statistics: Tools for Continuous Improvement,* 4th Edition, Air Academy Press, Colorado Springs, CO.

2. Myers, Raymond H., (1990) *Classical and Modern Regression with Applications*, 2nd Edition, Duxbury Press, Boston, MA.

3. Walpole, R.E., and Myers, R.H., (1985) *Probability and Statistics for Engineers and Scientists*, Macmillan Publishing Company, New York.

Chapter 5

Analysis of Experimental Data

5.1   Introduction

Once you collect data using well planned and highly disciplined experimental techniques, it can be assumed the data contains true knowledge of the process. The next step is to extract the knowledge from the data through graphical and statistical analyses. The purpose of this chapter is to expand upon the simple analysis techniques presented in Chapter 2 and incorporate the statistical methods discussed in Chapter 4.

The type of analysis technique you use will depend primarily on your experimental objective and the capability of your software.[*] In this chapter, we will discuss analysis techniques for the following design objectives: screening, modeling (also referred to as characterization), variance reduction, optimization, simple response surfaces, and robustness. Advanced topics on response surface methods and optimization are presented in Chapter 7.

If your objective is factor screening, i.e., reducing a large number of factors to a manageable subset of important factors, then resolution III designs such as those presented in Chapter 3 will typically be used. The analysis will most likely

------

[*]Most of the techniques discussed in this chapter can be performed on the accompanying software DOE KISS (student version) or DOE PRO from Air Academy Associates.

be a graphical approach such as plotting the averages for each effect, a Pareto diagram of the absolute value of the half effects ($|\Delta/2|$), or a normal probability plot. Analysis of Variance (ANOVA) and/or regression could also be utilized to get a statistical evaluation of important factors.

Modeling a process will usually require at least a resolution IV or V design, depending on your prior knowledge of interaction effects. To determine which terms appear in the final model will require a Pareto of $|\Delta/2|$ values and/or regression as a minimum. Other techniques such as ANOVA and normal probability plots might also be used to provide additional insight into the data.

Selecting the analysis technique(s) for variance reduction and robustness is similar to that for screening or modeling. The primary difference is that we are now graphing the standard deviation or modeling the variation through $\hat{s}$ or $\ln \hat{s}$ equations instead of modeling the average with the $\hat{y}$ equation.

5.2　　　Analyzing a 2-Level Design

Assume that you have collected the data for an 8-run Taguchi design matrix of 7 factors, each at 2 levels, as shown in Table 5.1. Since this is a resolution III design, it is likely you are using it as a screening design; however, we will use the data to discuss several analysis techniques.

	Factors							Response
Run	A	B	C	D	E	F	G	y
1	−	−	−	−	−	−	−	18
2	−	−	−	+	+	+	+	20
3	−	+	+	−	−	+	+	12
4	−	+	+	+	+	−	−	10
5	+	−	+	−	+	−	+	12
6	+	−	+	+	−	+	−	14
7	+	+	−	−	+	+	−	19
8	+	+	−	+	−	−	+	21

Table 5.1 L_8 Design Matrix for 7 Factors at 2 Levels

If this design is intended as a trouble-shooting exercise to find the best settings for the factors, the simplest analysis technique is to look at the response values for the experimental runs and select the response that best satisfies the objective (i.e., minimum, maximum, or closest to a target). This method, referred to as **pick the winner**, has obvious advantages: quick, easy, insensitive to interaction aliasing, and it already contains a confirmation result. Despite these advantages, there also exist several disadvantages: factor importance is unknown, a modeling equation is not available, factor settings will not necessarily minimize cost, and several possible combinations have been ignored. For example, if a small response is desired, then the "pick the winner" factor settings from Table 5.1 correspond to the highs (+) and the lows (−) of run number 4. However, the 7 factors at 2 levels produce a total of 128 different combinations and we've only considered 8. To look into the remaining 120 possibilities, we can analyze the marginal effects through average effect plots, Pareto diagrams of half effects, normal probability plots, ANOVA, and/or regression. Some of these procedures will not provide tests of significance of the data in Table 5.1 unless there is some replication. This problem will be discussed as we examine each of these methods.

To analyze the data by way of **average effect plots**, consider Figure 5.1. Assuming we want to minimize the response, the average effect plots in Figure 5.1 indicate the best settings would be $A_-\ B_+\ C_+\ D_-\ E_+\ F_-\ G_-$. The predicted average response based on these factor settings is computed as follows:

$$\hat{y} = \overline{\overline{y}} + \frac{\Delta_A}{2}A + \frac{\Delta_B}{2}B + \ldots + \frac{\Delta_G}{2}G$$

$$= 15.75 + \frac{1.5}{2}A_- - \frac{.5}{2}B_+ - \frac{7.5}{2}C_+ + \frac{1}{2}D_- - \frac{1}{2}E_+ + \frac{1}{2}F_- + \frac{1}{2}G_-$$

$$= 15.75 - .75 - .25 - 3.75 - .5 - .5 - .5 - .5$$

$$= 9$$

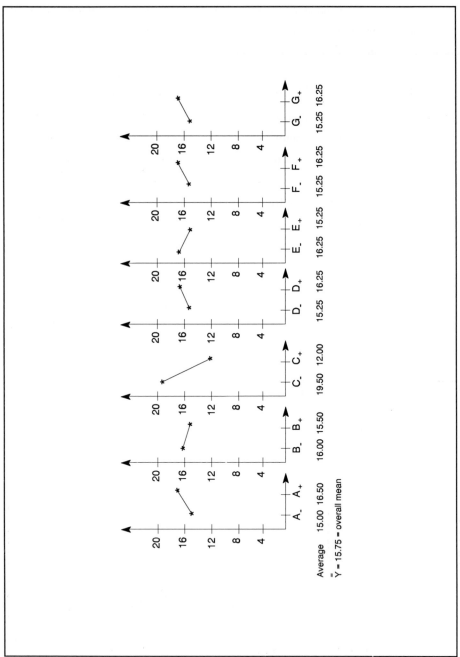

Figure 5.1 Average Effect Plots for Table 5.1 Data

This is an improvement over run #4; however, this estimate could be overly optimistic if some of the included effects occurred only due to chance and in reality are not significantly different from zero.

One way to assess the significance of effects is to look at a **Pareto diagram** of the absolute value of all half effect values found in the equation for $\hat{y}$ (see Figure 5.2). It appears that the only factor which stands out beyond what might be considered the noise level is factor C. The remaining effects could be considered to be different from zero due only to experimental error. This being the case, the best factor settings would consist of C_+ and the remaining factors set at levels which are based on: (1) economics, (2) convenience, (3) status quo, and/or (4) providing less variability in the response. The new predicted average response would be:

$$\hat{y} = 15.75 - 3.75C_+ = 12$$

which is slightly different from run #4 but probably a more realistic prediction of the average. Since only factor C appears to be significant, production costs may be reduced through more flexibility in selecting levels for the remaining factors.

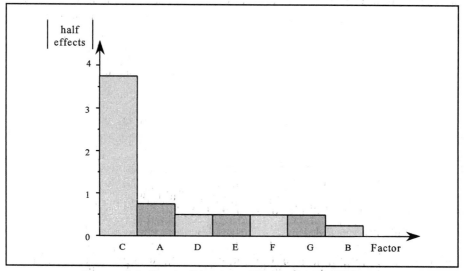

Figure 5.2 Pareto Diagram of Table 5.1 Half Effects

Not all Pareto diagrams provide as clear a picture of effect significance as Figure 5.2. Another approach is to use a **normal probability plot**. This method is especially useful when designs are saturated and there are no degrees of freedom for error. Since the design in Table 5.1 has n = 8 runs and k = 7 factors, there are no degrees of freedom for estimating error and, therefore, ANOVA and regression results with significance tests are not possible without some sort of modification. In these situations, the normal probability plot [5] or the Pareto diagram are the recommended techniques. To complete a normal probability plot:

(1)   For each factor and interaction, calculate the difference in average response for high and low levels (i.e., calculate the column effects:

$$\overline{A}_+ - \overline{A}_-, \ \overline{B}_+ - \overline{B}_-, \ ... , \overline{G}_+ - \overline{G}_-, \ \overline{A \times B}_+ - \overline{A \times B}_-, \text{ etc.}).$$

(2)   Assign a rank, i, to the differences in (1) starting from lowest to highest (i = 1, 2, . . . , M). M = the number of effects, i.e., the number of columns in the design matrix.

(3)   Convert each effect rank to a percentile using $p = \dfrac{100 \ (i - 0.5)}{M}$.

(4)   Scale the horizontal axis of normal probability plot paper to satisfy the values found in (1).

(5)   On normal probability plot paper, plot p versus values found in (1).

(6)   Draw a straight line through the majority of points which appear to line up (discounting any obvious "outliers"). When establishing the line, give priority to effects close to zero.

(7)   Any point which deviates considerably from the straight line represents a potentially significant effect.

Using the data set in Table 5.1, steps (1) through (7) result in the following:

	Factor	A	B	C	D	E	F	G
(1)	**Effect**	1.5	−.5	−7.5	1.0	−1.0	1.0	1.0
(2)*	**Rank**	7	3	1	5	2	5	5
(3)	*p*	92.9	35.7	7.1	64.3	21.4	64.3	64.3

(4) (5) (6)   See Figure 5.3

(7)   The only factor which appears significant is C. Thus, if our objective for the design in Table 5.1 is factor screening, the Pareto diagram and the normal probability plot both indicate that only factor C is important.

     Graphical analyses provide a visual understanding of the relative importance of each effect, but they do not provide a quantitative measure of confidence in our conclusion. To estimate this confidence, statistical methods such as **ANOVA** or **regression** must be used. To use ANOVA or regression, there must exist some degree of freedom for error. This is best accomplished by replicating the entire design or, more efficiently, replicating at one point. If runs are costly, replicating at one point is more economical, but the question is, at which point? To answer this question, you must first determine if there is any need to address non-linear effects. If so, you should replicate at the center point (0) and, thus, be able to graphically test for curvature and estimate error. If non-linear effects are not reasonable to test, then replicate at the level chosen for the best response. This replication serves as a way to estimate error and also provides confirmation runs.

     To demonstrate the use of ANOVA and regression, we will use the data from Table 5.1 with an additional replicate for each of the 8 runs. The new data set appears as shown in Table 5.2

---

*When two or more effects have equal values, their ranks become the average of what would otherwise be assigned. For example, the ranks for D, F, and G would be 4, 5, and 6, but since all effect values are the same, each receives an average rank of 5.0.

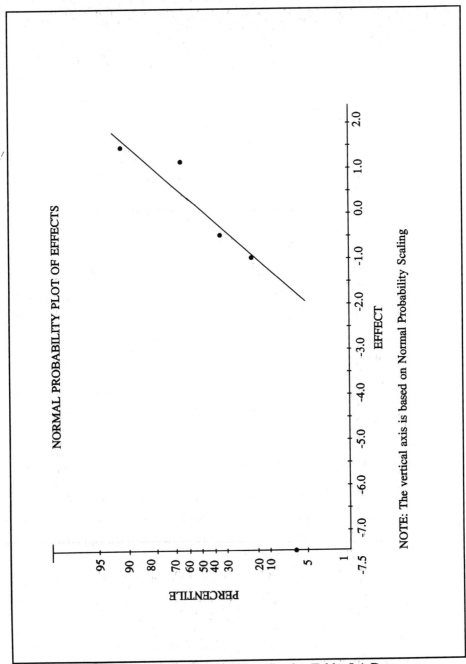

Figure 5.3   Normal Probability Plot for Table 5.1 Data

Run			Factors					Response		Average	Variance	
	A	B	C	D	E	F	G	y_1	y_2	$\bar{y}_r$	s_r^2	df_r
1	−	−	−	−	−	−	−	18	23	20.5	12.5	1
2	−	−	−	+	+	+	+	20	18	19.0	2.0	1
3	−	+	+	−	−	+	+	12	13	12.5	0.5	1
4	−	+	+	+	+	−	−	10	14	12.0	8.0	1
5	+	−	+	−	+	−	+	12	11	11.5	0.5	1
6	+	−	+	+	−	+	−	14	9	11.5	12.5	1
7	+	+	−	−	+	+	−	19	25	22.0	18.0	1
8	+	+	−	+	−	−	+	21	20	20.5	0.5	1

	A	B	C	D	E	F	G	
Avg −	16.0	15.625	20.5	16.625	16.25	16.125	16.5	$\bar{\bar{y}} =$
Avg +	16.375	16.75	11.875	15.75	16.125	16.25	15.875	16.1873
Δ	.375	1.125	−8.625	−.875	−.125	.125	−.625	
MSB_{effect}	.563	5.063	297.563	3.063	.063	.063	1.563	$F_c =$ F(.95,1,8)
F_{effect}	<1	<1	43.68	<1	<1	<1	<1	= 5.32

$$MSE = \frac{\sum (df_r)s_r^2}{\sum (df_r)} = \frac{1(12.5) + 1(2.0) + 1(.5) + \ldots + 1(18) + 1(.5)}{1 + 1 + 1 + 1 + 1 + 1 + 1 + 1} = 6.813$$

Table 5.2 Analysis of the Data from Table 5.1 with an Additional Replicate

Recall from Chapter 4 that 2-level designs will result in $MSB_{effect} = \frac{N}{4}(\Delta)^2$ where N is the total number of response values (for our example in Table 5.2, N = 16). Using the MSB_{effect} formula, the reader should verify the MSB_{effect} values in Table 5.2. The critical F value in Table 5.2 was obtained from Appendix F using $F_C = F(1 - \alpha, 1, \sum(df_r))$.

The results in Table 5.2 can be displayed in a standard ANOVA table as shown in Table 5.3. The % contribution column is calculated by the source sum of squares (SS) divided by the total sum of squares. To interpret the P column, use $(1 - P)100\%$ as the confidence in that term being important. The ANOVA results indicate that factor C is the only significant factor. The reader should realize that the preciseness of the P value is based on the normality of the response. Because response data is not always normally distributed [2], we recommend that ANOVA be used in conjunction with the previous approaches of Pareto charts, normal probability plots, size of the half effects, and prior knowledge. If you encounter severe non-normal data it can often be transformed using $\ln Y$, $\sqrt{Y}$, $1/Y$, or $1/\sqrt{Y}$ to normalize it. The analysis can then be performed on the transformed data.

ANOVA TABLE						%
Source	SS*	df	MS	F**	P***	Contrib.
A	.563	1	.5625	<1	.781	.155
B	5.063	1	5.0625	<1	.414	1.397
C	297.563	1	297.5625	43.68	.000	82.100
D	3.063	1	3.0625	<1	.521	.845
E	.063	1	.0625	<1	.926	.017
F	.063	1	.0625	<1	.926	.017
G	1.563	1	1.5625	<1	.645	.431
Error	54.5	8	6.8125	–		15.037
Total	362.438	15				

Table 5.3 Standard ANOVA Table for Table 5.2 Data Results

---

*For orthogonal 2-level designs, $SS_A = N \cdot b_A^2$ where N is the total number of runs (in this case N = 16) and b_A is the regression coefficient for factor A obtained when you regress the N response values on the design matrix of (+)'s and (−)'s. SS_A is also equal to $MSB_A \cdot df_A$. Other SS values are found in a similar fashion.

**See Appendix M for Rules of Thumb on significant F values.

***See Appendix M for Rules of Thumb on significant P values.

Another analysis approach is modeling using least squares regression. If we entered the 16 coded combinations from Table 5.2 into a software package and regressed the response value against factors A through G, the output would appear as shown in Table 5.4.

Dependent Variable: Y	N: 16		Multiple R: .922		Squared Multiple R: .850	
Adjusted Squared Multiple R: .718				Standard Error of Estimate: 2.610		
Variable	Coeff	Std Error	Std Coef	Tolerance	T	P(2 Tail)*
Constant	16.188	0.653	0.000	1.0000000	24.808	0.000
A	0.188	0.653	0.039	1.0000000	0.287	0.781
B	0.563	0.653	0.118	1.0000000	0.862	0.414
C	-4.313	0.653	-0.906	1.0000000	-6.609	0.000
D	-0.438	0.653	-0.092	1.0000000	-0.670	0.521
E	-0.063	0.653	-0.013	1.0000000	-0.096	0.926
F	0.063	0.653	0.013	1.0000000	0.096	0.926
G	-0.313	0.653	-0.066	1.0000000	-0.479	0.645
Analysis of Variance						
Source	Sum-of-Squares	df	Mean-Square	F-Ratio	P	
Regression	307.938	7	43.991	6.457	0.009	
Residual	54.500	8	6.813			

Table 5.4 Regression Output for Table 5.2 Example Data

For an orthogonal 2-level design, the coefficients for each factor are equal to

[response average at "+1" - response average at "-1"] ÷ 2;

thus, regression analysis has the same results as the (Δ/2) method. Reviewing Tables 5.3 and 5.4 indicates that both regression and ANOVA find only factor C

---

*See Appendix M on Rules of Thumb for significant P(2 Tail) values.

to be significant. In fact the ANOVA and regression results are identical (compare the P columns for each effect). The added benefit of regression is that we obtain the coefficients for the prediction equation (see the variable and coefficient columns in Table 5.4).

The prediction equation derived from Table 5.4 is:

$$\hat{y} = 16.188 - 4.313C.$$

Since our objective is to minimize the response, setting C at the high level (+1 orthogonal coded value) results in a predicted average response of 11.875. Assuming we had some degree of confidence in this model, the next step would be to make as many confirmation runs as possible (use 4 to 20 as a minimum) and determine the confirmation run average $\overline{CR}$. To test if the confirmation runs produced anticipated results you could conduct a t-test where

$$H_0: \mu = 11.875 \quad \text{and} \quad H_1: \mu \neq 11.875 \quad \text{using} \quad t_0 = \frac{\overline{CR} - 11.875}{S_{CR} / \sqrt{n_{CR}}} \quad \text{and}$$

$t_c = t\left[1 - \frac{\alpha}{2}, n_{CR} - 1\right]$ (see Chapter 4 for steps required for a t-test). If you fail to reject H_0, you have confirmed your results. However, if you reject H_0, i.e., conclude $H_1: \mu \neq 11.875$, you must conduct an organized investigation of what may have caused the difference between the confirmation run average and the predicted average response. The following is a partial list of what could have caused the difference.

(1) Factors not included in the experiment varying in an uncontrolled and non-random fashion.

(2) Poor experimental discipline (i.e., settings were not made correctly, response values read incorrectly, no randomization, etc.).

(3) Interactions aliased with main effects.

(4) Model inadequacy: (i) a second-order model is appropriate, but only first-order model is estimated due to a two-level type design; (ii) too few terms in the model; (iii) too many terms in the model.

(5)   Something changed between the initial experiment and the confirmation runs (i.e., different type of raw material, different ambient conditions, different machines, etc.).

(6)   Excessive variation in the data coupled with an inadequate number of response values per experimental run.

At this point you would want to review items (1) through (6) above and try to rule out as many as possible. Unfortunately, if you cannot rule out items (1) or (2), you may not be able to recover without redoing the entire experiment. If you suspect item (4) as the primary cause, you'd simply play with the prediction equation or add runs to allow for non-linear modeling (see CCD in Chapter 3). Provided all factors were set at the extremes during confirmation, i.e., either all (+) or all (−) settings, you can rule out non-linearity as a possible cause of not confirming. Investigating item (5) requires some group brainstorming; whereas recovery from item (6) necessitates more replicates. In our example, seven factors were placed in an eight-run design which results in a resolution of III. Recall from Chapter 3 that resolution III implies that main effects are aliased with two-way interactions. Thus, there is a strong possibility that any problem with confirmation runs in our example is associated with a strong two-way interaction, item (3). The next section will discuss the importance of including two-way interactions in the design matrix.

5.3   Importance of Modeling Interactions

How important is it to model interactions? This question can be addressed by way of a simple example. Consider the design below.

Run	A	B	A·B
1	+	+	+
2	+	−	−
3	−	+	−
4	−	−	+

Assume the researcher de-emphasizes interactions and is also interested in evaluating factor C. The 4-run design above could be used to test the A, B, and C main effects by aliasing A·B with factor C. Now, consider the following results:

Run	A	B	C	y
1	+	+	+	9
2	+	−	−	6
3	−	+	−	6
4	−	−	+	9

The average effect plots on the following page indicate no A or B effect, but a substantial C effect.

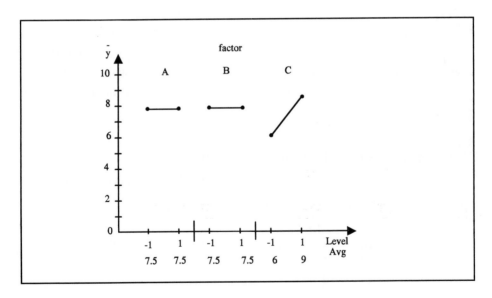

If "larger is better" then the researcher would suggest setting C at the high level and A and B settings to be based on economic considerations. For the sake of discussion, let's set A low and B high. Thus, the recommended settings are $A_{(-)}$ $B_{(+)}$ $C_{(+)}$ with a predicted average response of 9. We now make 5 confirmation runs with the following results.

Run	y
1	6
2	4
3	5
4	5
5	7
$\overline{CR}$	5.4
S_{CR}	1.14

$H_0: \mu = 9$ (confirmed)

$H_1: \mu \neq 9$ (did not confirm)

$\alpha = .05$

$t_0 = \dfrac{5.4 - 9}{1.14/\sqrt{5}} = -7.06$

$t_C = t(.975, 4) = 2.776$

Since $|t_0| > t_C$, we would reject H_0, i.e., conclude H_1. What has gone wrong? The problem in this hypothetical example is that we originally assumed the A· B effect was not applicable and aliased factor C with A· B. In our analysis, the A and B effects appeared unimportant, indicating that where we set them will not affect the response. However, if A· B is important and C is not, then a true representation of A and B will appear as shown in the following graphic.

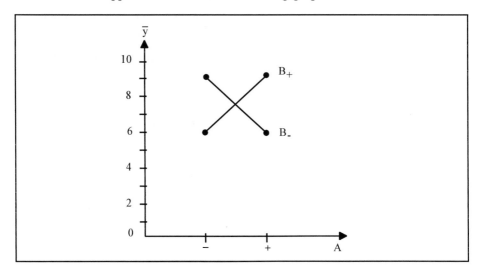

In this case, the settings for A and B cannot be made independently. For example, if A is set at (+), then y is maximized by setting B_+. However, if A is set at (−), then y is maximized by setting B_-. The results for the combination A_-B_+ in our confirmation runs is now predicted to be about 6.0 which is close to the 5.4 obtained.

The point to be made is that one should not ignore factor interactions just to simplify the design and to minimize the number of experimental runs. Even though it is true in most applications that strong interactions occur infrequently, the researcher should carefully consider whether or not to analyze interactions (see the Anodize case study from Boeing in Chapter 8 which further demonstrates the importance of identifying interactions). If you choose not to evaluate interactions, at least be aware of ways to recover from problems similar to the one presented above, i.e., if you use a R_{III} design which provides confusing results, you can fold it over to get a R_{IV}, separating 2-way interactions from the main effects. See Chapter 3 for more information on a foldover design.

5.4 Identifying Dispersion Factors (Factors Which Shift the Response Variability)*

Thus far, this analysis chapter has only discussed the identification of factors that shift the average response (i.e., location effects). Recent publicity of the Taguchi approach to parameter design for robust products and processes has emphasized the need to also identify which factors contribute to changes in variability of the response. The use of Taguchi's loss function described in Chapters 1 and 6 provides the motivation for experiments to focus on target values instead of engineering tolerances or specifications. The average loss for any product is based upon the product variability and how far the process average

---

*Much of this section is taken from the Schmidt, S. R. and Boudot, J. R. paper presented at the 1989 Rocky Mountain Quality Conference and the 1989 ORSA/TIMS Annual Conference [12].

deviates from a designated target. This being the case, a set of objectives for a designed experiment should be as follows:

(1) Identify factors which shift the average response (location effects) and then select those factor settings which minimize the difference between the average response and the desired target value (see Figure 5.4a).

(2) Identify factors which contribute to changes in the response variability (dispersion effects) and then select the factor settings which minimize response variability (see Figure 5.4b).

(3) Identify factors which shift the average response and the response variability and then select the factor settings which best increase the quality of the product or process (see Figure 5.4c).

(4) Identify factors which have no effect on the response and then set these factors at levels that result in lower costs and/or faster throughput (see Figure 5.4d).

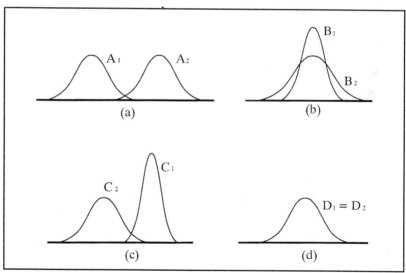

Figure 5.4 Objectives of a Designed Experiment

The bottom line is that the engineer needs to know which factors or knobs he/she can use to adjust the response average, which ones will reduce the variance, and which ones do not make a difference with regard to the average or the variance.

To accomplish the four previously stated objectives, competing strategies have emerged. Since methods for finding location effects have already been discussed, the remainder of this section will concentrate on the following three commonly used methods for identifying dispersion effects: (1) signal-to-noise ratios, (2) s or ln s, and (3) residual analysis for replicated or unreplicated designs.

Taguchi's three signal-to-noise ratios [13] are derived from loss functions associated with the following experimental goals: (1) maximize the response, (2) minimize the response, and (3) adjust the response to a specified target or nominal value. The resulting signal-to-noise ratios are shown below.

$$(1) \quad S/N_L = -10 \ Log_{10} \ \frac{1}{n_r} \sum_1^{n_r} \left(\frac{1}{y_i^2} \right) \quad \text{for maximizing the response.}$$

$$(2) \quad S/N_S = -10 \ Log_{10} \ \frac{1}{n_r} \sum_1^{n_r} \left(y_i^2 \right) \quad \text{for minimizing the response.}$$

$$(3) \quad S/N_N = 10 \ Log_{10} \ \frac{1}{n_r} \left(\frac{S_m - V_e}{V_e} \right) \quad \text{for a target response}^*$$

$$\text{where } S_m = n_r \overline{y}^2, \quad V_e = \frac{\sum_1^{n_r} y_i^2 - \frac{1}{n_r} \left(\sum_1^{n_r} y_i \right)^2}{n_r - 1}$$

and n_r is the number of replicates per experimental run (or row).

$^*S/N_N$ is also proportional to $10 \ Log_{10} \left(\dfrac{\overline{y}^2}{S^2} \right)$

The ln s method uses the natural logarithm of the standard deviation associated with responses for each experimental condition or run. To demonstrate the use of signal-to-noise ratios and ln s, consider the simple example presented in Table 5.5.

	Factors						
Run	A	B	C	y_1	y_2	$\bar{y}_r$	s_r
1	1	1	2	20	30	25	7.071
2	2	1	1	9	11	10	1.414
3	1	2	1	26	24	25	1.414
4	2	2	2	15	5	10	7.071

Table 5.5 Example Used to Demonstrate S/N Ratio and Ln s Analysis

If you look at the row changes in $\bar{y}$ and s and how these changes are correlated with certain design columns, you should be able to see that the A factor adjusts the mean response, the C factor adjusts the variability of the response and the B factor has no effect at all. A graphic display of each factor's effect on y appears in Figure 5.5.

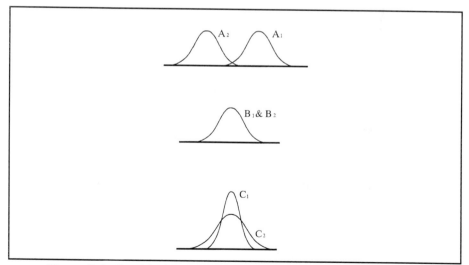

Figure 5.5 Graphical Representation of Factor Effects in Table 5.5

If you find the S/N_L values for each row in Table 5.5 and then calculate marginal averages of S/N_L for each factor, you will obtain the results shown in Table 5.6. Using the analysis results in Table 5.6, the S/N_L strategy identified the correct location effect (A), but not the dispersion effect (C).

	Factor		
	A	B	C
Avg (−1)	27.69	23.65	23.90
Avg (+1)	18.20	22.24	21.98
Δ	−9.49	−1.41	−1.92

Table 5.6 Analysis of Table 5.5 Data Using S/N_L

Had we analyzed the data using S/N_S the marginal averages would appear as shown in Table 5.7.

	Factor		
	A	B	C
Avg (−1)	−28.05	−24.09	−24.00
Avg (+1)	−20.51	−24.47	−24.55
Δ	7.54	−.38	−.55

Table 5.7 Analysis of Table 5.5 Data Using S/N_S

The marginal mean analysis for S/N_S also identified the location effect (A), but not the dispersion effect (C).

Using S/N_N to analyze the data, the marginal averages are shown in Table 5.8. In this case, the marginal mean analysis for S/N_N identified both A and C as important factors, but not until the subsequent use of $10 \log_{10}(S_m)$ in another marginal table can you determine which factor is a location effect and which is a dispersion effect.

	Factor		
	A	B	C
Avg (-1)	17.96	13.98	20.97
Avg (+1)	10.00	13.98	6.99
Δ	-7.96	0	-13.98

Table 5.8 Analysis of Table 5.5 Data Using S/N_N

The marginal mean analysis using ln s is shown in Table 5.9.

	Factor		
	A	B	C
Avg (-1)	1.15	1.15	.35
Avg (+1)	1.15	1.15	1.95
Δ	0	0	1.60

Table 5.9 Analysis of Table 5.5 Data Using Ln s

The analysis with ln s clearly indicates that factor C is a dispersion factor. Thus, given the type of data presented, the use of ln s appears to be an easier and more effective method for locating dispersion effects. Using ln s as the preferred approach for finding dispersion effects can also be verified by referring to the example data in Table 5.2. To analyze the variability effects of factors A through G, you can perform an analysis similar to the regression used for identifying location effects; however, the s^2 column of Table 5.2 is used as the response. As previously demonstrated, you could also use s, ln s, or ln s^2 as the response. The best models will be built on ln s and s because they are less skewed than s^2. Using an orthogonally coded design matrix (i.e., low = -1 and high = $+1$) and s^2 in Table 5.2 as the response, the regression results appear in Table 5.10.

Dependent Variable: s^2 N: 8   Multiple R: 1.000   Squared Multiple R: 1.000

Adjusted Squared Multiple R: 1.000 — Standard Error of Estimate: 0.000

Variable	Coeff	Std Error	Std Coef	Tolerance	T	P(2 Tail)
Constant	6.813	0.000	-	1.0000000	-	-
A	1.063	0.000	-	1.0000000	-	-
B	−0.063	0.000	-	1.0000000	-	-
C	−1.438	0.000	-	1.0000000	-	-
D	−1.063	0.000	-	1.0000000	-	-
E	0.313	0.000	-	1.0000000	-	-
F	1.438	0.000	-	1.0000000	-	-
G	−5.938	0.000	-	1.0000000	-	-

Table 5.10 Using Table 5.2 Data to Model the Variance, s^2

The R^2 value of 1.0 and the 0 values for standard error are due to a perfect overfit (i.e., the number of effects are one less than the number of runs). This model is obviously inappropriate for significance tests, but because of the orthogonal design, the coefficients can be used to estimate effects. The coefficients for each factor are computed from average variance at the factor high ($+$) minus the average variance at the low ($-$), i.e., these coefficient values are equivalent to factor half effects using s^2 as the response $((\bar{s}_+^2 - \bar{s}_-^2) \div 2)$. For factor A, the average run variance for levels ($-$) and ($+$) are 5.75 and 7.875, respectively. Thus, $(\bar{s}_{A+}^2 - \bar{s}_{A-}^2) \div 2$ is identical to the regression coefficient. You can now use a Pareto diagram or normal probability plot to determine which factors (if any) have substantial dispersion effects. Based on the coefficients in Table 5.10, there appears to be only a G dispersion effect. The negative sign indicates that there is lower dispersion at the G = +1 setting.

If you prefer to use ln s as the dependent variable, the results appear in Table 5.11. The same conclusion is made either way; however, the distribution of ln s is less skewed than that of s^2. Since ln s or s analysis are less sensitive to occasional extreme values, they are the preferred responses.

Dependent Variable: ln s	N: 8		Multiple R: 1.000		Squared Multiple R: 1.000		
Adjusted Squared Multiple R: 1.000				Standard Error of Estimate: 0.000			
Variable	Coeff*	Std Error	Std Coef	Tolerance	T	P(2 Tail)	
Constant	0.540	0.000	-	1.0000000	-	-	
A	-0.036	0.000	-	1.0000000	-	-	
B	-0.092	0.000	-	1.0000000	-	-	
C	-0.137	0.000	-	1.0000000	-	-	
D	-0.036	0.000	-	1.0000000	-	-	
E	0.082	0.000	-	1.0000000	-	-	
F	0.137	0.000	-	1.0000000	-	-	
G	-0.713	0.000	-	1.0000000	-	-	

Table 5.11 Using Table 5.2 Data to Model Ln s

In conclusion, the complete analysis of the example data in Table 5.2 indicates that you set factor C at the high (+1) for minimizing the response average and factor G at the high (+1) for minimizing the response variation. The remaining factors are set as dictated by economics, status quo, or convenience. At this point you would want to make several (4 to 20) confirmation runs to confirm the previous results and determine a crude estimate of the process

---

*See Appendix M for Rules of Thumb on significant terms for s and ln s models.

capability index, C_{pk}, for the optimal settings. The actual number of confirmation runs you make is dependent on the cost per run, the time per run and whether or not the run will generate production (or saleable products). The 4 to 20 suggestion is simply a rough Rule of Thumb. You should do enough to feel comfortable with your estimate of the average and the standard deviation. If the confirmation runs did confirm, but did not produce desired results, you might want to see Chapter 7 on how to change the experimental region.

5.5   Variance Reduction Case Study

To avoid the aliasing problem we had in modeling the data in Table 5.2, a resolution IV (or higher) design is recommended. In the example that follows, only 3 factors have been identified for experimentation. Provided the linearity assumption is valid and interactions are possible, a 3-factor (2-level) full factorial is a good design choice. This example is intended to reinforce the previous analysis procedures and to demonstrate one way to handle conflicts in optimizing both the $\hat{y}$ and $\hat{s}$ models.

In a metal casting process for manufacturing jet engine turbine blades, the experimental objective is to determine optimal factor settings to minimize part shrinkage (i.e., difference between mold size and part size) and minimize shrinkage variability. The factors to be tested are mold temperature (A), metal temperature (B), and pour speed (C). The coded test matrix together with the response values and analysis for this experiment are shown in Table 5.12. The

$\ln \dfrac{s_{(1)}^2}{s_{(-1)}^2}$ shown in Table 5.12 provides an alternative method for determining the

significance of terms for the $\hat{s}$ or $\ln \hat{s}$ prediction equation. Montgomery [9] has

shown that $\ln \dfrac{s_{(1)}^2}{s_{(-1)}^2}$ has an approximate standard normal distribution for terms that

do not belong in the variation prediction equation. Therefore, the statistical test for each column in the test matrix becomes:

	Factors and Interactions							Replicated Response Values				
Run	A	B	AB	C	AC	BC	ABC	y_1	y_2	y_3	$\bar{y}$	s
1	–	–	+	–	+	+	–	2.22	2.11	2.14	2.157	0.057
2	–	–	+	+	–	+	+	1.42	1.54	1.05	1.337	0.225
3	–	+	–	–	+	–	+	2.25	2.31	2.21	2.257	0.050
4	–	+	–	+	–	+	–	1.00	1.38	1.19	1.190	1.190
5	+	–	–	–	–	+	+	1.73	1.86	1.79	1.793	0.065
6	+	–	–	+	+	–	–	2.71	2.45	2.46	2.540	0.147
7	+	+	+	–	–	–	–	1.84	1.76	1.70	1.767	0.070
8	+	+	+	+	+	+	+	2.27	2.69	2.71	2.557	0.248
Avg y@ -1	1.735	1.957	1.945	1.993	1.522	1.975	1.913					
Avg y@ +1	2.164	1.942	1.954	1.906	2.377	1.924	1.986					
Δ	0.429	-.015	0.009	-.087	0.855	-.051	0.073					
MSB	1.104	0.001	0.001	0.045	4.386	0.016	0.032					
F	44.36	0.040	0.040	1.810	176.2	0.640	1.280					
Avg s@ -1	0.138	0.131	0.113	0.060	0.145	0.130	0.116					
Avg s@ +1	0.132	0.139	0.157	0.210	0.125	0.140	0.154					
Δ	-.006	0.008	0.044	0.150	-.020	0.010	0.038					
$\ln \frac{s^2_{(1)}}{s^2_{(-1)}}$	-.090	0.120	0.660	2.500	-.290	-.150	0.570					

$\bar{\bar{y}} = 1.950$ $\bar{s} = .135$

$MSE = .02489$

$F(.95, 1, 16) = 4.49$

$Z(.975) = 1.96$

Table 5.12 Turbine Blade Casting Example

(1) $H_0: \sigma^2_{(+1)} = \sigma^2_{(-1)}$ (i.e., the term representing the design column does not go into the variation prediction equation).

$H_1: \sigma^2_{(+1)} \neq \sigma^2_{(-1)}$ (i.e., the term representing the design column does go into the variation prediction equation).

(2) Select α (.1, .05, .01, .001).

(3) Compute $\ln \dfrac{s^2_{(1)}}{s^2_{(-1)}}$.

(4) Using Appendix D, find Z(a) where a $= (1 - \alpha/2)$.

(5) If $\left| \ln \dfrac{s^2_{(1)}}{s^2_{(-1)}} \right| \leq$ Z(a) conclude H_0, the column effect is insignificant.

If $\left| \ln \dfrac{s^2_{(1)}}{s^2_{(-1)}} \right| >$ Z(a) conclude H_1, the column effect is significant.

(See Appendix M for simple Rules of Thumb.)

If you can accept $\alpha = .05$, then the rule for determining if a column has a significant effect can be simplified to the following:

(i) If $\left| \ln \dfrac{s^2_{(1)}}{s^2_{(-1)}} \right| \leq 1.96$ conclude H_0.

(ii) If $\left| \ln \dfrac{s^2_{(1)}}{s^2_{(-1)}} \right| > 1.96$ conclude H_1.

Based on the information provided in Table 5.12, the two equations for modeling the turbine blade casting process are:

(i)* $\hat{s} = .135 + .075C$

(ii) $\hat{y} = 1.950 + .2145A + .4275A \cdot C - .0435C$

Optimizing the standard deviation ($\hat{s}$) first indicates that factor C (pour speed) should be set at the low level (coded level -1). Now, turning to the $\hat{y}$ equation and substituting a value of -1 for C, we obtain

$$\hat{y} = 1.950 + .2145A + .4275(A)(-1) - .0435(-1) \quad \text{or}$$

$$\hat{y} = 1.9935 - .213A.$$

Thus, minimizing $\hat{y}$ requires factor A to be set at the high level or coded value of 1. The resulting predicted percent shrinkage for $A_1B_?C_{-1}$ will be

$$\hat{y} = 1.950 + .2145(1) + .4275(1)(-1) - .0435(-1) = 1.7805.$$

Since factor B was not important for shifting the average or the variability, it should be set at an appropriate level to reduce costs.

A more in-depth examination of the $\hat{s}$ and $\hat{y}$ equations in this example indicates that, by minimizing $\hat{s}$ first, we have obtained a suboptimal value for $\hat{y}$. If we had instead minimized $\hat{y}$ first, we would have set A at the low level and C at the high level. This type of problem can potentially occur when the $\hat{s}$ and $\hat{y}$ equations contain one or more of the same factors. When both $\hat{s}$ and $\hat{y}$ cannot both be optimized, a tradeoff must take place. Usually, the engineer/researcher will weigh the importance of adjusting the average versus minimizing the variance and use the two equations ($\hat{s}$ and $\hat{y}$) to obtain the best settings.

---

*Ln s could have also been used to model variation. See Appendix M for Rules of Thumb on significant terms in s and ln s models.

An alternative is to use the loss function discussed in Chapter 1.   To compute average loss for each run we will use

$$\overline{L} = k\ (\hat{s}^2 + (\hat{y} - T)^2).$$

For our example, the target percent shrinkage (T) is zero; therefore,

$$\overline{L} = k\ (\hat{s}^2 + \hat{y}^2).$$

Using this equation, we can construct a table (see Table 5.13) to aid in selecting appropriate factor-settings. Thus, the best settings to minimize average loss are $A_{-1}\ B_?\ C_1$. The B factor will be set based on cost and/or convenience. Of the different strategies discussed, the best one depends on the objective of the experiment.

Factors				
A	C	$\hat{s}$	$\hat{y}$	$\overline{L}$
−1	−1	.060	2.2065	k(4.8722)
−1	+1	.210	1.2645	k(1.6431)*
+1	−1	.060	1.7805	k(3.1738)
+1	+1	.210	2.5485	k(6.5389)
* minimum average loss				

Table 5.13 Tradeoff Analysis Using Average Loss

5.6   Residual Analysis for Replicated or Unreplicated Designs

A third strategy for detecting dispersion effects allows for a reduction in experimental runs and/or an increase in design resolution through utilization of an unreplicated design. It is presented as a modification of the Box-Meyer method [6]. A modified version of their procedure applied to a 2-level design is presented below. This procedure can be used for replicated or unreplicated designs.

(1) Use a R_{IV} or higher design to avoid confounding interaction and dispersion effects.

(2) Fit the best model to the data (i.e., $\hat{y} = \ldots$).

(3) Compute the residuals (i.e., $y_i - \hat{y}_i$).

(4) For each level of each factor, compute the standard deviation of the residuals (i.e., s_+, s_-).

(5) Compute the difference between the two standard deviations (i.e., $s_+ - s_-$) for each factor.

(6) Rank order the differences found in (5).

(7) Use a Pareto chart, a normal probability plot or $\ln\dfrac{s_{(+)}^2}{s_{(-)}^2}$ to determine the important dispersion effects. (Also see Appendix M for Rules of Thumb.)

(8) For factors with important dispersion effects, determine the settings for least variation. Set location effect factors to optimize the response average.

(9) If a factor setting for minimizing variance differs from the setting that optimizes the location effect, use a trade-off study to determine which setting is most crucial to the product quality.

(10) Set all other factors based on economics, status quo, or convenience.

The example used to illustrate this method contains four factors. The data is generated from the following simulation model:

$$Y_i = 80 + 2A + 6B + 4A{\cdot}B + (Z_i\sigma)$$

(where $\sigma = 3 + 2C$ and $Z_i \sim N(0, 1)$). Therefore, you anticipate location effects for A, B, and A·B, a dispersion effect for C, and no effect for a fourth factor D. The generated data appears in Table 5.14. Notice that the run order has been altered from previously shown 16-run designs. The point to be made here is that starting with – or + is immaterial in generating the orthogonal matrix.

Run	A	B	C	D	y
1	+	+	+	+	94.50
2	+	+	+	−	102.15
3	+	+	−	+	95.91
4	+	+	−	−	95.66
5	+	−	+	+	76.45
6	+	−	+	−	65.85
7	+	−	−	+	72.79
8	+	−	−	−	72.75
9	−	+	+	+	83.75
10	−	+	+	−	76.15
11	−	+	−	+	79.91
12	−	+	−	−	80.15
13	−	−	+	+	85.00
14	−	−	+	−	74.25
15	−	−	−	+	77.38
16	−	−	−	−	78.08

Table 5.14 Design Matrix and Response Values for Residual Analysis Example

Using least squares regression on the data in Table 5.14 will generate the following prediction equation:

$$\hat{y} = 81.921 + 2.591A + 6.606B + 5.941A \cdot B$$

The computer output from the least squares regression procedure, shown in Table 5.15, shows only the A, B, and AB effects. These were the only location effects from the full model (output not shown) that were found to be significant.

Dependent Variable: y	N: 16	Multiple R: .939	Squared Multiple R: .881			
Adjusted Squared Multiple R: .851				Standard Error of Estimate: 3.927		
Variable	Coeff	Std Error	Std Coef	Tolerance	T	P(2 Tail)
Constant	81.921	0.982	0.000	1.0000000	83.438	0.000
A	2.591	0.982	0.262	1.0000000	2.635	0.022
B	6.606	0.982	0.670	1.0000000	6.724	0.000
A·B	5.941	0.982	0.603	1.0000000	6.056	0.000

Analysis of Variance

Source	Sum-of-Squares	df	Mean-Square	F-Ratio	P
Regression	1370.034	3	456.678	29.609	0.000
Residual	185.032	12	15.423		

Table 5.15 Computer Output for Regression Modeling of Table 5.14 Data

Next, using this model, the predicted values ($\hat{y}$) and the residuals ($y - \hat{y}$) are calculated and displayed in Table 5.16.

Run	y	$\hat{y}$	$y - \hat{y}$
1	94.50	97.06	−2.56
2	102.15	97.06	5.09
3	95.91	97.06	−1.15
4	95.66	97.06	−1.4
5	76.45	71.96	4.49
6	65.85	71.96	−6.11
7	72.79	71.96	.83
8	72.75	71.96	.79
9	83.75	79.99	3.76
10	76.15	79.99	−3.84
11	79.91	79.99	−.08
12	80.15	79.99	.16
13	85.00	78.68	6.32
14	74.25	78.68	−4.43
15	77.38	78.68	−1.30
16	78.08	78.68	−.60

Table 5.16 Response Values, Predicted Values, and Residuals for Table 5.14 Data

The standard deviation of the residuals for each factor level appears in Table 5.17.

	Factor			
	A	B	C	D
$s_{(+)}$	3.675	3.039	5.035	3.190
$s_{(-)}$	3.596	4.148	.905	3.530
$\ln \dfrac{s_{(+)}^2}{s_{(-)}^2}$	.043	−.622	3.432	−.203

Table 5.17 Standard Deviation of Residuals for Each Factor in
Table 5.14 Data
(Also see Appendix M for Rules of Thumb.)

Using the decision rules for $\left| \ln \dfrac{s_{(+)}^2}{s_{(-)}^2} \right|$ from Section 5.4, it is evident that the only

significant dispersion effect is associated with factor C, i.e.,

$[Z \left[1 - \dfrac{\alpha}{2} \right] = Z(.975) = 1.96] < 3.432$ implies we are at least 95 % confident

that factor C is a variance-shifting factor.

5.7 Robust Design Analysis

Robust designs are a type of designed experiment which identifies input factor settings that result in products and/or processes robust (or insensitive) to noise factors (sources of uncontrolled variation). Taguchi refers to this type of experimentation as parameter design (for more details see Chapter 6 and the Chapter 8 case study, Applying Experimental Design Techniques to Operating Systems and Software). To provide the reader with an introduction to robust design and how the analysis is conducted, consider the following example from an injection molding process.

Statement of the Problem

An injection molding process engineer has been experiencing problems associated with part shrinkage which occurs after curing. This shrinkage problem has contributed to increased part variability and customer dissatisfaction.

Objective

Identify factors which contribute to variability in part shrinkage and determine the best settings to minimize shrinkage.

Response

Part shrinkage is the amount of measured deviation from the desired part size.

Some of the brainstorming in the form of a cause and effect or fishbone diagram is shown in Figure 5.6. The asterisked factors are those determined to be uncontrollable in production; however, with special care, they can be varied in the experiment. These 3 factors are referred to as noise or uncontrollable factors (sources of uncontrolled response variation) and will eventually appear in a separate design array called the outer array. The phrase "uncontrolled factors" indicates that these factors are difficult or impossible to control during production or operation. Their use in the design matrix implies that with additional effort, we can control them during the experiment. A list of controllable and uncontrollable factors is shown in Table 5.18.

Controllable Factors	Uncontrollable Factors
A: Cycle Time	H: Ambient Temperature
B: Mold Temperature	I: Moisture Content
C: Holding Pressure	J: Percent Regrind
D: Gate Size	
E: Cavity Thickness	
F: Holding Time	
G: Screw Speed	

Table 5.18 List of Controllable and Uncontrollable Factors

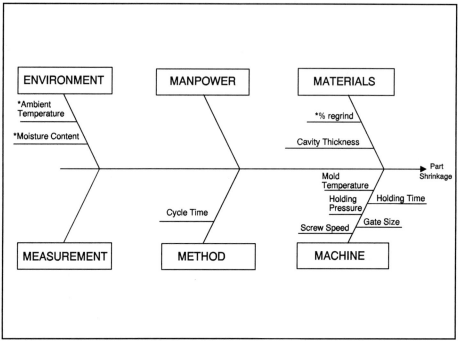

Figure 5.6 Cause and Effect Diagram for Injection Molding Example
(* denotes a noise factor)

Since there are seven controllable factors to be tested, the experimenter has chosen an L_8 screening matrix shown in Table 5.19. Combinations of the uncontrollable factors will be used to generate replicates of the L_8 inner array matrix. Notice that the outer array matrix for the uncontrolled factors is also orthogonal. This feature is not a requirement; rather, the combinations for the uncontrolled factors should represent the most extreme conditions of the noise variables. The design strategy as shown in Table 5.19 is to find a set of controllable factor conditions which will minimize the response variation due to uncontrolled factors.

							J	–	+	+	–		
		Factors					I	–	+	–	+		
							H	–	–	+	+		
Run	A	B	C	D	E	F	G	y_1	y_2	y_3	y_4	$\bar{y}_r$	s_r
1	–	–	–	–	–	–	–	2.6	2.7	2.8	2.8	2.725	.096
2	–	–	–	+	+	+	+	0.8	3.0	0.8	3.2	1.950	1.330
3	–	+	+	–	–	+	+	3.6	1.0	3.3	0.9	2.200	1.449
4	–	+	+	+	+	–	–	2.5	2.4	2.5	2.3	2.425	.096
5	+	–	+	–	+	–	+	3.5	3.6	3.5	3.5	3.525	.050
6	+	–	+	+	–	+	–	2.6	4.7	3.6	1.5	3.100	1.369
7	+	+	–	–	+	+	–	4.5	2.4	2.7	5.1	3.675	1.328
8	+	+	–	+	–	–	+	2.4	2.5	2.3	2.4	2.400	.082

Table 5.19 Robust Design Matrix with Response Values

The marginal mean analyses of the signal ($\bar{y}$) and the noise (ln s) appear in Tables 5.20 and 5.21, respectively.

Analysis of $\bar{y}$:

	Factor						
	A	B	C	D	E	F	G
Avg –1	2.325	2.825	2.687	3.031	2.606	2.769	2.981
Avg +1	3.175	2.675	2.812	2.469	2.894	2.731	2.519
Δ	.850	–.150	.125	–.562	.288	–.038	–.462

Table 5.20 Marginal Mean Analysis for Table 5.19 Data

Analysis of ln s:

	A	B	C	D	E	F	G
				Factor			
ln s Avg –	−1.01	−1.19	−1.07	−1.17	−1.04	−2.55	−1.02
ln s Avg +	−1.23	−1.05	−1.16	−1.06	−1.19	.313	−1.21
Δ	−.22	.14	−.09	.11	−.15	2.863	−.19

Table 5.21 Ln s Analysis for Table 5.19 Data

The Δ values in Tables 5.20 and 5.21 clearly indicate that factors A, D, and G shift the average, (i.e., these are the location effects), and that factor F shifts the variability, (i.e., it is a dispersion factor). Therefore, to minimize shrinkage and the variability of shrinkage across the uncontrolled factors, the best settings are A_ B_? C_? D_+ E_? F_ G_+. Factors with a question mark are set at levels based on economics, status quo, and/or convenience. If our assumptions in using this design are correct, these settings will minimize shrinkage and make the process robust to uncontrolled variations in percent regrind, ambient temperature, and moisture content. As always, confirmation runs should be performed.

As a final note on the comparison of dispersion factor identification methods, this set of data was originally analyzed using signal-to-noise "smaller is better," S/N_S. The results shown in Table 5.22 again failed to identify factor F as an important factor. Had we used signal-to-noise "nominal is best," S/N_N, the results shown in Table 5.23 do point out the importance of factor F; however, more cumbersome analysis is required to determine if F is a location or dispersion factor. In conclusion, the recommended procedure of analyzing the signal ($\bar{y}$) and the noise (ln s) or (s) separately is simple and straightforward.

Analysis Using S/N$_S$:

	Factor						
	A	**B**	**C**	**D**	**E**	**F**	**G**
Avg –	–7.90	–9.29	–8.78	–9.90	–8.70	–8.74	–9.60
Avg +	–10.20	–8.77	–9.28	–8.21	–9.36	–9.33	–8.43
Δ	–2.3	.52	–.50	1.69	–.66	–.59	1.17

Table 5.22 S/N$_S$ Analysis for Table 5.19 Data

Analysis Using S/N$_N$:

	Factor						
	A	**B**	**C**	**D**	**E**	**F**	**G**
Avg –	16.03	19.12	17.65	19.63	17.29	30.87	18.28
Avg +	20.57	17.48	18.94	16.96	19.30	5.72	18.32
Δ	4.54	–1.64	1.29	–2.67	2.01	–25.15	.04

Table 5.23 S/N$_N$ Analysis for Table 5.19 Data

5.8   Variance Reduction Using Multi-Variate Charts*

A semiconductor company had been experiencing problems with lot-to-lot, wafer-to-wafer, and within wafer variation of thickness values from a coating process. An engineering team narrowed the potential contributing factors down to temperature and time. Since only two factors are in question, the team decided to conduct a 2-factor (2-level) full factorial experiment with time (A) settings of 20 and 30 minutes, and temperature (B) settings of 50°C and 70°C. The design matrix, response values, and prediction equations are shown in Table 5.24. To

---

*The multivariate chart approach is credited to Dorian Shainin [2].

minimize wafer-to-wafer and lot-to-lot variation we would minimize the following equation:

$$\hat{s}_{\bar{y}} = 1.648 - .6675B$$

Setting factor B at $(+1)$, i.e., setting temperature at $70°C$, will minimize wafer-to-wafer and lot-to-lot variation. Fortunately, setting factor B at $(+1)$ also minimizes the equation for within wafer variation which is measured as within wafer range, R.

$$\hat{R} = 6.062 - 2.73B.$$

To achieve the desired target of 200 microinches, we would perform the following manipulations.

(1) $\hat{y} = 200.9 - 2.46A + 1.04AB - .04B$ (set $\hat{y}=200$ and $B=+1$)

(2) $200 = 200.9 - 2.46A + 1.04A(1) - .04(1)$

(3) $1.42A = 200.9 - 200 - .04$ (solve for A)

(4) $A = .86/1.42 = .606$

(5) Actual time $= \bar{f} + \dfrac{d_i}{2}x = 25 + \dfrac{(30-20)}{2}(.606) = 28.03$ minutes

The optimal settings of time $= 28.03$ minutes and temperature $= 70°C$ are closely related to experimental run number 4. Notice how the multivariate charts in Figure 5.7 indicate that a high time and high temperature setting will closely approximate the desired optimal. Having completed the experiment and found the best conditions, you would next confirm these findings and monitor production using control charting techniques and/or continued multivariate charting [2].

Response Values:
Wafer Average Thickness and (Wafer Range) in micro inches

Run	A	B	AB	LOT 1				LOT 2				LOT 3				Ȳ	Sȳ	R̄
				W1	W2	W3	W4	W1	W2	W3	W4	W1	W2	W3	W4			
1	−	−	+	204	201	206	204	207	204	205	206	203	205	201	207	204.4	2.021	(8.667)
				(10)	(6)	(7)	(8)	(8)	(10)	(6)	(9)	(9)	(11)	(10)	(10)			
2	−	+	−	201	203	202	202	203	201	203	203	204	202	202	201	202.2	0.965	(3.667)
				(3)	(6)	(3)	(3)	(5)	(3)	(2)	(5)	(4)	(5)	(2)	(3)			
3	+	−	−	194	200	197	199	201	194	197	196	197	199	194	201	197.4	2.610	(8.917)
				(8)	(10)	(7)	(9)	(9)	(10)	(7)	(10)	(9)	(11)	(9)	(8)			
4	+	+	+	199	200	198	199	199	201	200	198	200	199	201	199	199.4	0.996	(3.000)
				(2)	(3)	(4)	(5)	(2)	(3)	(2)	(4)	(2)	(3)	(6)	(2)			
Avg Y−	203.3	200.9	201.9															
Avg Y+	198.4	200.8	199.8															
Δ/2	−2.46	−.04	1.04															
F_e	90.52	0.03	16.25															
Avg R−	6.167	8.792	5.833															
Avg R+	5.958	3.333	6.292															
Δ/2	−.104	−2.73	−.229															

Wafer-to-Wafer and Lot-to-Lot Models:

$$\hat{y} = 200.9 - 2.46A + 1.04AB - 0.04B$$
$$\hat{S}_{\bar{y}} = 1.648 - 0.6675B .04B$$

Within Wafer Variance Model:

$$\hat{R} = 6.062 - 2.73B$$

Table 5.24 Design Matrix and Analysis Summary for
Multivariate Chart Example

Figure 5.7 Multivariate Charts [2] for Each Run in Table 5.24
(Vertical bars represent within wafer range. Open dots are the wafer
averages. The three sections represent the three lots.)

5.9   Three-Level Modeling Example

Assume you have 4 factors (A, B, C, D) to be tested and your objective is to build a non-linear model for the average response ($\hat{y}$) and the standard deviation ($\hat{s}$).   Based on the information presented in Chapter 3, you could use a Box-Behnken design of 27 runs or a central composite design which will also have 27 runs, assuming you choose 3 center points. Both of these designs will produce a similar non-linear model.   To choose which design is the best one for your needs, you should review the advantages and disadvantages of each type in Chapter 3. Suppose you elect to conduct the central composite design (CCD) shown in Table 5.25.

Notice that the α value for the design in Table 5.25 has been set equal to 1. A CCD with $\alpha = 1$ is known as a central composite face (CCF) design and it will not be perfectly orthogonal or rotatable — two desired statistical properties of a design matrix (see Chapter 3).   The degraded orthogonality can be seen in the Tolerance column of Table 5.26. If the Tolerance values get too low, there will be excessive amounts of partial confounding in the design. Most experimenters prefer to see a minimum Tolerance of .5. In this case the Tolerance is slightly less than .5, indicating potential partial confounding problems associated with the quadratic terms.   If the Tolerance values become unacceptable in a central composite face design, it will be necessary to use the recommended α value from Chapter 3. Let us assume that we have discussed the degraded orthogonality of our design and have decided that it is still the design of choice. The design matrix and data are displayed in Table 5.25.

	A	B	C	D	y_1	y_2	$\bar{y}$	$\hat{s}$
1	1	1	1	1	1.904	1.711	1.807	0.136
2	1	1	1	−1	1.256	1.310	1.283	0.038
3	1	1	−1	1	1.146	1.189	1.167	0.030
4	1	1	−1	−1	0.702	0.673	0.687	0.021
5	1	−1	1	1	1.658	1.723	1.690	0.046
6	1	−1	1	−1	0.892	0.885	0.888	0.005
7	1	−1	−1	1	0.043	0.078	0.060	0.025
8	1	−1	−1	−1	0.847	0.799	0.823	0.034
9	−1	1	1	1	0.409	0.255	0.332	0.109
10	−1	1	1	−1	0.321	0.211	0.266	0.078
11	−1	1	−1	1	0.324	0.529	0.426	0.145
12	−1	1	−1	−1	0.179	0.312	0.245	0.094
13	−1	−1	1	1	0.453	0.306	0.379	0.104
14	−1	−1	1	−1	0.255	0.399	0.327	0.102
15	−1	−1	−1	1	0.257	0.118	0.187	0.098
16	−1	−1	−1	−1	0.191	0.266	0.228	0.053
17	0	0	0	0	0.328	0.488	0.408	0.113
18	0	0	0	0	0.434	0.512	0.473	0.055
19	0	0	0	0	0.341	0.438	0.389	0.069
20	−1	0	0	0	0.247	0.311	0.279	0.045
21	1	0	0	0	0.986	0.943	0.964	0.030
22	0	−1	0	0	0.436	0.428	0.432	0.006
23	0	1	0	0	0.412	0.465	0.438	0.037
24	0	0	−1	0	0.391	0.385	0.388	0.004
25	0	0	1	0	0.526	0.611	0.568	0.060
26	0	0	0	−1	0.362	0.426	0.394	0.045
27	0	0	0	1	0.589	0.433	0.511	0.110

Table 5.25 Data from a Central Composite (Face) Design

The data from Table 5.25 has been analyzed using a standard statistical package similar to the student version in the back of the text. The results of the ŝ model are shown in Table 5.26. Using the Rule of Thumb on P(2 Tail), the only significant terms in the ŝ model are A, D, and D^2. The 3-dimensional response surface (Figure 5.8) and contour plots (Figure 5.9) indicate that a minimum ŝ of .024 occurs for A = +1.000 and D = -.333 coded values.

Regression Output for Std Dev, Rsq = 0.677022				F = 1.796728
Effect	Coeff	t	Tolerance	P(2 Tail)
Constant	0.05182	4.11818		0.00143
A	-0.0257	-3.1922	1.00000	0.00774
B	0.01202	1.49361	1.00000	0.16110
C	0.00967	1.20074	1.00000	0.25302
D	0.01858	2.30874	1.00000	0.03957
A^2	-0.0004	-0.0196	0.42857	0.98462
B^2	-0.0167	-0.7834	0.42857	0.44855
C^2	-0.0061	-0.2853	0.42857	0.78025
D^2	0.03953	1.85655	0.42857	0.08808
AB	0.00296	0.34687	1.00000	0.73470
AC	0.00712	0.83352	1.00000	0.42083
AD	0.00066	0.07766	1.00000	0.93938
BC	0.00155	0.18120	1.00000	0.85923
BD	0.00694	0.81281	1.00000	0.43216
CD	0.00473	0.55395	1.00000	0.58978

Table 5.26 ŝ Model Analysis

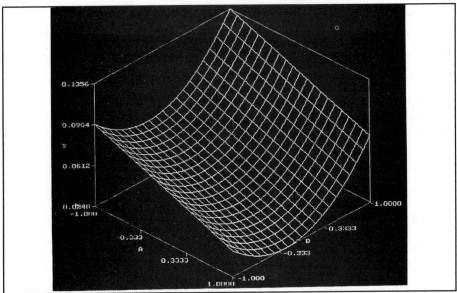

Figure 5.8 Response Surface Plot from the ŝ Model

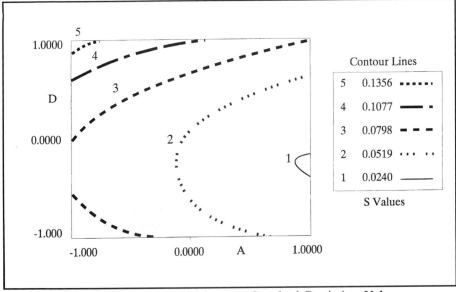

Figure 5.9 Contours of Constant Standard Deviation Values

The next step is to model $\hat{y}$, the objective of which is a targeted response greater than .3. The analysis in Table 5.27 indicates that several terms have P(2 Tail) values less than .100 and thus should be in the model. The 3 dimensional response surface plot, interaction plot and contour plot for A and D are shown in Figures 5.10, 5.11, and 5.12, where factors B and C have been held constant at -1. Since the objective for $\hat{y}$ is greater than or equal to .3, the settings of A and D that minimized $\hat{s}$ will also satisfy $\hat{y}$.

The combined analyses of $\hat{y}$ and $\hat{s}$ indicate that the coded settings of A = 1.000, B = -1.000, C = -1.000 and D = $-.333$ will produce $\hat{y}$ = .493 and $\hat{s}$ = 0.024.

The lower spec of .3 for the response will then result in an estimated C_{pk} of $\hat{C}_{pk} = \dfrac{\min(US - \hat{y}, \hat{y} - LS)}{3\hat{s}} = \dfrac{.493 - .3}{3(.024)} = 2.68$ which is very satisfactory.

Regression Output for Yi, Coded		Rsq = 0.881912		F = 20.80447
Effect	Coeff	t	Tolerance	P(2 Tail)
Constant	0.43184	9.34283		0.00000
A	0.37228	12.5925	1.00000	0.00000
B	0.09094	3.07626	1.00000	0.00454
C	0.18489	6.25399	1.00000	0.00000
D	0.07886	2.66753	1.00000	0.01237
A^2	0.18574	2.37468	0.42857	0.02440
B^2	-0.0007	-0.0097	0.42857	0.99232
C^2	0.04224	0.54004	0.42857	0.59329
D^2	0.01649	0.21083	0.42857	0.83449
AB	0.08347	2.66191	1.00000	0.01254
AC	0.16966	5.41053	1.00000	0.00001
AD	0.04909	1.56566	1.00000	0.12828
BC	-0.0515	-1.6433	1.00000	0.11110
BD	0.07503	2.39283	1.00000	0.02342
CD	0.09922	3.16420	1.00000	0.00364

Table 5.27 $\hat{y}$ Model Analysis

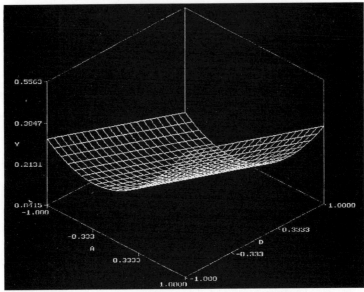

Figure 5.10 Response Surface Plot from the ŷ model (B = −1, C = −1)

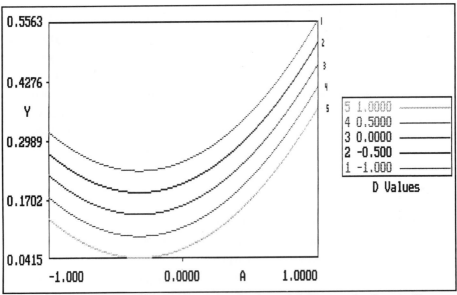

Figure 5.11 Interaction Plot from the ŷ model (B = −1, C = −1)

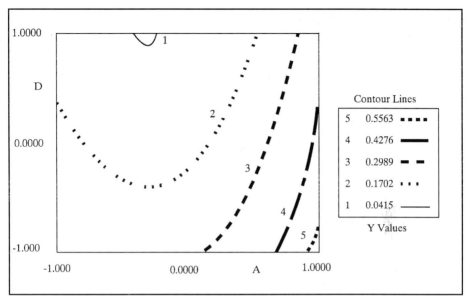

Figure 5.12 Contour Plot from the y model (B = -1, C = -1)

The reader is also referred to Chapter 8, "*Identifying a Plasma Etch Process Window*," for an industrial example of a 3-level design.

5.10   Three-Level Design Analysis and Comparison

This case study will use a simulation process to present 2nd order modeling from 3-level designs. "Plate 2" is a computer program that simulates a process used to train engineers in Design of Experiments. In this problem, students are asked to optimize an "auto bumper" plating process using thickness as a quantitative response with time, temperature, percent nickel, ph, and percent phosphorous as the independent variables. From background information, it is believed that all of the above independent variables are important in impacting thickness. Additionally, there is concern about strong interactive, as well as non-linear, factors in the process. Crudely following the sequence of steps suggested in Chapter 1, we will address this problem.

Statement of the Problem

We are chartered with characterizing a new process (the plating process). This will be accomplished by developing a mathematical model for predicting plating thickness which will be used for optimizing various products requiring different plating thicknesses. An empirical model will provide the information required for adjusting the settings of the important independent variables in order to obtain the necessary thickness for the various products.

Objectives

(1) Determine the effect of time, temperature, percent nickel, ph, and phosphorous on plating thickness.

(2) Develop a "good" mathematical model which characterizes the plating process.

QUALITY CHARACTERISTICS

Response	Type	Anticipated Range	How to measure?
Thickness	Quant.	0 - 2000	profilometer

FACTORS

Factor	Type	Control or Noise	Range	Interactions?
Time	Quant.	Control	4 - 12	possible
Temperature	Quant.	Control	16 - 32	possible
Nickel	Quant.	Control	10 - 18	possible
ph	Quant.	Control	2 - 10	possible
Phosphorous	Quant.	Control	1.40 - 3.88	possible

Appropriate Experiment Design

To resolve the stated problem, different strategies will be demonstrated as described in the following cases. The data has been analyzed using RS software from BBN Software Products, Inc.

CASE I:

A 2-level fractional factorial of resolution V is used to estimate all main effects and 2-factor interactions. The design matrix with the response values is shown in Table 5.28. The computer output for the mathematical model is shown in Table 5.29.

The lack of standard error estimates, t-values, and significance estimates is a result of using all the degrees of freedom for effects and, thus, none remain for estimating the experimental error. As previously demonstrated, a Pareto diagram for the effects can be used to separate the important (few) effects from the trivial (many) effects. The resulting Pareto diagram for the half-effects appears in Figure 5.13.

Run #	Time	Temp	Nickel	ph	Phos	Thickness
1	4.00	16.00	10.00	2.00	3.88	113
2	12.00	16.00	10.00	2.00	1.40	756
3	4.00	32.00	10.00	2.00	1.40	78
4	12.00	32.00	10.00	2.00	3.88	686
5	4.00	16.00	18.00	2.00	1.40	87
6	12.00	16.00	18.00	2.00	3.88	788
7	4.00	32.00	18.00	2.00	3.88	115
8	12.00	32.00	18.00	2.00	1.40	696
9	4.00	16.00	10.00	10.00	1.40	99
10	12.00	16.00	10.00	10.00	3.88	739
11	4.00	32.00	10.00	10.00	3.88	10
12	12.00	32.00	10.00	10.00	1.40	712
13	4.00	16.00	18.00	10.00	3.88	159
14	12.00	16.00	18.00	10.00	1.40	776
15	4.00	32.00	18.00	10.00	1.40	162
16	12.00	32.00	18.00	10.00	3.88	759

Table 5.28 Design Matrix (Actual Values) and Response (Thickness)
for Plating Example, Factorial Portion of the CCF

	Term	Coefficient	Std Error	T-Value	Significance
1	1	420.937500	-	-	-
2	~T	318.062500	-	-	-
3	~TE	−18.687500	-	-	-
4	~N	21.812500	-	-	-
5	~P	6.062500	-	-	-
6	~PHO	0.187500	-	-	-
7	~T·TE	−7.062500	-	-	-
8	~T·N	−6.062500	-	-	-
9	~T·P	1.437500	-	-	-
10	~T·PHO	3.812500	-	-	-
11	~TE·N	8.937500	-	-	-
12	~TE·P	2.437500	-	-	-
13	~TE·PHO	−9.937500	-	-	-
14	~N·P	15.187500	-	-	-
15	~N·PHO	12.312500	-	-	-
16	~P·PHO	−10.437500	-	-	-

No. cases = 16   Resid. df = 0   Cond. No. = 1

~ indicates factors are transformed

Table 5.29 *RS/Discover* Computer Output for Coded Data in Table 5.28

Based on the Pareto diagram of half-effects, it is obvious that time is an important factor. Nickel and temperature are ranked second and third, followed by everything else. At the outset, we discussed the possibility of non-linear effects. We will now tackle the question of non-linearity through the use of replicated centerpoints. This will be accomplished by complementing our 16 original data points with 4 centerpoints. For a non-simulated scenario, sequential application of the factorial and then the center points can be problematic, especially if there is a possibility that the process has shifted significantly. Therefore, the centerpoints are sometimes added during the 2-level experimental phase. The centerpoint data is as follows:

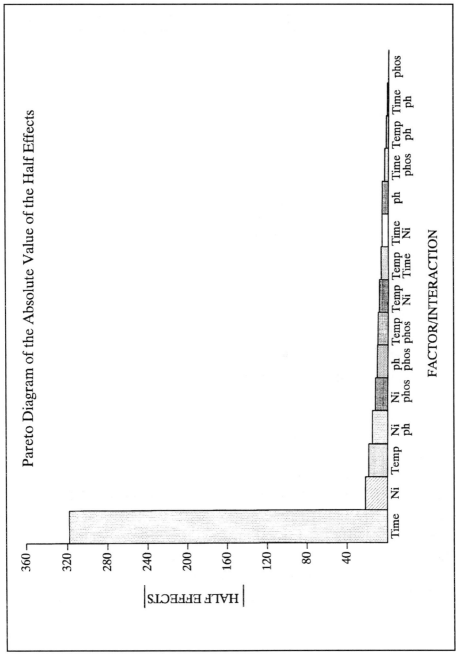

Figure 5.13 Pareto Diagram of Half-Effects Using Table 5.28 Data

Run #	Time	Temperature	Nickel	ph	Phos	Thickness
1	8	24	14	6	2.64	351
2	8	24	14	6	2.64	373
3	8	24	14	6	2.64	353
4	8	24	14	6	2.64	321

Analysis of the combined 20 data points is shown in Table 5.30.

Summary of ANOVA

Source	df	SS	MS	F-Ratio	Significance*
1 Total (Corr.)	19	1662583			
2 Regression	15	1644873	109658	24.77	0.0035
3 Linear	5	1632409	326482	73.74	0.0005
4 Non-linear	10	12464	1246	0.28	0.9527
5 Residual	4	17710	4427		
6 Lack of fit	1	16331	16331	35.53	0.0094
7 Pure error	3	1379	460		

$$R^2 = 0.9893 \qquad \text{Adj } R^2 = 0.9494$$

Model obeys hierarchy. The sum of squares for linear terms is computed assuming nonlinear terms are removed.

$F(1, 3)$ as large as 35.53 is a rare event $\Rightarrow$ unlikely that model is correct.

Table 5.30 *RS/Discover* Computer Output for Table 5.28 Data
with Centerpoints

Since the "lack of fit" is significant, we need to determine which of our factors are non-linear and develop a mathematical model. As discussed in Chapter 3, an efficient way to accomplish a second-order model is through the use of a Box-Wilson (central composite) design. Using $\alpha = (n_F)^{1/4}$ to determine the coded value for the axial points, we obtain $\alpha = 2$ for $n_F = 16$. The lows and highs from the fractional factorial portion of the experiment are, however, already

---

*Significance for this software is the same as P(2 Tail) discussed earlier.

at the extreme conditions. Therefore, $\alpha = 1.0$ is used to construct a Central Composite Face (CCF) design.

Since ph and phosphorous were screened out during the factorial portion of the experiment, we will not run α-points for these factors. Based on the above considerations, the resultant supplemental runs with associated thickness data are:

Run #	Time	Temp	Nickel	Thickness
1	12	24	14	736.1
2	4	24	14	96.0
3	8	32	14	328.9
4	8	16	14	303.5
5	8	24	18	358.2
6	8	24	10	347.7

By combining the 16 runs from the factorial portion of the design with the four centerpoints and the 6 runs from the axial portion of our design, we obtain the following regression table.

	Term	Coefficient	Std Error	T-Value	Significance
1.	1	342.7	11.747		
2.	T	318.3	7.092		
3.	TE	−15.2	7.092		
4.	N	19.9	7.092		
5.	T·TE	−7.1	7.522	−0.94	0.36
6.	T·N	−6.1	7.522	−0.81	0.43
7.	TE·N	8.9	7.522	1.19	0.25
8.	T^2	80.1	18.047	4.44	0.00
9.	TE^2	−19.7	18.047	−1.09	0.29
10.	N^2	17.0	18.047	0.94	0.36
	No. cases = 26		$R^2 = 0.9923$		RMS Error = 30.1
	Resid. df = 16		Adj R^2 = 0.9880		Cond. No. = 5.68

Table 5.31 *RS/Discover* Computer Output for Table 5.28 Data
with Centerpoints and Axial Points

Eliminating the unimportant terms leaves us with the final regression table shown below.   The RS software uses backward elimination [10] to remove unimportant terms.

	Term	Coefficient	Std Error	T-Value	Significance
1	1	342.0	10.498		
2	T	318.3	6.998		
3	TE	−15.2	6.998	−2.17	0.04
4	N	19.9	6.998	2.85	0.00
5	T^2	78.4	12.618	6.21	0.00
	No. cases = 26		R^2 = 0.9902	RMS Error = 29.69	
	Resid. df = 21		Adj R^2 = 0.9883	Cond. No. = 3.30	

Table 5.32 *RS/Discover* Final Model for Table 5.28 Data (Using a CCF)

The prediction equation generated from the orthogonally coded CCF design is:

$$\text{predicted thickness} = 342.0 + 318.3(\text{time}) - 15.2(\text{temp}) + 19.9(\text{nickel}) + 78.4(\text{time})^2$$

Step XI from Chapter 1 includes making some confirmation runs to verify predicted results. Utilizing *RS/Discover* software, a 95% confidence interval for the mean response when time = 4, temp = 16, and Nickel = 14 is calculated to be [72.9, 161.7]. This interval provides us something to judge our confirmation runs against. Ten confirmation runs at the above settings resulted in a mean value of 115.5 and a standard deviation of 20.4. Since the mean of our confirmation runs falls within our confidence interval, the model appears to be adequate.

Let's stop briefly and contrast our derived mathematical model with the equation in our simulation package.   The actual uncoded equation in our simulation package is:

$$\text{Thickness} = 5(\text{time})^2 - 2.3(\text{temp}) + 4.0(\text{nickel})$$
$$\text{with a standard deviation} = 40$$

When temperature is 16, time is 4, and nickel is 14, the actual model would generate an average value of 99.2 with a standard deviation of 40. The graphic below contrasts the true population distribution with the model (or empirical) distribution for the 10 confirmation runs at temp = 16, time = 4, and nickel = 14.

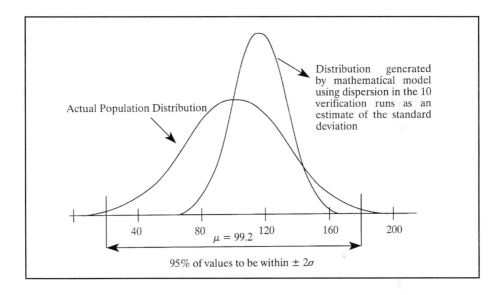

As we can see from the above graphic, our model is off by approximately 20 units on the average and we are nearly a factor of 2 off on the standard deviation. How can we improve on the mathematical model we determine? Large samples are required. A general rule of thumb is: For every term (excluding the constant) in our final prediction equation, we need roughly 10 or more samples. Using this guideline, with the 4 non-constant terms in our model, 40 or more samples in the original experiment would be desired to obtain a "good" model. We used only 26. To estimate standard deviation, the confirmation runs should have 15 or more data points. We used only 10. It is recognized, however, that oftentimes sample size is restricted due to time, money, and resource constraints. In these cases we must simply recognize the limitations of our results.

CASE II:

Instead of using a Central Composite Design to provide us with an empirical model, we could have used other approaches as well. For example, after the screening design was completed, we could have made use of a D-optimal design, a Box-Behnken design, or a Taguchi design.

Consider a computer generated D-optimal design capable of estimating all linear, quadratic, and 2-way interaction terms for the three most important factors. The minimum number of runs required in this case is 10. The D-optimal design generated by *RS/Discover* is 12 runs, but we ran replicates (3) on the last run. Our 14-run design along with the applicable thickness data is shown in Table 5.33, and the results are shown in Table 5.34.

Run	Time	Temp	Nickel	Thickness
1	12	32	10	717
2	4	16	10	136
3	12	16	18	787
4	4	32	18	78
5	4	16	18	87
6	4	32	10	80
7	12	16	10	760
8	8	24	10	282
9	8	32	14	318
10	4	24	14	96
11	12	32	18	682
12	12	24	14	747
13	12	24	14	764
14	12	24	14	742

Table 5.33 D-optimal Design (Actual Values) and
Response (Thickness) for Plating Example

From our summary ANOVA table, it appears we have an acceptable fit with the model, but the "T-values" for numerous terms in our coefficient table indicate we can get rid of some terms with little loss in R^2. Our final coefficient table using backward elimination is shown in Table 5.35.

Term	Coefficient	Std Error	T-Value	Significance
1. 1	314.786500	17.517042		
2. ~T	322.500000	6.491690		
3. ~TE	−26.625000	7.612182		
4. ~N	−7.375000	7.612182		
5. ~T^2	111.312500	17.517042	6.35	0.0031
6. ~T·TE	−10.375000	7.612182	−1.36	0.2446
7. ~T·N	5.375000	7.612182	0.71	0.5191
8. ~TE^2	29.937500	17.517042	1.71	0.1626
9. ~TE·N	−1.875000	7.612182	−0.25	0.8176
10. ~N^2	−40.062500	17.517042	−2.29	0.0841

No. cases = 14	R^2 = 0.9986	RMS Error = 21.53
Resid. df = 4	Adj R^2 = 0.9953	Cond. No. = 7.392

~indicates factors are transformed

Summary of ANOVA

Source	df	SS	MS	F-Ratio	Significance
1. Total (Corr.)	13	1283423			
2. Regression	9	1281569	142397	307.20	0.0000
3. Linear	3	1259538	419846	905.70	0.0000
4. Non-linear	6	22030	3672	7.92	0.0325
5. Residual	4	1854	464		
6. Lack of fit	2	1588	794	5.97	0.1435
7. Pure error	2	266	133		

$$R^2 = 0.9986$$

$$\text{Adj } R^2 = 0.9953$$

Model obeys hierarchy. The sum of squares for linear terms is computed assuming nonlinear terms are first removed. F(2, 2) as large as 5.971 is not a rare event ⇒ no evidence of lack of fit.

Table 5.34 *RS/Discover* Computer Output for D-optimal Design for the Plating Example

Term	Coefficient	Std Error	T-Value	Significance
1 1	311.470588	17.084202		
2 ~T	323.657143	6.874237		
3 ~TE	−22.941176	8.053570	−2.85	0.0173
4 ~T²	107.586555	18.415349	5.84	0.0002
No. cases = 14		R² = 0.9957	RMS Error = 23.48	
Resid. df = 10		Adj R² = 0.9944	Cond. No. = 5.328	
~ indicates factors are transformed				

Table 5.35 Final Model for D-optimal Design

The prediction equation from the orthogonally coded D-optimal design is:

predicted thickness = 311.47 + 323.66(time) − 22.94(temp) + 107.59(time)^2

Had we chosen a Box-Behnken design for 3 factors it would appear as shown in Table 5.36.

Run	Time	Temp	Nickel	Thickness
1	4	16	14	129
2	12	16	14	792
3	4	32	14	94
4	12	32	14	702
5	4	24	10	37
6	12	24	10	738
7	4	24	18	133
8	12	24	18	714
9	8	16	10	339
10	8	32	10	302
11	8	16	18	319
12	8	32	18	344
13	8	24	14	364
14	8	24	14	342
15	8	24	14	404

Table 5.36 Box-Behnken Design (Actual Values) and Response (Thickness) for Plating Example

Fitting the coded data with a regression model provided the statistics shown in Table 5.37.

	Term	Coefficient	Std Error	T-Value	Significance
1.	1	370.000000	16.123999		
2.	~T	319.125000	9.873892		
3.	~TE	−17.125000	9.873892		
4.	~N	11.750000	9.873892		
5.	~T^2	69.375000	14.533976	4.77	0.0050
6.	~T·TE	−13.750000	13.963792	−0.98	0.3700
7.	~T·N	−30.000000	13.963792	−2.15	0.0844
8.	~TE^2	−10.125000	14.533976	−0.70	0.5171
9.	~TE·N				
10.	~N^2				

No. cases = 15 R^2 = 0.9954   RMS Error = 27.93

Resid df = 5   Adj R^2 = 0.9872   Cond. No. = 4.522

~indicates factors are transformed

Summary of ANOVA

Source	df	SS	MS	F-Ratio	Significance
1. Total (Corr.)	14	851473.7			
2. Regression	9	847574.0	94174.9	120.70	0.0000
3. Linear	3	818176.8	272725.6	349.70	0.0000
4. Non-linear	6	29397.2	4899.5	6.28	0.0311
5. Residual	5	3899.8	780.0		
6. Lack of fit	3	1923.7	641.2	0.65	0.6535
7. Pure error	2	1976.0	988.0		

R^2 = 0.9954   Adj R^2 = 0.9872

Model obeys hierarchy. The sum of squares for linear terms is computed assuming nonlinear terms are first removed. $F_{(3, 2)}$ as large as 0.649 is not a rare event ⇒ no evidence of lack of fit.

Table 5.37 *RS/Discover* Computer Output Using Box-Behnken Design for Plating Example

From the summary ANOVA, it appears the model is adequate. From the coefficient table, non-important terms can now be consolidated. The result is the coefficients table shown in Table 5.38.

	Term	Coefficient	Std Error	T-Value	Significance
1	1	363.769231	14.624921		
2	~T	319.125000	10.763650		
3	~N	11.750000	10.763650		
4	~T²	70.153846	15.796720	4.44	0.0016
5	~T·N	−30.000000	15.222100	−1.97	0.0802
6	~N²	−33.096154	15.796720	−2.10	0.0656

No. cases = 15 R^2 = 0.9902   RMS Error = 30.44

Resid. df = 9   Adj R^2 = 0.9848   Cond. No. = 3.513

~ indicates factors are transformed

Table 5.38 Final Model for Box-Behnken Design

Thus, the prediction equation generated from the orthogonally coded Box-Behnken design is:

predicted thickness = 363.8 + 319.1(time) + 11.75(nickel) + 70.15(time)² − 30(time × nickel) − 33.10(nickel)²

To model the desired linear, quadratic, and 2-factor interactions, the appropriate 3-level Taguchi design is the L_{27}. In this instance, it is equivalent to the 3^3 full factorial.

The design and the generated thickness data are shown in Table 5.39. After fitting the coded data with the appropriate model and eliminating unimportant terms, our final coefficients table is shown in Table 5.40.

Run #	Time	Temp	Nickel	Thickness
1	4.00	16.00	10.00	113
2	4.00	16.00	14.00	152
3	4.00	16.00	18.00	147
4	4.00	24.00	10.00	65
5	4.00	24.00	14.00	53
6	4.00	24.00	18.00	130
7	4.00	32.00	10.00	83
8	4.00	32.00	14.00	40
9	4.00	32.00	18.00	94
10	8.00	16.00	10.00	339
11	8.00	16.00	14.00	303
12	8.00	16.00	18.00	381
13	8.00	24.00	10.00	348
14	8.00	24.00	14.00	342
15	8.00	24.00	18.00	420
16	8.00	32.00	10.00	327
17	8.00	32.00	14.00	255
18	8.00	32.00	18.00	322
19	12.00	16.00	10.00	745
20	12.00	16.00	14.00	780
21	12.00	16.00	18.00	740
22	12.00	24.00	10.00	772
23	12.00	24.00	14.00	769
24	12.00	24.00	18.00	755
25	12.00	32.00	10.00	735
26	12.00	32.00	14.00	726
27	12.00	32.00	18.00	757

Table 5.39 Full Factorial Design (Actual Values) and Response (Thickness) for Plating Example

Term	Coefficient	Std Error	T-Value	Significance
1 1	321.407407	13.192185		
2 ~T	327.888889	7.225657		
3 ~TE	−20.055556	7.225657	−2.78	0.0016
4 ~N	12.166667	7.225657		
5 ~T^2	87.888889	12.515205	7.02	0.0802
6 ~N^2	24.055556	12.515205	1.92	0.0656

No. cases = 27	R^2 = 0.9902	RMS Error = 30.66
Resid. df = 21	Adj R^2 = 0.9879	Cond. No. = 4.391

~ indicates factors are transformed

Table 5.40 Final Model for Full Factorial (L_{27}) Design

The prediction equation generated from the orthogonally coded L_{27} design is:

$$\text{predicted thickness} = 321.4 + 327.9(\text{time}) - 20.06(\text{temp}) + 12.17(\text{nickel})$$
$$+ 87.89(\text{time})^2 + 24.06(\text{nickel})^2$$

Let's now summarize the models we obtained with the various designs:

Design Type	R^2	Model # of Runs	Total # of Runs (including screening portion)	Model (factors are orthogonally coded)
D-optimal	.9957	14	34	$311.5 + 323.6(\text{time}) + 22.94(\text{temp}) + 107.6(\text{time})^2$
Box-Behnken	.9902	15	35	$363.8 + 319.1(\text{time}) + 11.75(\text{nickel}) + 70.15(\text{time})^2 - 30.0(\text{time} \cdot \text{nickel}) - 33.10(\text{nickel})^2$
CCF	.9902	26	26	$342.0 + 318.3(\text{time}) - 15.2(\text{temp}) + 19.9(\text{nickel}) + 78.4(\text{time})^2$
L_{27}	.9902	27	47	$321.4 + 327.9(\text{time}) - 20.06(\text{temp}) + 12.17(\text{nickel}) + 87.89(\text{time})^2 + 24.06(\text{nickel})^2$

The difference in prediction models is primarily due to large variation in the data coupled with small sample sizes. The actual simulation equation for the uncoded factors is:

$$\text{Thickness} = 5.0(\text{time})^2 - 2.3(\text{temp}) + 4.0(\text{nickel})$$

The program works as follows: You choose a specific uncoded value for time, temp, and nickel and enter it into the program. The simulation will generate a response based on the equation above plus a normal error distribution with a $\sigma = 40$.

One way to test the effectiveness for each of the various designs is to obtain cross-validated R^2 values. Using the simulation equation without the added error, you can obtain the expected value for all 27 combinations (see Table 5.41). Those 27 values will serve as the true response value in the crossvalidation. The cross-validated R^2 is obtained by correlating these true response values with the values obtained from the prediction equation for each design.

Design Type	R^2
D-Optimal	.996
Box-Behnken	.998
CCF	.999
L_{27}	.998

From the above table we see that the cross-validated R^2 for all of the models is quite good, with the best R^2 in this situation going with the CCF design. A major reason for the large R^2 terms for each model is that the thickness values against which the predicted values are correlated are actual values, without any noise, calculated directly from the base equation. This was done to provide a fair comparison of the prediction ability of each model.

	Actual Model			Predicted Values from Models			
Run	Time	Temp	Nickel	Thickness	D-Opt	B/B	CCF	L_{27}
1	4	16	10	83.2	118.3	39.9	97.3	113.4
2	4	16	14	99.2	118.3	114.8	117.3	101.5
3	4	16	18	115.2	118.3	123.5	137.3	137.7
4	4	24	10	64.8	95.4	39.9	82.1	93.2
5	4	24	14	80.8	95.4	114.8	102.1	81.4
6	4	24	18	96.8	95.4	123.5	122.1	117.6
7	4	32	10	46.4	72.5	39.9	66.9	73.2
8	4	32	14	62.4	72.5	114.8	86.9	61.4
9	4	32	18	78.4	72.5	123.5	106.9	97.6
10	8	16	10	323.2	334.4	318.9	337.3	353.4
11	8	16	14	339.2	334.4	363.8	357.2	341.5
12	8	16	18	355.2	334.4	342.4	377.2	377.7
13	8	24	10	304.8	311.5	318.9	322.1	333.3
14	8	24	14	320.8	311.5	363.8	342.0	321.4
15	8	24	18	336.8	311.5	342.4	362.0	357.6
16	8	32	10	286.4	288.5	318.9	306.9	313.2
17	8	32	14	302.4	288.5	363.8	323.8	301.4
18	8	32	18	318.4	288.5	342.4	346.9	337.6
19	12	16	10	723.2	765.7	738.2	733.9	769.1
20	12	16	14	739.2	765.7	753.0	753.9	757.2
21	12	16	18	755.2	765.7	701.7	773.8	793.5
22	12	24	10	704.8	742.7	738.2	718.7	749.1
23	12	24	14	720.8	742.7	753.0	738.7	737.2
24	12	24	18	736.8	742.7	701.7	758.6	773.4
25	12	32	10	686.4	719.8	738.2	703.5	729.0
26	12	32	14	702.4	719.8	753.0	723.5	717.1
27	12	32	18	718.4	719.8	701.7	743.4	753.4

Table 5.41 Summary of 3-Level Design Analysis Strategies for
the Plating Example

5.11 A Monte Carlo Simulation for Comparing Dispersion Factor Identification Methods [12] (optional)

The three most commonly used strategies to identify dispersion factors are signal-to-noise ratios, ln s, and analysis of dispersion effects from residuals. Several papers have been written discussing the appropriateness of these strategies from a theoretical standpoint. This section is not intended to add to the collection of theoretical arguments; rather, it is our purpose to put these strategies to the test using simulated data. For purposes of this section, four modeling scenarios are presented and the competing strategies are compared based on $k = 6$ input factors and 32 available experiments (or resources). The designs chosen to satisfy the number of factors and limitations of resources are (1) a Taguchi L_8 (2^{6-3} fractional factorial) with a sample size of four per run, (2) a Taguchi L_{16} (2^{6-2} fractional factorial) with a sample of two per run, and (3) a Taguchi L_{32} (2^{6-1} fractional factorial) with one response per run.

As you can see, the resolution of the designs will increase as you increase the number of combinations. The L_8 design is of resolution III indicating that the 6 main effects will be aliased with 2-way interactions. The resolution of the L_{16} design is IV indicating that the mains are free and clear of 2-way interactions, but each 2-way interaction is aliased with another 2-way interaction. The last design, L_{32}, has resolution VI which means that 2-way interactions are no longer aliased with other 2-way interactions. Ideally, the researcher prefers the highest design resolution possible, but he/she is constrained by limited resources and number of factors plus the assumed need to replicate each run to estimate dispersion effects.

Description of Models Tested

It is infeasible to model every different relationship of the response with the input factors while also incorporating several different types of error. However, since the problems encountered in industry are not always associated with additive error relationships having normally distributed error, it is important to include in this section some of the more common models.

ADDITIVE ERROR MODEL:

This model is defined as

$$y = f(\underline{x}_L, \underline{x}_B) + \epsilon(\underline{x}_D, \underline{x}_B)$$

where $\underline{x}_L$ is the set of factors which produce only location effects,

$\underline{x}_D$ is the set of factors which produce only dispersion effects, and

$\underline{x}_B$ is the set of factors which produce both a location and a dispersion effect.

For this model, the strategies will be tested using normally and exponentially distributed error.

The actual model for the simulation is $y = w_0 + w_1 x_1 + w_2 x_2 + w_3 x_1 \cdot x_2 + \epsilon$ where the w_i are generated randomly and each $w_i \geq 3\sigma$. In addition,

$$\epsilon = \begin{cases} zM\sigma & \text{if } x_3 = +1 \\ z\sigma & \text{if } x_3 = -1 \end{cases}$$

where z has a standardized normal distribution in one simulation and an exponential distribution ($\lambda = 0.5$) in the other. Therefore, x_3 is clearly the only dispersion effect and the amount of dispersion is controlled by the multiplier $M = 2, 3, 4,$ and 5.

MULTIPLICATIVE ERROR MODEL:

This model is defined as

$$y = f(\underline{x}_L, \underline{x}_B) + \epsilon(\underline{x}_D, \underline{x}_B)$$

In this case the actual simulator model is

$$y = (w_0 + w_1 x_1 + w_2 x_2 + w_3 x_1 \cdot x_2) \cdot \epsilon \text{ where}$$

$$\epsilon = \begin{cases} (4 + z)\sigma M & \text{for } x_3 = +1 \\ (4 + z)\sigma & \text{for } x_3 = -1 \end{cases}$$

and z has a standardized normal distribution. The values of M tested were 3 and 5.

EXPONENTIAL MODEL:

The exponential model appears as

$$y = \exp(w_0 + w_1 x_1 + w_2 x_2 + w_3 x_1 \cdot x_2 + \epsilon) \qquad \text{where}$$

$$\epsilon = \begin{cases} zM\sigma & \text{for } x_3 = +1 \\ z\sigma & \text{for } x_3 = -1 \end{cases}$$

and z has a standard normal distribution. The values of M tested were 3 and 5. A summary of the different tests is displayed in Table 5.42.

Strategies	Designs	Modeling Scenarios
(1) Signal-to-Noise Ratios - Smaller is better - Larger is better - Nominal is better	(1) 4 replications of L_8	(1) $y = f(\underline{x}) + N(0, 1)$
	(2) 2 replications of L_{16}	(2) $y = f(\underline{x}) \cdot N(4, 1)$
	(3) 1 replication of L_{32}	(3) $y = \exp[f(\underline{x}) + N(3, .01)]$
(2) $\bar{y}$ plus ln s		(4) $y = f(\underline{x}) + \exp(\lambda = .5)$
(3) Modified Box-Meyer		

Table 5.42 Summary of Different Simulation Tests

RESULTS:

The simulations were run 50 times for each strategy, model, and design type. The difference in average marginal S/N_L, S/N_S, S/N_N, and ln s was used to identify the dispersion factors for the replicated designs. The difference in average residual standard deviations was used to identify dispersion factors for the unreplicated designs. The results for each model follow:

ADDITIVE ERROR MODEL: (Error term is normally distributed.)

In the September 1988 IIE transactions, Pignatiello [11] discusses this model and concludes that maximizing S/N_N is accomplished by maximizing $\bar{y}$ and minimizing s^2. This is verified in that the results for S/N_N are very similar to ln s.

As stated in Box [5], S/N_L and S/N_S are based on ideas about location only and this is verified in that S/N_S and S/N_L never identified the dispersion effect. The results for this model are summarized in Figure 5.14. Starting with the L_8 designs, you can see that S/N_N and ln s are about equally efficient at correctly detecting x_3 as a dispersion effect. This result is discussed by Gunter [7] where he indicates that more information is required to identify dispersion effects than location effects. The L_{32} unreplicated method had very impressive results while providing much more information on interactions. Out of curiosity, an unreplicated L_{16} was run which again verified Gunter's statement above. As previously stated, S/N_L and S/N_S were ineffective methods for identifying the dispersion effect.

ADDITIVE ERROR MODEL: (Error term exponentially distributed.)
    This test produced the following results:

| | | | \multicolumn{5}{c}{Percent Correctly Detected} | | | |
| | | | | | Strategy | | |
			S/N_S	S/N_L	S/N_N	Ln s	Mod B-M
	L_8	M = 3	0%	0%	30%	72%	N/A
Design	L_8	M = 5	0%	0%	75%	90%	N/A
Type	L_{32}	M = 3	N/A	N/A	N/A	N/A	46%
	L_{32}	M = 5	N/A	N/A	N/A	N/A	50%

Table 5.43   Results for Additive Error Model

    For this method ln s was slightly better than S/N_N and the modified Box-Meyer approach was not as effective as before. The ineffectiveness of S/N_S and S/N_L is still confirmed in this type model.

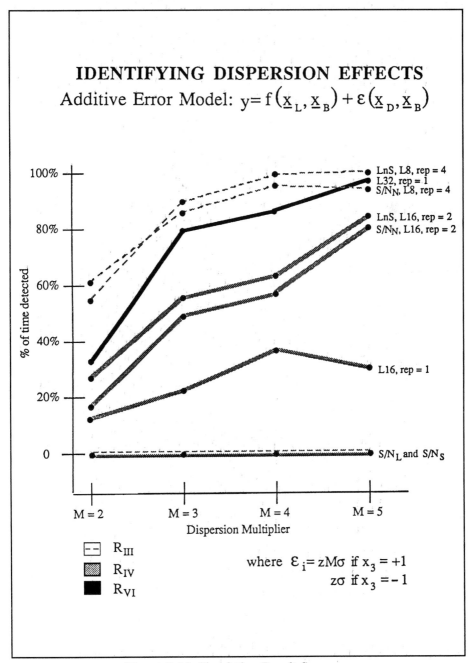

Figure 5.14 Simulation Result Summary

MULTIPLICATIVE ERROR MODEL:

Pignatiello [11] discussed this model and concluded that S/N_N and ln s would produce similar results. The simulation analysis provided a confirmation to his conclusion. The results are shown below.

| | | | Percent Correctly Detected | | | | |
| | | | | | Strategy | | |
			S/N_S	S/N_L	S/N_N	Ln s	Mod B-M
	L_8	M = 3	100%	100%	50%	45%	N/A
Design	L_8	M = 5	100%	100%	90%	90%	N/A
Type	L_{32}	M = 3	N/A	N/A	N/A	N/A	0%
	L_{32}	M = 5	N/A	N/A	N/A	N/A	0%

Table 5.44 Results for Multiplicative Error Model

The success of S/N_L and S/N_S was anticipated because, for the multiplicative error model, as the error increases, so does the response. Thus, any dispersion effect is also a location effect. Therefore, the success of S/N_L and S/N_S can be attributed to identifying location effect and not the dispersion effect.

EXPONENTIAL MODEL:

This test produced the results shown below.

| | | | Percent Correctly Detected | | | | |
| | | | | | Strategy | | |
			S/N_S	S/N_L	S/N_N	Ln s	Mod B-M
	L_8	M = 3	0%	0%	96%	35%	N/A
Design	L_8	M = 5	96%	96%	98%	98%	N/A
Type	L_{32}	M = 3	N/A	N/A	N/A	N/A	0%
	L_{32}	M = 5	N/A	N/A	N/A	N/A	22%

Table 5.45 Results for Exponential Model

Note the difficulty of all strategies except S/N_N in detecting a modest ($M = 3$) dispersion effect. However, S/N_S, S/N_L, and ln s performed comparably to S/N_N in detecting large dispersion effects. As with the multiplicative error model, it was apparent that the exponential model had transformed non-homogeneous variables into location effects, enabling S/N_S and S/N_L to better detect the dispersion factors.

Of course, if the exponential model $y = \exp(w_0 + w_1 x_1 + w_2 x_2 + w_3 x_1 x_2 + \epsilon)$ is anticipated, then a transformation using the natural logarithm would produce $\ln(y) = w_0 + w_1 x_1 + w_2 x_2 + w_3 x_1 x_2 + \epsilon$ which would yield the same results as the additive model. The modified Box-Meyer approach again had only marginal success with the nonlinear model; however, a log transformation as described above would compensate for this ineffectiveness.

CONCLUSIONS

The analysis of dispersion effects due to non-homogeneity of variance has become increasingly important as experimenters strive to optimize a response while minimizing its variance. Taguchi, Box, and others have suggested several techniques for analyzing dispersion effects within an experimental design, either simultaneous with or in addition to location effect analysis. This section compared the dispersion effect detection capabilities of five techniques (three Taguchi signal-to-noise ratios, a response standard deviation analysis, and a modified Box-Meyer technique) by way of a Monte Carlo simulation.

Simulation results clearly indicate that the standard deviation of responses for different factor combinations must typically change by at least a factor of 3 ($M \geq 3$) to be detected. This is consistent with the non-homogeneity threshold described by Keppel [8] for classical ANOVA analysis. Even with such a large dispersion effect, only S/N_N and ln s proved to be consistent detectors over the range of simulation models. Although comparable performers, the authors prefer using ln s over S/N_N due to its ease of interpretation. Because S/N_N simultaneously detects location and dispersion effects, it may be difficult to attribute the cause (either dispersion or location) to a particular factor. However, a traditional analysis of the average response for location effects in conjunction with the ln s analysis for dispersion effects dispels any question as to cause. The modified Box-Meyer method is as good as S/N_N and ln s for comparable number of runs if the underlying process model is linear, and it has the obvious advantages and efficiencies of an unreplicated design. However, its use should be questioned for process models which are truly unknown and possibly nonlinear unless the appropriate transformation is made. Taguchi's S/N_S and S/N_L are not recommended due to their inability to identify dispersion factors.

Sample size was also found to be a crucial part in finding dispersion factors. Historically, researchers have used 16 data points as a Rule of Thumb minimum[*] for estimating a variance (or standard deviation). The 16 run Box-Meyer approach suffered in performance because each set of residuals had 8 values, not enough for a good estimate of variance. The 32 run Box-Meyer approach performed well with 16 residuals per low or high for variance calculations. Thus, sample size for experimentation could be simply tied to the above rule of thumb for variance calculations instead of the rigorous approach in Chapter 4. Using 16 data points as the "ideal" minimum for estimating an average variance (at a factor low or high) in replicated designs would suggest the minimum number of replicates for the following 2-level designs:

2-Level Design	Minimum # of Replicates for Residual Analysis	Modeling Ln s
L_4	8	9
L_8	4	5
L_{16}	2	3
L_{32}	1	2

---

[*]See Appendix M for Rules of Thumb.

Realize that when costs, time, resources, etc., prevent you from achieving the above recommended number of replicates, you still have a chance of detecting dispersion factors.   However, the chances for detection are approximately proportional to the ratio of your choice in replicates with the ideal, i.e., replicating an L_8 only 2 times results in an approximate 2/5 (or 40%) chance of detecting a real dispersion factor when using ln s.   Another obvious conclusion is that the residual analysis approach will require less in the way of resources.

Problem Set

1. Given the following settings and results:

A	B	C	D = ABC	y_1	y_2	$\bar{y}$	s	ln s
−	−	−		50	70			
−	−	+		47	54			
−	+	−		44	35			
−	+	+		54	71			
+	−	−		57	53			
+	−	+		43	48			
+	+	−		42	39			
+	+	+		60	57			

a)   Complete the rest of the table.

b)   Find the prediction equation from these results. Include only significant main and 2-way interaction effects.

c)   What settings will minimize the response and the variance of the response?

2. An engineer needs to improve turbine blade quality by reducing thickness variability around a target of 3mm. The brainstorming session identified 4 variables that are likely to affect the thickness. Those variables and their ranges of values are:

	−1	1
Metal Temperature	20°C	22°C
Mold Temperature	3°C	5°C
Pour Time	1 sec	3 sec
Vendor	A	B

The engineer had been limited to 20 experimental runs, so he decided to run a full factorial of 16 runs, saving 4 runs to confirm his results.

Run #	Metal Temp	Mold Temp	Pour Time	Vendor	Thickness
1	20	3	1	A	2.314
2	20	3	1	B	2.307
3	20	3	3	A	2.321
4	20	3	3	B	2.233
5	20	5	1	A	4.204
6	20	5	1	B	4.260
7	20	5	3	A	4.221
8	20	5	3	B	4.198
9	22	3	1	A	2.687
10	22	3	1	B	2.748
11	22	3	3	A	2.697
12	22	3	3	B	2.728
13	22	5	1	A	4.769
14	22	5	1	B	4.868
15	22	5	3	A	4.787
16	22	5	3	B	4.849

a) Do the full analysis on this data so the engineer has a prediction equation. Look for main effects, interactions, and dispersion effects.

b) What settings should the engineer use to meet the goal of 3mm thickness with minimum variability?

3. A statistical process control analyst is trying to determine which factors have a major effect in a circuit board etching process. She wants to pare down the seven factors brought out in the brainstorming session to maybe two or three. The seven factors are:

	LOW	HIGH
A: Resist Thickness	.1mm	.5mm
B: Develop Time	80 sec	90 sec
C: Develop Concentration	3.1 : 1	2.7 : 1
D: Exposure	200	240
E: Develop Temperature	19°C	23°C
F: Circuit Line Thickness	1mm	3mm
G: Rinse Time	5 sec	10 sec

She ran the following L_8 design to find the significant effects.

Run	A	B	C	D	E	F	G	Response
1	−	−	−	−	−	−	−	74.48
2	−	−	−	+	+	+	+	70.07
3	−	+	+	−	−	+	+	75.71
4	−	+	+	+	+	−	−	68.79
5	+	−	+	−	+	−	+	84.98
6	+	−	+	+	−	+	−	81.57
7	+	+	−	−	+	+	−	84.03
8	+	+	−	+	−	−	+	80.97

a) Find the factors which affect the response variable.

b) If the goal is to maximize the response, what settings should be used? Predict what the response should be.

c) The analyst ran confirmation runs to see if she had all the effects under control. If the mean of those runs is 86, is there any problem? What if the mean were 90? 72?

4. Refer to the Chapter 8 Case Study on an Injection Molded Plastic Part, pages 8-23 through 8-27. In an attempt to get the width of the part to 9.380, the experimenter decided to look for interactions between the top four main effects on width:

    D: Mold Temperature
    A: Injection Velocity
    E: Hold Pressure
    B: Cooling Time

Run	D	A	E	B = DAE	y_1	y_2
1	–	–	–		9.34150	9.34160
2	–	–	+		9.36914	9.36916
3	–	+	–		9.34666	9.34664
4	–	+	+		9.36801	9.36809
5	+	–	–		9.36790	9.36800
6	+	–	+		9.34933	9.34937
7	+	+	–		9.36680	9.36690
8	+	+	+		9.35444	9.35446

Analyze his results and create a predictive equation for width. What are the best settings to reach the 9.38 goal for width?

5. As a reliability engineer, you have been asked to weed out infancy failures in component-populated printed circuit boards. The four factors of interest are:

		LOW	HIGH
A:	Stress Temperature	80°C	125°C
B:	Thermo Cycle Rate	5°C/min	20°C/min
C:	Humidity	15%	95%
D:	g level for a 10 minute sinusoid random vibration	3	6

The response is the number of electrical defects per board which has 1000 bonds.

Given the following design matrix and response data, determine the optimal screening method. (The more failures found, the better.)

	A	B	C	D	y_1	y_2	y_3
1	–	–	–	–	9	17	12
2	–	–	+	+	21	37	42
3	–	+	–	+	29	35	38
4	–	+	+	–	17	10	15
5	+	–	–	+	32	41	33
6	+	–	+	–	21	17	19
7	+	+	–	–	12	14	18
8	+	+	+	+	33	27	47

6. A metal casting process for manufacturing turbine blades has four controllable factors:

Metal Temperature	Pour Speed
Mold Temperature	Raw Material

The blades must be 3mm thick; however, the ambient temperature causes the blades to expand and contract. The following experiment was run with ambient temperature in the outer array:

A Metal Temp	B Mold Temp	C Pour Speed	D Raw Material	Temp L	Temp M	Temp H
+	+	+	+	3.06	3.14	3.06
+	+	–	–	3.01	3.00	3.05
+	–	+	–	2.81	2.81	2.80
+	–	–	+	2.80	2.88	3.01
–	+	+	–	2.62	2.61	2.62
–	+	–	+	2.61	2.66	2.72
–	–	+	+	2.42	2.43	2.52
–	–	–	–	2.42	2.43	2.41

a)   Find a combination of settings that will produce the required thickness and is also robust to ambient temperature.

b)   Plot the interactions of:

A with Ambient Temperature
B with Ambient Temperature
C with Ambient Temperature
D with Ambient Temperature

c)   Use the plots from b) to verify the optimal conditions found in a).

Bibliography

1. Barker, Thomas B., (1985) *Quality by Experimental Design*, Marcel Dekker.

2. Bhote, Keki R., (1988) *World Class Quality*, American Management Association, New York.

3. Box, G. E. P., William G. Hunter, and J. Stuart Hunter, (1978) *Statistics for Experimenters*, John Wiley and Sons, Inc., New York.

4. Box, G. E. P., (1988) "Signal-to-Noise Ratios, Performance Criteria, and Transformations," *Technometrics*, Vol. 30, No. 1.

5. Box, G. E. P. and R. D. Meyer, (1986) "Report No. 1, Studies in Quality Improvement: Dispersion Effects From Fractional Designs," Center for Quality and Productivity Improvement, University of Wisconsin, Madison, WI.

6. Box, G. E. P. and R. D. Meyer, (1986) "Report No. 2, Studies in Quality Improvement: An Analysis for Unreplicated Fractional Factorials," Center for Quality and Productivity Improvement, University of Wisconsin, Madison, WI.

7. Gunter, B., (1988) "Discussion: Signal-to-Noise Ratios, Performance Criteria, and Transformations," *Technometrics*, Vol. 30, No. 1.

8. Keppel, G., (1982) *Design & Analysis, A Researcher's Handbook,* 2nd ed., Prentice-Hall, Englewood Cliffs, NJ.

9. Montgomery, Douglas C., (1990) *Design and Analysis of Experiments*, 3rd ed., John Wiley and Sons, Inc., NY.

10. Myers, Raymond H., (1989) *Classical and Modern Regression with Applications*, 2nd ed., Duxbury Press, Boston, MA.

11. Pignatiello, J. J., (1988) "An Overview of the Strategy and Tactics of Taguchi," *IIE Transactions*, Vol. 20, pp. 247-254.

12. Schmidt, S. R. and J. R. Boudot, (1989) "A Monte Carlo Simulation Study Comparing Effectiveness of Signal-to-Noise Ratios and Other Methods for Identifying Dispersion Effects," Rocky Mountain Quality Conference and 1989 ORSA/TIMS Annual Conference.

13. Taguchi, G., (1987) *System of Experimental Design*, Kraus International Publications, White Plains, NY.

Chapter 6

Taguchi Philosophy, Design, and Analysis

6.1   Introduction*

An analysis of Japanese success in the American marketplace and their ability to produce quality products is a matter of current interest. The Japanese did in fact turn around a reputation of manufacturing junk to one of manufacturing quality products that can and still do obtain higher market value than many similar U.S. products. The tools used by the Japanese to accomplish this remarkable turnaround are focused at the product design phase versus quality inspection during manufacturing. By designing quality into the product from its conception, fewer resources are required to produce a quality product. The use of Taguchi Methods in parameter design, known as Quality Engineering, typically results in increased productivity, improved quality, and increased efficiency. The designs used by Taguchi were developed primarily by British and American statisticians as fractional-factorials, Plackett-Burman, Latin Square, and mixed designs. The idea of orthogonality has been around a long time; however, Taguchi's approach is different in that he (1) stresses only a few basic designs most commonly used in

---

*This chapter contains much of the information found in Warsavage and Schmidt [12].

industry, (2) provides a table of these designs, and (3) avoids statistical rigor by providing a cookbook approach to the analysis.

The Taguchi approach to quality control emanates from the following definition of quality: "The Quality of a product is the (minimum) loss imparted by the product to society from the time the product is shipped" [11]. From this definition, a loss function is developed that translates any deviation of a product's quality characteristic from its target value into a financial measure. Any deviation from the target value, either through variability (noise) or by a deviation of the average from target, translates into increased loss. The process of reducing variability and moving the mean to the target is based on designed experiments and emphasizes a phase of product and process design that Taguchi calls parameter design.

Despite much criticism of the Taguchi approach to designed experiments by traditional statisticians, the approach continues to expand. The expansion may be due to the simplicity associated with the presentation of the material by the American Supplier Institute and through the use of success stories such as those from Ford Motor Company, ITT, Digital Equipment Corporation, Xerox, Nippon Telephone and Telegraph Company, Toyota, and Fuji Film. Historically, engineers would vary only one factor at a time or intuitively select the best combination of factor settings based on engineering judgement. This can be extremely inefficient and will rarely lead to an optimal design. Traditional statistics were certainly an alternative; however, the presentation was not always simple and due to its perceived complexity, non-statisticians were less inclined to use it. One positive result of the discussions surrounding the use of Taguchi methods has been a great awakening of interest in industry to numerous experimental design approaches. Many engineers are now (as compared to just a couple of years ago) actively learning how to apply various types of orthogonal arrays and analysis approaches. Yet with all the discussions and learning that is taking place, probably fewer than 10% of the experiments done in industry today utilize a "designed experiment" approach. Much remains to be done.

The remainder of this chapter is divided into four parts. Under the topic of Taguchi philosophy and methodology, the loss function and Taguchi's approach to experimental design are discussed. Section 6.3 addresses the Taguchi strategy and Section 6.4 gives a detailed example of a Taguchi design. In the last section, we provide a 12-point comparison of the classical and Taguchi approaches.

6.2   Taguchi Philosophy and Methodology

6.2.1   Loss Function

Specification limits are the means by which manufacturers determine the details of a manufacturing product and process. These limits are frequently used in such a way as to imply that if the quality characteristic of a part falls within limits, the product is satisfactory; if it falls outside of these limits, it is defective. In many cases, a target value may be given, but hitting the target has not attained an importance beyond just being within specification. Taguchi suggests that any deviation from target results in incremental loss. He models this philosophy through a loss function which Chernoff and Moses [1] address as a part of statistical decision theory. Taguchi's contribution is the application of the loss function to the measurement of quality.

The quadratic loss function (based on a Taylor Series approximation) will increase as the quality characteristic deviates on either side of the target (see Figure 6.1). It is also symmetric about the target. Loss is minimized by producing all items as close as possible to the target. The old practice of using only the specification limits to determine loss does not hold in practice because a product just within specification is more like an out-of-specification product than a product right on target. As an example, consider a rod built within specification, but at the upper limit, and the sleeve within specification, but at the lower limit. This situation will result in a product with less than optimal performance and reliability. The end result is loss to the purchaser which eventually will be passed on to the producer as loss of market share. Other types of loss functions may be argued to be more accurate in that a deviation to one side of the target may be more of a problem than a deviation to the other side. The important point is this: the development of a loss function results in evaluating costs associated with a deviation from the target and the simplicity of the quadratic loss function results in ease of implementation.

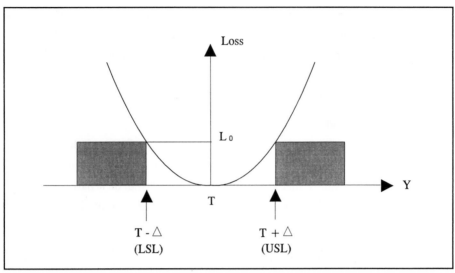

Figure 6.1 Taguchi's Quadratic Loss Function

An important aspect of the loss function is that it maps deviations from the target into a financial measure. Everyone understands money, and since it is a common measure, comparisons can be made between products and processes. Operationally, the average loss would be computed using a sample large enough to characterize the process measured on a critical quality characteristic. Using a quadratic loss function, the loss associated with each product would be computed as follows [4]:

$$L_i = k(y_i - T)^2$$

where　　　　　y_i is the quality characteristic of interest for product i,

T is the quality characteristic target, and

k is a constant that converts deviation to a monetary value.

To determine the value of k for a specific target, you need to estimate the loss at some measured value of the characteristic in question. Referring to Figure 6.1, assume you estimate the loss, L_0, associated with a point on the horizontal axis, y_0 (in this case y_0 = LSL). Then using $L_0 = k(y_0 - T)^2$ to represent the quadratic loss function and substituting known values for L_0, y_0, and T, you simply solve for k. For a specified target T (nominal is best) the average loss, $\bar{L}$, for n products is:

$$\bar{L} = k\sum_i \left[\frac{1}{n}\right](y_i - T)^2$$

where

n = sample size,

y_i = value of the critical parameter for product i, and

T = target value.

The decomposition of the average loss function for all n products is developed as follows:

$$\bar{L} = k\sum \frac{1}{n}(y_i - T)^2$$

$$= \frac{k}{n}\sum (y_i - \bar{y} + \bar{y} - T)^2$$

$$= \frac{k}{n}\sum [\ (y_i - \bar{y})^2 + (\bar{y} - T)^2 + 2(y_i - \bar{y})(\bar{y} - T)\ ]$$

$$= \frac{k}{n}\sum (y_i - \bar{y})^2 + k(\bar{y} - T)^2 + 2(\bar{y} - T)\frac{k}{n}\sum (y_i - \bar{y})$$

Since the last term results in a value of 0, $\bar{L} = k[\sigma^2_y + (\bar{y} - T)^2]$. Thus, average loss is partitioned into variance of the response and deviation of the average response from the target. To minimize loss, it is clear that not only do you minimize the average response from the target, but you must also minimize response variability.

One illustration of these points is cited by L. Sullivan [10] on the color intensity of television sets. Figure 6.2 illustrates the distribution of the product quality characteristic for Sony Japan and Sony U.S.A. Notice that Sony Japan was producing some products outside the specification limits, but Sony U.S.A. was not. Under conventional methods, Sony U.S.A. would be preferred. Yet, if given a choice, customers chose Sony Japan. A loss function can model this preference, whereas strict adherence to specification limits cannot.

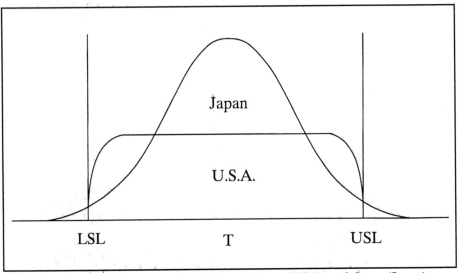

Figure 6.2 Performance Capability of Sony (USA) and Sony (Japan)

Ideally, you would want the variability of the product to appear as shown in Figure 6.3. When the majority of the response measurements occur near the target, the loss is reduced. The use of designed experiments will enable you to determine which factors shift the average (in order to minimize $(\overline{y} - T)^2$) and which factors shift the variability (in order to minimize σ_y^2).

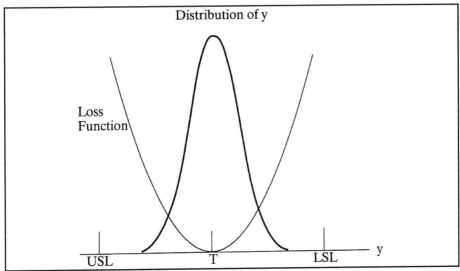

Figure 6.3 Idealized Distribution

Two quantitative measures which relate σ_y^2 to the specification limits are C_p and C_{pk}. $C_p = \dfrac{USL - LSL}{6\sigma}$ where $C_p < 1$ is unsatisfactory, $1.0 \le C_p \le 1.33$ is marginal, and $C_p > 1.33$ is desired. Since C_p assumes $\overline{y} = T$, which is not always the case, a different measure incorporating the product mean is C_{pk}. To calculate C_{pk}, we use $C_{pk} = \dfrac{\min(USL - \overline{y}, \overline{y} - LSL)}{3\sigma}$ where $C_{pk} < 1$ is unsatisfactory, $1.0 \le C_{pk} \le 1.33$ is marginal and $C_{pk} > 1.33$ is desired. The higher the C_{pk} value, the more the σ^2 portion of the loss function is minimized and the closer $\overline{y}$ is to T. Note that σ is typically estimated with s, the sample data standard deviation.

6.2.2   Quality Engineering

Currently there is a lot of emphasis placed on techniques to control the building of a product once it has entered the manufacturing phase. These techniques are referred to as statistical quality or process control. They were discovered and used by Shewhart in the 1920's and recently made popular by Deming's success with the Japanese. Taguchi refers to these techniques, as well as some innovations of his own, as **On-line Quality Control**.

One of Taguchi's major contributions is to direct attention to the contributions that statistics and quality engineering can make in the design phase of a product. These techniques are thus referred to as **Off-line Quality Control** [4]. Statistics and a quality oriented approach taken in the design phase can save large amounts of resources in terms of time and effort to manufacture a product. This approach also leads to a reduction of costly scrapping and reworking of the product.

6.2.3   Quality by Design

As previously discussed, the Taguchi approach emphasizes and elaborates on the design phase of a product's life. He subdivides the design phase into three steps or subphases: system design, parameter design, and tolerance design.

System Design: This is the inventive phase where engineers make the prototype. Historically, this is an area in which the U.S. has excelled to the point of leading the world in patents, whereas the Japanese have contributed little. Initially, the Japanese borrowed western product design and then manufactured it better; however, they are currently placing more emphasis on system design. In certain areas, the Japanese are surpassing the U.S.

Parameter and Tolerance Design: Taguchi argues that the Japanese have added or at least emphasized two additional steps between the prototype and the hand-off to manufacturing. The most important step is parameter design. In this step, Taguchi analyzes the effect of each of the controllable factors and determines the best factor settings in relation to the noise factors. Factors that most affect performance are determined, as well as factor settings that are optimal, in terms of performance, cost, ease of manufacture, and reliability. At this stage, one should avoid factor settings that are difficult to maintain and control. If those settings cannot be avoided, then tight tolerances must be set. The setting of those tolerances is addressed in the **tolerance design subphase**. The identification of critical factors and their optimal values are found by experimentation using designed experiment techniques. If tolerances must be determined, these are also determined experimentally.

6.2.4   Controllables and Noise

The factors that affect the performance of a product by increasing variability are referred to as noise or noise factors. Product reliability and ease of manufacture are the result of specifically designing the product to be resistant to noise. In contrast, those factors that are relatively easy to control are called controllables. Some factors can actually fall into both categories. For example, the temperature of a wave solder process can be set closer to 150 degrees than 200 degrees; however, even if 150 degrees is the target, the measured temperature may vary a few degrees within a run or between runs. In the first case, the temperature is considered a controllable factor and in the second, it is a noise factor. The Taguchi strategy investigates the effect of large and small changes. Another noise factor might be ambient humidity which may be systematically varied during experimentation in the parameter design phase, but remains uncontrollable in the operational phase of the process.

Taguchi's consideration of noise results in some changes in the way that an experimental design is set up and analyzed. Noise, a pseudonym for experimental error, is usually considered in statistical thinking as the element of unpredictability of the outcome and is assumed to be independent of the factor settings. In classical experimental design, you would assume that there are no errors in the factor settings. Any small errors, such as they are, are absorbed by randomization. Taguchi recommends that these errors or noise can be directly analyzed using experimental design techniques.

An example which illustrates some of Taguchi's philosophies is based on the characteristic function of a transistor (see Figure 6.4). Consider the following scenario:

(1)   Target output voltage is 100.

(2)   Gain has a large effect on output voltage and is non-linear.

(3)   The circuit engineer has many options in terms of choosing the gain setting.

One approach would be for the circuit engineer to set the transistor gain at 18 so as to obtain the desired output voltage of 100. However, small deviations from the target of 18 will result in relatively large variations in our output voltage. A better approach would be for the circuit engineer to select a transistor gain at a much flatter portion of the curve (for example, 30), thus minimizing the resulting variation (sometimes referred to as **transmitted variation**) in output voltage. By using some other component in the circuit (which has a linear effect on output voltage), the engineer can adjust the mean value from 120 to 100. Provided this reconfiguration did not increase the variability in the output voltage, the end result would be much improved.

In general, running transistors near the point where the slope of the output voltage is small may result in a higher quality product. This of course oversimplifies the problem, but it does illustrate a valuable point in that some factor settings can be used to reduce variability and others to obtain the response closest to the target.

Figure 6.4 Characteristic Function Plot

6.3   Taguchi Strategy of Experimentation

Taguchi's strategy is characterized by three elements: (1) orthogonal arrays based on fractional factorials, Plackett-Burman, or Latin Square experimental designs, (2) design modification that incorporates information about potential noise factors, and (3) analysis of a transformation called signal-to-noise ratio.

6.3.1   Saturated Fractional Factorial Experimental Designs

A fractional factorial experiment is saturated when the design only allows for the estimation of the main effects. For example, instead of estimating 4 main effects with 16 runs, it is possible to estimate 15 main effects independently of each other with those 16 runs. In this case, all the interactions are aliased with these main effects (see Chapter 3 for details on screening designs and Chapter 5 for screening design examples).

Fractional designs, when used as part of a screening process, are effective when there are large numbers of factors. Since large numbers of factors usually include some which do not affect the response measure, the screening process provides a way to remove from consideration (screen out) those unimportant factors as efficiently as possible. There are potentially hundreds of factors in most industrial processes and very little may be known about their effects. The screening process is, therefore, a way to start reducing the number of factors in a systematic fashion.

6.3.2   Designs That Assess Information About Noise

Randomization is the key to reaching conclusive results in a timely and resource efficient manner. Randomization in factorial designs implies that runs of all different factor combinations will be completed in a random order. There are two reasons for wanting to randomize: (1) the critical assumption of independence is usually ensured through random sampling and randomization, and (2) long-term trends and/or systematic noise factors not included in the design matrix will become part of the error by being spread in an equally likely fashion over the experimental conditions [6]. Since randomization in industrial experiments is frequently time consuming and often impossible, Taguchi does not emphasize the use of randomization, but he attempts to compensate for it by incorporating noise in the design. He, therefore, assumes all potential noise factors can be measured and will be included in the design. If this is not the case, there exists a strong possibility that the results will be ambiguous.

Taguchi's robust designs use two sets of factors: (1) an inner array of controllables, and (2) an outer array of noise factors. At each combination of the controllables, an outer array of noise factors is included. Thus, at each combination of the controllables, a separate estimate of the effect of the noise factors can be made. This is similar to a sensitivity analysis by J. S. Hunter, [3]. The primary objective of robust designs is to determine optimal settings for the controllable factors which will result in a desired response which is invariant (or insensitive) to noise.

6.3.3   Signal-to-Noise Ratio

Sensitivity to noise is analyzed through a signal-to-noise statistic which involves both location and dispersion effects. Many different estimates of signal to noise have been tested, but typically one of three are used. When a specified target is the criterion, a statistic related to the coefficient of variation is:

$$(S/N) \ = \ 10 \ \log_{10} \ \frac{\overline{y}^2}{S^2}$$

where

$$\overline{y} \ = \ \sum \frac{y_i}{n}$$

$$S \ = \ \sqrt{\frac{\sum (y_i - \overline{y}_i)^2}{n - 1}}$$

This transformation converts y to decibel units and is closely related to the conventional signal-to-noise ratio $\overline{y}/s$. The optimum factor levels for any process will be found by maximizing (S/N).

In the case of a zero or infinite target, the above formula is replaced by signal-to-noise ratios derived from the formula for average loss, resulting in:

$$(S/N) = -10 \log_{10} MSD$$

where $\qquad MSD = \dfrac{1}{n}\sum y_i^2$, for a zero target (smaller is better),

or $\qquad MSD = \dfrac{1}{n}\sum \dfrac{1}{y_i^2}$, for an infinite target (larger is better).

A typical Taguchi strategy involves analyzing one of these statistics with or without a graphical analysis of means.

6.4   Example of a Taguchi Experimental Design

In this section, the Taguchi strategy is illustrated with an example of a wave solder process optimization. In a printed circuit board assembly plant, parts are inserted either manually or automatically into a bare board with a circuit printed on it. After most of the parts are inserted, the board is put through a wave solder machine. This process is a means of connecting, both electrically and mechanically, all the parts into the circuit. Boards are placed on a conveyor, and taken through a series of steps. They are bathed in a flux mixture to remove oxide. To minimize warpage, the boards are preheated before the solder is applied. The soldering process takes place as the boards move across the wave of solder. After that, the board usually goes through a device similar to a dishwasher to remove any debris. Because of the many factors used in controlling the process, it is an excellent candidate for experimentation. For wave soldering processes which have not been studied statistically, it is not uncommon for 10-20% of the assembly work force to be involved in solder touchup.

The data used in this example are simulated to illustrate the Taguchi strategy. The objective of the experiment is to minimize the number of solder defects per million joints. Table 6.1 lists the factors and their settings for this example. The choice of factors to be used must be carefully considered in the brainstorming phase. Experts at other plants with different types of wave solder machines might choose a different set of factors. In our example, the inner array of controllables contained five factors. After brainstorming, it was decided to test each factor at two levels and to include an interaction between solder pot temperature and conveyor speed.

A Designed Experiment for Wave Solder
An Example of the Use of Orthogonal Arrays

Controllable Factors	Levels	
	Low	High
(1) Solder Pot Temperature (S)	480°F	510°F
(2) Conveyor Speed (C)	7.2 ft/m	10 ft/m
(3) Flux Density (F)	.9°	1.0°
(4) Preheat Temperature (P)	150°F	200°F
(5) Wave Height (w)	0.5"	0.6"

Noise Factors

(1) Product Noise	Assembly #1, Assembly #2
(2) Conveyor Speed Tolerance	−0.2, +0.2 ft/m
(3) Solder Pot Tolerance	−5°F, +5°F

Table 6.1 Factors and Settings for Wave Solder Example

The factors in the noise array are selected as well. Because several different types of assemblies are run through this wave solder process, two different types of assemblies were used. The objective is to find one setting for the wave solder process that is suitable for both types of assemblies. The design will also indicate if assembly type interacts with any of the controllables. In addition to product noise, both the conveyor speed and solder pot temperature will be moved around the initial setting given by the controllables array. This is because it is difficult to set the conveyor speed with any degree of accuracy and it is also difficult to maintain solder pot temperature.

Eight runs will be used to test the effects of the five controllables in a Taguchi L_8 design (see Table 6.2a). Notice that for each factor, there are four runs at the low setting and four runs at the high setting. This balancing is a property of the orthogonality of the design matrix. Table 6.2b lists the array of noise factor conditions to be run at each of the eight settings of the controllables. These four conditions were generated using a Taguchi L_4 design. The combination of the inner and outer arrays results in each run of the controllables being repeated over the four conditions of the noise factors.

Controllables Design Inner Array				
Run	Solder Pot Temperature	Conveyor Speed	Flux Density	Preheat Temperature	Wave Height
1	510	10.0	1.0	150	0.5
2	510	10.0	0.9	200	0.6
3	510	7.2	1.0	150	0.6
4	510	7.2	0.9	200	0.5
5	480	10.0	1.0	200	0.5
6	480	10.0	0.9	150	0.6
7	480	7.2	1.0	200	0.6
8	480	7.2	0.9	150	0.5

Table 6.2a Runs Used in Wave Solder Example

	Outer Array			
At each combination of the inner array, an outer array of noise factors is run.				
	Run			
Parameter	1	2	3	4
Product Noise	Assembly #1	Assembly #1	Assembly #2	Assembly #2
Conveyor Tolerance	−0.2	+0.2	−0.2	+0.2
Solder Tolerance	−5	+5	+5	−5

Table 6.2b Noise Factors for Wave Solder Example

Table 6.3 presents the results of the experiment. The two far right columns contain the average and signal-to-noise ratio for each of the eight settings of the controllables or inner array. Because the objective is to minimize the response solder defects per million, the formula for signal to noise is

$$S/N = -10 \log \left[\frac{1}{n} \sum y_i^2 \right].$$

An inspection of the results in Table 6.3 reveals that run number 4 maximized (S/N) and therefore, our experimental best result is S = 510, C = 7.2, F = .9, P = 200 and W = .5. It is also possible that some of the factors are more influential than others. To determine which factors are the most important, marginal averages are computed by averaging the four results at a particular level of a factor. For example, the first four runs of the inner array are with the solder pot temperature set at the higher level (510). These sixteen results (4 runs of the inner array, each with 4 runs of the outer array) are combined together to obtain an average of 170 solder defects per million.

Combined Inner and Outer Arrays
Results

Noise Factors	Run			
	1	2	3	4
Product Noise	#1	#1	#2	#2
Conveyor Tolerance	−.2	+.2	−.2	+.2
Solder Tolerance	−5	+5	+5	−5

Controllable Factors   Run

Run	Sold	Con	Flux	Pre-heat	Wave	1	2	3	4	Mean	S/N
1	510	10.0	1.0	150	0.5	194	197	193	275	215	−46.75
2	510	10.0	0.9	200	0.6	136	136	132	136	135	−42.61
3	510	7.2	1.0	150	0.6	185	261	264	264	244	−47.81
*4	510	7.2	0.9	200	0.5	47	125	127	42	85	−39.51
5	480	10.0	1.0	200	0.5	295	216	204	293	252	−48.15
6	480	10.0	0.9	150	0.6	234	159	231	157	195	−45.97
7	480	7.2	1.0	200	0.6	328	326	247	322	305	−49,76
8	480	7.2	0.9	150	0.5	186	187	105	104	145	−43.59

* Experimental Best Result

Table 6.3 Experimental Results for Wave Solder Example

The completed analysis is presented in Table 6.4 and displayed graphically in Figure 6.5. Analysis of the table and graph indicates that low flux density, high solder pot temperature and low wave height reduced the number of solder defects. It was this combination that influenced run number four. The effect of conveyor speed is marginal; however, economic considerations, such as throughput, might influence the decision to run the conveyor at a higher speed. Preheat temperature also appears insignificant with economic considerations favoring the lower temperature. Also note from Figure 6.6 that the anticipated interaction between conveyor speed and solder pot temperature does not appear important. The lines are nearly parallel, indicating that the effect (or Δ) for conveyor speed is the same at each level of solder pot temperature. From the S/N column of Table 6.4, a predicted best result would be S = 510, C = 10.0, F = .9, P = 150, and W = 0.5, where C and P were set based on economic considerations. This combination was never tested in the experiment, thus the term predicted best result To predict S/N at these settings, regression analysis was used to generate the following prediction equation (only the significant effects of S and F appeared significant enough to be placed in the model).

$$S/N = -45.52 + 1.35(S) -2.60(F)$$

To optimize our S/N equation we should set (S) and (F) at coded values of (+1) and (−1) respectively. Thus the best, predicted value for our S/N equation is 41.57. Although the S/N value for the predicted best result is a little smaller than

Parameter	Level	Mean	S/N
Solder Pot Temperature	480	225	−46.87
	510	170	−44.17*
Conveyor Speed	7.2	195	−45.17
	10.0	200	−45.87
Flux Density	0.9	140	−42.91*
	1.0	255	−48.11
Preheat Temperature	150	200	−46.03
	200	194	−45.01
Wave Height	0.5"	174	−44.50*
	0.6"	220	−46.54

* These are the optimum level settings for each significant factor. Factors without an asterisk are not significant and their levels can be based on other considerations.

Table 6.4 Analysis of Results for Wave Solder Example

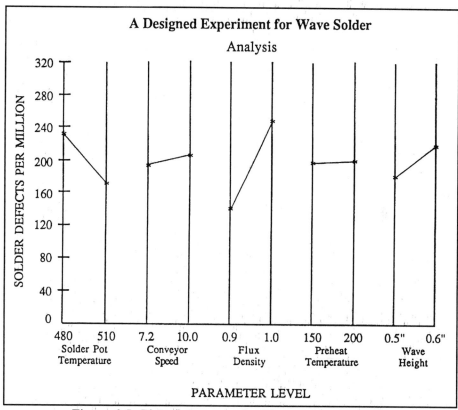

Figure 6.5 Plot of Averages for Wave Solder Example

that of the experimental best result the predicted best warrants further investigation because it could possibly cost less to produce. Confirmatory runs must now be conducted to confirm our prediction.

The use of S/N to simultaneously analyze the mean and variability provides a simple approach to the analysis. The reader is referred, however, to Chapter 5 where S/N analysis did not appear sensitive to information on variability. The combining of $\bar{y}$ and s in a S/N fashion may not always be the ideal approach. One should consider separate analysis of $\bar{y}$ and s (or ln s) to supplement S/N analysis (see discussion in Chapter 5).

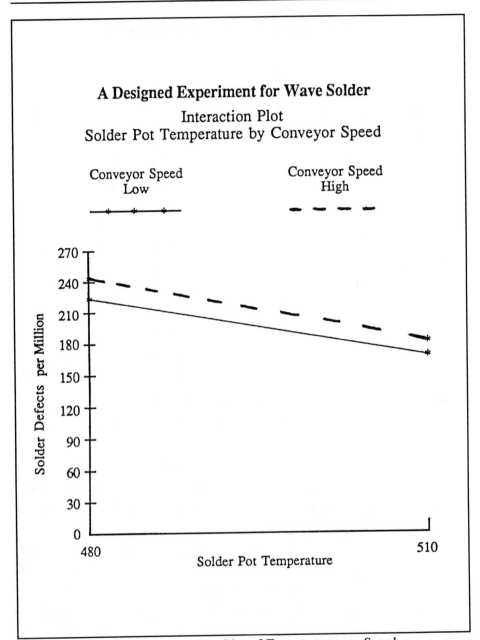

Figure 6.6 Interaction Plot of Temperature vs. Speed
for Wave Solder Example

6.5 Comparison of Classical and Taguchi Approaches

Classical	Taguchi

1. Brainstorming (as discussed in Chapter 1). Select responses based on customer needs.

1. Brainstorming (as discussed in Chapter 1). Try to select responses that do not have interactive inputs.

2. Effects to be tested depend on screening or modeling strategy. For screening estimate only linear effects and for modeling typically include linear 2-way interactions and non-linear (or quadratic) effects.

2. Effects to be tested usually consist of only linear and quadratic effects. Taguchi methods can incorporate 2-way interactions, but priority is given to the other effects for simplicity.

3. Determine desired levels of α and β in order to estimate the appropriate sample size to measure average shifting effects. For recent Rules of Thumb on sample size for average and variance shifting effects, see Appendix M.

3. Discussions about α and β normally are not addressed. Sample size, n, is found based upon the number of effects to be tested from (2) and the number of levels of each factor. Little guidance is provided on numbering of replicates.

4. Build an orthogonal design (usually R_{III} for screening and R_V or higher for modeling) to

4. Orthogonal designs are tabled and, because of de-emphasis of interactions, are usually

Classical	Taguchi

satisfy (2) and (3) above. (See Appendix M or inside front cover for Rules of Thumb)

screening designs or R_{III}. Normally the L_{12}, L_{18} and L_{36} are emphasized.

5. Randomize the order of the runs to spread all unmeasured noise factors evenly across the other effects. This step ensures internal validity of the experiment and allows for causality analysis. In other words, the effects of any uncontrolled variable will not bias our analysis.

5. Randomization is de-emphasized. Factors are placed in up to 4 groups based on their difficulty level to be reset during experimentation. The hardest factors to reset typically are assigned to columns with only one change of levels (for 2-levels designs) and the easiest to reset are assigned to alternating 1, 2 values. Thus, randomization is not enforced, making the experiment vulnerable to confounding of unsuspecting noise factors. Advocates of Taguchi will argue that if all important controllable and noise factors are in the design, then the confounding problem is resolved. Classical experimenters will still be concerned about factors which were not included due to

<table>
<tr><td>Classical</td><td>Taguchi</td></tr>
</table>

	incomplete brainstorming or inability to be measured.
6. Run the experiment and collect the data. Exercise extreme discipline.	6. Run the experiment and collect the data. Exercise extreme discipline.
7. Perform hypothesis tests on all desired effects and classify as significant or not significant using the t or F test. Use replication at some or all points to estimate error and/or pool higher order interaction sums of squares for the error. Recent emphasis is on the use of normal probability plots or Pareto charts for unreplicated designs. See Chapter 2 for simple modeling and Appendix M for Rules of Thumb on significance.	7. Usually, hypotheses tests are not emphasized; instead, a graphical analysis is conducted or the S/N ratio is used. If ANOVA is conducted, the error estimate is based on the pooling of insignificant sums of squares. Rules for pooling for sums of squares are not rigorous. F values are tested for significance per classical procedures; however, it is recommended that final unpooled F ratios greater than 2 not be ignored [4].
8. Build a simplified mathematical model for $\hat{y}$ and $\hat{s}$ to (i) allow you to predict the output for any settings of the inputs (ii) form prediction intervals (iii) estimate the optimal response	8. Important effects are determined graphically to select a "paper champion" or the "experimental champion" based on the best $\bar{y}$ or largest S/N. Prediction equations are generated, but the

Classical	Taguchi

through the use of Response Surface Methodology, and (iv) use the model for sensitivity and tolerance analysis.

modeling process is inferior to classical techniques. Prediction intervals are not calculated. The mathematical model is used for prediction, but not to locate the optimal as in Response Surface Methodology.

9. If the optimal response lies outside the sample region, conduct another experiment in the direction of the optimal. (See Chapter 7)

9. No iterative experimentation is used.

10. Find the optimal through an iterative experimental procedure.

10. The true optimal is not really found. Rather, settings for the best response over the experimental region are based on experimental and/or paper champions.

11. Set the process knobs at the optimal settings and go on-line. Recent emphasis is also on confirmation runs and the prediction of performance i.e., predicting C_p or C_{pk}.

11. The importance of confirmation runs is stressed. Confirmatory runs are made prior to going on-line.

| Classical | Taguchi |

12. Assumptions include normality, independence and equal variability. The F test is robust to minor violations of normality and homogeneity especially for large samples balanced for each experimental condition. If you are concerned about large violations, you can use a Pareto chart approach, i.e., rank order effects based on F values and select the larger ones as being important. More emphasis today is on blending science and engineering knowledge, common sense, and statistics when drawing conclusions.

12. Assumptions are similar to the classical method. Independence may be a problem due to lack of randomization. Unequal variance is accounted for by the transformation used in S/N calculations.

In addition to the 12 points previously discussed, the Taguchi method stresses simplicity. The idea of using tabled orthogonal arrays and marginal mean analysis is very appealing to engineers. It is obviously an easy way to get started in designing good experiments. However, engineers and scientists should not be content to limit themselves to the basics. Eventually they will want to progress to the more sophisticated designs and analysis techniques discussed in Chapters 3, 4, and 5 and summarized inside the front cover or in Appendix M. Regardless of which approach you use, make sure you have also included a plan to document your results.

Probably the most important contributions of the Taguchi approach are in the area of the loss function and robust designs. The signal-to-noise statistic does not appear to be the best metric for finding dispersion effects as pointed out in Chapter 5. In addition, the Taguchi approach will infrequently find the optimal; it is widely considered to be 60-80% effective. The Taguchi advocates will maintain that further improvements are so costly that more sophisticated efforts are not cost efficient. One could argue that this is true in many applications; however, there are applications that require more fine tuning in finding the optimal. For 80-100% effectiveness, the classical method of designed experiments and advanced Response Surface Methodology (RSM) techniques should be used.

Finally, design of experiments should not be an either/or approach (i.e., either Classical or Taguchi). Your objective should be to use the most powerful tools available to gain the necessary knowledge of your process (or activity). Just as you would never consider playing a major golf tournament with only a putter, you should never limit yourself to only one experimentation approach.

Problem Set

1-6. Redo the problems from Chapter 5 using the appropriate S/N formula for those questions. Compare the results using S/N and the results you found originally.

7.   When using signal-to-noise, how do you find the optimal levels for a process?

8.   Name 3 advantages of Taguchi's methodology. Name 3 shortcomings.

9.   Why does Taguchi use a quadratic approximation for the loss function? Into what components does the function partition ?

10.   When does a factor belong in an outer array?

Bibliography

1. Chernoff, H. and L. E. Moses, (1959) *Elementary Decision Theory*, John Wiley and Sons, Inc.

2. Fisher, R. A., (1966) *Design of Experiments*, 8th ed., Hafner, Macmillan Publishing Company.

3. Hunter, J. S., (1985) "Statistical Design Applied to Product Design," *Quality Progress*, Vol. 17, No. 4.

4. *Introduction to Quality Engineering, 5-Day Seminar Course Manual*, (1987) American Supplier Institute, Inc.

5. Kackar, Raghu N., (1986) "Taguchi's Quality Philosophy: Analysis and Commentary," *Quality Progress*, Dec. 1986.

6. Kirk, Roger E., (1982) *Experimental Design*, Brooks/Cole Publishing Co.

7. Lindquist, E. F., (1953) *Design and Analysis of Experiments in Psychology and Education*, Boston, Haughton Mifflin.

8. Norton, D. W., (1952) "An empirical investigation of some effects of non-normality and heterogeneity of the F-distribution," unpublished doctoral dissertation, State University of Iowa.

9. Rogan, J. C. and H. J. Kesselman, (1977) "Is the ANOVA F-test robust to variance heterogeneity when sample sizes are equal? An investigation via a coefficient of variation," *American Educational Research Journal* 1977, Vol 14, pp. 493-498.

10. Sullivan, L. P., (1984) "Reducing Variability: A New Approach to Quality," *Quality Progress*, Vol. 15, No. 4.

11. Taguchi, G., (1986) *Introduction to Quality Engineering*, Asian Productivity Organization.

12. Warsavage, B. and S. R. Schmidt, (1987) "Taguchi Philosophy and Methodology" presented at an IBM Technical Symposium.

Chapter 7

Optimization and Advanced Response Surface Methods

7.1   Introduction and Overview

If you have used the design and analysis techniques discussed in Chapters 3 and 5 and found factor settings that provide a satisfactory response, then there may be no need for further investigation.   Unfortunately, using the simplest optimization techniques may not always lead to the most desirable results. When your best response does not prove satisfactory, what can you do? There are four things to consider in selecting the next plan of action:

(1)   Were all potential factors and associated 2-way interactions included in the design?

(2)   Is there too much noise in the data?

(3)   Did you select the appropriate number of levels for each factor?

(4)   Did you select the appropriate range for each factor?

If your problem is related to (1) or (2) you will need to refer back to your process flow and cause/effect diagrams to insure you have considered all important factors for experimentation plus procedures to hold most of the remaining variables constant, thereby, reducing noise. To counter excessive noise, you can add more replicates to each run to reduce the impact of noise. If important factors and/or interactions were left out, you will need to perform more experiments. Once you have encountered problems such as these that point to inadequate initial brainstorming, the consequences involved in correcting the problems should convince you to invest more time in future brainstorming.

   If (3) and (4) are contributors to your problem, then you probably need to select a better method to optimize the response. This chapter is designed to compare 3 methods of finding the optimal input factor settings: (1) one-at-a-time designs, (2) orthogonal arrays (Taguchi methods), and (3) response surface methodology (RSM). Although the first two methods are simple and easy to implement, they typically will produce results that are $60-80\%$ of the total improvement to be made. If you need $80-100\%$ of the total improvement, you need another method. To illustrate this point, consider the example that follows.

   Given a hypothetical process with 2 input factors, f_1 = temperature (°F) and f_2 = pressure (psi), as shown below, the objective is to optimize some measure of percent yield, y. Figures 7.1a, 7.1b, 7.2a, and 7.2b present the one-at-a-time approach. In Figure 7.1a, the researcher uses brainstorming information to select f_2 = 150 as a place to start his investigation. He sets the pressure knob at 150 psi and varies the temperature over its range to find the optimal temperature setting, which appears to occur at 200°F with a corresponding yield of 85.3%.

f_1 Temperature (°F)	f_2 Pressure (psi)
$100 \leq f_1 \leq 400$	$100 \leq f_2 \leq 200$

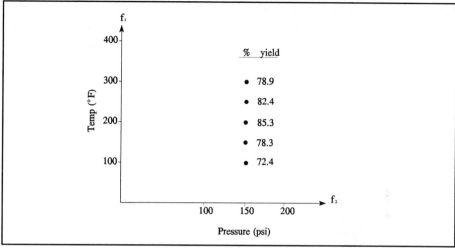

Figure 7.1a   Plot of Pressure and Temperature vs. Yield with
Pressure Held Constant

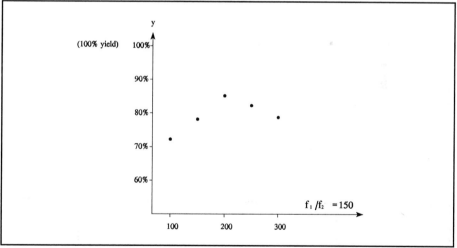

Figure 7.1b   Plot of Yield vs. Temperature with Pressure = 150 psi

The temperature knob is now fixed at 200°F and pressure is varied over its range
as shown in Figure 7.2a and 7.2b. These figures reveal that the best X_2 setting
(given f_1 = 200) is f_2 = 150. At this point, the researcher would conclude his

best results occur at (f_1 = 200, f_2 = 150) and he should make several confirmation runs to ensure the results produce a desired average yield and tolerable variance.

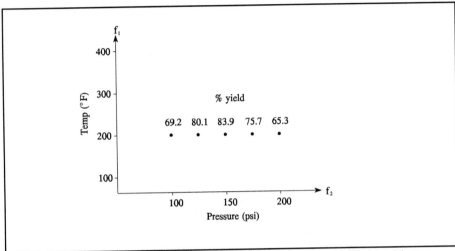

Figure 7.2a Plot of Pressure and Temperature vs. Yield with
Temperature Held Constant

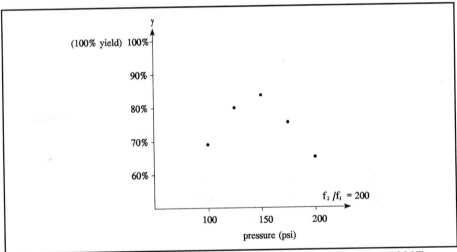

Figure 7.2b Plot of Yield vs. Pressure with Temperature = 200°F

Unfortunately, if you choose the one-at-a-time approach and your results are less than desirable, it is confusing as to how you might improve. In this example, the response contours (not known to the experimenter) are shown in Figure 7.3. The one-at-a-time approach will use up large amounts of resources and if results are unsatisfactory, there is no efficient way to seek improvements.

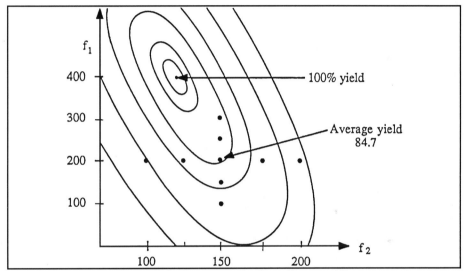

Figure 7.3 Response Surface Contour Plot

Using the Taguchi approach, or any other orthogonal array approach, without advanced RSM would consist of the following:

(1)   The researcher would select an experimental region over which he chooses to investigate the response.

(2)   Since the input factors in this example are both continuous, either a 2- or 3-level design would typically be used to evaluate the response over the experimental region.

(3)   Using analysis techniques as described in Chapter 5, a best setting for each of the factors is obtained.

Referring back to the example, assume the experimental region is decided to be $\{100 \leq f_1 \leq 200$ and $125 \leq f_2 \leq 175\}$. The results for a 3-level design appear in Figure 7.4. Based on average marginals, the best settings would be $f_1 = 200$ and $f_2 = 150$ which will produce an estimated average response of approximately 84.3.

Figure 7.4 Yield Results from 3-Level, 2-Factor Design

Obviously, the failures of the one-at-a-time and the Taguchi approaches associated with this example are due to a poor choice of factor values over which to investigate. As contrived as this data is, the point to be made is that this unfortunate choice of factor ranges can occur with old and new products even from the most experienced researchers, resulting in less than the desired results. When this situation arises, a sequential approach is needed to efficiently lead you in the direction of the optimal. The previous two methods do not have this flexibility and, if modified to produce an optimum search, they will typically be inefficient. What is recommended is an approach based on RSM.

7.2   Response Surface Methodology

The basic idea associated with response surface methodology and experimental design is to build an efficient design to determine if the experimental region contains the optimum. If the optimum lies in the experimental region, build a mathematical model to locate it, or else continue experimenting along a gradient of steepest ascent (direction of greatest improvement) to find the center of a new experimental region to be tested. At this new point, build another efficient design in the new experimental region and test whether it contains the optimum. Continue the process until the optimum is located. In order to conduct the process just described, we must be able to represent y as a function of the input factors. If the true statistical relationship is represented by $y = h(x_1, x_2, ..., x_k)$, our objective is to estimate $h(x_1, x_2, ..., x_k)$ with as few experimental runs as possible. To accomplish this task, a Taylor Series approximation of $h(x_1, x_2, ..., x_k)$ is recommended. In most cases, a second-order Taylor Series is a sufficient approximation. For k input factors, the second order Taylor Series approximation for $h(\underline{x})$ about $\underline{c} = (c_1, c_2, ..., c_k)$ is given by

$$h(\underline{x}) = h(\underline{c}) + \sum_1^k (x_i - c_i)\frac{\partial h}{\partial x_i}\Bigg|_{\underline{x} = \underline{c}} + \sum_1^k \frac{(x_i - c_i)^2}{2!}\frac{\partial^2 h}{\partial x_i^2}\Bigg|_{\underline{x} = \underline{c}}$$

$$+ \sum_{i>j}\sum (x_i - c_i)(x_j - c_j)\frac{\partial^2 h}{\partial x_i \partial x_j}\Bigg|_{\underline{x} = \underline{c}} + R(\underline{x}),$$

where $R(\underline{x})$ is the remainder or error term.

For coded data, i.e., $_i = 2\left(\dfrac{f_i - \overline{f}_i}{d_i}\right)$ it is advantageous to build the Taylor series about $\underline{c} = \underline{0}$. Thus,

$$h(\underline{x}) = h(\underline{0}) + \sum_{1}^{k} (x_i) \frac{\partial h}{\partial x_i}\bigg|_{\underline{x} = \underline{0}} + \sum_{1}^{k} (x_i^2) \frac{\partial^2 h}{\partial x_i^2}\bigg|_{\underline{x} = \underline{0}}$$

$$+ \sum_{i>j}\sum (x_i)(x_j) \frac{\partial^2 h}{\partial x_i \partial x_j}\bigg|_{\underline{x} = \underline{0}} + R(\underline{x})$$

If the second order model is a good fit, then $R(\underline{x})$ should be relatively small for any $\underline{x}$ in the experimental region. Notice that we will not be using calculus to build our Taylor Series type model. Rather, by properly designing our experiment, the modeling process will actually build a Taylor Series type model.

The Taylor Series type model from our experiment can be arranged such that it resembles the second order polynomial model below. The weights, b_0, b_i, b_{ii}, and b_{ij} are easily obtained from a multivariable regression software package.

$$y = b_0 + \sum_{1}^{k} b_i x_i + \sum_{1}^{k} b_{ii} x_i^2 + \sum_{i>j}\sum b_{ij} x_i x_j + \underline{e}$$

$$b_0 = h(\underline{0}) \qquad\qquad b_i = \frac{\partial h}{\partial x_i}\bigg|_{\underline{x} = \underline{0}}$$

where

$$b_{ii} = \frac{\partial^2 h}{\partial x_i^2}\bigg|_{\underline{x} = \underline{0}} \qquad\qquad b_{ij} = \frac{\partial^2 h}{\partial x_i \partial x_j}\bigg|_{\underline{x} = \underline{0}}$$

$\underline{e}$ is the error term.

The error term $\underline{e}$ can be broken into two parts:

(1) pure error   —   estimated through replication.

(2) lack of fit   —   estimated by the difference in the model error and pure error.

    —   This error indicates whether the model type (linear, quadratic, etc.) is appropriate.

Using the central composite design to accomplish RSM, the replications required to estimate pure error will occur at the center point. Consider the previous example where x_1(high) = 200, x_1(low) = 100, x_2(high) = 175, and x_2(low) = 125. The following table shows a first-order design with replicated center points, where x_i is the coded value of f_i.

f_1	f_2	x_1	x_2	$x_1 x_2$	y
200	175	+1	+1	+1	75.9
200	125	+1	−1	−1	82.1
100	175	−1	+1	−1	70.1
100	125	−1	−1	+1	69.7
150	150	0	0	0	75.6
150	150	0	0	0	76.2

The next step is to test for model adequacy. Using regression and a 0.10 significance level, the first-order model which fits the data is

$$\hat{y} = 75.9 + 4.55x_1 - 1.45x_2 - 1.65x_1 x_2$$

The computer output shown in Table 7.1 indicates non-significant lack of fit for a quadratic term.

Dependent Variable: y   N: 6   Multiple R: .999   Squared Multiple R: .998

Adjusted Squared Multiple R: .991   Standard Error of Estimate: 0.424

Variable	Coeff	Std Error	Std Coef	Tolerance	T	P(2 Tail)
Constant	75.900	0.300	0.000	1.0000000	253.000	0.003
x_1	4.550	0.212	0.888	1.0000000	21.449	0.030
x_2	−1.450	0.212	−0.283	1.0000000	−6.835	0.092
$x_1 \cdot x_2$	−1.650	0.212	−0.322	1.0000000	−7.778	0.081
quadratic term	−1.450	0.367	−0.163	1.0000000	−3.946	0.158

Analysis of Variance

Source	Sum-of-Squares	df	Mean-Square	F-Ratio	P
Regression	104.913	4	26.228	145.713	0.062
Residual	0.180	1	0.180		

Table 7.1 Regression Analysis Table for Initial Experimental Region

The insignificant quadratic lack of fit indicates that the optimum lies outside of the initial experimental region. Therefore, the next step is to find a gradient vector which will point us toward the optimum response. This is obtained by taking the derivative of the first order model with respect to each of the factors. The gradient vector, g, is really a vector perpendicular to contour lines developed from our model. However, using the calculus approach

$$ g \ = \left(\frac{\partial \hat{y}}{\partial x_1} , \ \frac{\partial \hat{y}}{\partial x_2} \right) \Bigg|_{\substack{x_1 = 0 \\ x_2 = 0}} = (4.55, \ -1.45). $$

To experiment along the gradient in reasonable increments, you divide g by the smallest absolute value of the components of g. In this case, we divide by 1.45, indicating that the first new experiment, R_1, should be conducted at (3.14, -1). Subsequent experiments, R_i, will take place at coordinates found by adding 3.14 to the x_1 component and subtracting 1 from the x_2 component. Experimentation along the gradient direction should continue until curvature is detected or until you reach the factor limitations. Then determine the estimated coordinates of the minimum or maximum along the gradient, make these coordinates the center of a new experimental region, conduct a new designed experiment and test for quadratic lack of fit. If the quadratic term is significant, add axial points to estimate the 2nd order model and locate the stationary point.

For our example, the experimental results for R_1, R_2, and R_3 are 91.0, 89.4, and 77.0, respectively. The best response along the gradient appears to occur at R_1; therefore, let the new experimental center point be (3.14, -1). See Figure 7.5. The new design matrix and corresponding response values are shown below.

f_1	f_2	x_1	x_2	y
357	150	4.14	0.0	87.0
357	100	4.14	-2.0	83.1
257	150	2.14	0.0	75.7
257	100	2.14	-2.0	86.9
307	125	3.14	-1.0	91.0
307	125	3.14	-1.0	90.1

To convert x_i back to the original values, use the formula

$$x_i = 2\left(\frac{f_i - \overline{f}_i}{f_{max} - f_{min}} \right)$$

For example, $x_1 = 4.14$ results in $4.14 = \dfrac{2(f_i - 150)}{100}$

which simplifies to $f_1 = 357$.

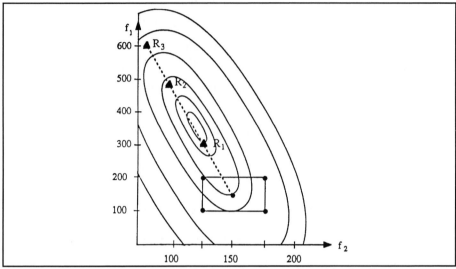

Figure 7.5 Response Surface Contour Plot with Gradient

To simplify the model building and analysis for this new experimental region, the design matrix will be altered to reflect R_1 as the center of the experimental region. This is accomplished by the following data table,

w_1	w_2	y
+1	+1	87.0
+1	−1	83.1
−1	+1	75.7
−1	−1	86.9
0	0	91.0
0	0	90.1

where,
$$w_i = 2 \left(\frac{f_i - \bar{f}_i}{f_{i\,max} - f_{i\,min}} \right)$$

$\bar{f}_i$ = mean for f_i over the new experimental region

$f_{i\,max}$ = maximum value of f_i over the new experimental region

$f_{i\,min}$ = minimum value of f_i over the new experimental region

i.e.,

$\bar{f}_1$ = 307, $f_{1\,max}$ = 357, $f_{1\,min}$ = 257

$\bar{f}_2$ = 125, $f_{2\,max}$ = 150, $f_{2\,min}$ = 100

Table 7.2 contains the least squares regression analysis of the data.

Dependent Variable: y N: 6 Multiple R: .999 Squared Multiple R: .997						
Adjusted Squared Multiple R: .987 Standard Error of Estimate: 0.636						
Variable	Coeff	Std Error	Std Coef	Tolerance	T	P(2 Tail)
Constant	90.550	0.450	0.000	1.0000000	201.222	0.003
w_1	1.875	0.318	0.299	1.0000000	5.893	0.107
w_2	−1.825	0.318	−0.291	1.0000000	−5.735	0.110
$w_1 \times w_2$	3.775	0.318	0.602	1.0000000	11.864	0.054
quadratic	−7.375	0.551	−0.679	1.0000000	−13.381	0.047
Analysis of Variance						
Source	Sum-of-Squares	df	Mean-Square	F-Ratio	P	
Regression	156.908	4	39.227	96.857	0.076	
Residual	0.405	1	0.405			

Table 7.2 Regression Analysis Table for Final Experimental Region
(without α points)

Since the quadratic term is significant at the .05 level, it is decided to add axial points to build the full second order model. The value for α (the axial point distance from the center) is

$$\alpha = (n_F)^{1/4} = (4)^{1/4} = 1.414,$$

where n_F equals the number of runs in the factorial portion of the design. Therefore, the completed data set is shown next.

w_1	w_2	y
1	1	87.0
1	−1	83.1
−1	1	75.7
−1	−1	86.9
0	0	91.0
0	0	90.1
1.414	0	85.4
−1.414	0	77.6
0	1.414	80.5
0	−1.414	77.4

The computer output for the fitted model is shown in Table 7.3. A graphical display is shown in Figure 7.6.

Variable	Coeff	Std Error	Std Coef	Tolerance	T	P(2 Tail)
Constant	90.549	2.135	0.000	1.0000000	42.412	0.000
w_1	2.316	1.068	0.399	1.0000000	2.170	0.096
w_2	-0.365	1.068	-0.063	1.0000000	-0.342	0.750
$w_1 \times w_2$	3.775	1.510	0.460	1.0000000	2.501	0.067
$(w_1)^2$	-3.787	1.412	-0.546	.8164322	-2.681	0.055
$(w_2)^2$	-5.063	1.412	-0.730	.8164322	-3.584	0.023

Dependent Variable: y   N: 10   Multiple R: .930   Squared Multiple R: .865

Adjusted Squared Multiple R: .695   Standard Error of Estimate: 3.019

Analysis of Variance

Source	Sum-of-Squares	df	Mean-Square	F-Ratio	P
Regression	232.776	5	46.555	5.107	0.070
Residual	36.465	4	9.116		

Table 7.3 Regression Analysis Table for Final Experimental Region
(with α points)

Using a significance level of .10, the simplified model is

$$\hat{y} = 90.549 + 2.316w_1 + 3.775w_1 \times w_2 - 3.787(w_1)^2 - 5.063(w_2)^2.$$

The stationary point is found by differentiating $\hat{y}$ with respect to both factors and solving simultaneously for w_1 and w_2.

$$(1) \qquad \frac{\partial \hat{y}}{\partial w_1} = 2.316 + 3.775w_2 - 7.574w_1 = 0$$

$$(2) \qquad \frac{\partial \hat{y}}{\partial w_2} = \qquad -10.126w_2 + 3.775w_1 = 0$$

Solving for w_1 in (2) results in $w_1 = 2.682w_2$.
Substituting this for w_1 in (1) results in

$$3.775w_2 - 7.574(2.682)w_2 = -2.316 \quad \text{or} \quad w_2 = 0.140.$$

Now, substituting this value in (2) results in $w_1 = .375$. Thus, our stationary point is estimated to be at $(w_1, w_2) = (.375, .140)$.

Define B as the second-order matrix shown below:

$$B = \begin{bmatrix} b_{11} & \dfrac{b_{12}}{2} \\ \dfrac{b_{21}}{2} & b_{22} \end{bmatrix} \quad \text{where} \quad$$

b_{11} is the coefficient of $(w_1)^2$

b_{22} is the coefficient of $(w_2)^2$

$b_{12} = b_{21}$ is the coefficient of $(w_1 \times w_2)$

The eigenvalues of B (computations not shown) are -6.419 and -2.433. Since both eigenvalues are negative, the stationary point $(w_1, w_2) = (.375, .140)$ is a maximum. The actual-value coordinates of the stationary point are $(325.75, 128.5)$. See Figure 7.6 for a summary of the experimentation used to find the optimum.

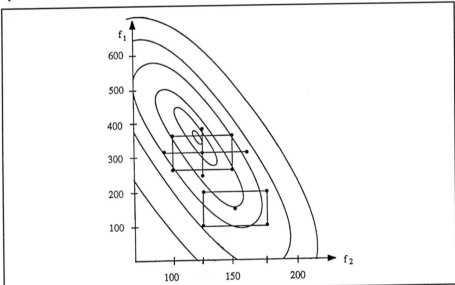

Figure 7.6 Response Surface Contour with Initial and Final
Experimental Regions

The following table of possible eigenvalues summarizes the characteristics of stationary points.

	Eigenvalue Sign	Stationary Point
(1)	all positive	minimum
(2)	all negative	maximum
(3)	some negative, some positive	saddle point
(4)	all of one sign with one or more close to zero	rising or falling ridge

For a more indepth study of RSM, see Box and Draper [1], Montgomery [2] or Myers [3].

Problem Set

1.   When should you use Response Surface Methodology? What findings in ANOVA indicate that you should use RSM?

2.   What is the gradient vector and what does it have to do with RSM?

3.   Given that you have searched along the gradient and have selected a new experimental region, how do you know this area contains a stationary point? How can you tell if that stationary point is a: maximum?
                                                                                         minimum?
                                                                                         saddle point?
                                                                                         a ridge?

4.   State the main reason for using coded variables in regression analysis.

5.   If the gradient vector from a 2-term model is:

$$g = (-2.67, 1.33)$$

Where should you conduct the first experiment for RSM? Find the second test values as well. If the second test value yields the maximum, what values should you use in a L_4 design?

Bibliography

1.   Box, G. E. P. and N. R. Draper, (1987) *Empirical Model Building and Response Surfaces*, John Wiley and Sons, Inc., New York.

2.   Montgomery, Douglas C., (1984) *Design and Analysis of Experiments*, 2nd ed., John Wiley and Sons, Inc., New York.

3.   Myers, Raymond H., (1976) *Response Surface Methodology*, Virginia Polytechnic Institute.

Chapter 8

Case Studies

8.1 Introduction

This chapter contains a collection of real industrial case studies, simulated studies, and training type case studies. The complexity of these studies ranges from the fairly simple to the state-of-the-art. The studies have not been altered by the authors of this text and thus, it is very possible that many of us would have conducted a similar study using a different approach. The purpose of this chapter is not to critique each case study, but rather to document the types of applications that exist. If the reader has serious concerns about any particular study, please share these concerns with the authors by writing or calling:

AIR ACADEMY ASSOCIATES
1155 Kelly Johnson Blvd., Suite 105
Colorado Springs, CO   80920
(719) 531 - 0777 ✧ (719) 531 - 0778 (FAX)
e-mail: aapa@airacad.com
website: www.airacad.com

The long range plan is to add more case studies to this text and develop a second volume dedicated to applications. If you would like to publish your case study in the next printing of this text, please contact us.

Results of a First Run Fractional Factorial for the TRE 10X Photo Process on CMOS III Gate Policy

Submitted by: J. Merritt
Honeywell
Colorado Springs, CO

Background

At the time of this experiment, the author's process was generating an unacceptable rate of rework on the CMOS III 10X stepper process because of poor resolution. In an effort to evaluate this process and determine which process parameters have the most influence on poor resolution, a 2^{8-4} fractional factorial was devised.

Experimental Objective

To detect the relative magnitude of first order effects and to screen out less significant factors.

	Controlled Factors	"low" (-1)	"center" (0)	"high" $(+1)$
(A)	Resist Thickness	12,400 Å	13,700 Å	15,000 Å
(B)	Softbake Time	20 min	30 min	40 min
(C)	Softbake Temperature	95°C	100°C	105°C
(D)	Develop Concentration	3.1:1	3.0:1	2.7:1
(E)	Develop Temperature	19°C	21°C	23°C
(F)	Develop Time	80 sec	90 sec	100 sec
(G)	Focus	430	480	530
(H)	Exposure	200	220	240

Response

The response is a quantitative Critical Dimension (CD). The CD was measured at five points across the wafer for each experimental combination. For this data, the sample mean and sample standard deviation were calculated.

Experimental Array

Design #	A	B	C	D	E	F	G	H
1	−	−	−	−	−	−	−	−
2	+	−	−	−	+	+	+	−
3	−	+	−	−	+	+	−	+
4	+	+	−	−	−	−	+	+
5	−	−	+	−	+	−	+	+
6	+	−	+	−	−	+	−	+
7	−	+	+	−	−	+	+	−
8	+	+	+	−	+	−	−	−
9	−	−	−	+	−	+	+	+
10	+	−	−	+	+	−	−	+
11	−	+	−	+	+	−	+	−
12	+	+	−	+	−	+	−	−
13	−	−	+	+	+	+	−	−
14	+	−	+	+	−	−	+	−
15	−	+	+	+	−	−	−	+
16	+	+	+	+	+	+	+	+
*17	0	0	0	0	0	0	0	0

NOTE: Multiple centerpoints were run so as to obtain an experimental error estimate. This estimate was used to sort out important from unimportant effects for both the mean and standard deviation.

Experimental Conduct

The experiments were all conducted during a 10 hour period. Randomization was conducted where it was possible.

Calculation of Effects

Factor	Description	Effect (in order of absolute magnitude)
A	Resist Thickness	.504
H	Exposure	−.401
B	Softbake Time	.389
F	Develop Time	−.386
A × C	Resist Thickness × Softbake Temp	−.369
A × B	Resist Thickness × Softbake Time	−.359
C	Softbake Temp	.309
E	Develop Temp	−.301
B × C	Softbake Time × Softbake Temp	−.129
G	Focus	−.121
D × E	Develop Concentration × Develop Temp	.111
C × D	Softbake Temp × Develop Concentration	.069
A × D	Resist Thickness × Develop Concentration	−.059
D	Develop Concentration	.039
B × D	Softbake Time × Develop Concentration	.024

Table 1 Mean CD

Calculation of Effects

Factor	Description	Effect (in order of absolute magnitude)
C	Softbake Temp	−.117
A	Resist Thickness	.109
B × C	Softbake Time × Softbake Temp	−.090
H	Exposure	−.084
A × B	Resist Thickness × Softbake Time	−.070
D	Develop Concentration	.066
C × D	Softbake Temp × Develop Concentration	−.063
F	Develop Time	−.052
E	Develop Temp	−.037
A × D	Resist Thickness × Develop Concentration	−.036
G	Focus	.032
B	Softbake Time	.024
D × E	Develop Concentration × Develop Temp	.021
A × C	Resist Thickness × Softbake Temp	−.016
B × D	Softbake Time × Develop Concentration	.014

Table 2 Standard Deviation of CD

Analysis (Mean CD)

From the variation observed in the centerpoints, it was determined that any effect greater than 0.30 would be considered significant (see Table 1). The major surprise for this characteristic was that the Developer Concentration (Factor D) appeared to have little or no effect on the mean CD.

Analysis (standard deviation of CD)

The data in Table 2 indicates that Softbake Temperature (Factor C) may be a dispersion effect. The sign of the effect indicates smaller variability may result at the higher softbake temperature. Technically, this appears readily explainable since a lower softbake temperature is less efficient in removing solvents from the photoresist.

Conclusions

The experiment was successful in screening out major mean effect factors and interactions, as well as identifying a potential dispersion reduction factor.

Reduction of Measurement Device Variability
Using Experimental Design Techniques

Submitted by: Dale Owens
                      Kurt Manufacturing
                      Minneapolis, Minnesota

Background

Kurt Manufacturing, headquartered in Minneapolis, Minnesota, is a precision machine shop of considerable size and expertise. Kurt's clients include the nation's elite from aerospace, communications, and computers. Starting in the mid-1980's, Kurt embarked upon an aggressive program to implement gage studies and Process Control techniques across their entire operation. Findings from initial gage studies indicated that in certain situations, the inherent variability of some measurement devices was "large" in relation to the specified engineering tolerances. Kurt's own interval criteria is to use only measurement systems which have an inherent variability of less than 10% of the engineering tolerance.

In what follows, we will learn how Kurt made use of a simple orthogonal array so as to come up with a gaging process with vastly reduced variability.

Case Study

When setting up a new process with dedicated gaging, it is not only important to have a capable machine, but a capable gaging device as well. At Kurt, our goal is to have a gage R&R (repeatability and reproducibility) of less than 10% of specification and a process C_{pk} greater than 1.33.

An initial gage R&R study on our device in question yielded a gage error of 80.5% of the characteristic tolerance. Obviously, this was found to be far from acceptable. Accordingly, a team composed of the operator, supervisor, manufacturing engineer, quality engineer, and statistical facilitator met together to determine the variables and levels for a designed experiment.

Extensive brainstorming resulted in the following variables and levels:

(1) Type of indicator

> level 1 = current situation = .0001" increment indicator
> level 2 = .00005" indicator

(2) Indicator support

> level 1 = existing special stand
> level 2 = height stand

(3) Inspection movement

> level 1 = keep the part stationary and move the indicator
> level 2 = keep the indicator stationary and move the part

(4) Type of inspection master

> level 1 = existing part master
> level 2 = use gage blocks

(5) Indicating method

> level 1 = bring the indicator straight into the part
> level 2 = sweep the indicator into the part from the side

Since this was strictly a screening design, it was decided not to test for 2-factor or higher order interactions. After some discussion, the experimental goal was established as "to reduce the amount of gage error." Byproducts of meeting this overall goal would result in reduced cost through a reduction in inspection time and more accurate and precise measurement readings. The experimental objectives of the experiment included determining the effect of factors (1) through (5) on repeatability. Repeatability was quantified through the use of the sample standard deviation.

The team decided to make use of a 2^{5-2} fractional factorial (Taguchi L8). The test plan called for checking one part five times in succession per inspection setup. This approach was used for all eight experimental combinations. The experimental setup was as follows:

Exp #	Increment	Indicator Support	Movement	Master	Method	Result*
1	.0001	tree	part	part	straight	.00004
2	.0001	tree	part	gage	sweep	.00001
3	.0001	stand	indic	part	straight	.00001
4	.0001	stand	indic	gage	sweep	.00005
5	.00005	tree	indic	part	sweep	.00000
6	.00005	tree	indic	gage	straight	.00002
7	.00005	stand	part	part	sweep	.00003
8	.00005	stand	part	gage	straight	.00003

*NOTE: *Result - sample standard deviation of five readings*.

By simply "eyeballing" the results, we find that combination #5 is the best. Verification runs by the company showed that this particular combination produced a gage R&R value of 15.2%, a big improvement over previous settings. Efforts were taken by the company to further decrease the gage error.

Corrosion Study

The following data was provided to us by an engineer wishing to study an inhibitor of corrosion on a metallic alloy. The questions to be answered in this study were: Is time an important factor? Do corrosion thicknesses change over time? To answer these questions, samples were assessed at zero month, one month, 2 month, and 3 month intervals. Values shown in Table 1 are amounts of material remaining.

Zero Months	One Month	Two Months	Three Months
99	104	108	96
114	77	107	68
85	80	69	98
*222	110	sample destroyed	105
127	84	89	104
124	90	124	90

Table 1

NOTE: The value of 222 in column "Zero Months" was discarded for technical reasons. Investigation revealed it was not representative. Because of some additional confusion, only five sample values were available in the "Two Months" column.

Two approaches were taken to answer the question posed. The first approach was to provide a graphical representation of the data. It appears in Figure 1. From this chart we see no clear cut answer to the question. The second assessment approach, a one-way ANOVA, shown in Table 2, resulted in the following statistics:

Source	df	SS	MS	F	"P Value"
Factor	3	1134.9	378.3	1.41	.274
Error	18	4844.3	269.1		
Total	21	5979.2			

Table 2

As we can see from Table 2, at a significance level of .05, there is insufficient evidence to reject the hypothesis that the sample came from populations with equal means.

Based upon the two above analysis approaches, it was concluded there was no major change in corrosion thickness over time.

Figure 1 Corrosion Thickness vs. Replication

Linear Motor Design*

Submitted by: Karen Cornwell
                           Digital Equipment Corporation

Case Study

The force constant typically drops off approximately 10% on the ends of the motor. To reach optimum seek-time, a more linear distribution is required.

Objective

Determine the effect various end cap and motor configurations have on the force constant and leakage flux values.

Analysis

The experiment was conducted using Taguchi's performance parameter (the signal-to-noise ratio). This made it possible to look at the force constant at both the mean level and variation around the mean. The analysis indicates that the 0.180 groove end cap and standard motor provide the best results.

Results

The improved configuration represents a major improvement over the standard configuration.

Introduction

In an attempt to improve the force constant linearity, the end cap was modified with a groove which created an air gap between the magnet and end cap. In the current configuration, the end cap contacts the magnets. With the addition of the air gap, the magnetic lines of flux will be channeled more toward the center pole rather than through the end cap. The same principle was also attempted with

---

*Originally presented at the Rocky Mountain Quality Conference, June 1987

a groove cut at the front of the motor. There is a limited amount of space available for an air gap or additional steel.

The Taguchi analysis technique was employed in developing an optimum motor/end cap configuration which resulted in the maximum force constant with the least variability.

The force constant is measured using a load cell. Current is applied to the coil in the forward and reverse directions. The resulting force is measured on a strain gage. The force constant is simply the measured force divided by the applied current.

Experimental Objective

Determine the effect of end cap, front motor cut, and center cut on the dependent variables: force constant and leakage flux.

Identification of Dependent Variable(s)

Force Constant

Leakage Flux

Identification of Independent Variables

Noise	Control
Tester Repeatability and Positional Location	A. End Cap
	1. .100 depth groove
	2. .155 depth groove
	3. .180 depth groove
	4. .200 depth groove
	5. .220 depth groove
	6. 44D Stainless Steel
	7. 303 Stainless Steel
	8. Standard
	B. Front Cut Motor
	C. Center Cut Motor

Setup of O.A.

We progressed through several experimental stages in conducting the experiment. Briefly, the stages (with applicable results) were as follows:

Stage 1:

We believed that several of the end cap configurations would not produce acceptable results for leakage flux. Because of this, we first tested the leakage flux at six levels for end cap. Results of this experiment were as follows:

End Cap	Leakage Flux (must be < 25)
standard	16
.155	17
.180	18
.200	23
.225	31
standard (303 stainless)	58

From this series of experiments, we found only the first four to have an acceptable leakage flux.

Stage 2:

Using the information from Stage 1, we set up an L8 with three control factors. Factor settings for the experiment were:

	Variable	Level 1	Level 2
(A)	end cap	standard	.180
(B)	front	standard	.120
(C)	center	standard	.080

The experimental design with applicable results was:

Run	A	B	AB	C	AC	BC	ABC	S/N*(of force constant)
1	1	1	1	1	1	1	1	11.97
2	1	1	1	2	2	2	2	10.97
3	1	2	2	1	1	2	2	11.08
4	1	2	2	2	2	1	1	10.22
5	2	1	2	1	2	1	2	13.12
6	2	1	2	2	1	2	1	12.10
7	2	2	1	1	2	2	1	12.35
8	2	2	1	2	1	1	2	11.23

*NOTE: *Signal-to-noise* $= 20 \log\left[\dfrac{y^2}{s^2}\right]$, *where y^2 and s^2 were calculated from*

8 data points.

MARGINAL MEAN RESULTS

FACTOR	LEVEL 1	LEVEL 2	|Δ|
A	11.06	12.20	1.14
B	12.04	11.22	.82
AB	11.63	11.63	0.00
C	12.13	11.13	1.00
AC	11.60	11.67	.06
BC	11.64	11.60	.01
ABC	11.66	11.60	.06

From the "Marginal Mean Results" on page 8-15, it appears that end cap at 0.180 cut is best, a standard front is best, and a standard center is best.

NOTE: *Leakage flux was acceptable for all combinations tested, i.e., the average was well below the upper specification of 25.*

Stage 3:

In Stage 3, we decided to panic a bit. Perhaps we had been premature in going to 2 levels instead of 4 levels on the end cap. Because of this, we set out to conduct a 4 x 3 full factorial using end cap and motor configuration as factors. The design, with resulting S/N ratios, was as follows:

		MOTOR		
		Center Cut	**Front Cut**	**Standard**
	standard	10.79	11.08	11.83
END	.100	11.24	11.66	12.65
CAP	.155	12.11	12.32	12.69
	.180	12.10	12.35	13.17
	.200	12.09	12.11	12.45

A graphical comparison of the results is provided in Figure 1. As is apparent from the graph, the standard motor with the 0.180 groove is the best performer.

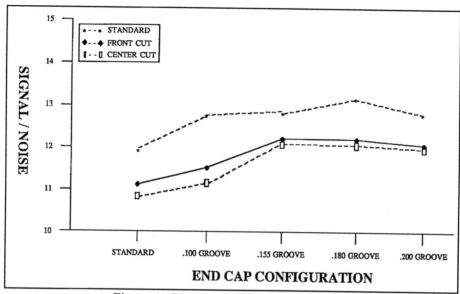

Figure 1 S/N vs. End Cap Configuration

Stage 3a:

After extensive discussion, we decided to re-analyze the raw data using the standard deviation. Our data table appeared as follows:

END CAP	CENTER CUT	FRONT CUT	STANDARD
standard	.041	.039	.030
.100	.037	.034	.025
.155	.031	.028	.025
.180	.031	.026	.022
.200	.031	.026	.026

Graphically, these results were as follows:

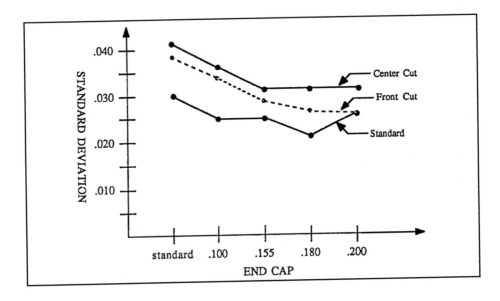

Analysis using the standard deviation indicates that we have the least dispersion with a 0.180 end cap and a standard cut motor.

Stage 4:

We conducted verification testing of the predicted best combinations. Results obtained were in agreement with our predicted best combination.

Conclusions

Based upon the above findings, it is recommended to go with the standard motor and the 0.180 end cap.

Factorial Experiment for Bottom Electrode Stress

Submitted by: Doug Sheldon
Ramtron Corporation

Abstract

Test Lot XXX was run using a 2-level, 3-factor orthogonally designed experiment to characterize Bottom Electrode film stress. The results show that current factor A thickness (200) is too thin for good, controllable adhesion and must be made thicker.

Factors and Levels

Factor	"Low"	"High"
A	200	1500
B	500	4300
C	no	yes

Responses: Stress and Resistance

Experimental Stages

Stage 1:

Stress and resistance measurements were taken with A and C at all possible combinations (providing us with a 2-level, 2-factor, full factorial design).

Stage 2:

Adding factor B (3rd process step) at 2-level provided us with a 2^3 full factorial design.

Data and Analysis

Stage 1:

Table 1 lists the applicable results after deposition of factor A.

	Factor		Stress			Resistance		
Run	A	C	x_1	x_2	$\bar{x}$	y_1	y_2	$\bar{y}$
1	200	no	−10.00	−1.26	−5.63	40	101.0	70.50
2	1500	no	1.31	1.23	1.27	6	9.5	7.75
3	200	yes	7.76	14.40	11.08	111	82.5	96.75
4	1500	yes	1.47	2.43	1.95	7	10.5	8.75

Table 1

Parameter	Stress Effect	Resistance Effect
A	−1.115	−75.375
C	8.690	13.625
AC	−8.015	−12.625

For "stress," note that the AC interaction effect is the second largest of the three. This means the effect of factor A on stress cannot be considered independently of C. Figure 1 provides a graphical representation of the interaction. At thin values of factor A, the stress varies greatly. At thick values for A, there is less difference in stress regardless of whether C is in or out of the experiment.

For resistance, the most significant factor is A. As factor A thickness increases to 1500, the resistance drops an order of magnitude. The fact that the thin factor A values have a large mean with relatively high dispersion indicates a good factor A layer may not be forming. Resistance values with factor A set at 1500 seems to have much less dispersion. The effect of having factor C present is to increase resistance.

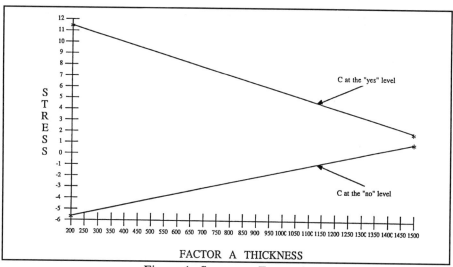

Figure 1   Stress vs. Factor A

Stage 2:

Once factor A was deposited and measured, factor B was deposited and stress was measured. The applicable data with resultant effects are shown in Tables 2a and 2b, respectively.

	Factors			Response
Run	A	B	C	Stress
1	200	500	no	1.67
2	1500	500	no	2.30
3	200	4300	no	0.85
4	1500	4300	no	2.20
5	200	500	yes	0.53
6	1500	500	yes	2.20
7	200	4300	yes	0.72
8	1500	4300	yes	2.70

Table 2a

Analysis	
Parameter	Effect
A	1.41
B	$-.06$
C	$-.22$
AB	.26
AC	.42
BC	.40
ABC	$-.10$

Table 2b

From Table 2b, only factor A appears to stand out. The interaction terms, as well as factors A and C, appear to be of no importance. Figure 2 provides a graphical representation of the effects.

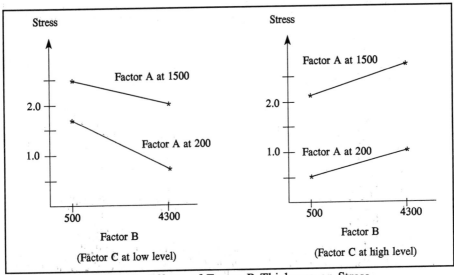

Figure 2 Effects of Factor B Thickness on Stress

Conclusions

The current thickness of factor A (200) appears to be too stressful and prone to great dispersion. It is recommended the setting for factor A be increased to 1500. With a setting of 1500 for factor A, the occurrence of factor C is not required. If, however, we go to 1500 for factor A, we will need to reduce factor B. The lower limit of factor B would be where possible factor A interdiffusion into factor B becomes a problem. Results from previous tests indicate it is possible to run with very thin layers of factor B.

Experimental Design on an Injection Molded Plastic Part

Background

Shrinkage is a major concern with some injected molded parts. Typically, a molded die for a part will be built oversize to allow for part shrinkage after it has been produced. In the following situation, a new die had been produced for the production of a part. What are the proper process settings to produce a product with the proper dimensions?

Objective

Determine the effect of injection velocity, cooling time, barrel zone temperature, mold temperature, hold pressure, and back pressure on shrinkage of part XXXX (as defined by the 9.380 and 14.500 dimensions).

Brainstorming

The brainstorming process lasted approximately four hours. Attenders included a process engineer, process supervisor, quality control inspector, quality control supervisor, and a person with some rudimentary training in experimental design. Barrel zone temperatures was a tough one. Barrel zones actually had six different control points. Some discussion was conducted as to whether we should first run a series of experiments about zone temperatures. After a series of discussions, it was concluded to look at only a "low" and "high" group of settings for barrel zone temperatures in the screening design chosen.

Design Type

An 8-run fractional factorial at 2 levels (Taguchi L8) was chosen. The controlled factors with applicable levels chosen were:

	Control Factor	Low	High
A	Injection Velocity	1.0	3.1
B	Cooling Time	40 sec	50 sec
C	Barrel Zones	"low temps"	"high temps"
D	Mold Temperature	100	150
E	Hold Pressure	200	1100
F	Back Pressure	50	150

We decided to look only for main effects in this particular design.

Test Plan

We chose to run all experiments during the same shift using raw materials from the same lot and containers. We also chose to use only "fresh" raw materials. No reground product was to be used. Additionally, our test procedure called for cycling in the settings, running until these settings were stable at the specified settings, running 10 shots, and then running the parts for the experiment. Five experimental parts were measured at each design setting with applicable dimensional measurements recorded. Since factor D was a factor which was very time consuming to change, we assigned it to the first column of our array, then randomized the run order within the 2 levels of factor D.

Design and Resultant Data

Design #	D	C	A	B	E	F	Length
1	1	1	1	1	1	1	14.5000, 14.5005, 14.5000, 14.5000, 14.5005
2	1	1	1	2	2	2	14.5075, 14.5090, 14.5070, 14.5065, 14.5065
3	1	2	2	1	1	2	14.5045, 14.5050, 14.5045, 14.5045, 14.5045
4	1	2	2	2	2	1	14.5100, 14.5105, 14.5105, 14.5110, 14.5105
5	2	1	2	1	2	1	14.5105, 14.5110, 14.5105, 14.5120, 14.5100
6	2	1	2	2	1	2	14.5045, 14.5055, 14.5065, 14.5050, 14.5050
7	2	2	1	1	2	2	14.5150, 14.5140, 14.5155, 14.5150, 14.5145
8	2	2	1	2	1	1	14.5055, 14.5065, 14.5055, 14.5055, 14.5060

Design #	D	C	A	B	E	F	Width
1	1	1	1	1	1	1	9.3575, 9.3560, 9.3570, 9.3585, 9.3590
2	1	1	1	2	2	2	9.3650, 9.3640, 9.3640, 9.3640, 9.3645
3	1	2	2	1	1	2	9.3545, 9.3545, 9.3545, 9.3550, 9.3540
4	1	2	2	2	2	1	9.3630, 9.3630, 9.3625, 9.3635, 9.3635
5	2	1	2	1	2	1	9.3555, 9.3560, 9.3560, 9.3555, 9.3560
6	2	1	2	2	1	2	9.3580, 9.3565, 9.3550, 9.3540, 9.3545
7	2	2	1	1	2	2	9.3600, 9.3580, 9.3585, 9.3590, 9.3585
8	2	2	1	2	1	1	9.3565, 9.3560, 9.3565, 9.3565, 9.3560

Analysis

Table of Average Marginals
(14.5 - nominal, length)

Level	D	C	A	B	E	F
1	14.5057	14.5059	14.5070	14.5076	14.5040	14.5068
2	14.5092	14.5089	14.5078	14.5072	14.5109	14.5080
|Effect|	.0035	.0030	.0008	.0004	.0069	.0012

Table of Average Marginals
(9.380 - nominal, width)

Level	D	C	A	B	E	F
1	9.3599	9.3583	9.3593	9.3567	9.3560	9.3582
2	9.3566	9.3582	9.3573	9.3598	9.3605	9.3583
|Effect|	.0033	.0001	.0020	.0031	.0045	.0001

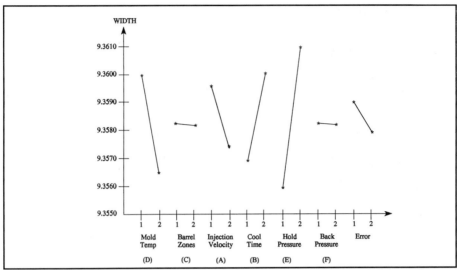

Figure 1 Plot of Average Marginals for 9.380 ± .005 (nom) Dimension

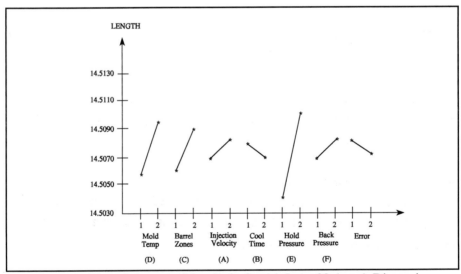

Figure 2 Plot of Average Marginals for 14.5 ± .02 (nom) Dimension

Interpretation of Results

LENGTH	WIDTH
Best settings for 14.5 (nominal)	Best settings for 9.380 (nominal)
Mold temp (1)	Mold temp (1)
Barrel zones (1)	Barrel zones (don't care)
Hold pressure (1)	Hold pressure (2)
Inj. velocity (don't care)	Inj. velocity (1)
Cool time (don't care)	Cool time (2)
Back pressure (don't care)	Back pressure (don't care)

The only major conflict exists on hold pressure. Because of problems with holding the 9.380 (nom) dimension, we will select hold pressure at level 2.

NOTE: *Under no operating conditions during the experiment were we able to meet the specification on the 9.380 dimension.*

Best conditions appear to happen at mold temperature (1), barrel zones (1), hold pressure (2), injection velocity (1), and cool time (2). Back pressure, with little impact, is best set at level 1.

Findings from Verification Run

Twenty-five verification runs were conducted at the "best conditions" as predicted by our experiment. Under no conditions were we able to meet the criteria established for the 9.380 (nominal).

Conclusions

Based upon the above, it was determined to conduct a redesign of the injection die.

Consider the Metric:
Dealing with Mixtures and Temperature Gradients
(Are you defining your inputs sensibly?)

Submitted by: Jack E. Reece, Ph.D.
                      Honeywell

Introduction

Engineers and scientists eager to try new tools to improve their abilities to solve processing problems readily adapt to the matrix approach to experimentation required by statistical experimental design. Once a person understands the structures of the factorial, fractional-factorial, or various response surface design matrices, setting up an experiment involving common input factors such as time, temperature, power, catalyst type, etc., becomes almost second nature.

However, some particularly troublesome processes for experimental design are those involving combinations of reactive ingredients in a fixed or flowing environment. One example includes the mixture of gases used to deposit a coating or to etch a given layer selectively in the semiconductor industry. Another might be a batch or flow process in a chemical industry involving a mixture of reactants in either a static or flow reactor.

Most texts on experimental design devote most of their attention to the justification of the technique of experimental design and to the statistics and analysis of results. Few devote other than passing reference to assisting the user in defining the metric for these cases. The experimental design literature has generally treated these mixture experiments as a special class and devised special treatments for them (see, for example, "Experiments with Mixtures" by John A. Cornell, New York, Wiley, 1981). The transformations of input metrics described below can accommodate a wide variety of experimental situations.

Dealing with Mixtures

1. Experiments with Photographic Emulsions

Conventional photographic emulsion formulations contain light sensitive elements, chemical and photo sensitizers, binders, and other materials. Developing an image in the systems requires treating the exposed film with an external developing chemistry.

At one time the author conducted product development research on a family of photographic emulsions whose entire chemistry (including the developers) resided in the light sensitive coating. Developing an image in an exposed coating for these materials required only heat.

The developing chemistry required a hindered phenolic reducing agent (A) and another agent generally considered a "co-developer" or "development accelerator" (B). The system worked best when the emulsion contained an excess of the phenolic compared to the accelerator. However, too much or too little of the phenolic in the emulsion gave unacceptable image properties; too much of the combination might cause the entire unexposed area to darken upon development; too little gave very slow development with poor image density. Proper balance of these ingredients allowed the "tuning" of the emulsion for specific photographic applications.

In attempting to optimize the performance of this emulsion for a particular application, the author decided to use the amount (or concentration) of the two materials in a fixed batch and their ratio as factors in the design rather than to manipulate the amount of each in the emulsion separately. This approach produced two equations with two unknowns:

$$A + B = \text{Total weight in fixed batch (X1)}$$
$$A/B = \text{Ratio (X2);} \quad A = B \times X2$$

$$B \times X2 + B = X1$$
$$B = X1/(1 + X2)$$
$$A = X1 - B$$

Where A and B are the amounts of developer and are actually added to the emulsion, and X1 and X2 are the factors included in the design matrix. Using this approach obviously requires knowledge of the likely functional limits of each factor.

The previous relationships provide for the following 2-factor matrix in X1 and X2; solving the previous algebraic expressions provides the actual amounts of each reactant to add:

X1	X2
-1	-1
$+1$	-1
-1	$+1$
$+1$	$+1$

Thus, the design is orthogonal in X1 and X2, but not with respect to the weights of each ingredient actually added to each batch.

The logic of this approach is that it considers the stoichiometry (chemistry) of the system and includes the concentration of these two reactions and their "pseudo-molar" ratios. The practical significance of the approach is that it provided the author a convenient metric for managing these two variables and optimizing the emulsion characteristics.

2. Experiments Involving Gas Flows
a. Two Reactive Gases

An extension of the approach described above provides a means for controlling two reactive gases flowing into a reaction chamber to provide a product or to modify (etch, for example) a layer on a substrate. In this example, G1 and G2 have flows measured in SCCM (Standard Cubic Centimeters per Minute) or equivalent units. A logical approach to characterizing this system would use a range of the actual flow of each gas in the design matrix.

Based on his experience in optimizing photographic emulsions, the author prefers to use a "chemical" approach to this problem. The usefulness of the method has been demonstrated in actual practice, in that models prepared in this fashion have shown smaller residual sums of squares, less lack of fit sums of squares, and fewer outlier or maverick points in an investigation. Models prepared this way better explain the relationships between inputs and outputs in the system being investigated.

The equations are adaptations of those shown on Page 8-29:

$$G1 + G2 = \text{Total Flow (TF)}$$
$$G1/G2 = \text{Ratio (R)}$$

TF and R become the factors in the experiment, and one calculates the flow settings used to investigate the process as demonstrated above.

b. More than Two Gases (All Reactive)

Additional gases necessarily complicate the algebra necessary to solve for the actual flow. The example below assumes we must explore three gas flows, all of which play a reactive part in the process.

$$G1 + G2 + G3 = \text{Total Flow (TF)}$$
$$G1/G2 = \text{Ratio1 (R1)}; \quad G1 = G2 \times R1$$
$$G3/G2 = \text{Ratio2 (R2)}; \quad G3 = G2 \times R2$$

$$G2 \times R1 + G2 + G2 \times R2 = TF$$
$$G2 = TF(1 + R1 + R2)$$

Choosing one of the gases as the denominator in both of the ratio expressions simplifies the algebra somewhat; one convenient approach uses the major component of the gas mixture as that denominator.

c. Several Reactive Gases and a Carrier (or Diluent)

A further complication occurs when one of the gases in a muixture of three or more is a diluent or carrier. Here we can conveniently use the ratios of reactants as well as the *concentration* of reactants in the stream:

$$G1 + G2 + G3 + D = \text{Total flow (TF)}$$
$$G1/G2 = \text{Ratio 1 (R1)}; \; G1 = G2 \times R1$$
$$(G1 + G2 + G3)/TF = \text{Concentration (R3)}$$

$$(G2 \times R1 + G2 + G2 \times R2)/TF = R3$$
$$G2 = R3 \times TF(1 + R1 + R2)$$

Using the expression for the concentration of the reactants both simplifies the algebra and probably better describes the relationship among inputs and outputs.

3. Chemical Reactions in Batches

Consider the case in which a chemist wishes to design experiments to study the reaction of two species in a vessel to produce a particular product. In this case, the vessel has a finite capacity, commonly expressed in volumetric or weight measure. Most chemists prefer to consider reaction components in terms of *mole ratios* of the reactants based on some presumed equation for the chemistry of the process. The moles of reactant X1 = k_1 x wt X1; moles of X2 = k_2 x wt X2, where k_1 and k_2 are constants whose values depend on the respective molecular weights of X1 and X2. The mole ratio X1/X2 = k_1 x wt X1/k_2 x wt X2 or k_3 x (wt X1/wt X2). The mole ratio X1/X2 = k_1 x wt X1/k_2 x wt X2 or k_3 x (wt X1/wt X2). Therefore, one may use the raw weights of each reactant, understanding that multiplying that ratio of weights by k_3 would yield the actual mole ratio.

Quite commonly, chemical reactions in a batch reactor require an inert solvent or diluent - note that similarity between this problem and the one described in the Section 2. So, another concern is the concentration of the reactive species in the reactions mixture. Assume for this example that the maximum weight allowed in this reaction vessel is 1,0-00 units. Then:

$$X1 + X2 + S = 1000$$
$$X1/X2 = Ratio1\ (R1);\ X1 = X2 \times R1$$
$$(X1 + X2)/1000 = Concentration\ (R2)$$
$$(X2 \times R1 + X2)/1000 = R2$$
$$X2 = R2 \times 1000/(1 + R1)$$
$$X1 = X2 \times R1$$
$$S = 1000 - (X1 + X2)$$

Thus, the controlled factors in this experiment are the ratios of the reactants to each other and their concentration in the system.

Dealing with Temperature Gradients

Figure 1 represents a tube furnace similar to those used in the semiconductor industry to prepare coatings on silicon wafers. Among the variables one might manipulate in characterizing the performance of this device are the three temperature zones B1, B2, and B3. In this particular furnace configuration reactive gases enter through the ports J in the end cap C, and proceed down the furnace to the exhaust pump outlet B. Because the process depletes the reactive species in the gas flow, "tilting" the furnace temperature from low to high from the entry to the exit is a common practice.

One might decide to investigate this process using entry temperatures that average 20 degrees cooler than the center temperatures, and exit temperatures 20 degrees warmer:

Entry	790 to 810
Center	810 to 830
Exit	830 to 850

Figure 1 LPCVD Tube Furnace Configuration

A. Quartz Furnace Tube
B. Heating Zones
C. Metal Sealing Flange
D. Quartz Carrier for Wafer Boats
E. Quartz Flow Baffles
F. Quartz "Boats" for Wafers

G. Inert Ballast Gas Inlet
H. Outlet to Vacuum System and Exhaust
I. Closed-loop Pressure Sensor
J. Reactive Gas Mixture Inlet (from Mixing Manifold)

The 2-level, 3-factor matrix that results is:

Entry	Center	Exit
790	810	830
810	810	830
790	830	830
810	830	830
790	810	850
810	810	850
790	830	850
810	830	850

While the general trend of the temperature settings is higher toward the exit in each run, the distribution of settings produced contains center temperatures equal to entry and center temperatures equal to exit. The author suggests an alternative metric: consider the center zone the key or master zone and *offset* the other zones with respect to it.

Entry	-30 to -10
Center	810 to 830
Exit	10 to 30

Entry	Center	Exit
$-30(780)$	810	10(820)
$-10(800)$	810	10(820)
$-30(780)$	830	10(840)
$-10(800)$	830	10(840)
$-30(780)$	810	30(840)
$-10(800)$	810	30(840)
$-30(780)$	830	30(860)
$-10(800)$	830	30(860)

Now each run contains an upward temperature bias from entry to exit; the severity differs from run to run. The design is orthogonal in the offset settings, but not in the actual temperatures run.

Summary

The metrics described in the preceding sections ultimately have provided useful regression models describing the behavior of one or more outputs as a function of these inputs. These metrics by no means represent all those an investigator might select to use.

Experiments and the time necessary to conduct them are expensive. Experiments using the matrix approach of statistical experimental design such as factorials, etc., can make our investigations much more cost effective and provide more reliable information than investigations done by some other method. But the engineer or scientist can improve on the results obtainable from a statistical experimental design if she or he will take the time to step back from a problem and consider alternative, possible novel, metrics for input variables before starting the experiments. Conversion of factors to a new metric *after* all runs are complete may give useful information, but the non-orthogonality of the matrix resulting in that case may cast doubt on the validity of any regression models derived.

Minimizing Variability

Submitted by: Jack E. Reece, Ph.D.
Honeywell Corporation

Introduction

The techniques of statistical experimental design evolved from the work led by Sir Ronald A. Fisher during World War I at Rothamstad Agricultural Experiment Station in London as he and his colleagues strived to increase crop yields. Generally scientists and engineers using these methods concentrated most commonly on the "location" of a process, trying to increase yields, decrease byproducts, or understand a process enough to set and control inputs to maintain a desired result or combination of results [1,2].

Particularly in recent years the widespread teaching of so-called "Taguchi Methodology" has focused attention not only on the location of a process, but also on its dispersion or variability [3]. Within that methodology are a number of metrics designed to describe variability and location in one parameter. Classical statisticians have attacked these metrics, creating considerable confusion among "lay practitioners" of the art of statistical experimental design [4].

Engineers tend to focus on the "bottom line". What interests them are methods they can exploit to find answers in terms they and their managers can understand. This paper discusses the analysis of the results of a designed experiment in the semiconductor industry in which the investigators sought to understand the factors controlling coating uniformity during a batch process [5].

The study was successful using one of the "Performance Statistics" described on pages 8-38 through 8-42. Having obtained access to an advanced experimental design package called *RS/Discover 2.0*[1], the author elected to re-examine that data to compare predictions of optimal regions from several such statistics.

---

[1]*RS/Discover is marketed by BBN Software Products Corporation, Cambridge, MA. It supports a variety of "performance statisitics" and contains a sophisticated multiple regression utility.

The Process

Figure 1 (Page 8-34) illustrates the furnace used to prepare the samples for the experimental study.

One inserts a batch of silicon wafers mounted on edge in the boats "F" on the quartz carrier "D" and places them in the furnace. The objects "E" on the carrier are quartz baffles which provide turbulent flow through the furnace and which have thermal mass to preheat the reactive gases before they contact the wafers. From a mixing manifold (not shown) the gas mixture containing two reactive species enters at the positions "J" in the end cap "C" and passes through the furnace tube "A" to the exhaust pump and scrubbing system at "H". Heaters around the tube at "B1", "B2", and "B3" heat the gas stream to about 800 degrees centigrade. Metering an inert ballast gas at the port "G" controls the pressure.

In characterizing the process the engineers manipulated six variables: CLP (Closed Loop Pressure), CTMP (temperature of zone B2), Z1OFF (offset of zone B1 with respect to zone B2), Z3OFF (offset of zone B3 with respect to zone B3), TF (total flow of reactive gases), and RAB (ratio of gas A to gas B). Because they expected complex, nonlinear relationships among the input and outputs of the process, they elected to run a Box-Wilson Central Composite design requiring 53 trials to allow estimation of curvilinear effects. Each trial contained 12 pilot wafers (among the rest used for ballast); and each boat contained 3 pilots. The engineers gathered 10 thickness measurements from each wafer in each run, for a total of 6360 thickness measurements.

The experimental objective was to identify a process window which provided reasonable deposition rates with minimum variation within or among wafers.

Performance Statistics

1. Taguchi's "Target is Best"

As stated above, the experimental objective is to get the process on target with minimum variability. Taguchi suggests a parameter called "Target is Best" or SN_T for this purpose. In mathematical terms this parameter is:

$$10 \log_{10} (\bar{y}^2/s^2)$$

where $\bar{y}$ is the average value observed in each run, and

s is the observed standard deviation in each run.

Taguchi uses the parameter to define the optimum operating point for the process in that the combination of factor settings which maximizes this function defines the optimum. Classical statisticians object to the use of this statistic because it blends the mean with a measure of variability. The absence of close correlation between the variability and the mean causes this statistic to lose information compared to separate measurements of mean and variability in that case.

The form of the statistic reflects Taguchi's conversion of these numbers to "decibel" readings because of his engineering background. This statistic is actually a variant of the classical coefficient of variation statistic:

$$CV = s/\bar{y}$$
$$10 \times \log_{10}(CV) = 10 \times \log_{10}(s/\bar{y})$$
$$-20 \times \log_{10}(CV) = 20 \times \log_{10}(\bar{y}/s)$$
$$= 10 \times \log_{10}(\bar{y}^2/s^2)$$

Thus, Taguchi's "Target is Best" statistic is the same as $-20 \times \log_{10}CV$.

2. Performance Statistic Based on Range

A common metric used for expressing the "uniformity" of a variety of different kinds of measurements within a group of measurements in the semiconductor industry is:

$$UNIF = range/(2 \times \bar{y}) \times 100$$

This is also a variant of the coefficient of variation CV in that for small samples one may estimate the standard deviation by dividing the range by 4:

$$\text{UNIF} = (2/2) \times (\text{range}/2) \times (1/\bar{y}) \times 100$$
$$= 2 \times (\text{range}/4) \times (1/\bar{y}) \times 100$$
$$= 2 \times s/\bar{y} \times 100 = 2 \times \text{CV} \times 100$$

The author recommends against using the statistic based on the range because of the influence outlier-prone data can have - the standard deviation is far more robust to outliers than the range. Because the natural limit of either statistic is 0, one should transform the response to preclude negative predictions before attempting to prepare a regression model; common transforms are the logarithm, square root, inverse, inverse square root, etc.

3. Performance Statistic Based on the Standard Deviation

Dr. George Box of the University of Wisconsin, Madison, and his colleagues have published a series of papers dealing with modeling variability independent of the mean or location. They recommend:

$$-\log 10(S)$$

At once you can see that this is a very simple metric, requiring the calculation of the standard deviation of each set of observations and finding the logarithm of each standard deviation. The negative sign causes this function to follow the "Target is Best" statistic in that one seeks to maximize its value. The results presented below show how well it performs for the data used here in which the standard deviation appears proportional to the mean.

Another form of this statistic is somewhat more difficult to generate:

1. Find the logarithm of each observation in each group.
2. Determine the standard deviation of each group of logarithms of the observations.
3. Find the logarithm of that standard deviation.

For comparison purposes, the author generated this metric and called it "LGSDLGTHK."

4. Other Performance Statistics

All of the metrics described so far assume that the distribution of the observations for each run is normal, although the transforms in each minimize the impact of non-normality. In the particular case described in this paper, only 17 of the sets of observations of the 53 indicated a normal distribution using the Pearson's Chi Square test. Therefore, metrics independent of the nature of the distribution of the observations can be quite useful in some cases.

One such statistic is the InterQuartile Range (IQR). If one orders observations from low to high and defines the end of the first quartile (25% of the observations) and the end of the third quartile (75% of the observations), the difference between the two numbers is the IQR. The range defined by the IQR represents 50% of the observations. Recall that $+/-$ 1 standard deviation from the mean in a normal distribution represents about 2/3 of the area under the normal curve or 2/3 of the data. Thus, one may model the IQR of a group of data or its $-\log_{10}$ and obtain results analogous to the metric recommended by George Box and associates.

Another statistic independent of the distribution is the Median. The median is the middle value in the ordered data when the count of the data is odd; or the average of the two middle values in the ordered data when the count is even. The ratio of the IQR to the Median provides a metric quite analogous to the coefficient of variation and is robust to large departures from normality. Another approach to the problem described here went as follows:

> Each run contained 4 wafer boats which held three wafers on which we made 10 measurements of thickness.

> Because the distribution of the measurements tended to be non-normal, calculating the mean of each boat helped normalize the distribution.

> If we then calculated the standard deviation among the boat means and created a form of the coefficient of variation using the standard deviation and the grand mean of each run, we had a metric describing the relative spread of the data in each boat in each run:

CV' = STDEV (MEANS OF BOATS 1 TO 4 PER RUN)/GRAND MEAN
PER RUN

To convert this metric to a measure relatable to the $+/-$ percent variation around a central value, we used the following:

$$BTUNIF = 2 \times CV' \times 100$$

This contrived metric is the one the author used initially in helping engineers model the process discussed here. An important part of its usefulness was that it gave "uniformity" values engineers could understand. It and those based more directly on the coefficient of variation translate to a $+/-$ variation about a central point.

About the Software

BBN Software Products Corporation markets *RS/Discover*™ and its companion multiple regression analysis package *MULREG*™. Together these packages provide a comprehensive environment for the design, analysis, and interpretation of statistical experimental designs. In its current form, Version 2.0, *RS/Discover* supports experiments having inner and outer arrays and calculates a variety of performance statistics automatically when requested by the user.

The *MULREG* package associated with *RS/Discover 2.0* is extremely useful in that it incorporates the Box-Cox system of response transformations and can advise the user on which is statistically appropriate. Furthermore, it supports both least squares and robust (bisquare) regression and can transform inputs based on variance inflation factors associated with a particular factor's settings. To give each metric the best opportunity possible to describe the variability seen in this study the author used every utility the *MULREG* routines indicated were statistically appropriate.

The author completed the initial study of this process using *RS/Discover 1.2* and its accompanying multiple regression utilities. These products were the state-of-the-art for menu-driven, user-friendly statistical packages in 1987, but did not support the sophistication and convenience available in 1989. Nor did they support experiments with outer arrays.

One could consider the 120 thickness measurements in each experiment part of an outer array. While they are not true "experiments" or "replicates", they provide an estimate of how the thickness of material deposited varied from one end of the furnace to another. The author created a suitable experiment within the software, then inserted the data from the previous experiment into the tables created. The system calculated the performance statistics "Target-is-Best", IQR, and LogS.

The software also allows the user to define custom performance statistics whose metric is more familiar in his environment. Examples of these include the "BTUNIF", IQR/Mean, UNIF, and LGSDLGTHK discussed above. In these cases the system provides space for them in the worksheet generated, but the user must calculate or insert them.

Consider the Data

Space considerations prohibit the display of the entire data base. Rather, Tables 1 and 1a show the conditions of each run, the mean thickness observed in each wafer boat, the grand mean thickness for each run (as well as the median), standard deviation among boats, minimum and maximum thicknesses, standard deviation of each run, IQR, and standard deviation of the logarithm base 10 of thicknesses within each run.

Figure 2 plots the observed means of the boats versus the run number. The error bars on this graph represent the minimum and maximum thickness observed. From it one can see that normally the wafers decreased in thickness as boat number increased (boat 1 was near the entrance; boat 4 near the exit); but not always. Some of the runs had very tight distribution of thicknesses —the purpose of regression modeling is to find the simplest expression relating the input factors to the observed distribution of thicknesses.

The Regression Models

We intend to model variability independent of the mean or median of the response. The "Correlation Matrix" on Page 8-44 shows the linear correlation coefficients between the mean thickness value observed and the metric created to estimate variability. Note the $-\log_{10}(S)$ statistic correlates at about -0.45 with the mean —this translates to an R^2 of about .20. With this amount of data, the correlation is probably statistically significant. Note that those metrics involving a ratio of the standard deviation to the mean correlate less well with the mean. We can decide the importance of this correlation when we compare how well each statistic identifies the region of minimum variability in the process. (Incidentally, the previous results using older software identified such a region, and confirming runs proved it.)

Correlation Matrix
Location and Variability Metrics

Term	Mean Thickness
Mean Thickness	1.000
Median Thickness	0.996
$-$ Log 10 (S)	-0.449
IQR	0.432
2 * SD/Mean * 100	0.243
Target is Best	-0.196
Log 10 (SD(LOG 10 (THICK))	0.190

EXP NUM	CLP	TOT FLOW	RAB	CTMP	Z1 OFF	Z3 OFF
1	425	350	5	790	-20	20
2	478	287	4	786	-24	24
3	372	287	6	794	-16	24
4	478	287	6	794	-24	24
5	372	413	6	786	-24	16
6	478	413	6	794	-16	24
7	425	350	5	790	-20	20
8	478	413	6	786	-24	24
9	372	287	4	794	-24	24
10	478	413	4	786	-16	24
11	372	413	6	794	-24	24
12	425	350	5	790	-20	30
13	425	350	5	790	-20	20
14	372	413	4	794	-16	24
15	372	287	6	786	-16	16
16	478	413	6	794	-24	16
17	478	287	4	794	-16	24
18	425	350	7	790	-20	20
19	372	287	4	786	-24	16
20	478	287	4	786	-16	16
21	372	287	4	786	-24	24
22	300	350	5	790	-20	20
23	372	287	6	794	-24	16
24	425	350	5	790	-20	20
25	425	350	5	790	-20	10
26	425	350	5	800	-20	20
27	478	413	4	794	-16	16
28	425	350	5	790	-20	20
29	478	287	6	786	-24	16
30	478	287	6	786	-16	24
31	478	413	4	794	-24	24
32	478	287	4	794	-24	16
33	425	350	3	790	-20	20
34	372	413	4	794	-24	16
35	425	350	5	790	-30	20
36	372	287	4	786	-16	24
37	372	413	4	786	-16	16
38	425	350	5	790	-16	20
39	478	287	6	794	-16	16
40	372	287	4	794	-16	16
41	372	413	4	786	-24	24
42	478	413	4	786	-24	16
43	372	413	6	786	-16	24
44	425	350	5	790	-20	20
45	425	500	5	790	-20	20
46	478	413	6	786	-16	16
47	372	413	6	794	-16	16
48	425	350	5	790	-10	20
49	425	350	5	780	-20	20
50	425	350	5	790	-20	20
51	425	200	5	790	-20	20
52	425	350	5	790	-20	20
53	550	350	5	790	-20	20

Table 1 Factor Settings

EXP NUM	BOAT 1 AVG	BOAT 2 AVG	BOAT 3 AVG	BOAT 4 AVG	BOAT STDEV	GRND MIN	GRND MED	MIN	MAX	GRAND STDEV	IRQ	STDEV LOG 10 (THK)
1	995	942	901	895	46.71	933.2	923.5	851	1084	49.42	70.0	0.023
2	968	935	919	908	24.67	932.6	922.0	852	1090	44.16	51.5	0.020
3	893	809	749	732	72.28	795.7	781.5	690	936	68.49	115.0	0.037
4	933	872	797	811	68.27	853.2	859.0	705	1017	69.32	102.5	0.036
5	1003	945	884	868	59.37	925.0	911.0	813	1138	76.24	107.0	0.035
6	1183	1070	996	976	94.16	1056.1	1034.0	903	1401	109.81	156.5	0.044
7	972	925	884	880	43.98	915.4	911.0	837	1053	47.68	63.0	0.022
8	963	951	932	944	15.46	947.4	938.0	886	1052	38.18	55.5	0.017
9	950	920	892	903	29.27	916.2	911.0	853	1019	35.39	49.0	0.017
10	1152	1092	1072	1129	41.67	1111.5	1112.5	1009	1301	64.15	96.0	0.025
11	951	931	909	916	21.02	927.1	922.5	867	1004	31.84	47.5	0.015
12	951	919	905	936	23.43	927.8	922.0	872	1037	36.23	53.5	0.017
13	992	928	888	894	52.50	925.4	916.5	838	1177	58.77	70.5	0.027
14	1105	1056	1032	1077	37.42	1067.4	1063.0	980	1198	49.20	75.0	0.020
15	786	716	658	634	63.95	698.5	682.5	611	843	63.05	102.5	0.039
16	1096	1031	946	904	74.78	994.3	986.5	865	1199	85.66	140.0	0.037
17	1055	936	872	871	92.99	933.7	908.5	812	1312	92.84	118.0	0.041
18	876	824	759	730	58.88	797.0	796.0	694	921	63.85	110.5	0.035
19	880	857	821	809	29.76	841.4	843.5	771	957	37.25	58.0	0.019
20	974	890	823	796	75.61	870.7	860.0	757	1091	76.80	110.0	0.038
21	764	739	707	707	28.66	729.0	727.5	673	801	32.26	50.0	0.019
22	890	865	835	843	27.13	858.2	858.5	798	925	30.03	49.5	0.015
23	863	805	729	689	67.00	771.4	762.5	655	916	72.17	129.0	0.041
24	983	932	889	882	46.85	921.5	914.5	842	1061	50.06	64.0	0.023
25	1072	978	883	820	94.68	938.1	919.0	771	1201	105.02	165.0	0.048
26	1018	956	901	883	58.67	939.6	925.5	849	1097	60.70	93.5	0.028
27	1119	1061	1010	1009	54.74	1049.7	1047.0	951	1195	58.65	85.5	0.024
28	902	877	855	870	23.38	876.1	879.5	819	949	28.00	43.0	0.014
29	741	716	675	656	33.16	697.2	696.0	635	775	37.93	62.5	0.024
30	743	697	667	676	38.49	695.6	691.0	639	802	36.60	53.5	0.023
31	1103	1104	1102	1154	0.85	1115.8	1107.0	1049	1270	46.71	66.0	0.018
32	918	890	846	829	36.26	870.7	870.5	794	965	41.35	67.5	0.021
33	1038	1013	993	1029	22.64	1018.4	1018.0	951	1121	38.40	55.0	0.016
34	1111	1102	1078	1089	16.95	1094.7	1092.5	1019	1184	38.94	63.0	0.015
35	889	896	880	882	8.06	886.5	884.5	841	941	22.19	32.0	0.011
36	1095	945	861	846	118.00	936.8	906.5	781	1322	116.43	151.5	0.051
37	994	955	926	940	33.76	953.9	947.5	890	1072	40.07	59.0	0.018
38	1223	1052	930	884	147.00	1022.4	981.5	816	1470	153.40	210.0	0.063
39	855	772	689	654	82.87	742.7	722.5	627	919	83.39	147.5	0.048
40	988	909	648	832	70.08	894.0	883.0	793	1092	69.27	100.5	0.033
41	1403	1310	1222	1212	90.58	1286.6	1273.5	1099	1642	120.95	183.0	0.040
42	1434	1309	1179	1111	128.00	1258.1	1245.0	1009	1714	156.65	232.0	0.053
43	1234	1052	929	884	153.00	1024.9	991.5	809	1503	160.26	231.0	0.065
44	1205	1028	899	837	153.00	992.2	944.0	771	1426	157.88	231.0	0.067
45	1040	1029	1016	1047	12.34	1033.0	1035.0	971	1143	34.21	54.0	0.014
46	864	828	789	783	37.37	815.9	817.0	749	903	39.43	60.0	0.021
47	943	894	847	827	47.87	877.8	872.5	792	1009	51.18	82.5	0.025
48	887	831	791	803	48.54	828.1	822.5	751	937	44.65	66.5	0.023
49	781	771	758	781	11.39	772.8	774.0	727	829	22.30	34.0	0.013
50	881	857	829	838	26.37	851.0	851.0	790	914	29.31	47.0	0.015
51	649	613	578	575	35.53	603.4	598.0	551	677	33.24	52.0	0.024
52	1222	1051	930	871	147.00	1018.2	977.0	817	1423	150.69	217.5	0.062
53	1222	1046	926	873	149.00	1016.7	969.0	802	1428	151.53	227.0	0.062

Table 1a Results

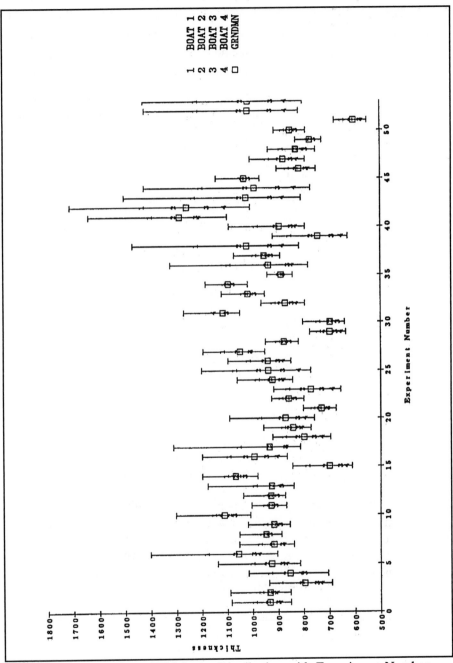

Figure 2 Variation of Thickness Distribution with Experiment Number

Table 2 summarizes the regression models derived in this study. Note the similarity in structure for locations as well as variation effects. Note also that two interactions appear to control the variability independent of the location.

In performing this analysis we postulated a full second order Taylor Series Model and refined it using stepwise regression which obeyed hierarchy. (Hierarchy rules require that one remove a main effect term only when all higher-order terms involving it have been removed.) This contrasts directly with the methods advocated by G. Taguchi. The quotes below are from a recent paper he published in the ASI Journal [6]:

"...In order to prove that there are no interactions, one assigns only main effects to the orthogonal array and finds the optimum conditions. One makes the assumption that there is not interaction and estimates the process average. If there is little interaction, the estimated value of the process average and its confirmatory experiment value will match well. If the effect of interactions is large, these two values will differ greatly and it becomes necessary to take additional measures, such as redoing the experiment on site or bracing oneself for the necessity of improving the design according to marketplace information, or redesigning by rethinking the quality characteristic so that there will no longer be interactions..."

"...The author suggests that, even if there are interactions, only main effects should be assigned to an orthogonal array. The process average is then estimated, and one compares this process average with the value of the confirmatory experiment..."

"...When an interaction is significant, this indicates that the optimum level of a control factor will differ, depending on the levels of other control factors. This also means the optimum conditions we have found will not be useful downstream where conditions differ. This is true because if there are interactions between control factors, there will also be interactions with the manufacturing scale and interactions between laboratory and marketplace conditions. Experience shows that optimum conditions determined after finding interactions do not stand up under large-scale production conditions..."

"... The greatest general difference between Taguchi Methods and traditional design of experiments is the belief that it is useless and counterproductive to assign interactions, whether or not there are interactions. Orthogonal arrays should be used to help us discover experimental failures when interactions exist..."

How in the world anyone would be able to predict a significant interaction is beyond this author. Furthermore, how do we define the "quality characteristic" variability so that interactions are no longer important? The author's experience is that interactions among control factors remain important regardless of the scale of the production. Perhaps new adjustments become necessary if one scales up a process to much larger equipment or output, but control factor interactions do NOT disappear.

This general ignoring of interactions in both optimization and some screening experiments, the general disregard for the importance of randomizing experimental tests, and the tendency to "pick the winner" among existing experimental conditions are the THREE main problems with the "Taguchi Method". Twenty years of doing experimental designs has shown this author that interactions are common, that processes drift with time while we probe them, and that experiments done to satisfy a matrix seldom contain the optimum point.

Modeling a collection of outputs as functions of inputs requires multiple regression analysis. Box and Draper [7] have recently reminded us that all regression models are wrong, but that some might be useful. How useful depends on their ability to predict responses within the limits of our experimental space. Table 3 shows the results of optimization runs done within the software using all the metrics for variability studied here. Regardless of their complexity, they seem to converge on the same general experimental region. Are the differences observed of any consequence?

Within the *MULREG* program a utility allows us to predict responses from our models, given a set of input parameters. Table 4 summarizes several of these. The ranges of values chosen for use in generating Table 4 represent the ranges seen from the various optimization runs. The confidence intervals show that the minor differences observed among the metrics are inconsequential.

Term	MEAN	MEDIAN	TARGET IS BEST	-LOG10 (STDEV)	LOG10 STDEV	2*SD/MN *100	2*SDEV/GMN *100	THKIQR	IQR/MED *100	-LOG10 IQR
1	0.0329	0.0331	23.939	-1.757	-1.565	0.290	2.302	0.112	0.335	-1.932
~CLP	-0.0045	-0.0046	-0.241	-0.060	0.014	-0.018	0.142	-0.006	-0.017	-0.046
~TF	0.0027	0.0028	1.042	-0.054	-0.055	0.023	-0.498	-0.005	0.027	-0.052
~RAB	-0.0009	-0.0009	-1.781	0.002	0.091	-0.024	0.340	-0.004	-0.040	-0.032
~CTMP	0.0007	0.0009	0.148	-0.040	-0.005	-0.011	0.169	-0.011	-0.020	-0.060
~ZN1	-0.0010	-0.0010	-3.227	-0.163	0.155	-0.061	0.709	-0.022	-0.072	-0.155
~ZN3			1.271	0.025	-0.066	0.020	-0.373	0.011	0.041	0.075
~CLP*CTMP			-9.923	-0.327	0.482	-0.095	1.121	-0.037	-0.103	-0.312
~TF*CTMP	0.0018	0.0017	6.607	0.337	-0.329	0.090	-0.771	0.041	0.107	0.355
~RAB*CTMP	-0.0021	-0.0020	-4.884	-0.312	0.247	-0.083	0.818	-0.038	-0.096	-0.327
~ZN1*ZN3			-8.469	-0.324	0.409	-0.091	0.815	-0.045	-0.126	-0.368
TRANSFORM	INVSQRT	INVSQRT	NONE	NONE	NONE	INVSQRT	LOG	INVSQRT	INVSQRT	NONE
VAR. WT.	NONE	NONE	RAB	CTMP	RAB	CTMP	CTMP	CTMP	CTMP	CTMP
ADJ R**2	0.8156	0.8573	0.462	0.296	0.462	0.450	0.502	0.403	0.568	0.403
REL. PRES	0.7760	0.8200	0.314	0.089	0.313	0.282	0.373	0.236	0.449	0.183

Table 2   Regression Models — Location and Variability

Factor, Resp.; Formula	Range	Initial Setting	TARGET IS BEST	-LOGS	LOG10SD LOGTHK	2*SD/MN *100	2*SDBT/GMN *100	THKIQR	IQR/MED *100	-LOG10 IQR
Factors										
CLP	300 to 550	425	300.440	300.000	300.000	300.010	307.260	300.120	300.140	300.000
TOTFLW	200 to 500	350	500.000	499.620	500.000	491.000	490.320	-498.780	499.260	499.990
RAB	3 to 7	5	3.006	3.001	3.240	3.003	3.014	3.001	3.002	3.019
CTMP	780 to 800	800	799.920	800.000	799.940	799.990	799.930	799.990	799.950	800.000
ZN10FF	-30 to -10	-30	-29.230	-28.601	-30.000	-29.184	-29.900	-29.993	-29.874	-29.999
ZN30FF	10 to 30	30	27.622	28.630	25.538	26.821	26.339	29.824	27.031	29.749
Resp.										
BTUNIF			0.067	0.064	0.078	0.078	0.083	0.046	0.065	0.046
IQRMED			1.270	1.250	1.330	1.326	1.347	1.135	1.265	1.137
LGIQR			-0.522	-0.502	-0.576	-0.568	-0.590	-0.386	-0.517	-0.389
LSDLTH			-3.258	-3.281	-3.207	-3.207	-3.176	-3.404	-3.263	-3.403
SDMNUN			1.803	1.784	1.871	1.869	1.899	1.642	1.791	1.644
THKTGT			58.248	58.701	57.231	57.213	56.596	61.256	58.367	61.222
THKIQR			12.803	12.582	13.396	13.335	13.579	11.371	12.743	11.396
THKLOGS			-0.429	-0.415	-0.468	-0.465	-0.486	-0.315	-0.420	-0.317

Table 3 Optimization Summary — Variability Metrics

Predictions and 95% simultaneous confidence intervals for mean responses of IQRMED using model TKIQRMD
RAB = 3, CTMP = 800, CLP = 300, TF = 490

ZN3		ZN1 = −30	ZN1 = −29	ZN1 = −28
	Lower	0.657995	0.677811	0.698022
25	Predicted	1.382611	1.427586	1.474793
	Upper	4.563445	4.741089	4.938965
	Lower	0.535777	0.561639	0.588548
30	Predicted	1.145710	1.195903	1.249468
	Upper	3.963159	4.089269	4.238291

Predictions and 95% simultaneous confidence intervals for mean responses of IQRMED using model TKIQRMD
RAB = 3, CTMP = 800, CLP = 300, TF = 500

ZN3		ZN1 = −30	ZN1 = −29	ZN1 = −28
	Lower	0.640432	0.659353	0.678640
25	Predicted	1.353949	1.397522	1.443234
	Upper	4.541693	4.719994	4.918446
	Lower	0.523517	0.548327	0.574109
30	Predicted	1.124059	1.172821	1.224826
	Upper	3.931724	4.059521	4.210180

Table 4 Predictions from Regression Models (IQR/MED × 100)

Another utility uses the regression models to create contour plots showing us graphically the region of minimum variability predicted by each. Figures 3 and 4 are the contour plots for all the metrics discussed here. Note how all converge to one region.

Figure 3

Figure 4

Summary

What then is the engineer to do? This study has shown that where some correlation exists between the variability of a process and the location of a process, one may use that metric to describe that variability that makes the most sense in his environment and is the most convenient.

Without doubt the $\log_{10}(S)$ statistic advocated by Dr. George Box is adequate to the task described here. In this case, however, because engineers in the semiconductor industry are now more familiar with metrics connoting a $+/-$ variation around a central value, we elect to use any one of the several generally based on the coefficient of variation. If one employs a "fudge-factor" to generate values similar to those found using the standard deviation, that metric based on the IQR and the Median is probably the one of choice.

If you wish to interpret decibel values from the SN_T metric, then do so. Most importantly, however, instead of the highly-fractionated or saturated designs advocated by Taguchi, use more conservative 1/2-, or 1/4-fraction factorials so that you can estimate interactions. Randomize experimental trials to protect yourself and your results from nuisance variation due to a change in the process as you probe it. Finally, model your responses using multiple regression to help identify optimal regions without trying to "pick a winner" among the experimental trials actually done.

About the Author

Dr. Jack E. Reece received the Ph.D. in Organic Chemistry from the University of N. Carolina in 1966. He has practiced statistical experimental design techniques since 1970 in all areas of process development and manufacturing in both the manufacture of specialty coatings and semiconductor devices. Currently he is a Senior Principal Process Engineer for a major semiconductor manufacturer and defense contractor. In that position he assists in the integration of applied statistics methods, including experimental design and SPC, in all manufacturing stages and trains co-workers to use the techniques.

References for the Minimizing Variability Case Study

1. G.E.P. Box et al., *Statistics for Experimenters*, John Wiley & Sons, New York, 1978.

2. D.C. Montgomery, *Design and Analysis of Experiments*, John Wiley & Sons, New York, 1984.

3. See for example: G. Taguchi, and M.S. Phadke, "Quality Engineering Through Design Optimization," Conference Record Vol. 3, IEEE Globecom 1984 Conference, Atlanta, GA, pp. 1106-1113; G. Taguchi, and Y. Wu, "Introduction to Off-Line Quality Control," Central Japan Quality Control Association 1986; and G. Taguchi, "Introduction to Quality Engineering: Designing Quality into Products and Processes," Asian Productivity Organization, UNIPUB, Kraus International Publications, White Plains, NY. See also M.S. Phadke, "Quality Engineering Using Design of Experiments," Proceedings of the American Statistical Association, section on Statistical Education, 1982, pp. 11-20.

4. G.E.P. Box, "Studies in Quality Improvement: Signal to Noise Ratios, Performance Criteria, and Transformation," Report No. 26, University of Wisconsin Center for Quality and Productivity Improvement, Madison, WI, 1987.

5. J.E. Reece, and M.A. Good, "Process Variability and Response Surface Methodology," Proceedings of the 1988 Rocky Mountain Quality Conference, Denver, CO.

6. G. Taguchi, "The Development of Quality Engineering," ASI Journal, Vol. 1, 1988.

7. G.E.P. Box, and N.R. Draper, *Empirical Model-Building and Response Surfaces*, John Wiley & Sons, New York, 1987.

Computer Network Performance Analysis[*]

Submitted by: Lt Col Mark J. Kiemele
Tenure Associate Professor
United States Air Force Academy

Introduction

The verification and validation of simulation models is a difficult task [8, 18]. In the absence of real data from the system being simulated, the task seems even close to insurmountable. This study addresses the prospective validation of a special kind of computer network. This network is capable of transmitting voice and data simultaneously and is thereby labeled an integrated network. Since such a network does not now exist in real form, a computer software simulation, hereafter called the simulator, is used to perform the verification and validation.

Verification of simulator performance with regard to voice/data transaction arrival patterns was accomplished by Clabaugh [3]. He found that the simulator behavior was within a 95% confidence interval of the true means, where the true means were determined by the use of standard Poisson generators and user run time specifications. Clabaugh also provides additional verification of simulator output by configuring the simulator to correspond to an analytic model and comparing the output results [3]. These verification efforts apply to the modified model as well, since none of the modifications impact Clabaugh's verification work. This study documents another step taken toward the verification and validation of the integrated network simulator.

The motivation for this analysis stems not only from the desire to move in the direction of model verification, but also from the need to obtain and analyze performance data from an integrated circuit/packet-switched computer network, for such data is scarce. Fortunately, these two encompassing goals do not diverge.

---

[*]Reprinted from the author's dissertation, "Adaptive Topological Configuration of an Integrated Circuit/Packet-Switched Computer Network," Texas A&M University, 1984.

Goals and Scope of the Analysis

The specific goals of the analysis are as follows:

(1)   Design an experiment whereby the performance data can be efficiently and economically obtained.

(2)   Determine the effective ranges of network parameters for which realistic and acceptable network performance results.

(3)   Investigate the sensitivity of performance measures to changes in the network input parameters. Specifically, determine how network performance is affected by changes in the network traffic load, trunk line or link capacity, and network size.

The scope of the analysis is restricted to those parameters that are closely related to the network topology and the workload imposed on that topology. With regard to performance sensitivity to workload and link capacity, the following four parameters are investigated:

CS: Circuit Switch Arrival Rate (voice calls/min)

PS: Packet Switch Arrival Rate (packets/sec)

SERV: Voice Call Service Rate (sec)

SLOTS: Number of Time Slots per Link (a capacity indicator)

The sensitivity to network size is also investigated by varying the number of nodes and links in a network. In all cases, the performance measures observed are mean packet delay (MPD), fraction of voice calls blocked (BLK), and average link utilization (ALU).

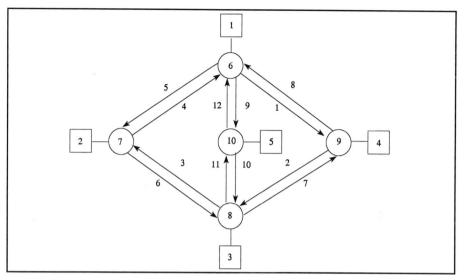

Figure 1   10-Node Network Topology

Design of Experiment

The 10-node network topology shown in Figure 1 was considered sufficiently complex to provide practical performance data without exhausting the computing budget. The circuit switch (circular) nodes correspond to the major computing centers of Tymshare's TYMNET, a circuit-switched network [14]. The trunk lines are full-duplex (FDX) carriers, each of which is modeled by two independent, half-duplex (HDX) channels.

The preliminary sizing analysis [13] indicates that the effective range of input parameters to be investigated is as follows:

CS: 0-8 (calls/min at each node)
PS: 0-600 (packets/sec at each node)
SERV: 60-300 (sec/call)
SLOTS: 28-52 (a link capacity indicator)

The fixed parameter settings were such that each slot represents a capacity of about 33 Kbits/sec.

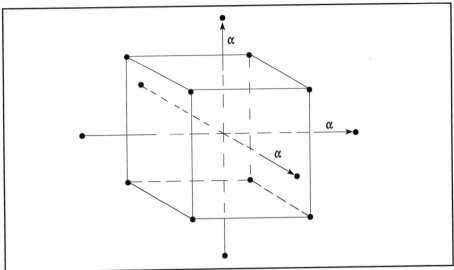

Figure 2 Central Composite Design (k=3)

The experimental design selected for this analysis is a second order (quadratic), rotatable, central composite design [10, 11]. Such a design for k (number of parameters) = 3 is illustrated in Figure 2. This design was chosen because it reduces considerably the number of experimentation points that would otherwise be required if the classical 3^k factorial design were used. The "central composite" feature of the design replaces a 3^k factorial design with a 2^k factorial system augmented by a set of axial points together with one or more center points. A "rotatable" design is one in which the prediction variance is a function only of the distance from the center of the design and not on the direction. Myers [10] shows that the number of experimental points required for k = 4 is 31. That is, there are 16 factorial points, 8 axial points, and 7 replications at the center point. In order for the design to be rotatable, Myers [10] shows that α must be chosen as $(2^k)^{1/4}$. In this case, α, the distance from the center point to an axial point, is 2. The seven replications at the center point will allow an estimate of the experimental (pure) error to be made; thus, a check for model adequacy is possible [12]. Myers [10] recommends seven replications at the center point simply because it results in a near-uniform precision design. That is, it is a design for which the precision on the predicted value ŷ, given by

$$\frac{N \; var \; (\hat{y})}{\sigma^2}$$

where N = total number of experimental points, and
σ^2 = the error variance,

is nearly the same at a distance of 1 from the center point as it is at the center point [10].

In this design, each parameter is evaluated at five different levels. The five levels for each of the four parameters are as follows:

CS	0	2	4	6	8
PS	0	150	300	450	600
SERV	60	120	180	240	300
SLOTS	28	34	40	46	52

The center point is defined as CS / PS / SERV / SLOTS = 4 / 300 / 180 / 40.

Analysis of the results of these 31 runs indicated that the original range of data was too large. Ten of these experimental data points resulted in network performance that would be totally unacceptable, e.g., a MPD of more than 10 seconds. These "outliers" were eliminated and 14 additional points in the moderate to heavy loading of the network were added. Two observations were taken at 8 of these 14 additional points for the purpose of estimating the error variance at points other than the center of the design. Table 1 shows the 43 observations that form the data set used in the subsequent regression analysis. The first 21 observations in Table 1 represent those data points retained from the original 31-run design. The remaining 22 observations represent the data points used to augment the original design.

OBS	CS	PS	SERV	SLOTS	MPD	ALU	BLK
1	2	150	120	34	0.117	0.262	0.000
2	2	150	120	46	0.116	0.194	0.000
3	2	150	240	34	0.113	0.372	0.000
4	2	150	240	46	0.116	0.275	0.000
5	2	450	120	34	0.430	0.553	0.016
6	2	450	120	46	0.260	0.396	0.000
7	2	450	240	34	0.433	0.676	0.031
8	2	450	240	46	0.449	0.481	0.003
9	6	150	120	34	0.339	0.520	0.012
10	6	150	120	46	0.115	0.386	0.000
11	6	450	120	46	0.462	0.594	0.016
12	0	300	180	40	0.123	0.231	0.000
13	4	0	180	40	0.000	0.325	0.000
14	4	300	180	52	0.153	0.427	0.000
15	4	300	180	40	0.350	0.561	0.016
16	4	300	180	40	0.527	0.571	0.005
17	4	300	180	40	0.502	0.561	0.011
18	4	300	180	40	0.316	0.581	0.020
19	4	300	180	40	0.381	0.562	0.017
20	4	300	180	40	0.798	0.608	0.036
21	4	300	180	40	0.616	0.560	0.007
22	5	400	210	43	1.613	0.739	0.070
23	5	400	210	43	2.754	0.746	0.082
24	5	400	150	37	1.656	0.725	0.067
25	5	400	150	37	1.345	0.725	0.084
26	5	400	150	43	0.192	0.615	0.027
27	5	400	210	40	4.814	0.785	0.106
28	5	400	210	40	5.624	0.782	0.125
29	4	400	210	37	1.604	0.747	0.084
30	4	400	210	37	2.730	0.745	0.101
31	4	400	210	40	2.440	0.695	0.063
32	4	400	210	40	1.199	0.687	0.057
33	4	400	180	40	0.910	0.648	0.035
34	5	400	180	40	1.970	0.722	0.074
35	5	400	180	40	1.130	0.731	0.077
36	5	400	180	37	4.084	0.775	0.110
37	5	400	180	37	4.019	0.778	0.139
38	4	400	180	37	4.817	0.712	0.057
39	4	400	180	37	1.135	0.701	0.065
40	3	400	210	37	0.599	0.652	0.017
41	3	400	210	43	0.293	0.541	0.003
42	3	400	150	37	0.454	0.570	0.003
43	3	400	150	43	0.280	0.476	0.005

Table 1 Experimental Data

Regression Analysis and Model Selection

The Statistical Analysis System (*SAS*) [15, 16, 17] has been used to perform a multiple regression analysis for each of the three performance measures. The regression variables are the four input parameters. *SAS* procedures REG [16] and RSREG [16] were used to analyze the data. The assumption of a quadratic response surface allows for the estimation of 15 model parameters, including the intercept.

Although the model selection procedure can be based on a number of possible criteria [4, 6, 9], the approach taken is as follows. RSREG was used first to check the full (quadratic) model for specification error (lack of fit test) and to determine significance levels for the linear, quadratic, and crossproduct terms. The models for MPD, BLK, and ALU exhibited a lack of fit that was significant at the .14, .05, and .82 levels, respectively. It was found, however, that these significance levels were heavily influenced by the observations at the extremes of the heavy-loading region. For example, by eliminating 13 of the observations in the heavy-loading range of data, the lack of fit for BLK could be raised to the .70 significance level. Hence, although the .05 level for the BLK model borders on statistical significance, the lack of fit was not deemed sufficient to justify a more complex model.

The RSREG results also indicated that some terms were insignificant and possibly could be deleted from the model. Using the RSREG results as a guide, several subsets of the full quadratic models were investigated using *SAS* procedure REG. In these models, all linear terms were retained because even though a lower order term in a polynomial model may not be considered significant, dropping such a term could produce a misleading model [6]. Several crossproduct and quadratic terms were deleted, however, and the final regression models selected are shown in Table 2. Despite their appearance of being insignificant, several of the crossproduct and quadratic terms were retained in the model simply because their retention resulted in a smaller residual mean square than if they had been deleted. The interpretation of these regression models is presented graphically in the following section in terms of a sensitivity analysis.

Sensitivity Analysis

This section presents data to describe how the various performance measures change with respect to changes in traffic load, link capacity, and network size. All of the graphs presented are obtained from the quadratic response surfaces (regression models) developed in the previous section.

```
DEP VARIABLE: MPD
                      SUM OF          MEAN
SOURCE        DF      SQUARES        SQUARE      F VALUE     PROB>F
MODEL         10     61.561986      6.156199      6.874      0.0001
ERROR         32     28.658932      0.895592
C TOTAL       42     90.220918
      ROOT MSE        0.946357     R-SQUARE      0.6823
      DEP MEAN        1.218093     ADJ R-SQ      0.5831
      C.V.           77.69169

                    PARAMETER      STANDARD     T FOR HO:
VARIABLE   DF       ESTIMATE        ERROR      PARAMETER=0    PROB > |T|

INTERCEP    1       12.556905      13.142988      0.955        0 3465
X1   CS     1       -1.654903       1.326989     -1.247        0.2214
X2   PS     1        0.014620       0.012349      1.184        0.2452
X3   SERV   1       -0.026635       0.010541     -2.527        0.0167
X4   SLOTS  1       -0.531166       0.597025     -0.890        0.3803
X11         1        0.232675       0.078122      2.978        0.0055
X12         1        0.001153381  0.0009976058    1.156        0.2562
X13         1        0.013125       0.00331975    3.954        0.0004
X14         1       -0.048214       0.025537     -1.888        0.0681
X24         1       -0.000395179  0.0003018693   -1.309        0.1998
X44         1        0.009018565   0.007302992    1.235        0.2259




DEP VARIABLE: ALU
                      SUM OF          MEAN
SOURCE        DF      SQUARES        SQUARE      F VALUE     PROB>F
MODEL         11      1.123895      0.102172    970.909      0.0001
ERROR         31      0.003262242  0.0001052336
C TOTAL       42      1.127158
      ROOT MSE        0.010258     R-SQUARE      0.9971
      DEP MEAN        0.581233     ADJ R-SQ      0.9961
      C.V.            1.764929

                    PARAMETER      STANDARD     T FOR HO:
VARIABLE   DF       ESTIMATE        ERROR      PARAMETER=0    PROB > |T|

INTERCEP    1        0.022183       0.150510      0.147        0.8838
X1   CS     1        0.068950       0.014713      4.686        0.0001
X2   PS     1        0.001859007  0.0001334741   13.928        0.0001
X3   SERV   1        0.001788392  0.0006903726    2.590        0.0145
X4   SLOTS  1       -0.0094644     0.006917257   -1.368        0.1811
X11         1       -0.000929364  0.0008468003   -1.098        0.2809
X13         1        0.0003897799  .00003664077   10.638        0.0001
X14         1       -0.00127589   0.0002929912   -4.355        0.0001
X24         1       -.0000259887   .00000329748   -7.881        0.0001
X33         1       -.0000245180   .00000146947   -1.669        0.1053
X34         1       -.0000214946   .00000932481   -2.305        0.0280
X44         1        0.0001583866  .00008483328    1.867        0.0714
```

Table 2 Regression Models

```
DEP VARIABLE:  BLK
                          SUM OF          MEAN
     SOURCE       DF      SQUARES        SQUARE       F VALUE        PROB>F
     MODEL        12      0.062669      0.005222451    26.743        0.0001
     ERROR        30      0.005858444   0.0001952815
     C TOTAL      42      0.068528
          ROOT MSE        0.013974      R-SQUARE       0.9145
          DEP MEAN        0.038163      ADJ R-SQ       0.8803
          C.V.           36.61764


                         PARAMETER      STANDARD     T FOR HO:
     VARIABLE     DF      ESTIMATE        ERROR      PARAMETER=O    PROB > |T|

     INTERCEP      1      0.197892       0.205845        0.961        0.3441
     X1   CS       1     -0.028562       0.020039       -1.425        0.1644
     X2   PS       1   0.0003439953    0.0002038975      1.687        0.1020
     X3   SERV     1   0.0001307622    0.0005157113      0.254        0.8016
     X4   SLOTS    1     -0.012008      0.00887909       -1.352       0.1863
     X11           1    0.006553606     0.00115578        5.670       0.0001
     X12           1    .00005438269   .00001502268       3.620       0.0011
     X13           1    0.0003339836   0.0000496682       6.724       0.0001
     X14           1    -0.00175778    0.0004020761      -4.372       0.0001
     X22           1    3.79530E-07    1.75917E-07        2.157       0.0391
     X24           1    -.0000158292   .00000450668      -3.512       0.0044
     X34           1    -.0000194136   .00001274852      -1.523       0.1383
     X44           1    0.0002843414   0.0001092323       2.603       0.0142
```

Table 2 (Continued)

1. Sensitivity to Traffic Load

The workload imposed upon an integrated network is described by the voice arrival rates at each circuit switch (CS) node and the data packet arrival rates at each packet switch (PS) node, as well as the length of service for each voice call (SERV). It is assumed that, for a given arrival at any particular node, all of the other nodes of the same type are equally likely to be the destination node for that arrival. That is, the workload is said to be uniformly distributed between node pairs. Unless otherwise stated, it is also assumed that SERV = 180 seconds. Besides these three parameters, the link capacity (SLOTS) parameter also affects the network traffic load. If the CS and PS arrival rates are also assumed to be fixed, then the smaller values of SLOTS represent a heavier network load while the larger SLOTS values correspond to a lighter network load.

Figures 3, 4, and 5 depict the MPD, BLK, and ALU performance measures, respectively, as a function of CS for four different SLOTS/PS combinations. Similarly, Figures 6, 7, and 8 show MPD, BLK, and ALU, respectively, as a function of PS for four load levels defined by combinations of

SLOTS and CS.   The performance measure sensitivity to the traffic load parameters CS, PS, and SLOTS (for SERV = 180) is seen by observing the relative slopes and ordinate values of the four curves in each of the Figures 3 - 8. For example, examination of the curves in Figures 3 and 6 indicates that, within the range of data shown, MPD is more sensitive to CS than to either SLOTS or PS. Additionally, the sensitivity of MPD to both CS and SERV is shown in the three dimensional plot of Figure 9, where the higher levels of one input parameter are seen to magnify the effects of the other parameter.

2. Sensitivity to Link Capacity

The trunk line carrying capacity is an important design parameter in integrated networks. Hence, in addition to the plots of the performance measures versus number of SLOTS, which are shown in Figures 10, 11, and 12, the confidence intervals for each of the performance measures are also presented. Figures 13, 14, and 15 give the 95% confidence limits for a mean predicted value of MPD, BLK, and ALU, respectively. The voice and data arrival rates for these graphs correspond to a fairly heavy traffic load (CS = 4, PS = 400, SERV = 180).

3. Sensitivity to Network Size

In addition to checking the performance of the simulator for varying traffic loads and link capacities on a fixed network topology, the analysis also examines a fixed traffic load on varying sized topologies. In particular, a throughput requirement of 2000 data packets per second with a voice call arrival rate of 20 calls per minute was imposed upon three different sized networks. A 10-node, 20-node, and 52-node network were each subjected to the fixed traffic load.

The 10-node network is the TYMNET topology shown previously in Figure 1. Six links interconnect the backbone nodes.

The 20-node network consists of 10 packet switches and 10 circuit switches. The 10 circuit switches forming the backbone of the network are 10 of the major computing centers in the CYBERNET network [14]. The nodes on the subnet are interconnected by 12 trunk lines. The backbone of the 20-node network is depicted in Figure 16.

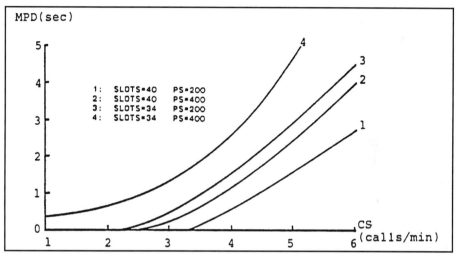

Figure 3   Mean Packet Delay (MPD) vs. Voice Arrival Rate (CS)

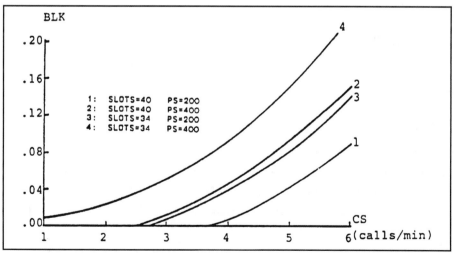

Figure 4   Fraction of Calls Blocked (BLK) vs. Voice Arrival Rate (CS)

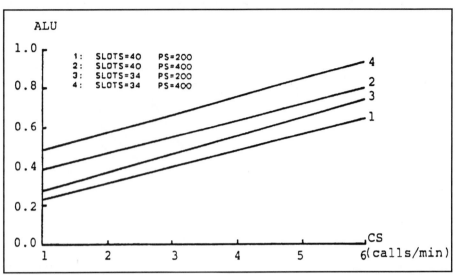

Figure 5 Average Link Utilization (ALU) vs. Voice Arrival Rate (CS)

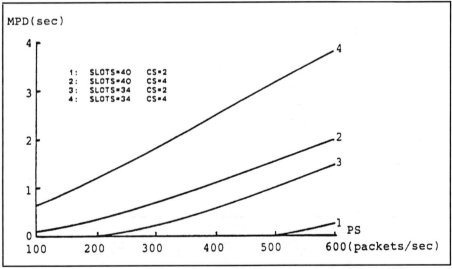

Figure 6 Mean Packet Delay (MPD) vs. Data Arrival Rate (PS)

Figure 7 Fraction of Calls Blocked (BLK) vs. Data Arrival Rate (CS)

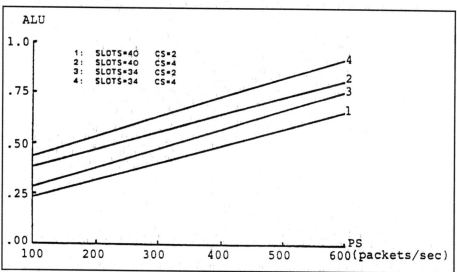

Figure 8 Average Link Utilization (ALU) vs. Data Arrival Rate (CS)

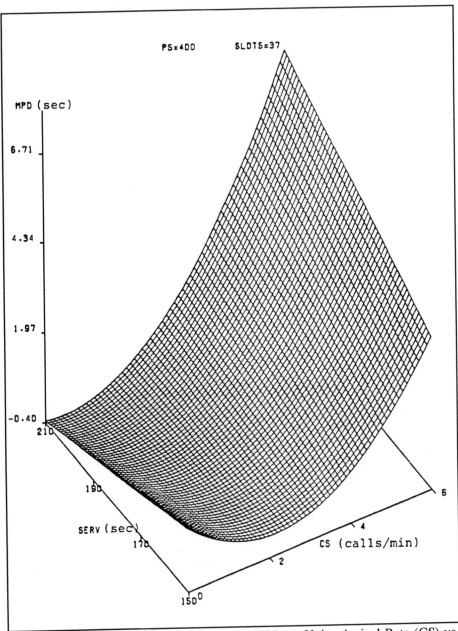

Figure 9 Mean Packet Delay (MPD) vs. Voice Arrival Rate (CS) vs.
Voice Service Time (SERV)

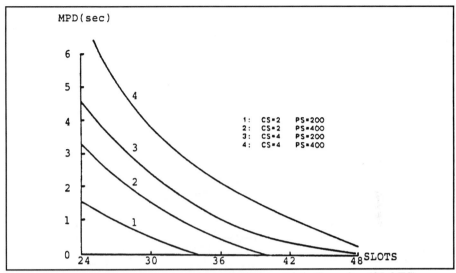

Figure 10 Mean Packet Delay (MPD) vs. Link Capacity (SLOTS)

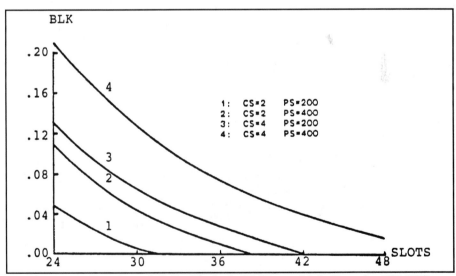

Figure 11 Fraction of Calls Blocked (BLK) vs. Link Capacity (SLOTS)

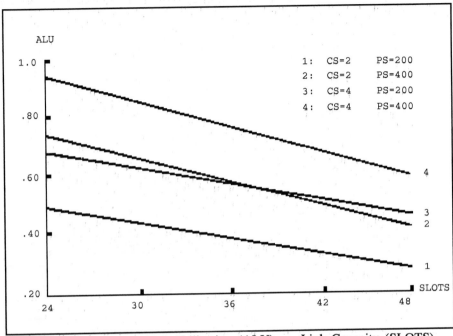

Figure 12   Average Link Utilization (ALU) vs. Link Capacity (SLOTS)

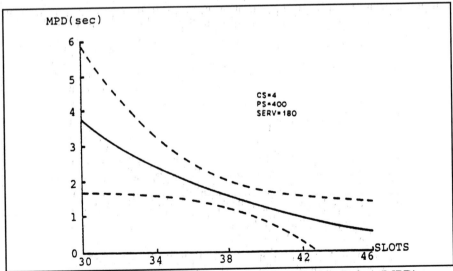

Figure 13 95% Confidence Limits for Mean Packet Delay (MPD)

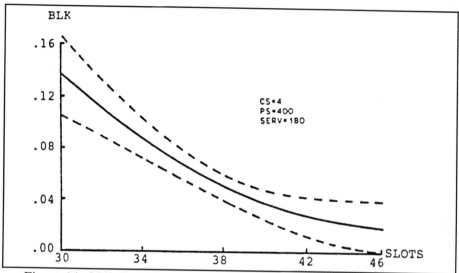

Figure 14 95% Confidence Limits for Fraction of Calls Blocked (BLK)

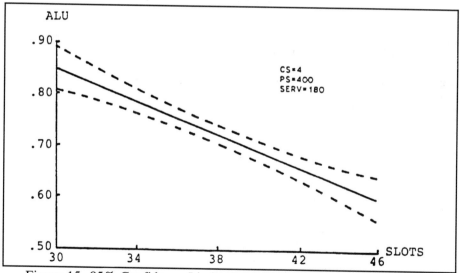

Figure 15 95% Confidence Limits for Average Link Utilization (ALU)

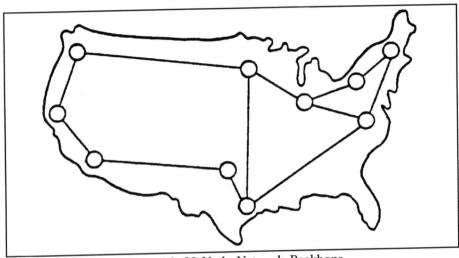

Figure 16   20-Node Network Backbone

The 52-node network is comprised of 26 packet switches and 26 circuit switches, with the circuit switch nodes corresponding to a 26-node substructure of the ARPANET network. This 26-node subset of ARPANET is commonly used in the literature for comparative analyses [1, 2, 5, 7]. The subnet is interconnected with 33 links. Figure 17 shows the communications subnet of the 52-node network.

All links in the three topologies have a fixed capacity of 40 slots. Additionally, each of the three communications subnets is node-biconnected (i.e., there are at least two node-disjoint paths between any pair of nodes).

Table 3 summarizes the results obtained when subjecting the three different network topologies to the given workload. The performance measures given for the 10-node network are averages obtained from a total of 10 simulation runs. Correspondingly, the data for the 20-node and 52-node networks are averages of 7 simulations and 1 simulation, respectively. Despite the fact that there is a reduction in MPD as the number of nodes increases, the relatively smaller decrease in MPD when going from the 20-node network to the 52-node network (as compared to going from the 10-node network to the 20-node network) results from the fact that the backbone of the integrated network is circuit-

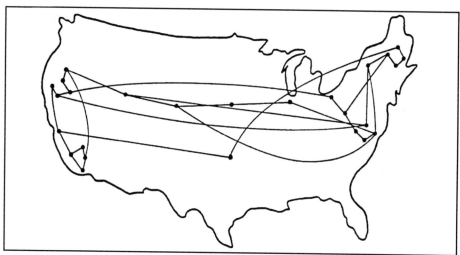

Figure 17 52-Node Network Subnet

switched. This implies that there is a fixed switching delay (assumed to be 50 ms in this study) at each node, and each packet incurs this delay at each intermediate node on its route. Hence, if the subnet link/node ratio remains fairly constant, the MPD is expected to decrease to a certain point and then begin to increase as the number of nodes in the network increases. This phenomenon is due to the fact that as more nodes are added to the network, a greater proportion of the delay can be attributed to switching delays, even though the queuing delay steadily decreases. However, as technology improves the switching delays, the effect of this phenomenon on total MPD is reduced.

Topology	Subnet Link/Node Ratio	MPD	BLK	ALU
10-Node	1.2	.987	.034	.654
20-Node	1.2	.451	.033	.416
52-Node	1.27	.313	.006	.289

Table 3 Sensitivity to Network Size

Summary

Simulation model validation is a never-ending-process, but a well designed sensitivity analysis of the simulator can increase the user's confidence in the model, as well as his knowledge of the model. In this regard, a sensitivity analysis is an important step in the direction of model validation. This chapter has outlined an approach to analyzing the performance characteristics of an integrated computer-communication network simulation model and has presented the results of the analysis.

The investigation has centered on the integrated network traffic load parameters of packet arrival rate (PS), voice call arrival rate (CS), and voice call service times (SERV), as well as the design parameters of link capacity (SLOTS) and network size. Network performance has been measured in terms of mean packet delay (MPD), a strict data measure; fraction of voice calls blocked (BLK), a pure voice criterion; and average link utilization (ALU), a gauge which tends to combine both the data and voice attributes of an integrated network.

The analysis has determined a range of input parameters for which second-order response surfaces can be used to adequately describe realistic network performance. This range is summarized as follows:

CS: 1-6 calls/min

PS: 100-600 packets/sec

SERV: 150-210 sec

SLOTS: 24-48 slots

In light of the fact that an integrated network with a circuit-switched subnet has not yet been implemented, the term "realistic" performance is admittedly dubious. However, performance criteria for current packet-switched and circuit-switched networks can and have been used as guidelines (e.g., a packet delay of 10 seconds is clearly unacceptable, for the message would automatically time out), although flexibility in the guidelines has been preserved so as to not stifle the range of model applicability.

Furthermore, the graphs presented in this chapter consistently support the theme that the performance measures are more sensitive to the CS (voice) arrival rate than to the other parameters investigated. The "slices" (i.e., graphs) of the response surfaces that correspond to an increased CS level generally have higher ordinate values and steeper slopes than the "slices" that correspond to an increased network loading due to variations in the other parameters. In essence, voice arrival rates tend to dominate the network in the sense that the virtual circuits established by successful call initiations provide the framework of paths by which data packets can "piggyback" the digitized voice.

Additionally, heavily loaded networks tend to intensify the effect of any parameter. Increasing the number of nodes in a network (along with a corresponding increase in the number of links) for a fixed traffic load will decrease link utilization rates and call blocking, but may or may not decrease mean packet delay, depending on what proportion of the delay is switching delay.

The simulation model analyzed in this research is a tool that can be used by integrated network designers and managers alike. As a result of this analysis, the user of the simulation model now has a more precise understanding of the relationship between integrated network performance and the network parameters that influence such performance. In light of this, the model is a more viable tool now than it was prior to the analysis.

References for The Computer Network Performance Analysis Case Study

1.   R.R. Boorstyn and H. Frank, Large-scale network topological optimization, *IEEE Trans. on Comm. Com-25*, Vol. 1, Jan 1977, pp. 29-47.

2.   W. Chou and D. Sapir, A generalized cut-saturation algorithm for distributed computer communications network optimization, *IEEE 1982 Int Conf. on Comm. (ICC-82)*, Philadelphia, PA, June 13-17, 1982, 4C.2.1-4C.2.6.

3.   C.A. Clabaugh, "Analysis of flow behavior within an integrated computer-communication network," doctoral dissertation, Texas A&M University College Station, TX, May 1979.

4.   N.R. Draper and H. Smith, *Applied Regression Analysis, 2nd ed.*, John Wiley & Sons, Inc., New York, 1981.

5.   H. Frank and W. Chou, Topological optimization of computer networks, *Proc. of IEEE 60*, Vol. 11, Nov 1972, pp. 1385-1397.

6.   R.J. Freund and P.D. Minton, *Regression Methods*, Marcel Dekker, Inc., New York, 1979.

7.   M. Gerla and L. Kleinrock, On the topological design of distributed computer networks, *IEEE Trans. on Comm. Com-25*, Vol. 1, Jan 1977, pp. 48-60.

8.   W.T. Graybeal and U.W. Pooch, *Simulation: Principles and Methods*, Winthrop Publishers, Inc., Cambridge, MA, 1980.

9.   D.C. Montgomery and E.A. Peck, *Introduction to Linear Regression Analysis*, John Wiley & Sons, Inc., New York, 1982.

10. R.H. Myers, *Response Surface Methodology*, Allyn and Bacon, Inc., Boston, MA, 1971.

11. T.H. Naylor (ed.), *The Design of Computer Simulation Experiments*, Duke University Press, Durham, NC, 1969.

12. B. Ostle and R.W. Mensing, *Statistics in Research*, 3rd ed., The Iowa State University Press, Ames, IA, 1975.

13. U.W. Pooch and M.J. Kiemele, "A simulation model for evaluating integrated circuit/packet-switched networks," presented at Winter Simulation Conf., Arlington, VA, Dec 12-14, 1983.

14. U.W. Pooch, W.H. Greene, and G.G. Moss, *Telecommunications and Networking*, Little, Brown and Company, Boston, MA, 1983.

15. SAS Institute Inc., *SAS User's Guide: Basics, 1982 Edition*, SAS Institute Inc., Cary, NC, 1982.

16. SAS Institute Inc., *SAS User's Guide: Statistics, 1982 Edition*, SAS Institute Inc., Cary, NC, 1982.

17. SAS Institute Inc., *SAS/Graph User's Guide, 1981 Edition*, SAS Institute Inc., Cary, NC, 1981.

18. R.E. Shannon, *Systems Simulation: The Art and Science*, Prentice-Hall, Inc., Englewood Cliffs, NJ, 1975.

Optimizing the Anodize Process and Paint Adhesion for Sheet Metal Parts

Submitted by: Tom Bingham
Supplier Improvement Manager
Boeing Commercial Airplanes

Introduction

Sheet metal used for aircraft parts must be finished in such a way that it is resistant to the problems associated with corrosion. The anodizing process and painting of sheet metal are important procedures to insure acceptable corrosion resistance. The problems addressed in this experiment were optimization of the anodizing process and maximization of paint adhesion in order to assure acceptable corrosion resistance on metals exposed to salt spray tests. One of the problems to be overcome was the dilemma caused by conflicting affects of seal times and temperature on anodizing results and paint adhesion. The historical solution was to balance the process so both salt spray tests and paint adhesion failures are in spec. This balancing required slowing down the anodize time and increasing the temperature. Both of these settings produced an expensive solution to the problem. Therefore, it was decided to obtain information through experimental design which could ultimately lead to a more economical solution.

Approach to the Problem

A team of qualified people from diverse backgrounds and levels of expertise was formed to brainstorm the problem. A fishbone diagram generated by the team yielded over 50 possible factors relevant to the anodize process and paint adhesion. The team then reduced the possible factors to 9 probable factors. Two experiments were then designed to model both paint adhesion and salt spray failures. The factors and the levels of interest are shown on the following page.

The response of interest for the anodize process is the number of pits that appeared on the finished metal product after two weeks of exposure to the salt spray test.

Experimental Variables

Factor	Test Values		
	1	2	
A: Deox Time	6 min	20 min	
B: Anodize Time	35 min	55 min	
C: Anodize Temp	91°F	99°F	Controlled
D: Free CrO₃	32%	4.7%	in Anodize
E: Seal Time	23 min	28 min	Process
F: Seal Temp	191°F	199°F	
G: Ph	3.2	3.8	
H: Primer Thickness	.3 mils	1.5 mils	Paint Adhesion
I: Topcoat Thickness	.3 mils	1.5 mils	Specific Factors

Results

The results indicated that anodize time and temperature had average response values as shown in the following graphic.

Salt Spray Pits

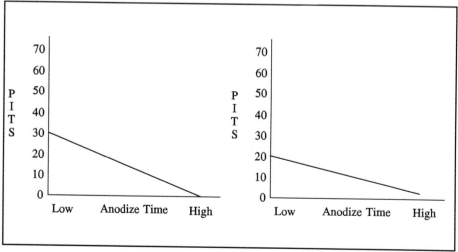

At first glance, it appeared that anodized time and temperature would play a moderate role in optimizing the anodize process. The optimal settings still seemed to favor high time and high temperature which were not the best for reducing costs. However, further investigation through the use of interaction graphs of time with CrO_3 and temperature with CrO_3 provided the following results.

Salt Spray Pits

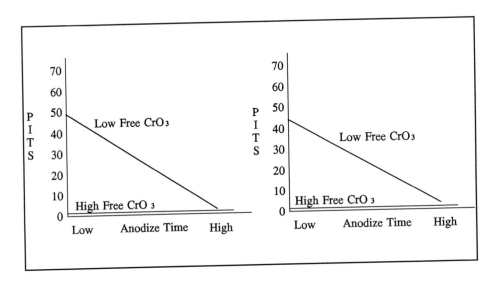

The interaction graphs revealed that when high free CrO_3 is used the time and temperature become unimportant for the anodize process. These two factors can now be set at levels which will speed up the process and reduce costs. The discovery of this very important interaction provided the knowledge required to produce a superior quality product at a lower cost.

Building Paper Helicopters

Submitted by: Cadet First Class Jim Hecker
Cadet First Class Phil Herre
United States Air Force Academy

Statement of the Problem

The purpose of this experiment is to find the best design and construction of a paper helicopter through the use of experimentation.

Objective of the Experiment

Find a combination of the input factors which will maximize the hangtime.

Start Date: 22 JAN 1989

End Date: 27 JAN 1989

Quality Characteristic

Our response variable will be hangtime − time in the air. Hangtime is a quantitative variable (seconds), with an anticipated range of one to four seconds. The person who launches (drops) each helicopter from a height of ten feet will also time the fall in order to cut down the variability of reaction time.

Factors

Three controllable variables are considered: 1) ratio of blade length to width, 2) type of paper, and 3) the number of paper clips attached.

The ratio of blade length to width will have two levels: 2:1 and 1:2 (length to width). Paper is also a categorical variable with two levels: filler paper and construction paper. The paper clip is a quantitative variable with either one or two as the levels of interest. We expect all the interactions to be worthy of investigation, except for the 3-way interaction of ratio-paper-clips.

The noise variables (uncontrollable and unavoidable) that may have an effect are:

1)   **wind or drafts** — To control for these, we performed the experiments in a closed room.

2)   **fall height** — The helicopter does not start to spin when released, creating a "fall height" where the helicopter does not spin. From observation, we assume the fall height to be random for each helicopter, therefore, not an important factor.

3)   **reaction time** — In timing the hangtime, there may be some variance caused by the reaction time of hitting the timer when the helicopter is released and when it hits the ground. To control for this problem, the same person launched the helicopters and timed the fall in order to minimize and randomize reaction time.

Experimental Design
       Due to the minimal cost, both in time and price, of a single run, we choose a full-factorial design for this experiment. A full-factorial has the advantage of testing all possible combinations of the factors and accounting for all possible interaction effects.

       Since we had three factors at two levels, this meant we needed 8 runs. In order to check our results and to have a variance interval, we duplicated each run, meaning there were 16 total runs.

       Here is how the design looked:

Run	Ratio	Paper	R−P	Clip	R−C	P−C	R−P−C	Y_1	Y_2	S^2
1	+	+	+	+	+	+	+	1.28	1.31	.00045
2	+	+	+	−	−	−	−	1.60	1.58	.00020
3	+	−	−	+	+	−	−	1.15	1.10	.00125
4	+	−	−	−	−	+	+	1.19	1.20	.00005
5	−	+	−	+	−	+	−	2.03	2.05	.00020
6	−	+	−	−	+	−	+	2.50	2.42	.00320
7	−	−	+	+	−	−	+	1.32	1.40	.00320
8	−	−	+	−	+	+	−	1.64	1.64	.00000

Ratio (length to width)		Paper		Clips	
+	1:2	+	filler	+	2
−	2:1	−	construction	−	1

Analysis Techniques

The primary method we used to analyze our results is a 2-way ANOVA table (ANalysis Of VAriance). We found the average value at each level of each factor, added the squared differences of each level from the grand mean (1.588) per factor, multiplied by the number that went into each average (4 runs each level × 2 runs) and divided by two. The error was calculated by adding the variances of each run. See Table 1 for the results.

Figure 1 is a graph of the marginal means − the averages at each level per factor. The main factors − ratio, paper, and clips − clearly stand out, as do the three 2-way interactions. The 3-way interaction is insignificant, as expected.

Figure 2 is a Pareto Diagram, which shows the half effects —half the difference of the averages. This plot is not as clear as the marginal means; however, it does show the importance of the main factors and the 2-way interactions in comparison to the 3-way interaction.

Since the variable per run was so small, it did not appear to be worthwhile to investigate dispersion effects.

Conclusions

Based on a simple technique for full-factorials called "pick-the-winner", supported by the marginal means plot, we concluded that the best combination is run six − 2:1 ratio, filler paper, one clip (see Figure 3). Five confirmation runs were made of this design, giving an average hangtime of 2.378 seconds and a variance of 0.00868. These ratios are much better than those from any other combination.

Recommendations

It is our professional opinion that in order to make a paper helicopter with the greatest hangtime, a light paper with a small weight attached to the bottom with the blades twice as long as wide will be the best design.

We recommend for anyone else who tries this experiment to use a greater release height in order to create more variance in the results. Also, a 3-factor, 3-level design would not be that hard to accomplish: simply create another ratio, like a 1:1 length to width; find a third kind of paper, like onion-skin typing paper; and a third level for the additional weight, i.e., 3 clips or even no clips.

	Ratio	Paper	Clip	R−P	R−C	P−C	R−P−C
Average +	1.301	1.846	1.455	1.477	1.630	1.543	1.578
Average −	1.875	1.330	1.721	1.705	1.546	1.634	1.599

Source	SS	df	MS	F_0
Ratio	1.318	1	1.318	1318
Paper	1.065	1	1.065	1065
Clip	0.283	1	0.283	283
R−P	0.219	1	0.219	219
R−C	0.028	1	0.028	28
P−C	0.033	1	0.033	33
R−P−C	0.002	1	0.002	2*
Error	0.009	8	0.001	−
Total	2.957	15	−	−

$$F_{CRIT} = F_{(0.05, 1, 8)} = 5.32$$
* Not significant because $F_0 < F_{CRIT}$

Table 1 ANOVA Table

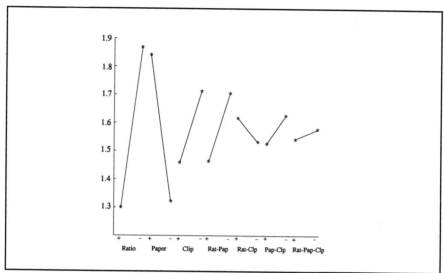

Figure 1   Marginal Means Plot

Figure 2   Pareto Diagram

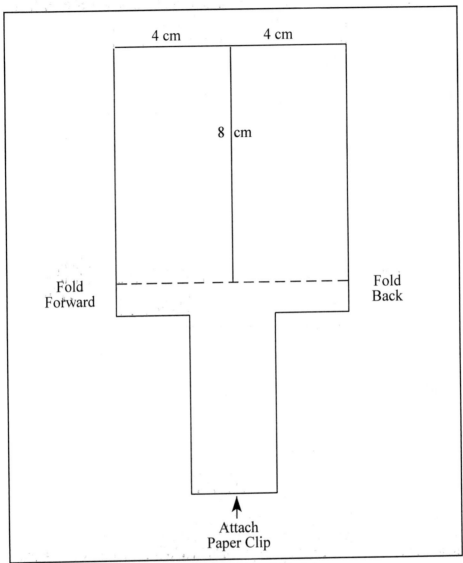

Figure 3 The Optimal Design

Acknowledgement:

Thanks to C.B. (Kip) Rogers for the original paper helicopter application for designed experiments.

Optimizing Lunar Concrete

Submitted by: Col. David O. Swint, Professor
United States Air Force Academy
Engineering Design Project
Presented to the American Concrete Institute, Committee 125

Lt. Anderies	Lt. Kleinsmith	Lt. Milam	Lt. Peddycord
Lt. Bennett	Lt. Krause	Lt. Mingus	Lt. Platt
Lt. Bramer	Lt. Leante	Lt. Neulander	Lt. Shackleford
Lt. Finn	Lt. Martin	Lt. Novak	Lt. Smith
Lt. Haggard	Lt. Meyer	Lt. Noyes	

Problem Statement

The Spring 1989 Engineering 410 class at the U.S. Air Force Academy took on the project of making lunar concrete. The task at hand was to design, fabricate, test, and analyze concrete mixtures, mixed under ambient and semi-evacuated environments which optimized and maximized engineering properties. The design, construction, and testing of the specimens were performed in accordance with American Society for Testing Materials [4] standards to ensure quality control and to minimize errors [12].

Background

Of the various types of lunar construction materials examined, concrete was found to be one of the most feasible [5, 6]. Dr. T. D. Lin has advocated the use of lunar soils and rocks as aggregates because of their quality physical characteristics [5, 6]. According to Dr. Lin, lunar materials contain sufficient amounts of silicate, aluminum, and calcium to produce the "cementitious material" required. Finally, when compared to three other major construction materials (aluminum alloy, mild steel, and glass), concrete requires less energy, and thus, less cost for its production. It is also temperature, radiation, and abrasion resistant [5, 6].

A foreseeable problem with lunar construction of concrete lies in the hydration process required in making the concrete [12]. Due to the evacuated conditions on the lunar surface, curing of the concrete would have to be accomplished in a pressure chamber to ensure that at least water vapor pressure is maintained. This would prevent the necessary amount of water for optimal concrete strength from being drawn out into a lower pressure environment.

Problem Solving Strategy

With the feasibility of lunar concrete as a construction material established, one is left with solving the problem of determining the best concrete to be used. To accomplish this task, the Engineering 410 group had to determine the best of 80 possible concrete combinations. The group used a D-optimal design to test 18 different combinations of cement, environment, plasticizer, wetting agent, and additive variables. From the results of the testing and the prediction equation, the characteristics of the remaining 62 combinations were estimated. The generation of these 18 test runs will be discussed in the section entitled "Generation of Test Runs."

Concrete Variables

Cement − Two types of cement were used for the experiment, Calcium Aluminate and Portland Type III. The Portland Type III contains approximately 65% calcium oxide, 23% silica, and 4% alumina [5]. Calcium Aluminate is a high alumina cement (38% calcium oxide, 10% silica, 52% alumina oxide) and approximates well to a cement that could be made from lunar soils. According to Dr. Lin, these proportions are satisfactory for the Portland cement to be considered a cementitious material [5].

Environment − The two mixing environments used during this experiment were ambient and semi-evacuated (mixed at 200 Torr) conditions. This was done to confirm the effect of mixing conditions on the resulting porosity of the concrete.

Plasticizer/Wetting Agent (admixture) − The concrete combinations were made with and without each of these variables. According to Dr. Lin, these variables would be two of the determining factors in the ultimate strength of the concrete produced.

Reinforcements − The reinforcements used were steel wire, aluminum wire, polymer, simulated lunar soil, and no additive. Though aluminum is known to react with the calcium oxide present in the cements, the group was asked to investigate its feasibility as an additive. If it was a feasible additive, using aluminum already in space would eliminate the cost of shipping it to the moon.

Generation of Test Runs

With the five factors and their possible variations mentioned above, the group had 80 possible combinations of ingredients to test (see Table 1). However, due to time and monetary constraints, it was not feasible to test all 80 combinations. After consulting with Lt. Col. Stephen R. Schmidt, the class used his work on Experimental Design theory for the project [8]. Lt. Col. Schmidt consulted Dr. Lin to insure that the most probable interactions, those between environment, plasticizer, and wetting agent, were included in the design generation. These interactions combined with the other variables were used to generate the 18 combinations of ingredients that would provide the information necessary to predict the properties of the other 62 combinations (see "Prediction Process" section). The actual combinations generated can be seen in Table 2.

Mixing Conditions

For each of the 18 combinations discussed above, one beam and three column specimens were produced. For every concrete batch made, certain mixing conditions were standardized.

Sand − The sand used in the concrete mixtures is Ottawa sand because its particle size distribution is similar to that of a lunar soil sample [6]. The grain size distribution was the same proportion as that used by Dr. Lin, 4% of #30 sieved sand, 31% of #40 sieved sand, 45% of #50 sieved sand, and 20% of #100 sieved sand (to the nearest percent). It was sieved according to ASTM procedures, C136 [6].

Water − As with Dr. Lin's experiment, the distilled water was used.

Mix − Though the recommended ASTM mixture for Portland cement is 2.75:1:0.485 of sand-cement-water, the sand-cement ratio was adjusted to achieve the same level of workability and attain the same water-cement ratio as the Calcium Aluminate cement [6]. The resulting ratio used by Dr. Lin was 1.75:1:0.485. Due to work from previous classes that produced different results, it was decided that a ratio likely to provide better results would be 1.75:1:0.435. This ratio was used in all concrete combinations poured.

Additives − The additives were added by volume after removing 2% of the sand from the mixtures. This was done because of the different densities of the various additives.

Mixing Time − Each specimen was mixed for 3 minutes before being placed in the molds. According to the Fall 1988 class, a 3-minute mix resulted in the strongest mix and removed a high percentage of air pockets from the specimens.

Curing Time − Dr. Lin's test results confirmed a 3- to 4-day cure of the Calcium Aluminate cement for optimal strength [6]. For this reason, the Engineering 410 class allowed the Calcium Aluminate cement mixes to cure for 4 days. The Portland cement cured for 7 days according to standard testing procedures. The first day of curing took place in the molds in a 100% humidity environment, the molds were removed after 24 hours, and the remaining days of curing took place in a 100% humidity environment also.

Dimensions for the cylinder and beam specimens are:
    Cylindrical Column: Diameter = 1"   Length = 2"
    Square Cross Section Beam: Side dimension = 1.57"   Length = 6.69"

Testing Conditions

From each test run, the specimens were tested for compressive and flexural strength. A 3-point load was placed on the beams to be tested (according to ASTM C 78-64) using the Satec Systems 120 WHVL Universal Testing Machine. The load rate used was 0.0025 in/min, consistent with the Spring and Fall 1988's load rate, and was applied until beam failure.

The columns were capped with sulfur (for uniform application of load), and strain gauges were applied with epoxy onto the columns after surface preparation. The wires were connected to a Vishay/Ellis 20 Strain Indicator and a load rate was set at 0.0012 in/min. The lower load rate (as compared with the beam loading) was done so the group could obtain the necessary information from the testing device. As with the beam, the columns were loaded to ultimate failure.

From these tests it was possible to determine maximum flexural strength (modulus of rupture), maximum compressive strength, modulus of elasticity, and Poisson's ratio.

Experimental Data

Two sets of specimens were prepared for each of the 18 experimental conditions shown in Table 2. Each specimen was tested for compressive strength resulting in two response values (A and B) for each row of the design matrix. The average compressive strength per experimental condition is displayed in Table 2 in the column labeled "Compressive Strength Actual."

Analysis

Once all the specimens were tested, the information from the testing was used to determine the performance of the specimens in terms of four different properties: maximum compressive strength, maximum flexural strength, Poisson's Ratio, and Young's Modulus of Elasticity. Maximum compressive strength is a measure of the compressive load per unit area a specimen can withstand before catastrophic failure occurs. Maximum flexural strength is the product of the bending moment times the distance from the neutral axis to the outer fibers divided by the area moment of inertia. Poisson's Ratio is the ratio of lateral strain to axial strain in the specimen. Young's Modulus of Elasticity is the ratio of stress versus strain which can be determined using the compressive load versus longitudinal deformation curve. It measures the stiffness of the specimen.

To determine the values for these four characteristics, a computer program, *CNCRT.PAS* was written. The input values for this program were: maximum compressive load, two points defining the load versus deformation curve, maximum flexural load, and the code defining the concrete mixture. The equations used in this program are as follows:

MAXIMUM COMPRESSIVE STRENGTH (see Table 2)

$$f_c' = P_{max} / \text{Area}$$

POISSON'S RATIO

$$\mu = \frac{\text{LATERAL STRAIN}}{\text{AXIAL STRAIN}}$$

FLEXURAL STRENGTH

$$f_r = \frac{My}{I} = \frac{Pl}{bh^3}$$

YOUNG'S MODULUS OF ELASTICITY

$$E = \frac{\text{STRESS}}{\text{STRAIN}} = \frac{\sigma}{\epsilon}$$

Using the program, these four properties for both runs of the 18 combinations were obtained.

Prediction Process

At this point in the experiment, the group decided to carry on only with the compressive strength for the prediction equation. The reason this was done is that, although standard mixing and testing procedures were followed, the group was not confident about the measurements for the three properties of Poisson's Ratio, maximum flexural strength, and Young's Modulus of Elasticity. The group was, however, confident about the maximum compressive strength because of the simple procedure used to obtain the results. Having the values necessary to determine the maximum compressive strength, the prediction equation below was used to estimate the value of the compressive strength for the other 62 untested combinations.

$$\hat{Y} = \bar{T} + (\bar{C}_i - \bar{T}) + (\bar{E}_i - \bar{T}) + (\bar{P}_i - \bar{T}) + (\bar{A}_i - \bar{T})$$
$$+ [(\bar{P_iW}_i - \bar{T}) - (\bar{P}_i - \bar{T}) - (\bar{W}_i - \bar{T})]$$
$$+ [(\bar{W_iE}_i - \bar{T}) - (\bar{W}_i - \bar{T}) - (\bar{E}_i - \bar{T})]$$
$$+ [(\bar{E_iP}_i - \bar{T}) - (\bar{E}_i - \bar{T}) - (\bar{P}_i - \bar{T})]$$

[8]

In this equation, " $\hat{Y}$ " represents the predicted compressive strength for any concrete sample with a given combination of ingredients. " $\bar{C}_i$ " represents the average response when C is at level i. The " i " subscript allows for variation of the type of cement used. For example, $\bar{C}_1$ represents the average response (f_c') for specimens made with Portland Type III cement, while $\bar{C}_2$ represents the average response (f_c') for samples made with Calcium Aluminate cement. The other factors are interpreted in a similar fashion. The marginal averages can be located in Table 3.

Just as the single factors provide marginal averages (average response), the double factor combinations provide information on interactions. An interaction is very similar to a marginal average except that instead of averaging the value of the property for experimental specimens containing one specific ingredient, one must average the value of the property for the experimental specimens containing the two ingredients in the interaction. For example, for "$\hat{Y}$" representing compressive strength, $\bar{E}_1\bar{P}_1$ is calculated by finding the average of the compressive strength for all of the 18 combinations containing both E_1 and P_1 (see Table 3). "$\bar{T}$" is the average value of compressive strength for all 18 of the experimental runs. Applying the prediction equation to any of the 80 possible combinations of ingredients, one would be able to determine the estimated compressive strength for a concrete specimen containing the given combination. For instance, if one wishes to determine the compressive strength for the specimen with the ingredients Calcium Aluminate (C_2), wetting agent (W_1), plasticizer (P_1), and no additives (A_1) mixed under ambient conditions (E_1), the equation would be as follows:

$$\hat{Y} = \bar{T} + (\bar{C}_2 - \bar{T}) + (\bar{E}_1 - \bar{T}) + (\bar{W}_1 - \bar{T}) + (\bar{P}_1 - \bar{T}) + (\bar{A}_1 - \bar{T})$$
$$+ [(\bar{P}_1\bar{W}_1 - \bar{T}) - (\bar{P}_1 - \bar{T}) - (\bar{W}_1 - \bar{T})]$$
$$+ [(\bar{W}_1\bar{E}_1 - \bar{T}) - (\bar{W}_1 - \bar{T}) - (\bar{E}_1 - \bar{T})]$$
$$+ [(\bar{E}_1\bar{P}_1 - \bar{T}) - (\bar{E}_1 - \bar{T}) - (\bar{P}_1 - \bar{T})]$$

In order to speed up the prediction process, two programs were written to implement the equation described above. The first program, *410NIT.FOR*, accepts the property values for the 18 runs and generates all the marginal averages, interactions, and $\bar{T}$ value for the given property, i.e., compressive strength (see Tables 2 and 3).

This data is then inserted into the text of the second program, *410.FOR*. This program iterates through all 80 combinations using the prediction algorithm above and gives the predicted $\hat{Y}$ value for each combination (see Table 1). In this way, it is possible to approximate the remaining, untested 62 conditions with the results of only 18 tests.

Optimal Combinations

Using the mathematical model, the group was able to predict the compressive strength for all of the 80 combinations. In determining the optimal combinations, the graphical results of the prediction algorithm were analyzed (see Figure 1). The high points marked the greatest value of compressive strength. From this graph, runs 61, 66, and 71 were of interest and chosen for confirmation testing (see Table 4).

Confirmation Tests

Once the optimal runs were chosen, they were remixed and tested to confirm the results with those of the prediction algorithm. The combinations of interest were as follows:

1)   **#61** − Calcium Aluminate cement, semi-evacuated mixing conditions, wetting agent, plasticizer, no reinforcement.

2)   **#66** − Calcium Aluminate cement, semi-evacuated mixing conditions, wetting agent, no reinforcement.

3)   **#71** – Calcium Aluminate cement, semi-evacuated mixing conditions, plasticizer, no reinforcement.

One mix, consisting of three cylinders and a beam, was done for each of the combinations of interest. The mixing and testing procedures were carried out exactly as they had been for the original 18 specimens.

The confirmation results (see Table 4) exceeded our prediction (from 647 psi to 833 psi). Although the difference in predicted and confirmation results is in a direction advantageous to our objective, the group is still concerned as to what has caused this difference. One strong possibility is that an important interaction was not included in this design. To overcome this problem it is recommended that a follow-on experiment be conducted allowing for all 2-way interactions to be estimated.

Conclusions

The purpose of this experiment was twofold. It was designed to find the optimum concrete mixture for lunar construction. It also used Lt. Col. Schmidt's prediction equation as a statistical tool in this process. The confirmation data supports the algorithm's predictions and thus shows that the design of experiments approach is a vital tool to reduce the experimenter's work. The group realizes that for actual lunar construction, many more inputs must be analyzed to obtain the optimal combination. However, the group also realizes that the use of the prediction equation is a must to decrease time and money while still obtaining the optimum concrete mixture.

Run #	C	E	W	P	A	VALUE
1	1	1	1	1	1	3862.417
2	1	1	1	1	2	2534.727
3	1	1	1	1	3	1314.186
4	1	1	1	1	4	2990.017
5	1	1	1	1	5	3126.379
6	1	1	1	2	1	4737.217
7	1	1	1	2	2	3409.527
8	1	1	1	2	3	2188.986
9	1	1	1	2	4	3864.817
10	1	1	1	2	5	4001.179
11	1	1	2	1	1	3574.465
12	1	1	2	1	2	2246.775
13	1	1	2	1	3	1026.234
14	1	1	2	1	4	2702.065
15	1	1	2	1	5	2838.427
16	1	1	2	2	1	3893.400
17	1	1	2	2	2	2565.710
18	1	1	2	2	3	1345.169
19	1	1	2	2	4	3021.000
20	1	1	2	2	5	3157.362
21	1	2	1	1	1	4762.402
22	1	2	1	1	2	3434.713
23	1	2	1	1	3	2214.712
24	1	2	1	1	4	3890.003
25	1	2	1	1	5	4026.365
26	1	2	1	2	1	5020.270
27	1	2	1	2	2	3692.580
28	1	2	1	2	3	2472.039
29	1	2	1	2	4	4147.870
30	1	2	1	2	5	4284.232
31	1	2	2	1	1	5064.096
32	1	2	2	1	2	3736.406
33	1	2	2	1	3	2515.865
34	1	2	2	1	4	4191.696
35	1	2	2	1	5	4328.059
36	1	2	2	2	1	4766.098
37	1	2	2	2	2	3438.408
38	1	2	2	2	3	2217.867
39	1	2	2	2	4	3893.698
40	1	2	2	2	5	4030.060

Table 1 Possible Combinations of Ingredients

Run #	C	E	W	P	A	VALUE
41	2	1	1	1	1	4962.018
42	2	1	1	1	2	3634.328
43	2	1	1	1	3	2413.787
44	2	1	1	1	4	4089.618
45	2	1	1	1	5	4225.980
46	2	1	1	2	1	5836.818
47	2	1	1	2	2	4509.128
48	2	1	1	2	3	3288.587
49	2	1	1	2	4	4964.418
50	2	1	1	2	5	5100.780
51	2	1	2	1	1	4674.066
52	2	1	2	1	2	3346.376
53	2	1	2	1	3	2125.835
54	2	1	2	1	4	3801.666
55	2	1	2	1	5	3938.029
56	2	1	2	2	1	4993.001
57	2	1	2	2	2	3665.311
58	2	1	2	2	3	2444.770
59	2	1	2	2	4	4120.601
60	2	1	2	2	5	4256.963
61	2	2	1	1	1	5862.004
62	2	2	1	1	2	4534.313
63	2	2	1	1	3	3313.773
64	2	2	1	1	4	4989.604
65	2	2	1	1	5	5125.966
66	2	2	1	2	1	6119.871
67	2	2	1	2	2	4792.181
68	2	2	1	2	3	3571.640
69	2	2	1	2	4	5247.471
70	2	2	1	2	5	5383.833
71	2	2	2	1	1	6163.697
72	2	2	2	1	2	4836.007
73	2	2	2	1	3	3615.466
74	2	2	2	1	4	5291.297
75	2	2	2	1	5	5427.659
76	2	2	2	2	1	5865.699
77	2	2	2	2	2	4538.009
78	2	2	2	2	3	3317.468
79	2	2	2	2	4	4993.299
80	2	2	2	2	5	5129.661

Table 1   Possible Combinations of Ingredients (Cont.)

Mixing Assignments

C_1 - Portland Type III cement

C_2 - Calcium Aluminate cement

E_1 - Ambient mixing

E_2 - Semi-Evacuated

W_1 - Wetting Agent - 0.07 ml

W_2 - No Wetting Agent

P_1 - Plasticizer - 1ml

P_2 - No Plasticizer

A_1 - No Reinforcement

A_2 - Steel

A_3 - Aluminum

A_4 - Polymer

A_5 - Simulated Lunar Soil

COMPRESSIVE STRENGTH (psi)

	Design Matrix					Response		Compressive Strength		Deviation
Run	C	E	W	P	A	A	B	Actual	Predicted	(%)
1	1	1	2	2	1	4132.94	3428.83	3780.89	3893.40	−2.98
2	2	1	1	1	1	5067.50	6042.79	5555.15	4962.02	+10.68
3	1	2	2	1	1	5207.55	4004.34	4605.95	5064.10	−9.95
4	2	2	1	2	1	6944.25	6009.69	6476.97	6119.87	+5.51
5	1	2	1	1	3	3101.61	2892.21	2946.91	2214.17	+24.86
6	1	1	1	2	3	2272.73	2639.43	2456.08	2188.99	+10.87
7	1	1	2	1	4	2317.30	4278.08	3297.69	2702.07	+18.06
8	1	2	2	2	5	3997.97	4507.27	4252.62	4030.06	+5.23
9	2	1	2	2	2	4010.00	3670.00	3840.00	3665.31	+4.55
10	2	2	1	1	4	4260.26	5458.38	4859.32	4989.60	−2.68
11	2	1	2	1	3	1914.95	1960.79	1937.87	2125.84	−9.70
12	2	1	1	2	5	4318.83	5672.28	4995.56	5100.78	−2.11
13	2	2	2	2	4	4870.00	4210.00	4540.00	4993.30	−9.98
14	2	2	2	1	2	5692.65	5692.65	5692.65	4863.01	+15.05
15	1	1	1	1	2	1706.14	2294.38	2000.26	2534.73	−26.72
16	2	2	1	1	5	3496.32	4219.52	3857.92	5125.97	−33.13
17	1	2	1	2	2	3808.30	3342.25	3575.28	3692.58	−3.28
18	2	2	2	2	3	2591.04	3179.28	2885.16	3317.47	−14.98

$$\overline{T} = 3975.35$$

Table 2 Actual Generations and Compressive Strength (psi)

MARGINAL AVERAGES

Variable	Compressive Strength
$\underline{C}_1$	3364.46
$\underline{C}_2$	4464.06
$\underline{E}_1$	3482.94
$\underline{E}_2$	4369.28
$\underline{W}_1$	4080.38
$\underline{W}_2$	3870.31
$\underline{P}_1$	3861.52
$\underline{P}_2$	4089.17
$\underline{A}_1$	5104.74
$\underline{A}_2$	3777.05
$\underline{A}_3$	2556.51
$\underline{A}_4$	4232.34
$\underline{A}_5$	4368.70

INTERACTIONS

Interaction	Compressive Strength
$\underline{W}_1, \underline{E}_1$	3751.76
$\underline{W}_1, \underline{E}_2$	4343.28
$\underline{W}_2, \underline{E}_1$	3214.11
$\underline{W}_2, \underline{E}_2$	4395.28
$\underline{E}_1, \underline{P}_1$	3197.74
$\underline{E}_1, \underline{P}_2$	3768.13
$\underline{E}_2, \underline{P}_1$	4392.55
$\underline{E}_2, \underline{P}_2$	4346.01
$\underline{W}_1, \underline{P}_1$	3843.91
$\underline{W}_1, \underline{P}_2$	4375.97
$\underline{W}_2, \underline{P}_1$	3883.54
$\underline{W}_2, \underline{P}_2$	3859.73

Table 3 Marginal Averages and Interactions

Figure 1 Compressive Strength

	Design Matrix					Response				Compressive Strength	
Test	C E W P A					1	2	3	Avg (psi)	Predicted	Dev
61	2 2 1 1 1					4588.6	8225.1	7270.2	6694.6	5862.0	832.6
66	2 2 1 2 1					6369.8	7109.8	7345.3	6941.6	6119.9	821.7
71	2 2 2 1 1					5765.3	8641.5	6026.2	6811.0	6163.7	647.3

Table 4 Confirmation Results

References for the Lunar Concrete Case Study

1. *Design and Control of Concrete Mixtures*, The Portland Cement Association, 12th ed., 1979.

2. ENGR 410 class, Fall 1988, Final Report.

3. ENGR 410 class, Spring 1988, Display.

4. Flexural Strength of Concrete (Using Simple Beam with Third-Point Loading), ASTM C 78-64, pp. 40-42.

5. T.D. Lin, "Concrete for Lunar Base Construction," The Portland Cement Association, 1987.

6. T.D. Lin, H. Love, and D. Stark, "Physical Properties of Concrete Made With Apollo 16 Lunar Soil Sample," Construction Technology Laboratories, 1987.

7. Stanley P. Rader, Maj. (USAF) Consultant, Fall, 1988, Spring, 1989.

8. Stephen R. Schmidt, and Robert G. Launsby, *Understanding Industrial Designed Experiments*, 3rd ed., 1992.

9. S.P. Shah, "Alternative Reinforcing Materials for Concrete Construction." *International Journal for Development Technology*, Vol. 1, 1983, pp. 3-15.

10. Student Manual for Strain Gauge Technology, Measurements Group, Inc., Raleigh, NC, 1983.

11. Stephen J. Simmerer, Capt. (USA), "Lunar Construction: The Countdown Begins." *The Military Engineer*, Jul 1987, pp. 354-357.

12. David O. Swint, Col. (USAF), Instruction and discussions. Spring, 1989.

Design of Experiments
Silane Doping in GaAs

Submitted by: Alan Arnholt
Steve Smith
Robert Kaliski
Statistics Seminar Project - University of Northern Colorado

Objective

In the growth of gallium arsenide (GaAs), a material used to fabricate electronic devices, the level of impurities in the crystal has a great influence on the electrical properties. The level of these trace impurities is known as the doping level. Since the doping level determines some of the electrical properties, we desire to have control of this level to tailor the electrical properties to the device or circuit requirements.

Response

For the operation of heterostructure bipolar transistors, there are two different doping levels of silicon in GaAs. These levels are 3×10^{16} cm^{-3} and 5×10^{18} cm^{-3}. The goal of this experimental design is to model doping level response, select the proper conditions to obtain the two doping levels, and then confirm the results. The response will be the actual doping level measured on a profiling instrument. This instrument can measure the doping level both accurately and precisely in a range of 10^{14} to 10^{19} cm^{-3}.

Factors

In the case of silicon doping in GaAs, three factors may influence the doping level: the silane flow, the arsine flow, and the trimethylgallium (TMGa) flow. There are other factors that could also modify the response, but the machinery is calibrated at these fixed levels. In addition, engineering is only interested in the perturbation of three factors and cannot afford to recalibrate the reactor. The three factors are quantitative and are highly controllable ($+/- 2\%$

of set level), and literature and engineering experience suggest that no interactions exist.

Experimental Constraints

The growth parameters selected are to be used in the growth of heterostructure in March. Hence, engineering prefers to have the data in three weeks (end of February) with 15 runs and a maximum of 30 runs including confirmation runs. Although cost is not a concern, each run requires four hours from start to completion.

Experimental Plan

Since engineering believes no interaction exists, we will use a fractional factorial design, the Taguchi $L_4(2^3)$, with two replications. The factors with both high and low levels are listed below.

Factors	− level	+	Total range
silane flow	15	50	0 - 100
TMGa flow	20	35	5 - 50
arsine flow	100	140	90 - 150

The analysis will include a Pareto chart (see Figure 1) of the three main effects and also a determination of variance (see Figure 2). A regression analysis will follow using linear terms. We plan to check the center point for validity of the linear model. If the linear model is inappropriate, we will then examine the response of the axial points and determine the appropriate regression equation. From the empirical regression, we finally will compute the correct levels and make at least two confirmation runs for each target level.

Figure 1 Pareto Diagram

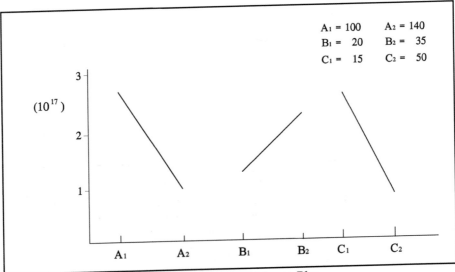

Figure 2 Marginal Means Plot

Experiment and Analysis

Results:

Run #	A=Arsine	B=TMGa	C=Silane	Y1	Y2	S	Ln (S)
1	−1	−1	−1	0.70	0.66	0.0282	35.578
2	−1	+1	+1	2.0	1.6	0.282	37.881
3	+1	−1	+1	4.0	3.7	0.212	37.593
4	+1	+1	−1	1.6	1.2	0.282	37.881

T = 1.93

Note: Response is in terms of 1×10^{17} cm^{-3}.

Avg. [Ln(S)] = 37.233

Marginal Means:

	A	B	C
Avg +	2.63	1.60	2.83
Avg −	1.24	2.27	1.04
d	1.39	−0.67	1.79

Factor C is the largest location effect.

	A	B	C
Avg Ln(S.D.) +	37.731	37.881	37.737
Avg Ln(S.D.) −	36.730	36.586	36.730
d	1.007	1.295	1.007

From our limited data, no factor appears to stand out as a dispersion reduction factor.

Test for Significance of Regression

1) H_0: $A = B = C = 0$

H_1: $A \neq 0$ or $B \neq 0$ or $C \neq 0$

2) a. F-test

b. $\alpha = 0.05$

c. DF = (3, 4)

d. CV for $F_\alpha = 0.05, 3, 4 = 6.59$

3) Calculations

$$X = \begin{bmatrix} 1 & -1 & -1 & -1 \\ 1 & -1 & 1 & 1 \\ 1 & 1 & -1 & 1 \\ 1 & 1 & 1 & -1 \\ 1 & -1 & -1 & -1 \\ 1 & -1 & 1 & 1 \\ 1 & 1 & -1 & 1 \\ 1 & 1 & 1 & -1 \end{bmatrix} \quad X' = \begin{bmatrix} 1 & 1 & 1 & 1 & 1 & 1 & 1 & 1 \\ -1 & -1 & 1 & 1 & -1 & -1 & 1 & 1 \\ -1 & 1 & -1 & 1 & -1 & 1 & -1 & 1 \\ -1 & 1 & 1 & -1 & -1 & 1 & 1 & -1 \end{bmatrix} \quad Y = \begin{bmatrix} 0.7 \\ 2.0 \\ 4.0 \\ 1.6 \\ 0.66 \\ 1.6 \\ 3.7 \\ 1.2 \end{bmatrix}$$

$$X'X = \begin{bmatrix} 8 & 0 & 0 & 0 \\ 0 & 8 & 0 & 0 \\ 0 & 0 & 8 & 0 \\ 0 & 0 & 0 & 8 \end{bmatrix} \quad (X'X)^{-1} = \begin{bmatrix} 0.125 & 0 & 0 & 0 \\ 0 & 0.125 & 0 & 0 \\ 0 & 0 & 0.125 & 0 \\ 0 & 0 & 0 & 0.125 \end{bmatrix}$$

$$\underline{B} = (X'X)^{-1}X'Y = \begin{bmatrix} 1.93 \\ 0.63 \\ -0.33 \\ 0.89 \end{bmatrix} \quad \underline{Y} = 1.93 + 0.69A - 0.33B + 0.89C$$

$$\underline{B}X'Y = \begin{bmatrix} 1.93 & 0.69 & -0.33 & 0.89 \end{bmatrix} \begin{bmatrix} 15.46 \\ 5.54 \\ -2.66 \\ 7.14 \end{bmatrix} = 40.892$$

$$Y'Y = 41.1756 \qquad \frac{(\sum y_i)^2}{n} = 29.876$$

Source	SS	DOF	MS	F_0
Regression Error Total	$S_R = \underline{B}X'Y - (\sum y_i)^2 / n =$ 11.016 $SS_E = Y'Y - \underline{B}X'Y =$ 0.2823 $SS_T = Y'Y - (\sum y_i)^2 / n =$ 11.299	$K = 3$ $n - k - 1 = 4$ $n - 1 = 7$	3.672 0.0707	51.93

4)   Reject H_0, $F_0 = 51.93 > F_\alpha = 0.05, 3, 4 = 6.59$

5)   We found evidence that at least A $\neq$ 0 or B $\neq$ 0 or C $\neq$ 0
      $R^2 = SS_R / SS_T = 0.975$

Test on Individual Regression Coefficients

1)   a. H_0: A = 0   b. H_0: B = 0   c. H_0: C = 0
      H_1: A $\neq$ 0 H_1: B $\neq$ 0 H_1: C $\neq$ 0

2)   a. t-test
      b. $\alpha = 0.05$
      c. DF = n - k - 1
      d. CV for $t_{0.05/2,4} = 2.776$

3)   Calculations

$$t_0 = \underline{B}_j / \sqrt{\sigma^2 c_{jj}}$$

a. $t_0 = 0.69 / \sqrt{(0.0707)(0.125)} = 7.339$

b. $t_0 = -0.33 / \sqrt{(0.0707)(0.125)} = -3.51$

c. $t_0 = 0.89 / \sqrt{(0.707)(0.125)} = 9.467$

4)   Reject H_0 for A, B, and C since $|t_0| > t_{\alpha/2, n-k-1}$
   a. $|t_0| = 7.339 > t = 2.776$
   b. $|t_0| = 3.51 > t = 2.776$
   c. $|t_0| = 9.467 > t = 2.776$

5)   We found evidence that A, B, and C are all non-zero.

From the experimental data, all three factors are significant while no factors have any greater influence on the variability. Since the variability of each factor is approximately equal, we will not concern ourselves with the standard deviation any further.

Using the half-effects of the marginal means, the regression equation is:

$$\underline{Y} = (1.93 + 0.69*A - 0.33*B + 0.89*C) \; 10^{17} \; cm^{-3}.$$

With two runs at the center point A = B = C = 0, the responses are 2.0 and 2.2 (10^{17} cm^{-3}). The average of the center points is 2.1. This value does not appear to be deviate enough from 1.93 to justify the addition of curvature to the regression model.

To reach the lower target of 0.3, we need to set A $= -1$ and B $= 1$ by inspection. Solving the regression equation for C, we obtain a coded value for C of -0.67. Uncoding this value by using the inverse of the standard coding transformation yields a silane setting of 21. Two confirmation runs were performed with the three factors set at the specified levels. The responses are 0.27 and 0.30 (10^{17} cm^{-3}) which are acceptable.

Conclusions

To achieve the lower target value, we recommend the arsine be set at 90, the TMGa set at 50, and the silane set at 21. Since the upper target cannot be met with the present machine configuration, we recommend the engineering staff make changes to the machine to achieve this doping level or change their target level.

Bonding Gold Wire to Gallium Arsenide Wafers: Overcoming Preconceived Notions

Submitted by: Randy Boudreaux
                      Cray Research, Inc.

Statement of the Problem

The opening of a new Cray plant in Colorado Springs posed an interesting problem. Gold wire bonding to Gallium Arsenide wafers had been optimized at the Wisconsin plant and, using the same parameters, should have optimized the process in Colorado. The solution was not to be that simple. Several months passed while "best guess" and "one-factor-at-a-time" experimentation were used to find an improved process. After hearing and reading the materials in this text, it was decided to try a designed experiment matrix as shown in Chapter 3. Several lessons learned from this first experiment are:

1. Proper experimental designs can be used with little training provided adequate guidance is available.

2. Do not consume all of your resources during the initial experiment —things may go wrong, or the information gained may lead to the need for further experimentation.

3. Preconceived notions are a barrier to process improvements, but they can be overcome through properly designed experiments.

4. Designed experimentation is a much more powerful tool than previously used "best guess" or "one-factor-at-a-time" approaches.

The Design

Prior to the actual experimentation, the brainstorming session generated five input factors to be tested along with two responses of interest as shown in Table 1.

INPUT FACTORS	LEVELS
Force	60, 150
Time	40, 100
Ultrasonics	40, 150
Temperature	125, 175
Z-Speed	50, 150
RESPONSE	**TYPE**
Bond/No Bond	Attribute
Bond Strength	Variable

Table 1

The L_8 design matrix and response values for Bond/No Bond are shown in Table 2.

TABLE 2

DESIGN MATRIX AND RESPONSE VALUES FOR BONDING EXPERIMENT

Run	Force	Time	U/S	Temp	Z-Speed	1	2	3	4	5	6	7	8	9	10	11	12	13	14	15	16	17	18	19	20	21	22	23	24
1	60	40	40	125	50	0	0	0	0	0	0	0	0	0	0	0	0	0	0	0	0	0	0	0	0	0	0	0	0
2	60	40	150	175	131	0	0	0	0	0	0	0	0	0	0	0	0	0	0	0	0	0	0	0	0	0	0	0	0
3	60	100	40	125	50	1	0	0	0	0	0	0	1	0	0	0	1	0	1	0	0	0	0	1	1	0	1	1	0
4	60	100	150	175	50	0	0	0	0	0	0	0	0	0	0	0	0	0	0	0	0	0	0	0	0	0	0	0	0
5	150	40	40	175	150	0	0	0	0	0	0	1	1	0	0	0	0	0	0	0	0	1	1	1	1	1	1	0	0
6	150	40	150	125	50	1	1	1	1	1	1	1	1	1	1	1	1	1	1	1	1	1	1	1	1	1	1	1	1
7	150	100	40	175	50	1	1	1	0	1	1	1	1	1	1	1	1	1	1	1	1	1	1	1	1	1	1	0	1
8	150	100	150	125	131	1	1	1	1	0	1	1	1	0	1	1	1	1	1	1	0	1	1	1	1	0	1	1	1

Results

As you can see, some errors were made in Z-speed settings for runs 2, 3, and 8. For runs 2 and 8, the problem was associated with machine limitations. Run 3 should have been set at 150, but was inadvertently set incorrectly. These problems will limit our ability to obtain unambiguous results for each factor's effectiveness; however, there still exists a very important finding from comparing the results for each of the 8 runs. For example, prior to collecting the data, it was generally accepted that runs 1, 2, and 6 would be poor performers and were only run because they showed up in the design matrix. Run 6 was obviously a surprise, among other things, preconceived notions had previously prevented this successful combination from being tested.

Conclusions

Not only did the experimental matrix provide the best combination of input factors, resulting in 100% good bonds, these bonds all exceeded the minimum bond strength requirements. In fact, the bond strength testing device could not knock them off. The knowledge gained in this one experiment was worth many times that of four months of one-at-a-time experiments.

Constant Velocity Joints on a Front Wheel Drive
Analysis of Multiple Responses from Designed Experiments

Submitted by: Don F. Wilson
                      Quinas, Inc.
                      Littleton, Colorado

Abstract

This case study presents methods of visualizing and analyzing multiple responses from a designed experiment. This case study is a product development example of elastomeric boots for constant velocity joints on front wheel drive automobiles. Methods presented include use of linear graphs to represent interactive factors, blocking to eliminate noise factors, use of percent contribution to identify strong relationships, analysis of non-parametric (qualitative) data, and compromises to balance performance of several key objectives.

Background

The use of constant velocity (CV) joints on front wheel drive automobiles presents a unique challenge for development of elastomeric products. The CV joints are permanently sealed with a boot to contain the grease and to prevent contaminants from entering. If undetected by the customer, the tearing or loosening of the elastomeric boot may require replacement of an entire CV joint, costing the customer in excess of $350.00 per CV joint.

The boot must be able to flex at angles up to 45 degrees; operate at speeds up to 1900 RPM without more than 7mm of radial growth; maintain resistance to attack by ozone, oxidation, and grease; perform at temperatures below -40 degrees and over 200 degrees; and provide cut resistance to road hazards.

At the time of this case study, the product dimensions had been determined by the customer and most of the key functional parameters of the CV boots had been met. The primary focus of this study was to improve the resistance of the boot to ozone attack, low temperature properties, and radial growth characteristics at high RPM, while maintaining the performance of previous rubber formulations on other functional tests.

Determining Objectives

The primary objective of the experiment was to substantially improve the dynamic ozone characteristics of the formulation without adversely affecting the other requirements:

Dynamic Ozone after Heat Aging

Low Temperature Brittleness

Radial Growth (Speed at 7mm Growth)   - Tested on CV assemblies

Functional Life Testing

Tensile Strength

100% Modulus

Elongation at Break   - Samples cut from parts

Compression Set Characteristics

Tear Resistance

Primary Objective

The customer tests the ozone resistance of the formulation on the CV boot, on an accelerated life test preceded by heat aging the samples to simulate internal deterioration (inner noise) of the product. The specification was to show no cracking under 7X magnification after 168 hours of running the CV joint at a 34 degree angle at 23 mph (350 RPM) in an atmosphere of 50 ppm of ozone. At the time of this study, minute cracks were visible at 168 hours (no cracks at 144 hours). Several attempts to "patch" the formulation with one-at-a-time changes had failed. Parts had been marginally passing the radial growth test and low temperature flex requirements. Any change to the rubber formulation might cause problems with these tests.

Measuring the Objective

The dynamic ozone test at the customer's facility could not be directly duplicated in the development lab because 1) the expense of the test equipment was high, 2) the time necessary to acquire the equipment would delay development, and 3) correlation problems between labs existed. To circumvent these problems, a "predictor" test had been developed in the supplier's development lab to show relative improvements in ozone performance of formulations.

Brainstorming Potential Causes / "Knobs"

The primary cause of deterioration of the rubber boot is the ozone level in the atmosphere. However, the intent of the experiment is not to control the root cause (ozone), but to make the boots "robust" or insensitive to the "outer" noise of ozone. The major factors to improve ozone resistance relate to either reducing the stress level of the rubber in operation or using protective chemicals in the formulation.

Prioritizing the Investigations

Since the product design had been finalized, and modulus requirements were necessary, the focus of the experiment was on modifying the rubber formulation to improve chemical resistance to ozone.

The rubber formulation consists of thirteen ingredients, but the chemist believed only four ingredients were potential major contributors to improving the ozone performance.

Selecting Factors for the Experiment

The four factors were to be characterized by both type and amount, creating seven Control Factors for the experiment:

Neoprene Type	(N)	Antioxidant Level	(XL)
Antiozonant Type	(Z)	Wax Type	(W)
Antiozonant Level	(ZL)	Wax Level	(WL)
Antioxidant Type	(X)		

One noise factor was considered to be dominant in the ozone testing –the variability of the ozone concentration from test to test.

Interactions of the three protective ingredients were considered to be strong possibilities, creating seven suspected interactions:

$Z \times ZL$ (Antiozonant Type × Antiozonant Level)

$X \times XL$ (Antioxidant Type × Antioxidant Level)

$W \times WL$ (Wax Type × Wax Level)

$Z \times X$ (Antiozonant Type × Antioxidant Type)

$Z \times W$ (Antiozonant Type × Wax Type)

$X \times W$ (Antioxidant Type × Wax Type)

$ZL \times XL$ (Antiozonant Level × Antioxidant Level)

Selecting Levels for Each Factor

Two types of neoprene (GNA, GW) were selected to investigate differences in performance on the three major responses: Dynamic Ozone, Low Temperature, and Radial Growth. Technical literature indicated that stress levels might be significantly different in the neoprene types to improve performance characteristics in all three areas.

The selection of the protective ingredients were based on technical literature and previous lab screenings. The levels of the antioxidant and antiozonant were selected to produce "adequate" to "improved" protection, but below the maximum solubility of the materials in the formulation. The levels of the wax types were varied from 0 to 4 parts per hundred (pph) of base polymer to examine a published theory that way may improve static ozone resistance, but inhibit dynamic ozone resistance.

Choosing the Appropriate Design

With the seven factors listed above and selecting two levels per factor, all combinations of seven factors at two levels each would require 128 formulations. Running the 128 trials would produce information on all interactions of the factors. However, resources for the experiment were limited. The dynamic ozone chamber in the development lab would test only 8 samples at a time, and each test required approximately 2 weeks (32 weeks for 128 combinations). The cost of testing each formulation was $250.00, or $32,000 for 128 combinations.

The minimum use of resources to examine the effects of each factor would be 8 trials, using a one-at-a-time strategy, or an L8 design (a fully saturated fractional factorial). A one-at-a-time strategy would not permit analysis of interactions and would be very sensitive to individual test results. The L8 Design would confound all 2-way interactions with the factor effects (3 2-way interactions with each factor), but factor effects would be based on averages of 4 data points each. Larger fractional factorial designs (L16, L32, L64) would reduce confounding of interactions at the expense of spending more resources. Selection of non-geometric orthogonal designs (L12, L20, L24, L28) would prevent total confounding of interactions with factors, but would not permit analysis of the interactions.

The chemist decided to select a design that would prevent confounding of the seven factors and seven interactions that were suspected, but allowing the other 14 potential interactions to be confounded with factors and/or interactions. Using the linear graph method developed by Genichi Taguchi, the experiment can be accomplished using the L16 design as shown in Figure 1.

Reviewing the Design

The linear graph does not show the confounding of 14 other potential 2-way interactions, or higher order interactions. Using an interaction table, the confounding situations in Table 1 were identified.

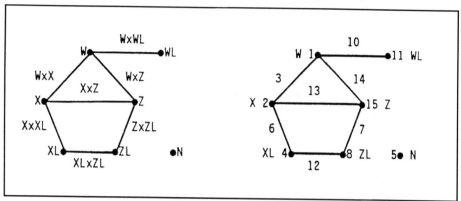

Figure 1 L16 Design

FACTOR			POTENTIAL 2-WAY INTERACTIONS
1	W	Wax Type	N×XL
2	X	Antiox Type	
3	W×X	Wax × Antiox	ZL×WL
4	XL	Antiox Lvl	W×N WL×Z
5	N	Neoprene	W×XL
6	X×XL		
7	Z×ZL		X×N
8	ZL	Antioz Lvl	
9	(not assigned)		W×XL X×WL
10	W×WL		X×ZL N×Z
11	WL	Wax Level	XL×Z
12	XL×ZL		
13	X×Z		N×ZL
14	W×Z		N×WL
15	Z	Antioz	XL×WL

Table 1 Confounding Situations

The 2-way interactions that were confounded were not considered to be any of the "vital few" effects that would significantly alter ozone performance.

A review of the formulation combinations showed all combinations to be feasible for the experiment:

Design #	WAX TYPE	Antiox Type	Antiox Level	Neoprene Type	Antioz Level	Wax Level	Antioz Level
1	Parafin	X1	1 PHR	GNA	1 PHR	0 PHR	Z1
2	Parafin	X1	1 PHR	GNA	3 PHR	4 PHR	Z2
3	Parafin	X1	3 PHR	GW	1 PHR	0 PHR	Z2
4	Parafin	X1	3 PHR	GW	3 PHR	4 PHR	Z1
5	Parafin	X2	1 PHR	GNA	1 PHR	4 PHR	Z2
6	Parafin	X2	1 PHR	GNA	3 PHR	0 PHR	Z1
7	Parafin	X2	3 PHR	GW	1 PHR	4 PHR	Z1
8	Parafin	X2	3 PHR	GW	3 PHR	0 PHR	Z2
9	Poly.	X1	1 PHR	GW	1 PHR	4 PHR	Z2
10	Poly.	X1	1 PHR	GW	3 PHR	0 PHR	Z1
11	Poly.	X1	3 PHR	GNA	1 PHR	4 PHR	Z1
12	Poly.	X1	3 PHR	GNA	3 PHR	0 PHR	Z2
13	Poly.	X2	1 PHR	GW	1 PHR	0 PHR	Z1
14	Poly.	X2	1 PHR	GW	3 PHR	4 PHR	Z2
15	Poly.	X2	3 PHR	GNA	1 PHR	0 PHR	Z2
16	Poly.	X2	3 PHR	GNA	3 PHR	4 PHR	Z1

Determining the Sequence of Experimenting

The test samples could be prepared in random sequence (to reduce the chances of a nuisance variable affecting the results). However, the variability of the dynamic ozone test (8 tests/setup) was likely to affect the test results substantially. In Table 2 we see the setup table sorted by test order. Factor 9 was used to block the order of testing to isolate the effect of the ozone test variation.

Trial Order	Design No.	Wax Type	Antiox Type	Antiox Level	Neoprene Type	Antioz Level	Test Block	Wax Level	Antioz Type
1	3	Parafin	X1	3 PHR	GW	1 PHR	1st	0 PHR	Z2
2	10	Poly.	X1	1 PHR	GW	3 PHR	1st	0 PHR	Z1
3	7	Parafin	X2	3 PHR	GW	1 PHR	1st	4 PHR	Z1
4	14	Poly.	X2	1 PHR	GW	3 PHR	1st	4 PHR	Z2
5	12	Poly.	X1	3 PHR	GNA	3 PHR	1st	0 PHR	Z2
6	5	Parafin	X2	1 PHR	GNA	1 PHR	1st	4 PHR	Z2
7	1	Parafin	X1	1 PHR	GNA	1 PHR	1st	0 PHR	Z1
8	16	Poly.	X2	3 PHR	GNA	3 PHR	1st	4 PHR	Z1
9	15	Poly.	X2	3 PHR	GNA	1 PHR	2nd	0 PHR	Z2
10	13	Poly.	X2	1 PHR	GW	1 PHR	2nd	0 PHR	Z1
11	6	Parafin	X2	1 PHR	GNA	3 PHR	2nd	0 PHR	Z1
12	11	Poly.	X1	3 PHR	GNA	1 PHR	2nd	4 PHR	Z1
13	9	Poly.	X1	1 PHR	GW	1 PHR	2nd	4 PHR	Z2
14	8	Parafin	X2	3 PHR	GW	3 PHR	2nd	0 PHR	Z2
15	4	Parafin	X1	3 PHR	GW	3 PHR	2nd	4 PHR	Z1
16	2	Parafin	X1	1 PHR	GNA	3 PHR	2nd	4 PHR	Z2

Table 2   L16 Design - Setup Table Sorted by Test Order

Performing Experiments/Obtaining Results

As anticipated, the dynamic ozone results (hours to first cracks observed) changed dramatically from the first test group to the second, averaging 342 hours and 174 hours, respectively. The chemist, on noting the difference, requested a supplier (DuPont) to perform a "second opinion" test at their laboratories, using a different apparatus. Results were compiled in Table 3:

Trial Order	Design No.	Dynamic Ozone	DuPont Ozone	Ozone Rating	Modulus 100%	Tensile PSI	Elong %	Hi Speed Spin	Low Temp
7	1	384	4	2B	380	1870	325	2250	−49
16	2	168	5	2B+	390	1930	350	2300	−47
1	3	360	5	2B+	300	2150	375	2100	−45
15	4	192	2	1B+	315	2070	350	2100	−46
6	5	288	7	3B	360	1970	300	2250	−48
11	6	168	5	2B+	420	1930	325	2350	−48
3	7	240	7	3B	300	2020	375	2050	−46
14	8	168	5	2B+	340	2060	400	2200	−47
13	9	168	5	2B+	295	2140	325	2150	−44
2	10	408	3	2B−	340	2180	375	2150	−46
12	11	216	2	1B+	370	2030	325	2150	−48
5	12	408	4	2B	415	2060	300	2300	−49
10	13	168	7	3B	310	2200	450	2100	−47
4	14	336	6	3B−	325	2170	425	2100	−45
9	15	144	7	3B	390	2080	300	2300	−48
8	16	312	6	3B−	395	2050	350	2250	−47

Table 3   Second Opinion Test

Analyzing Data

The data for each response was analyzed using *Design Cube* software. Analysis of Variance (ANOVA) tables and Level Effects reports were generated for each response (see Tables 4 and 5). The analysis for the Dynamic Ozone responses are shown in Table 4. To determine significant effects, factors and interactions with less importance were "pooled" into error terms.

The results on Dynamic Ozone in Table 5 show the TEST BLOCK (the difference between the first and second test groups) to be the largest effect on ozone performance (79.9% contribution), followed by ANTIOXIDANT TYPE (10.1%), and WAX LEVEL (3.6%). The pooling process may have created "significant" results out of insufficient information by pooling the smaller sources of variation preferentially. Thus, the Probability column (1−alpha) may not be accurate. Reviewing the tables of Level Means and Level Effects shows the preferred Antioxidant is X1. The preferred wax level is 0 PHR. The interaction plots in Figure 2 show a preference for 3 PHR of Antioxidant if X1 is selected.

The Dynamic Ozone results, however, were questionable, considering the large variation between tests. Therefore, a "second opinion" test was conducted by DuPont on a different method of dynamic ozone testing. All samples were run for 336 hours in the ozone chamber and removed. Unfortunately, the DuPont measurement criteria (Table 6) was not a continuous quantitative scale as with the "time to cracking" criteria. The data was qualitative, using a visual rating scale, from 1B− (very slight cracks under 7X magnification) to 3B+ (almost ready to break). The data from the DuPont Test was reviewed in two ways: 1) By sorting the trials from low to high ratings and 2) by assigning a numerical rating scale to the visual ratings (1B = 1 . . . 3B+ = 8).

```
                    CV BOOT COMPOUNDING STUDY        L16 DESIGN
Grp. Factor   Name       Level 1   Level 2       2 way interactions

I     1  W     WAX TYPE   PARAFIN   POLY.      2x3 4x5 6x7 8x9 10x11 12x13 14x15
II    2  X     ANTIOX.    X1        X2         1x3 4x6 5x7 8x10 9x11 12x14 13x15
II    3  WxX                                   1x2 4x7 5x6 8x11 9x10 12x15 13x14
III   4  XL    ANTIOX LVL 1 PHR     3 PHR      1x5 2x6 3x7 8x12 9x13 10x14 11x15
III   5  N     NEOPRENE   GNA       GW         1x4 2x7 3x6 8x13 9x12 10x15 11x14
III   6  XxXL                                  1x7 2x4 3x5 8x14 9x15 10x12 11x13
III   7  ZxZL                                  1x6 2x5 3x4 8x15 9x14 10x13 11x12
IV    8  ZL    ANTIOZ LVL 1 PHR     3 PHR      1x9 2x10 3x11 4x12 5x13 6x14 7x15
IV    9  B     TEST BLOCK 1st       2nd        1x8 2x11 3x10 4x13 5x12 6x15 7x14
IV    10 WxWL                                  1x11 2x8 3x9  4x14 5x15 6x12 7x13
IV    11 WL    WAX LEVEL  0 PHR     4 PHR      1x10 2x9 3x8  4x15 5x14 6x13 7x12
IV    12 XLxZL                                 1x13 2x14 3x15 4x8 5x9  6x10 7x11
IV    13 XxZ                                   1x12 2x15 3x14 4x9 5x8  6x11 7x10
IV    14 WxZ                                   1x15 2x12 3x13 4x10 5x11 6x8  7x9
IV    15 Z     ANTIOZ.    Z1        Z2         1x14 2x13 3x12 4x11 5x10 6x9  7x8
```

```
Trial  Design Factors                                         DYNAMIC
Order  Order  1 2 3  4 5  6 7  8 9  10 11 12  13 14 15          OZONE

  7      1    1 1 1  1 1  1 1  1 1   1  1  1   1  1  1           384
 16      2    1 1 1  1 1  1 1  2 2   2  2  2   2  2  2           168
  1      3    1 1 1  2 2  2 2  1 1   1  1  2   2  2  2           360
 15      4    1 1 1  2 2  2 2  2 2   1  1  1   1  1  1           192
  6      5    1 2 2  1 1  2 2  1 1   2  1  1   1  2  2           288
 11      6    1 2 2  1 1  2 2  2 2   1  1  2   2  1  1           168
  3      7    1 2 2  2 2  1 1  1 1   2  2  2   2  1  1           240
 14      8    1 2 2  2 2  1 1  2 2   1  1  1   1  2  2           168
 13      9    2 1 2  1 2  1 2  1 2   1  2  1   2  1  2           168
  2     10    2 1 2  1 2  1 2  2 1   2  1  2   1  2  1           408
 12     11    2 1 2  2 1  2 1  1 2   1  2  2   1  2  1           216
  5     12    2 1 2  2 1  2 1  2 1   2  1  1   2  1  2           408
 10     13    2 2 1  1 2  2 1  1 2   2  1  1   2  2  1           168
  4     14    2 2 1  1 2  2 1  2 1   1  2  2   1  1  2           336
  9     15    2 2 1  2 1  1 2  1 2   2  1  2   1  1  2           144
  8     16    2 2 1  2 1  1 2  2 1   1  2  1   2  2  1           312

            Grand Average   258.00    Averages    258.00    0.00
```

ERROR (e)	SOURCE	NAME	df	SUM OF SQUARES	MEAN SQ.	% CONTR.	F	Prob.
	1 W	WAX TYPE	1	2304.00	2304.00	1.54	18.67	.9960**
	2 X	ANTIOX.	1	14400.00	14400.00	10.12	116.67	.9998**
e	3 WxX		1	0.00	0.00	0.00		
e	4 XL	ANTIOX LVL	1	144.00	144.00	0.00		
e	5 N	NEOPRENE	1	144.00	144.00	0.00		
	6 XxXL		1	1296.00	1296.00	.83	10.50	.9858 *
e	7 ZxZL		1	144.00	144.00	0.00		
	8 ZL	ANTIOZ LVL	1	2304.00	2304.00	1.54	18.67	.9960**
	9 B	TEST BLOCK	1	112896.00	112896.00	79.91	914.67	.9999**
	10 WxWL		1	576.00	576.00	.32	4.67	.9341
	11 WL	WAX LEVEL	1	5184.00	5184.00	3.59	42.00	.9993**
e	12 XLxZL		1	144.00	144.00	0.00		
	13 XxZ		1	1296.00	1296.00	.83	10.50	.9858 *
e	14 WxZ		1	144.00	144.00	0.00		
e	15 Z	ANTIOZ.	1	144.00	144.00	0.00		
		Replications	0	0.00	0.00	0.00		
		Error	7	864.00	123.43	1.31		
		Total	15	141120.00		100.00		
		Standard Error		11.11				

ANOVA TABLE

Table 4 Analysis for the Dynamic Ozone Responses

```
                    CV BOOT COMPOUNDING STUDY
                           L16 DESIGN

                                   LEVEL      LEVEL           TOTAL
FACTOR    NAME       LEVEL         MEANS      EFFECTS         EFFECTS

 1 W     WAX TYPE   1 PARAFIN     246.00      -12.00          24.00**
                    2 POLY.       270.00       12.00

 2 X     ANTIOX.    1 X1          288.00       30.00         -60.00**
                    2 X2          228.00      -30.00

 3 WxX              1             258.00        0.00           0.00
                    2             258.00        0.00

 4 XL    ANTIOX LVL 1 1 PHR       261.00        3.00          -6.00
                    2 3 PHR       255.00       -3.00

 5 N     NEOPRENE   1 GNA         261.00        3.00          -6.00
                    2 GW          255.00       -3.00

 6 XxXL             1             249.00       -9.00          18.00 *
                    2             267.00        9.00

 7 ZxZL             1             261.00        3.00          -6.00
                    2             255.00       -3.00

 8 ZL    ANTIOZ LVL 1 1 PHR       246.00      -12.00          24.00**
                    2 3 PHR       270.00       12.00

 9 B     TEST BLOCK 1 1st         342.00       84.00        -168.00**
                    2 2nd         174.00      -84.00

10 WxWL             1             264.00        6.00         -12.00
                    2             252.00       -6.00

11 WL    WAX LEVEL  1 0 PHR       276.00       18.00         -36.00**
                    2 4 PHR       240.00      -18.00

12 XLxZL            1             261.00        3.00          -6.00
                    2             255.00       -3.00

13 XxZ              1             267.00        9.00         -18.00 *
                    2             249.00       -9.00

14 WxZ              1             255.00       -3.00           6.00
                    2             261.00        3.00

15 Z     ANTIOZ.    1 Z1          261.00        3.00          -6.00
                    2 Z2          255.00       -3.00

                               Grand Average   258.00
```

Table 5 Analysis for the Dynamic Ozone Responses (Cont.)

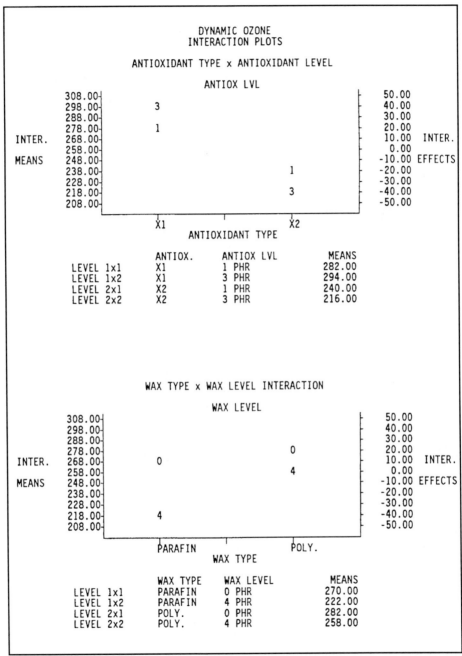

Figure 2 Dynamic Ozone Interaction Plots

The results from the DuPont Test in Table 6 tended to confirm Antioxidant X1 as the largest contributor in the formulation for ozone protection, but did not confirm the effects of the Wax Level. ANOVA Analysis using the Numerical rating scale appeared to show an effect from the ANTIOXIDANT TYPE (59.5% contribution), the ANTIOZONANT TYPE (9.5%), the ANTIOZONANT LEVEL (9.5%), and an ANTIOXIDANT TYPE / ANTIOZONANT TYPE interaction (9.5%).

```
                         DUPONT OZONE RESULTS
                         SORTED BY VISUAL RATING

                                                        |DUPONT   DUPONT
Des|WAX     ANTIOX ANTIOX NEOPRENE ANTIOZ TEST  WAX   ANTIOZ|VISUAL   NUMER.
No |TYPE    TYPE   LEVEL  TYPE     LEVEL  BLOCK LEVEL TYPE  |RATING   RATING

11 |POLY.   X1     3 PHR  GNA      1 PHR  2nd   4 PHR Z1    | 1B+       2
 4 |PARAFIN X1     3 PHR  GW       3 PHR  2nd   4 PHR Z1    | 1B+       2

10 |POLY.   X1     1 PHR  GW       3 PHR  1st   0 PHR Z1    | 2B-       3

12 |POLY.   X1     3 PHR  GNA      3 PHR  1st   0 PHR Z2    | 2B        4
 1 |PARAFIN X1     1 PHR  GNA      1 PHR  1st   0 PHR Z1    | 2B        4

 6 |PARAFIN X2     1 PHR  GNA      3 PHR  2nd   0 PHR Z1    | 2B+       5
 3 |PARAFIN X1     3 PHR  GW       1 PHR  1st   0 PHR Z2    | 2B+       5
 9 |POLY.   X1     1 PHR  GW       1 PHR  2nd   4 PHR Z2    | 2B+       5
 8 |PARAFIN X2     3 PHR  GW       3 PHR  2nd   0 PHR Z2    | 2B+       5
 2 |PARAFIN X1     1 PHR  GNA      3 PHR  2nd   4 PHR Z2    | 2B+       5

14 |POLY.   X2     1 PHR  GW       3 PHR  1st   4 PHR Z2    | 3B-       6
16 |POLY.   X2     3 PHR  GNA      3 PHR  1st   4 PHR Z1    | 3B-       6

 7 |PARAFIN X2     3 PHR  GW       1 PHR  1st   4 PHR Z1    | 3B        7
13 |POLY.   X2     1 PHR  GW       1 PHR  2nd   0 PHR Z1    | 3B        7
 5 |PARAFIN X2     1 PHR  GNA      1 PHR  1st   4 PHR Z2    | 3B        7
15 |POLY.   X2     3 PHR  GNA      1 PHR  2nd   0 PHR Z2    | 3B        7
```

Table 6 DuPont Ozone Results

Analysis of Multiple Responses

Each response can be analyzed separately by ANOVA to determine the effect of each factor on each response. To gain a better understanding of the relationships of each factor to each response, Multiple Response Summary Tables can be constructed to show a matrix of the factors and product characteristics on a single page. The Summary Tables may display the Percent Contribution, Level

Effects, or Level Means. Table 7 is the Percent Contribution Summary Table and shows the relative effect of each factor on each product characteristic and can be used to quickly identify the "vital few" factors, versus the "trivial many." The Percent Contribution Summary Table for this experiment shows the importance of antioxidant selection for Ozone Protection, and the importance of Neoprene selection for Modulus, Tensile, Elongation, High Speed Spin, and Low Temperature properties. The Wax selection appears important to maintain tensile strength. Factors of lesser significance are also identified.

The Percent Contribution Table does not indicate the preferred level for each factor –only the relative importance. Table 8 summarize information by Level Effects and shows which level is preferred for any particular response. The Level Effect is obtained by subtracting the Grand Average from the Mean Response of each Factor Level. Thus, a prediction can be made by adding the Level Effects for any combination of Factor Levels to the Grand Average. For example, the Tensile prediction for the combination of Poly Wax with GNA Neoprene would be: 2056.88 (Grand Average) + 56.88 (Poly Wax Level Effect) − 66.88 (GNA Neoprene Level Effect) = 2046.88, assuming other Factor effects to be insignificant.

Table 9 is the Level Means Summary Table and shows the preferred Levels of Neoprene Type to be GW Neoprene for Tensile and Elongation Properties, but GNA Neoprene for High Speed Spin and Low Temperature. The selection of Neoprene Type is dependent on the relative importance of the response criteria. For factors that have no substantial relationship to any responses, the factor level with the lower cost would be the choice.

CV BOOT COMPOUNDING STUDY
SUMMARY TABLE - % CONTRIBUTION

FACTOR	NAME	LEVEL	DYNAMIC OZONE	DUPONT OZONE	MODULUS 100%	TENSILE PSI	ELONG. %	HI SPEED SPIN	LOW TEMP BRITTLE.
1 W	WAX TYPE	1 PARAFIN 2 POLY.	1.63	0.00	.28	36.56	.52	.48	.79
2 X	ANTIOX.	1 X1 2 X2	10.20	59.52	.28	.11	8.38	.48	.79
3 WxX		1 2	0.00	2.38	.28	.75	8.38	.48	.79
4 XL	ANTIOX LVL	1 1 PHR 2 3 PHR	.10	2.38	.00	.75	2.09	1.93	.79
5 N	NEOPRENE	1 GNA 2 GW	.10	0.00	81.05	50.55	52.36	69.56	63.78
6 XxXL		1 2	.92	2.38	.05	2.76	.52	1.93	.79
7 ZxZL		1 2	.10	0.00	.05	2.34	13.09	1.93	7.09
8 ZL	ANTIOZ LVL	1 1 PHR 2 3 PHR	1.63	9.52	12.64	.00	2.09	7.73	0.00
9 B	TEST BLOCK	1 1st 2 2nd	80.00	2.38	.05	.04	0.00	1.93	0.00
10 WxWL		1 2	.41	2.38	.00	.53	.52	0.00	3.15
11 WL	WAX LEVEL	1 0 PHR 2 4 PHR	3.67	0.00	4.81	.99	.52	7.73	7.73
12 XLxZL		1 2	.10	0.00	.14	.22	.52	.48	3.15
13 XxZ		1 2	.92	9.52	.00	.11	.52	.48	3.15
14 WxZ		1 2	.10	0.00	.28	2.34	8.38	.48	0.00
15 Z	ANTIOZ.	1 Z1 2 Z2	.10	9.52	.05	1.95	2.09	4.35	3.15

☐ = STRONG CONTRIBUTOR TO RESPONSE

☐ = LESSER CONTRIBUTOR TO RESPONSE

Table 7 Percent Contribution

CV BOOT COMPOUNDING STUDY
SUMMARY TABLE - LEVEL EFFECTS

FACTOR	NAME	LEVEL	DYNAMIC OZONE	DUPONT OZONE	MODULUS 100%	TENSILE PSI	ELONG. %	HI SPEED SPIN	LOW TEMP BRITTLE.
1 W	WAX TYPE	1 PARAFIN	-12.00	0.00	-2.19	-56.88	-3.12	6.25	-.12
		2 POLY.	12.00⊙	0.00	2.19	56.88●	3.12	-6.25	.12
2 X	ANTIOX.	1 X1	30.00●	-1.25●	-2.19	-3.12	-12.50	-6.25	.12
		2 X2	-30.00	1.25	2.19	3.12	12.50⊙	6.25	-.12
3 WxX		1	0.00	.25	-2.19	8.12	12.50⊙	-6.25	.12
		2	0.00	-.25	2.19	-8.12	-12.50	6.25	-.12
4 XL	ANTIOX LVL	1 1 PHR	3.00	.25	-.31	-8.12	6.25	12.50	.12
		2 3 PHR	-3.00	-.25	.31	8.12	-6.25	-12.50	-.12
5 N	NEOPRENE	1 GNA	3.00	0.00	37.19	-66.88	-31.25	75.00●	-1.12●
		2 GW	-3.00	0.00	-37.19	66.88●	31.25●	-75.00	1.12
6 XxXL		1	-9.00	.25	.94	-15.62	-3.12	12.50	.12
		2	9.00	-.25	-.94	15.62	3.12	-12.50	-.12
7 ZxZL		1	3.00	0.00	.94	-14.38	15.62⊙	-12.50	-.38⊙
		2	-3.00	0.00	-.94	14.38	-15.62	12.50	.38
8 ZL	ANTIOZ LVL	1 1 PHR	-12.00	.50	-14.69	.62	-6.25	-25.00	0.00
		2 3 PHR	12.00⊙	-.50⊙	14.69	-.62	6.25	25.00⊙	0.00
9 B	TEST BLOCK	1 1st	84.00?	.25	-.94	1.88	0.00	-12.50	0.00
		2 2nd	-84.00?	-.25	.94	-1.88	0.00	12.50	0.00
10 WxWL		1	6.00	-.25	.31	-6.88	3.12	0.00	.25
		2	-6.00	.25	-.31	6.88	-3.12	0.00	-.25
11 WL	WAX LEVEL	1 0 PHR	18.00⊙	0.00	9.06	9.38	3.12	25.00⊙	-.50⊙
		2 4 PHR	-18.00	0.00	-9.06	-9.38	-3.12	-25.00	.50
12 XLxZL		1	3.00	0.00	-1.56	-4.38	-3.12	6.25	-.25
		2	-3.00	0.00	1.56	4.38	3.12	-6.25	.25
13 XxZ		1	9.00	-.50⊙	-.31	-3.12	-3.12	-6.25	-.25
		2	-9.00	.50	.31	3.12	3.12	6.25	.25
14 WxZ		1	-3.00	0.00	2.19	-14.38	-12.50	6.25	0.00
		2	3.00	0.00	-2.19	14.38	12.50⊙	-6.25	0.00
15 Z	ANTIOZ.	1 Z1	3.00	-.50⊙	.94	-13.12	6.25	-18.75	-.25
		2 Z2	-3.00	.50	-.94	13.12	-6.25	18.75	.25
	Grand Average		258.00	5.00	352.81	2056.88	353.12	2193.75	-46.88

● = STRONG PREFERENCE FOR LEVEL SETTING

⊙ = MILD PREFERENCE FOR LEVEL SETTING

Table 8 Level Effects

CV BOOT COMPOUNDING STUDY
SUMMARY TABLE - LEVEL MEANS

FACTOR	FACTOR	LEVEL	DYNAMIC OZONE	DUPONT OZONE	MODULUS 100%	TENSILE PSI	ELONG. %	HI SPEED SPIN	LOW TEMP BRITTLE.
1 W	WAX TYPE	1 PARAFIN	246.00	5.00	350.62	2000.00	350.00	2200	-47.00
		2 POLY.	270.00	5.00	355.00	2113.75	356.25	2188	-46.75
2 X	ANTIOX.	1 X1	288.00	3.75	350.62	2053.75	340.62	2188	-46.75
		2 X2	228.00	6.25	355.00	2060.00	365.62	2200	-47.00
3 WxX		1	258.00	5.25	350.62	2065.00	365.62	2188	-46.75
		2	258.00	4.75	355.00	2048.75	340.62	2200	-47.00
4 XL	ANTIOX LVL	1 1 PHR	261.00	5.25	352.50	2048.75	359.38	2206	-46.75
		2 3 PHR	255.00	4.75	353.12	2065.00	346.88	2181	-47.00
5 N	NEOPRENE	1 GNA	261.00	5.00	390.00	1990.00	321.88	2269	-48.00
		2 GW	255.00	5.00	315.62	2123.75	384.38	2119	-45.75
6 XxXL		1	249.00	5.25	353.75	2041.25	350.00	2206	-46.75
		2	267.00	4.75	351.88	2072.50	356.25	2181	-47.00
7 ZxZL		1	261.00	5.00	353.75	2042.50	368.75	2181	-47.25
		2	255.00	5.00	351.88	2071.25	337.50	2206	-46.50
8 ZL	ANTIOZ LVL	1 1 PHR	246.00	5.50	338.12	2057.50	346.88	2169	-46.88
		2 3 PHR	270.00	4.50	367.50	2056.25	359.38	2219	-46.88
9 B	TEST BLOCK	1 1st	342.00	5.25	351.88	2058.75	353.12	2181	-46.88
		2 2nd	174.00	4.75	353.75	2055.00	353.12	2206	-46.88
10 WxWL		1	264.00	4.75	353.12	2050.00	356.25	2194	-46.62
		2	252.00	5.25	352.50	2063.75	350.00	2194	-47.12
11 WL	WAX LEVEL	1 0 PHR	276.00	5.00	361.88	2066.25	356.25	2219	-47.38
		2 4 PHR	240.00	5.00	343.75	2047.50	350.00	2169	-46.38
12 XLxZL		1	261.00	5.00	351.25	2052.50	350.00	2200	-47.12
		2	255.00	5.00	354.38	2061.25	356.25	2188	-46.62
13 XxZ		1	267.00	4.50	352.50	2053.75	350.00	2188	-47.12
		2	249.00	5.50	353.12	2060.00	356.25	2200	-46.62
14 WxZ		1	255.00	5.00	355.00	2042.50	340.62	2200	-46.88
		2	261.00	5.00	350.62	2071.25	365.62	2188	-46.88
15 Z	ANTIOZ.	1 Z1	261.00	4.50	353.75	2043.75	359.38	2175	-47.12
		2 Z2	255.00	5.50	351.88	2070.00	346.88	2212	-46.62

Table 9   Level Means

Determining Settings

The original set of priorities indicated that the Dynamic Ozone, High Speed Spin, and Low Temperature Brittleness characteristics were the most important responses. Through use of the Summary Tables, the following formulation was derived:

MATERIAL	AMOUNT	TO IMPROVE:
Neoprene GNA		High Speed Spin, Low
Antioxidant X1	3 phr	Temperature
Antiozonant Z1	3 phr	Dynamic Ozone
Wax - Poly.	1 phr	(Dynamic Ozone ?)
		(Ozone / Tensile ?)

This combination had not been run as one of the 16 formulations in the experiment. Instead of "Picking the Winner" from among the 16 formulations, the analysis is predicting a different combination. Predictions of average performance with the above combination (conservative estimates using Grand Average + largest Level Effects):

Dynamic Ozone:	Improved, but cannot predict hours
DuPont Ozone:	$5.0 - 1.25 - .5 - .5 - .5 = 2.25 = 1B+$ rating
High Speed Spin:	$2194 + 75 + 25 =$ Approx 2300 RPM
Low Temp. Brittleness:	$-46.88 - 1.12 = -48$ deg. F.
Tensile:	$2057 + 57 - 67 = 2047$ PSI
Elongation:	$353 - 31 - 15.5 - 12.5 - 12.5 + 12.5 = 294\%$
Modulus:	$353 + 37 + 15 = 405$ PSI

Verifying Results

Results from the experiment trials should always be verified before action is taken. Mixing the above formulation gave results close to the prediction for Tensile, Elongation, Modulus, and High Speed Spin. The dynamic ozone improvements were predicted to be better than previous formulations, but could not be verified in running hours because of the large variability in testing. Samples of CV boots were sent to the customer for functional testing and all parts passed the performance tests as predicted.

Through Statistical Process Control, the Tensile, Elongation, and Modulus tests were monitored to maintain performance. Subsequent process improvements were introduced to improve consistency of the manufactured product.

The methods incorporated in this experiment have been used frequently in the development and production of CV boots and other rubber products. Through the use of Design of Experiments and Statistical Process Control, a better understanding of the chemistry of the formulation and manufacture of the boots was achieved.

A Strategic Defense Case Study

Submitted by: Peter L. Knepell
PhD, Operations Research
Logicon/RDA

Minutes of the First Meeting

"O.K. folks, here's the architecture we need to study," explained Capt. Kirk, an enthusiastic strategic defense expert. "Don't look at the picture too long because it might change any minute. Some things do remain constant:

- a two-tiered interceptor system (one is space-based and the other is ground-based),

- command centers,

- space-based and ground-based sensors to support the command centers and interceptors, and

- communications links.

"Look, this study is quite simple." "It's a simple matter of input/output, like knobs on a radio." His knobs were labeled:

Number of SBI	Space-Based Interceptors (SBI)
Number of GBI	Ground-Based Interceptors (GBI)
RVs	Reentry Vehicles (alias: blivits or bombs)
Decoys	The objects used to look like RVs
Pk	The one-shot probability of kill of an interceptor against a target
Trelease	The time it takes someone to give the command to shoot at missiles
Discrimination factor	A measure of how well real missiles are picked over decoys
Tdiscrim	The time it takes to track and discriminate an RV so that an interceptor can shoot at it

This stimulated a discussion on what measures will be used to gauge "system performance." The fundamental measure agreed upon was RVs destroyed. Then things got interesting and everyone got a chance to add knobs to the system.

"Not enough knobs," said Dee Coy, a renowned US expert on Russian threats through the year 2010. "You've got to consider the number of fast burn boosters and fast bussers." (Bussing is a term describing the act of casting several RVs off an ICBM in the proper direction –it's not a rapid transit system.)

"Don't forget the ASAT (anti-satellite) threat," piped in Dr. Larry Leber, a national labs scientist. "A good ASAT attack could 'smear' SBIs all over the galaxy."

"I have a new SBI design that will make all ASATs ineffective," countered Dr. S. B. Iacocca.

"Yeah, but what about SEOs?" remarked the consultant from ERUDITE Corporation.

"Yep, don't forget SEOs."   "SEOs are tricky."   "Oh no, SEOs!"

It got very quiet, especially after I asked, "What are SEOs?"

"Survivability Enhancement Options," someone said disdainfully.

"Whose?" I asked undaunted –after all, I was conducting the "simple input/output" study.

"The SBI's of course. You've got to consider these in your study," said Kirk.

"What exactly do the SBIs do?"

"THEY haven't decided yet. Anyway, if I knew, I couldn't tell you –it's classified above your level."

"When will THEY decide?"

"Soon, probably after you finish the study."

Minutes of the Second Meeting

I was much better prepared this time. "Gentlemen, this table shows all factors that you want me to consider.

PARAMETER	LEVELS	PARAMETER	LEVELS
ASAT Attack	4 types	Nr of Decoys	2 levels
SBI Design	2 types	Nr Fast-Burn	2 levels
Pk	2 levels	Boosters	
Quantity	4 levels		3 levels
		Trelease	
GBI Design	2 types	Discrimination	2 levels
Pk	2 levels	Factor	
Quantity	4 levels		2 levels
		Tdiscrim	

FULL FACTORIAL DESIGN =
$4 \times 2 \times 2 \times 4 \times 2 \times 2 \times 4 \times 2 \times 2 \times 3 \times 2 \times 2 = 49152$ cases

I am prepared to conduct a complete factorial analysis (49,152 cases). Since I can simulate an entire war in only 15 minutes, I'll be prepared to brief the results in 512 days (737,280 minutes) –as long as the computer doesn't crash."

"I want an answer in three months," boomed Capt. Kirk.

"O.K., how about a fractional factorial design where I'll tell you which factors were dominant in affecting system performance," I offered.

"No good," he said. "I need to make clear statements about the architecture THEY are designing. All factors are important!"

"All right, please tell me which factors are more likely to remain fixed while we vary everything else." The following list was developed:

SBI & GBI design & Pk	Trelease
Number of decoys	Tdiscrim
Number of fast-burn boosters	Discrimination factor

"I see your problem with the number of factors and I have a suggestion," offered Dee Coy. "If we composed two different threat scenarios, which differ by the number of fast-burn boosters and decoy composition, we could combine two of your factors into one called 'strategic threat.' Then we can measure the effects separately since they occur at different times in the ICBM flyout."

This was an excellent suggestion except for one consideration: "Could there be an interaction effect between the number of fast-burn boosters and decoy composition?" I asked.

Dee Coy was very accommodating: "Given this architecture, there is no reason to believe that there is an interaction. I might also suggest that the effects of Pk, Trelease, and Tdiscrim could be checked in just a few cases to give us a feel for their impact."

"So gentlemen, let's summarize this discussion. It appears that the following factors are foremost in your minds:

ASAT Threat	SBI Design
Number of SBI	GBI Design
Number of GBI	Strategic Threat

Then Dr.Leber had an interesting suggestion: "Why not see how the SBI design performs against the ASAT threats."

"But I want a complete tradeoff analysis for varying the numbers of SBIs and GBIs," said Kirk.

"Gentlemen, if we all agree that our main objective is to conduct in interceptor tradeoff analysis and our secondary objective is to see how changes in other parameters affect these trades, then I think I can have a study design very soon!"

Minutes of the Third Meeting

I opened the meeting by restating our objective: "If we all agree that our primary mission is to conduct a study on the tradeoffs (i.e., effects) involved with varying the number of SBIs and GBIs in the architecture, I think I have a proposition we all can live with. In particular, I will discuss a study that will support tradeoff analyses for 10 different scenarios and some other limited excursions with less than 200 simulations. With that number of runs, I can meet our schedule."

Here's what I presented: For each of the 10 scenarios, I have a "design space" representing different numbers of interceptors. I would conduct 18 simulations at the lattice points shown.

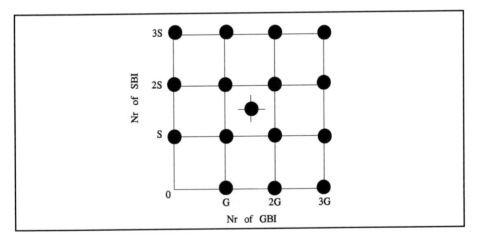

(I don't need to simulate the origin for obvious reasons.)

"There are only 16 lattice points shown. Why 18 runs?"

The reason: "The center point is simulated (replicated) three times to get a measure of variation. This will support the fit of a quadratic response surface. The contours of this surface (level curves) can then be plotted in the design space as shown."

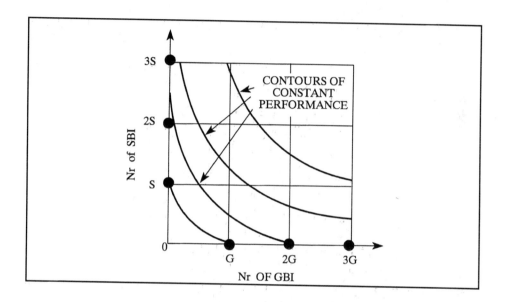

"What do these curves represent?" asked Capt. Kirk.

"One curve shows which combinations of SBIs and GBIs give a constant number of RVs killed. In fact, if you have some baseline number of RV kills that you want to compare the results to, I can generate that curve and highlight it in this display."

"So we could call these curves 'isomorts' for constant mortality," responded Kirk.

At this point, a vocal critic said: "This is statistical voodoo. You must replicate at all points at least 6 times and use the average to counteract variation."

This would increase the number of simulation runs to almost 1000! Discussions of statistical design, analysis of variance, etc. did not calm this persistent nay sayer. Nor did an appeal to the fact that these methods have been accepted by the most respected people in the field. After many unsuccessful attempts we entered into an interesting dialogue:

"You agree that there is a chance for variation at each point in the design space?" I asked.

"Of course," he replied.

"And you agree that with sufficient data, a representative response surface can be created."

"Yes."

"But your fear is that with only one replication for 15 of the points, the response could be too high or too low."

"Correct."

"Could the results at these points all come in too high?"

"Yes, but that would be rare."

"Well then, what would you expect to happen across 16 points in the design space?"

"I guess some would come in too high and some too low."

"Just like we would see with replications at one point?"

"Yes," he agreed.

"Well then, it seems like the 16 points will act like replications where some are too high, some are too low, but together, they should balance each other out to provide a close approximation of the true response surface."

"O.K., but how is this all put together to make a statistical statement?"

"That's voodoo –F statistics!" I answered cryptically.

Capt. Kirk looked at the data and, while he liked the contour plots, he suggested another display which is more appealing to his tastes.

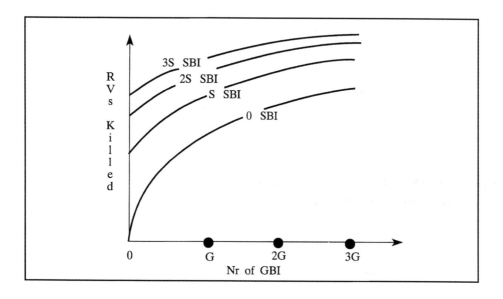

"These curves are essentially generated by cutting the response surface with four planes parallel to the GBI axis, where these planes cut through fixed levels of SBIs," I noted.

"Enough on display of data; what are your scenarios?" asked Kirk.

I suggested the following five cases of SBI type and ASAT threat be considered:

	SBI TYPE	
ASAT MIX	I	II
NO ASATS	CASE A	
FORCE 1	CASE B	
FORCE 2	CASE C	CASE D
FORCE 3		CASE E

The audience was in wide agreement. I then asked, "Which of these cases would represent a baseline, that we should examine while changing other factors?" For the first time a consensus choice was easy –Case D was chosen as a baseline for all other sensitivity studies. Then using Case D, I suggested that we run complete tradeoff studies while changing settings in:

Discrimination Factor	2 settings
Strategic Threat	2 settings
GBI Design	one change

Thus, with the five cases for ASAT-SBI factors and the five excursions noted above, I got agreement on the study design. Each scenario required 18 simulation runs for a total of 180 to provide tradeoff analyses for all scenarios. We then could conduct a handful of simulations to provide a feel for the effects of Pk, Trelease, and Tdiscrim.

Epilog

This study took six months, primarily due to a massive software development effort to simulate the ASAT threat and the SEOs. Results were widely appreciated for two reasons: people liked our display of results (the tradeoff contours and the tradeoff curves) and we answered a multitude of questions without confounding the results. Specifically, we:

- Characterized the effects and interactions when the number of interceptors was varied (tradeoff analysis)

- Provided a statement on the effects of:
    ASAT Attack
    Interceptor Design
    Strategic Threat
    Discrimination Factor

- Provided a feel for the effects of variations in:
    Pk
    Trelease
    Tdiscrim

What was not appreciated by the customer, but was fundamental to our success, was the importance of a study design to limit the number of simulations. Given time and resource constraints, we had to limit the number of simulation runs early and design our study around that limit. When the limit was discussed with the customer, he had a much greater appreciation of our need to use an efficient design as well as our need to limit the scope of the study. Future studies will now focus on some of the more salient results of our initial study.

Identifying a Plasma Etch Process Window

Submitted by:

Dr. Jack E. Reece, Sr. Principal Process Engineer
David Daniel, Process Engineering Supervisor
Honeywell Solid State Electronics Division

Ms. Robin Bloom, Sr. Process Engineer
United Technologies Microelectronics Center
Colorado Springs, CO

The Problem

Removing a top layer in the presence of an underlayer is a common requirement in the manufacture of integrated circuits. The process must etch the layer uniformly across the wafer and must maintain considerable selectivity to avoid damaging any underlayer. Furthermore, to minimize the risk of damaging other parts of the circuit, operators must have a convenient means for detecting the end point when the process has removed the target layer.

For approximately one year preceding the start of this study, four similar processes run in this equipment had given satisfactory etch rate uniformity and selectivity. In the process a relatively low-powered pulsed plasma activated an etchant gas at low pressure to remove the target layer. Unexpectedly, operators began having difficulty determining end points for the etch process, presumably due to the breakdown of some shielding or circuit that had protected the detector from the pulsing RF in the reactor. In addition, while the uniformity of the etch rate across a wafer had been marginally acceptable, design demands for new circuits required better control.

The machine vendor, working cooperatively with factory maintenance personnel, failed to fine an assignable mechanical or electrical cause for the malfunctions. After three weeks with the $200,000 machine offline, the best he could suggest was that the engineers devise new processes using higher-intensity pulsed cathode power to enhance etch rate uniformity. All parties had somewhat agreed that an unstable plasma due to the pulsing at low power contributed to the uniformity problems and that pulsing the plasma in general caused the end point detection interference. But the vendor could not show that he had eliminated the electrical interference with the photocell end point detector nor that the problem would not return.

When the engineers on site suggested using a steady-state, rather than pulsed cathode power, the vendor was adamant that the machine absolutely would not etch the layers uniformly under such conditions − it was designed to use the pulsed power. In fact, preliminary best guess probing experiments suggested that the vendor was probably correct.

Given that they were faced with devising a new process and that they had no real confidence that end point detection problems would cease to plague the machine's operation, the engineers elected to invest the time necessary to characterize its performance using fixed power first.

Experimental Approach
Factor Settings and Design Matrix

When restricted to constant power, the machine offered only three process control variables: power, pressure, and flow of etchant gas. Experience gained from the probing experiments and previous use of the equipment identified the logical limits for factor settings, so the engineers elected to use a Box-Wilson central composite design requiring 20 trials. Table 1 shows the factors and their limits for this experiment.

Factor, Units	Low	High
Pressure, mtorr	200	300
Power, watts	75	150
Etchant Flow, sccm	20	40

Table 1   Factor Settings

Simple 2-level factorial designs support linear or planar models; to characterize any curvature present in a response as a function of the inputs requires at least 3 levels for each factor. The Box-Wilson Central Composite Design [1, 4] augments the usual 2-level factorial array with *star points* and *center points* such that one explores 5 levels of each factor in relatively few trials.

Among alternatives to this design are a 3-level full factorial which requires 27 trials and the 3-level Box-Behnken, which requires 15 trials (including replication). We choose the central composite design as it provides 5 points for estimating curvilinearity and requires only 20 trials (including replication).

Modern experimental design software simplifies the tasks of creating and interpreting designs such as these. The limits for low and high indicated in Table 1 were truly limits, but the software used here allowed us to assign these points to the "star" positions while it constructed the 2-level matrix within them, saving the tedium of the arithmetic calculations for factor settings. Table 2 summarizes the design matrix generated.

Responses

Responses included the etch rates of layers 1 and 2, the uniformity of each etch, and the selectivity between layers 1 and 2. Measurements at several points on the wafer before and after each trial provided the etch rates. The ratio of etch rates during each trial provided the selectivity.

Etch rate measurements at several points on each wafer gave ranges of values for the rates suggesting that one or more of the factors was contributing a dispersion effect in the process. In addition the range of etch rate values observed was proportional to the mean etch rate. To describe the dispersion of etch rates we choose a relative of the *coefficient of variation*, in Equation (1), because it was familiar to those involved in the investigation.

$$\text{Percent Uniformity} = [(\text{Max} - \text{Min})/2 \times \text{Mean}] \times 100 \qquad (1)$$

# Reps	PRESSURE (mtorr)	POWER (watts)	GASFLOW (sccm)
1	220	90	24
1	280	90	24
1	220	135	24
1	280	135	24
1	220	90	36
1	280	90	36
1	220	135	36
1	280	135	36
1	200	113	30
1	300	113	30
1	250	75	30
1	250	150	30
1	250	113	20
1	250	113	40
6	250	113	30

Table 2 Design Matrix of Factor Settings

Calculations are simple and the result provides a concept of a $+/-$ variation about a central value. However, this measurement has a natural limit of 0, and regression models readily predict meaningless **negative** values. Also, range is sensitive to outliers, so a more robust metric that provides approximately the same answers substitutes 2σ for the range.

Taguchi [2] has published a variety of complex equations to model variability and calls them **signal-to-noise** ratios. His SN_T for "Target is Best" is also a variant of the coefficient of variation and works well when significant correlation exists between the standard deviation of a series of observations and the mean of those observations. Box, [3], advocates a simpler approach by determining the logarithm of the standard deviations of the observations and basing the analysis on that metric.

The software used in this study supports Box-Cox transformations of responses and performs the arithmetic totally transparent to the user. If the user takes full advantage of the scaling options available, she/he sees only normal response values and appropriate regression coefficients, regardless of the transformation used. Supported transformations include y^2, y, $\sqrt{y}$, $\log_e y$, $1/\sqrt{y}$), and $1/y$.

Randomization of experimental trials distributes environmental or equipment bias uniformly among all factor levels; and while the software supports orthogonal blocking, we elected not to use it in this study. The first 3 columns of Table 3 illustrate the random run order for this investigation, which was software-generated. The last 5 columns contain the observations.

Run	Prs	Pwr	Flow	Etch Rate 1	Layer 1 Unif	Etch Rate 2	Layer 2 Unif	Select
1	280	90	24	827	13.213	193	23.714	4.3
2	250	113	30	1417	8.395	446	9.099	3.2
3	250	113	40	1351	6.295	375	12.912	3.6
4	250	75	30	719	14.289	200	20.701	3.6
5	250	113	30	1365	10.304	372	7.907	3.7
6	220	90	36	978	9.099	279	11.508	3.5
7	280	135	36	1597	9.795	397	6.998	4.0
8	250	113	30	1475	6.607	405	10.399	3.6
9	250	113	20	1376	8.395	316	7.295	4.4
10	250	113	30	1370	8.091	348	9.795	3.9
11	220	90	24	905	12.106	256	11.092	3.5
12	220	135	24	2100	2.000	734	5.794	2.9
13	250	113	30	1400	6.295	367	21.106	3.8
14	220	135	36	1961	2.698	660	5.105	3.0
15	280	90	36	841	11.508	195	21.979	4.3
16	200	113	30	1824	0.800	551	7.096	3.3
17	250	150	30	2065	4.198	615	10.789	3.4
18	300	113	30	1085	11.298	245	25.704	4.4
19	280	135	24	1647	9.099	398	12.388	4.1
20	250	113	30	1449	5.297	462	10.990	3.1

Table 3 Randomized Worksheet and Results Etch Process Optimization

Results and Analysis

Each etch rate is the average of 5 measurements on each wafer, and the authors used the range of those in Equation (1) to model uniformity after transforming the values to the natural logarithm using software utilities. Selectivity is the ratio of the etch rates.

Backwards stepwise regression provided the Taylor expansion series coefficients displayed in Table 4. The authors prefer this technique because it efficiently provides the simplest (most parsimonious) model which adequately describes the response as a function of the factors. In this case we set software parameters such that the α risk associated with retaining a coefficient was 0.10. Analysis of Variance (ANOVA) and residual diagnostic routines within the software help the user decide on the adequacy of a model.

The software coded the predictor variables according to the relationship

$$(X_i - ((MAX\ X_i + MIN\ X_i)/2)\ /\ ((MAX\ X_i - MINX_i)/2)$$

which results in values of -1 and $+1$ for the low and high predictor values, respectively. Therefore, we may judge the relative importance of each factor on a response by the absolute value of its coefficient. For example, the models for LAYER1 ETCH and LAYER1 UNIF indicate that Pressure and Power are highly **coupled** or **interactive**. That is, the effect on the response of changing Pressure from its low to its high setting depends on the setting of Power.

Note that selectivity is a quadratic function of etchant gas flow while the other responses are independent of this factor. In Table 4, the coefficients for the uniformity models reflect our choice of a scaling option that displays approximate values for the raw metric regardless of the transformation. This makes interpretation of the model somewhat simpler than using logarithms. In UNIF2, the α risk associated with the PWR^2 coefficient was 0.11.

Model Term	Etch Rate 1	Log Unif 1	Log Unif 2	Selectivity Layer1/Layer2
Constant	1382.68	8.095	10.251	3.590
PRS	−278.96	7.795	6.232	0.696
PWR	737.89	−5.800	−5.804	−0.237
FLW	-----	------	------	−0.164
PRS^2	-----	−6.320	------	------
$PRS \times PWR$	−213.22	7.385	------	------
PWR^2	-----	------	3.106	------
FLW^2	-----	------	------	0.377
Adj-R^2	0.97	0.826	0.748	0.664
Rel. Press	0.95	0.605	0.581	0.491
RMS Error	71.16	2.503	2.496	0.269

Table 4   Regression Coefficients

Interpretation and Conclusions

An extremely convenient method for interpreting the Taylor series regression models is contour plotting. Unlike a conventional graph, this technique assigns a factor to each axis while holding any others at fixed values. The contour lines are lines of equal response, quite analogous to the contour intervals on a map.

Figures 1 to 6 are computer-generated plots based on the Taylor series models. The authors adjusted the maximum value of the Pressure axis to expand the scale of the figures and focus attention on the area of interest. The '★' on each figure represent actual data points. The first 4 figures add contours at etchant flow = 30 sccm – the center of the investigation. The box in Figures 1 to 4 indicates the best conditions found during probing experiments which seemed to support the vendor's contention that this machine would not operate properly using fixed cathode power. Figures 1 to 4 show that reducing the pressure and increasing the power improve uniformity and increase potential throughput at the cost of selectivity. Figure 5 shows that increasing the etchant flow to the limit studied improves selectivity, but not enough; Figure 6 shows that decreasing the etchant flow to the minimum investigated improves it further and provides an acceptable result. The box in Figure 6 represents a proposed process window.

An investigator must not forget that the models derived contain experimental error. The values indicated on a contour plot are not absolute, but have confidence intervals. An additional software utility permits estimation of these intervals. The final stage in any investigation requires predicting results and testing them in the process. Table 5 summarizes the predicted values and observed values for each response when the authors tested the modified process.

Figure 1 ETCH1
FLOW = 30

Figure 2 ETCH1, UNIF1
FLOW = 30

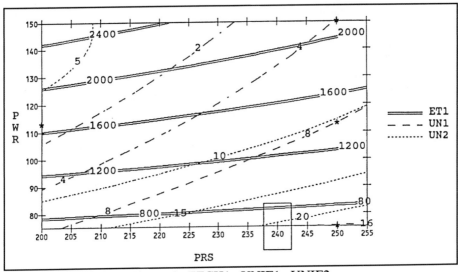

Figure 3 ETCH1, UNIF1, UNIF2
FLOW = 30

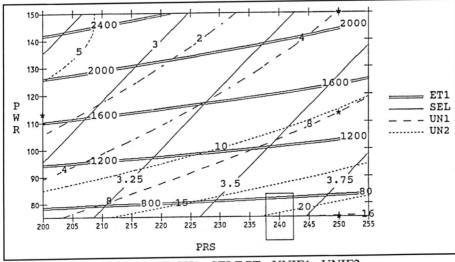

Figure 4 ETCH1, SELECT, UNIF1, UNIF2
FLOW = 30

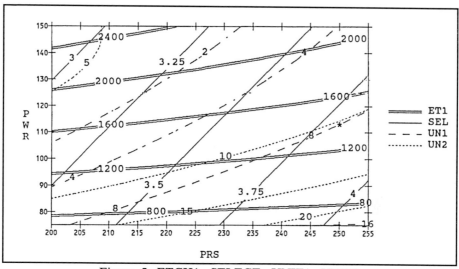

Figure 5 ETCH1, SELECT, UNIF1, UNIF2
FLOW = 40

Figure 6 ETCH1, SELECT, UNIF1, UNIF2
FLOW = 20

Factors	Etch Rate 1	Unif Layer1	Unif Layer2	Selectivity Layer1/Layer2
PRESSURE	230	230.00	230.00	230.0
POWER	115	115.00	115.00	115.0
GAS FLOW	20	20.00	20.00	20.0
Lower Bound	1477	3.30	6.00	3.0
RESPONSE →	1549	4.60	7.90	3.8
Upper Bound	1620	6.50	10.30	4.6
Observed	1517	4.60	7.70	3.5

Table 5 Predictions and Observed Values for Responses
in Etch Process (95% Confidence Intervals)

This investigation required four days and included 20 experiments, their analysis and interpretation, and the testing of predictions. It identified a new process window which satisfied all process requirements which the vendor swore could not exist. The new process has performed as expected for 30 months, requiring nothing more than regular preventative maintenance and reducing scrap and rework while increasing the throughput at this step. The new process unexpectedly replaced the four previous ones, giving acceptable results regardless of device type or layer.

The application of experimental design and the contour plots derived from the models generated provide a far superior understanding of process windows available to the process engineer than any other experimental method.

Attempting to build these models from other than the orthogonal arrays provided by statistical experimental design usually leads to mathematical correlations among supposedly independent factors and potentially misleading models.

Acknowledgements

The authors gratefully acknowledge the contributions of Mr. Tom Tjaden, Process Engineer, Honeywell Solid State Electronics Division, in helping to define the study and in interpreting its results and implications.

References for the Plasma Etch Case Study

1. G.E.P. Box, W.G. Hunter, and J.S. Hunter, *Statistics for Experimenters*, John Wiley & Sons, New York, 1978.

2. See for example: G. Taguchi, and M.S. Phadke, "Quality Engineering Through Design Optimization," Conference Record Vol. 3, IEEE Globecom 1984 Conference, Atlanta, GA, pp. 1106-1113. G. Taguchi, and Y. Wu, "Introduction to Off-Line Quality Control," Central Japan Quality Control Association, 1986. G. Taguchi, "Introduction to Quality Engineering: Designing Quality into Products and Processes," Asian Productivity Organization, UNIPUB, Kraus International Publications, White Plains, NY. See also M.S. Phadke, "Quality Engineering Using Design of Experiments," Proceedings of the American Statistical Association, section on Statistical Education, 1982, pp. 11- 20, .

3. G.E.P. Box, "Studies in Quality Improvement: Signal-to-Noise Ratios, Performance Criteria, and Transformation," Report No. 26, University of Wisconsin Center for Quality and Productivity Improvement, Madison, WI., 1987.

4. G.E.P. Box, and N.R. Draper, *Empirical Model-Building and Response Surfaces*, John Wiley & Sons, New York, 1987.

Applying Experimental Design Techniques
to Operating Systems and Software

Even though there are large numbers of documented designed experiments in the chemical and hardware segments of industry, few documented examples exist in the operating systems and software world. In view of the above, it is now fitting that we share the following case study with you.

A software engineer concerned with overall system performance when running a particularly large computer program decided to run a designed experiment. The experimental objective was to determine the best combination of operating system factors in order to minimize CPU time (regardless of the disk file scheme). Factors of interest to the experimenter were:

FACTOR	LOW	MIDDLE	HIGH
A. Buffered I/O Limit	−1	0	+1
B. "Q" Limit	−1	0	+1
C. Block Size	−1	0	+1
D. File Limit	−1	0	+1
E. Direct I/O Limit	−1	0	+1
F. Working Set Quota	−1	0	+1
G. Byte Limit	−1	0	+1
H. "Bf"	A		B

In addition, three different disk file schemes were appropriate to evaluate as a "noise factor" (X_1, X_2, X_3).

For the actual experiment, six different responses were considered worthy of consideration. In our example, however, we will only address one response (CPU time). Utilizing an 18 run computer-generated D-optimal design, coupled with a "Taguchi style" outer array, the design matrix with applicable data is shown in Table 1.

Exp #	A	B	C	D	E	F	G	H	X_1	X_2	X_3
1	-1	-1	-1	-1	-1	-1	-1	A	127	155	269
2	-1	0	0	0	0	0	0	A	118	150	285
3	-1	1	1	1	1	1	1	A	115	169	2321
4	0	-1	-1	0	0	1	1	A	41	163	2237
5	0	0	0	1	1	-1	-1	A	110	134	207
6	0	1	1	-1	-1	0	0	A	131	160	791
7	1	-1	0	-1	1	0	1	A	139	271	1358
8	1	0	1	0	-1	1	-1	A	127	145	216
9	1	1	-1	1	0	-1	0	A	112	139	579
10	-1	-1	1	-1	0	0	-1	B	40	69	144
11	-1	0	-1	0	1	1	0	B	56	125	715
12	-1	1	0	1	-1	-1	1	B	63	92	652
13	0	-1	0	1	-1	1	0	B	41	73	574
14	0	0	1	-1	0	-1	1	B	68	125	889
15	0	1	-1	0	1	0	-1	B	49	82	129
16	1	-1	1	0	1	-1	0	B	51	100	580
17	1	0	-1	1	-1	0	1	B	50	80	786
18	1	1	0	-1	0	1	-1	B	57	87	153

Table 1 Design Matrix

Analysis of Table 2 and Figure 1 indicate that G, H, and possibly F have an important impact on the mean response. An analysis of the data in Table 3 indicates that factor G may be a variance reduction factor with the low level being the best setting. Figure 2 provides us with a graphical representation of the relationship between disk file scheme and factor G. Of particular interest is the relatively constant low CPU time when G is at the low level, regardless of the disk file scheme being used.

	Low	Mid	High
A	337	339	279
B	363	265	327
C	333	276	347
D	315	321	319
E	252	331	373
F	247	290	418
G	128	288	540
H	417	---	220

Table 2 Average Marginals for
the Statistic "$\bar{x}$"

	Low	Mid	High
A	5.48	5.41	5.26
B	5.55	5.26	5.34
C	5.43	5.27	5.45
D	5.44	5.31	5.39
E	5.27	5.41	5.47
F	5.24	5.31	5.59
G	3.94	5.75	6.45
H	5.56	---	5.20

Table 3 Average Marginals for
the Statistic LnS*

* natural log of the sample std dev

Figure 1

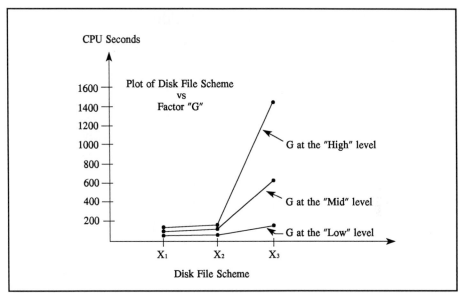

Figure 2

In summary, setting factor G at the low level will allow us to minimize CPU time.

Reducing Variability in a Machining Process:
An Industrial Application of Experimental Design

Submitted by: Maxcor   U.S. AIR FORCE ACADEMY
                    Gary Fenton   C1C James B. Hecker
                    John Hopkinson   C1C Phillip A. Herre
                    William Woodard   C1C Ray L. Plumley
                                           C2C Raymond X. Sagui

Background

An industrial bushing has a close tolerance requirement on the outer diameter of 14 mm $\pm$ .01 mm (.5512" $\pm$ .00039"). The machining process was producing approximately .6% of the product out of the tolerance limits. Some of the variability of the outer diameter was due to an "out-of-round" condition. A joint effort between Maxcor and the Air Force Academy was initiated to reduce the amount of runout (as measured by Total Indicator Reading - TIR) by utilizing a designed experiment approach to optimizing the process settings.

Experiment Objective

The objective of the experiment was to see an improvement in the roundness of the bushings. Roundness is defined as the condition on a surface of revolution, such as a cylinder, where all points on the surface intersected by any plane perpendicular to a common axis are equidistant from the center.

Quality Characteristics

The two ends of the bushing may be distinguished by the presence of a dimple near one end (see Figure 1). The manufacturing specifications required that the entire surface (composed of an infinite number of diameters) meet the 14 $\pm$ .01 mm condition. To meet this condition, both the taper and the out of roundness must be minimized. Thus, both of these were used as criteria measures (response variables).

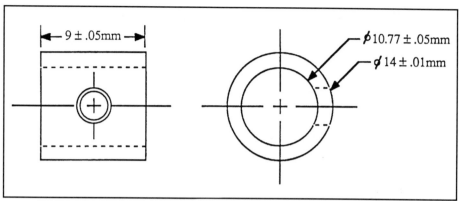

Figure 1   Outer Bushing Top View and Cross Section

The inspection equipment used a two point contact so that as the part was rotated, both maximum and minimum diameters could be recorded. Both ends of the bushing were inspected, generating four pieces of data for each part. These four data pieces were then used to mathematically calculate the roundness (maximum minus minimum for each end) and the taper (maximum of one end minus maximum of the other end).

Quality Characteristics

Response	Type	Anticipated Range
TIR, dimple	quantitative	0 - 0.0110 mm
TIR, non-dimple	quantitative	0 - 0.0110 mm
Taper	quantitative	−0.0075 - 0 mm

Factors

Perhaps the most difficult step of experimental design is deciding which factors should be included in the design. Many times there are so many potential factors that affect the results that the design becomes infeasible because of the size, or the time to run the entire design, or most common, the cost per experiment.

In this case study, a brainstorming session with the engineers brought out the possible factors that go into the manufacturing of a single bushing. This initial list consisted of almost twenty different factors shown in Figure 2.

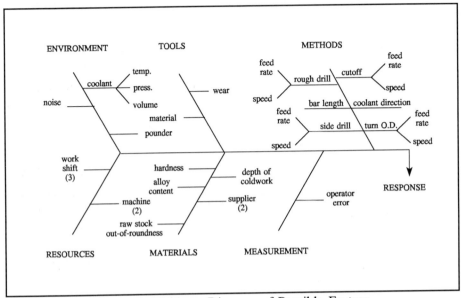

Figure 2 Fishbone Diagram of Possible Factors

Some factors, such as alloy content, stock roundness, and bar feeding length, were beyond control so they were not included in the experiment. Others, like machine, worker shift, and environment were controlled by randomization or holding them constant. After two weeks of weighing the priority and appropriateness of each factor, ten 2-level factors were chosen for the design, see Table 1.

Factor	Levels	Type
Hardness (HD)	Lo - Hi	Qualitative
Rgh Drill Feed Rate (RDFR)	.0015 - .0025	Quantitative
Rgh Drill Speed (RDS)	2000 - 3000 rpm	Quantitative
Sd Hole Drill Feed Rate (SHFR)	.07 - .17 ipm	Quantitative
Sd Hole Drill Speed (SHDS)	1800 - 2800 rpm	Quantitative
Turn O.D. Feed Rate (TFR)	.002 - .003	Quantitative
Turn O.D. Speed (TODS)	3000 - 5000 rpm	Quantitative
Cutoff Feed Rate (CFR)	.0002 - .0005	Quantitative
Cutoff Speed (CS)	2500 - 4000 rpm	Quantitative
Coolant Flow (CF)	Top - Bottom	Qualitative

Table 1   Experimental Factors

Choosing the Factors for the Experiment

The hardness of the raw stock was considered the most important factor to the Maxcor engineers. Because of all the drilling and cutting required to make a bushing, the drill bit can severely deform the metal if the metal hardness is low. The feed rate and speed factors were considered important because they cause movement of the raw stock during drilling and cutting. The place of coolant flow was also considered important because of affect on lubrication and cooling.

Run Limitations

The parts manufactured during the experiment were eventually to be sent to a customer, so it was important to select factor level settings that would not intentionally produce non-conforming products. The appropriate sample size for each run was selected by calculations based upon the historical standard deviation and the amount of shift that was to be detected, along with the alpha and beta risks that were reasonable (see Chapter 4).

Appropriate Experimental Designs

Due to Maxcor's familiarity with the Taguchi design, an L_{16} Taguchi sequence was used. Except for the factor of material hardness, which could not be easily changed due to equipment set-ups, the runs were randomized by making use of a random number table.

Experimental Implementation and Data Collection

Maxcor's technicians were instructed to set up the machinery in accordance with the randomized Taguchi L_{16} design. After each run, the quality characteristics were measured and recorded for each bushing. Throughout this entire process, management was present to make certain there was no deviation from the experimental design.

Analysis and Results

The analytical method used was a regression analysis on the three-response mean for each replicated run to determine main effects. For each of the three regressions, there was no significance at the 95% confidence level. Due to the small differences in factor settings, the resulting regression weights were not statistically significant. A set of "vital few" factors were suspected to exist despite the regression results. Therefore, it was necessary to rank order the regression weights to determine which factors appeared to be more significant than others. The coefficients of the orthogonally coded factors were used to set up the prediction equations; these coefficients are listed in Table 2.

A regression on the variance of the three responses for each replicated run was used to determine dispersion effects. Again, the regressions were not statistically significant. Likewise, these regression weights were used to determine the most likely significant factors. The coefficients of the orthogonally coded factors were used to set up the prediction equation (see Table 2).

Using the coefficients listed in Table 2, each factor was set to minimize the mean and variance of the response variables. This involved setting factors with a positive coefficient at the low level, and factors with a negative coefficient at the high level. This procedure was accomplished for each of the response variables. In some instances, the factor coefficients for the mean and variance responses were opposite in sign. Since Maxcor desired to emphasize decreasing the variance in the responses, it was decided to give precedence to the coefficients of the variance response. The best factor setting for optimizing the mean and variance can be seen in Table 3.

Factor	Y1	Y2	Y3	$S_1 2$	$S_2 2$	$S_3 2$
HD					0.00505	
RDFR			0.024		-0.00462	
RDS						
SHFR			0.030	0.00641		
SHDS	0.042	0.046	0.037		-0.00385	
TFR		0.031		0.00459	0.00466	
TODS			-0.036		0.00444	
CFR	-0.038			-0.00453	-0.00571	
CS				0.00498		
CF				-0.00526	0.00529	-0.00478

Table 2 Coefficients

Factor	Setting
HD	Low
RDFR	High
RDS	High or Low
SHFR	Low
SHDS	High
TFR	Low
TODS	Low
CFR	High
CS	Low
CF	High

Table 3 Optimal Settings

Confirmation Runs

Maxcor ran 200 confirmation runs at the recommended levels. To obtain an overall picture of the bushing, the TIR responses for both the dimpled and non-dimpled ends were combined. Although there was no significant change in the mean TIR response, the variation was decreased by 50% (see Figures 3 and 4). Although the confirmation runs produced parts with reasonable taper, there is no way to discern any improvement in this response due to the lack of historical data.

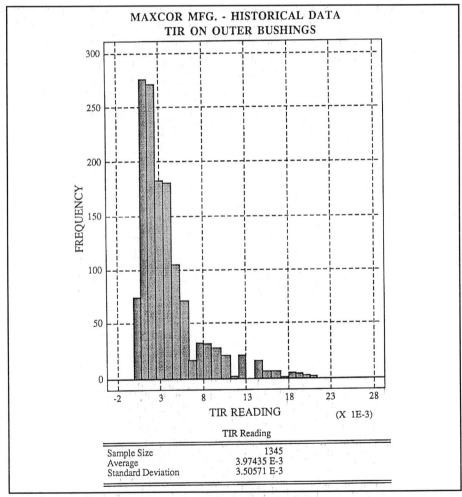

Figure 3 TIR on Outer Bushings

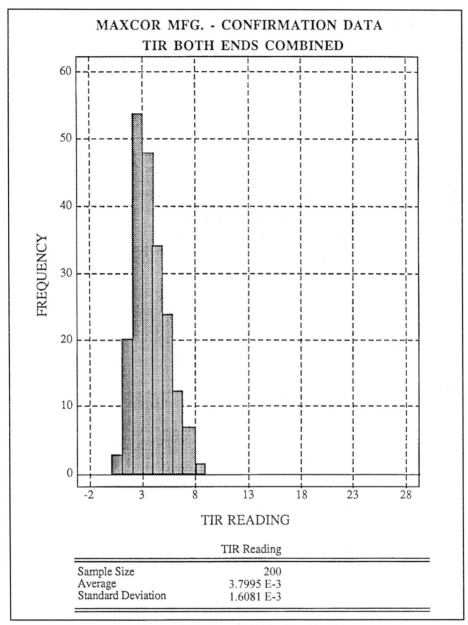

Figure 4 TIR Both Ends Combined

The Effect of the CE Rivet H Parameter on Head Protrusion

Submitted by: Tom Gardner
                    James D. Riggs

Problem Statement

Determine the effect of the CE rivet h parameter (see Figure 1) on head protrusion in skin panel rivet installations.

Objective

Use a designed experiment and knowledge gained from previous experimentation to quantify the effect of the h parameter on rivet protrusion. This information will be applied to determine the required countersink dimension for each value of h.

Background

The variation in head diameter of CE rivets has been found to be a major contributor to the variation in head protrusion in CE rivet skin installations. A request to rivet vendors to better control head diameter was rejected in favor of controlling the h parameter. Apparently, the head diameter is controlled by the h parameter. Examination of Figure 1 shows the geometric reasoning for this statement. A correlation analysis between h and head diameter showed marginal correlation, thus challenging the geometric foundation. This experiment is designed to establish the role of h on head protrusion.

Results

The h parameter does effect rivet protrusion, but the magnitude of the effect (0.0011") is much smaller than that of the combination of countersink and head diameter (0.0093"). See Figures 2 and 3. Figure 2 shows the expected increase in protrusion with an increase in h. Minimal correlation exists between h and the head diameter. The countersink required to control protrusion to 0.0020" can be determined from h only marginally. Head diameter is the factor

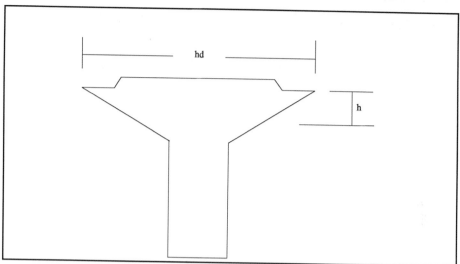

Figure 1 The CE Rivet H Parameter

from which the countersink should be determined.

This experiment shows that the countersink ranges over which the protrusion can be controlled to 0.0020" are different from the existing ranges shown below. This may be due to a change in rivet vendor — now Precision Form and formerly Allfast.

	CE 5	CE 6
Existing range	0.240" − 0.245"	0.290" − 0.295"
Experiment range	0.245" − 0.252"	0.292" − 0.294"

Figure 3 shows two effects. First, as the countersink increases, protrusion decreases as expected. Secondly, as the head diameter increases, the protrusion decreases when the countersink is small. When the countersink is larger, the protrusion increases as the head diameter increases. This effect is due to the mechanics compensating for small countersink diameter with increased pressure on the rivet gun.

Figure 2 Effect of H on Protusion

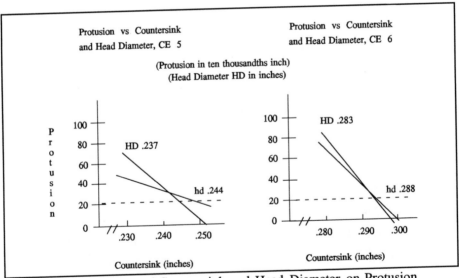

Figure 3 Effect of Countersink and Head Diameter on Protusion

Recommendations

This experiment used rivets from a variety of batches which gives no clue as to the within batch variation. Both head diameter and h need to be measured from within several batches. The within variation can be determined, as well as the correlation between head diameter and h. Also, the number of rivets that must be measured to find accurate averages for these factors can be found. If the correlation between head diameter an h is good enough, h can be used to characterize the batch and to determine the target countersink diameter. Otherwise, the head diameter must be used. In addition, if h can be correlated, recommendations to vendors for target h values and variances can be made.

The graphs in Figure 3 give countersink ranges which will produce protrusions for the different head diameters for both the CE 5 and CE 6 rivets. Batches (boxes) on the floor need to have average head diameters and variation measured to allow the graph to give the target countersink diameter for each box. The goal is to find two or three countersink targets for each rivet type (CE 5 or 6). This will permit utilization of two or three countersink cutters per rivet type.

Experiment Design

Previous experimentation in similar skin riveting installations shows the following factors, with levels, to be important.

Factor	Levels	Symbol
panel	0	
complexity	0	
shimming	0	
rivet head diameter	2	hd
h	3	csk
countersink diameter	3	h
rivet head protrusion	random	PROT

All samples are taken from a single panel in Acc 315 thus eliminating the panel-to-panel effect.

The complexity factor was eliminated as all the combinations of h and csk were available in the open (easy) areas of installation on the selected panel.

The shimming effect was eliminated by working only in areas where no shims are present.

The rivet head diameter, i.e., CE 5 and CE 6 is a fixed factor over which the experiment is blocked. The actual measurements are expected to be 0.228" to 0.248" and 0.279" to 0.299", respectively. The deviation from nominal for both the CE 5 (0.238") and CE 6 (0.288") rivets is used in the analysis.

The h parameter has the following levels within hd. The deviation from nominal (level 2 for each type) is used.

	Level	Actual
CE 5	1	0.0140"
	2	0.0148"
	3	0.0162"
CE 6	1	0.0209"
	2	0.0216"
	3	0.0230"

The countersink diameter is considered fixed within hd as follows. The deviation from nominal is used in the analysis and the nominal are the same as those of the head diameter.

	Level	Actual
CE 5	1	0.228"
	2	0.238"
	3	0.248"
CE 6	1	0.279"
	2	0.288"
	3	0.298"

The rivet head protrusion is the response variable. It has been shown that controlling this factor clearly reduces defects associated with skin panel installation. These include both rivet installation defects (such as exposed countersinks, gaps, shanked rivets, cracked butts) and skin defects (including scratches, tool marks, pillowing, canning).

Model

The informal model to be tested is:

$$
\begin{aligned}
\text{PROTRUSION} = \text{mean} &+ hd + h + csk + hd \times h + hd \times csk + h \times csk + h \times h \\
&+ csk \times csk + hd \times h \times csk + hd \times h \times h + hd \times csk \times csk \\
&+ \text{RESIDUALS}
\end{aligned}
$$

The anticipated ANOVA table is:

Source	df	FC	RC
hd	1	$18n\emptyset(hd)$	σ^2_R
h	2	$18n\emptyset(h)$	σ^2_R
csk	2	$18n\emptyset(hd)$	σ^2_R
hd $\times$ h	2	$9n\emptyset(hd \times h)$	σ^2_R
hd $\times$ csk	2	$9n\emptyset(hd \times csk)$	σ^2_R
h $\times$ csk	4	$3n\emptyset(h \times csk)$	σ^2_R
hd $\times$ h $\times$ csk	4	$n\emptyset(hd \times h \times csk)$	σ^2_R
REP	$18(n - 1)$	0	σ^2_R
Total	$18n - 1$		

ANOVA Table

Sample Size

To compute the number of samples needed per cell (h and csk in full factorial combinations) we first assume the df to be infinity, then recompute the sample size based on the previous estimate. The difference (in ten thousandths) we need to detect is 5, and 0_{-E} is equal to 3 (determined from past experience). This gives us:

$$t_{0.025}(\infty) \times 0_{-E} (2/n)^{1/2} = 5$$

$$=> \quad 1.96 \times 3 \, (2/n)^{1/2} = 5$$

$$=> \quad n = 2 \times ((1.96 \times 3)/5)^2$$

$$n = 2.77 \quad => \quad 3$$

If $n = 3$, we will take a total of 42 samples. Let df for $t = 41$.

$$=> \quad n = 2 \times ((2.02 \times 3)/5)^2$$

$$n = 2.94 \quad => \quad 3$$

The sample size needed is 3 per cell, yielding a total of 42 samples.

Randomization

Randomization is by runs within a block. Random numbers are generated and assigned to the factorial runs within a block. See Figure 4 for a diagram of the normal order of installation-first CE 6's, CE 5's, CE 6's, and then CE 5's.

CE 6	CE 5	CE 6	CE 5	

Block   1   2   3   4

Four blocks containing either CE 5 or CE 6 rivets are laid out on the 65-45800-139 skin panel in ACC 315. The order of installation is from left to right in the diagram.

Figure 4 Test Panel Layout

Within each block, the randomized run combinations are as follows:

Block 1

Run	h	c
1	1	2
2	2	1
3	2	3
4	3	3
5	2	2
6	3	2
7	3	1
8	1	3
9	1	1

Block 2

Run	h	c
1	2	1
2	3	2
3	3	1
4	2	3
5	1	2
6	2	2
7	3	3
8	1	1
9	1	3

Block 3				Block 4		
Run	h	c		Run	h	c
1	2	2		1	2	3
2	1	1		2	1	2
3	2	3		3	2	1
4	1	2		4	1	1
5	3	1		5	3	3
6	2	1		6	2	2
7	1	3		7	3	1
8	3	2		8	1	3
9	3	3		9	2	1

Data Analysis

The experimental data collected, listed in Table 1, are severely reduced from the plan above. None of the blocks are balanced. Block 1 is the only block with no duplicate runs.

The data for csk, h, and hd are the differences of the specification nominal from the actual readings. The length (len) of the rivet is taken from the part number rather than from actual measurements as previous experimentation shows high correlation between the measured length and the length obtained from the part number. The mechanics suggested two factors that could (but did not) show a sequence effect. Tape is applied to the surface of the rivet gun which interfaces with the rivet head. As rivets are installed, this tape wears thus diminishing a cushioning effect. The other factor is the firmness of the structure into which the rivets are installed. As the installation proceeds from left to right (see Figure 4), the structure becomes progressively less firm.

Multiple regression was used to analyze the data. The factors which contribute to the variability of the rivet head protrusion are, countersink, head diameter, and the h parameter, as shown in Figures 2 and 3 on Page 8-170.

	Prot	csk	h	Len	hd	ce	Run
	35.0	-8.0	0.0	5.0	-1.5	6.0	2
	2.0	11.0	0.0	5.0	-2.5	6.0	3
Factor 1	18.0	10.0	1.5	5.0	0.0	6.0	4
	14.0	2.0	0.2	5.0	-4.5	6.0	5
	44.0	0.0	1.4	6.0	-1.0	6.0	6
	47.0	0.0	-0.7	5.0	1.5	5.0	2
Factor 2	23.0	10.0	-0.4	5.0	1.5	5.0	7
	6.0	10.0	-0.6	5.0	0.5	5.0	7
	39.0	0.0	-0.5	5.0	-0.5	6.0	1
	82.0	-10.0	-0.7	5.0	-3.0	6.0	2
	1.0	9.0	-0.5	5.0	-1.0	6.0	3
	0.0	10.0	0.1	5.0	-3.0	6.0	3
Factor 3	90.0	-10.0	1.0	5.0	-1.5	6.0	5
	63.0	-7.0	0.1	5.0	-2.0	6.0	6
	83.0	-8.0	0.8	7.0	-4.0	6.0	6
	41.0	0.0	0.8	7.0	-4.5	6.0	8
	2.0	11.0	1.0	7.0	-4.5	6.0	9
	25.0	0.0	-0.2	5.0	2.0	5.0	2
	40.0	0.0	-0.7	5.0	1.0	5.0	2
	66.0	-8.0	0.4	5.0	2.5	5.0	3
	40.0	-7.0	-0.8	5.0	-0.5	5.0	4
Factor 4	28.0	10.0	1.4	6.0	5.5	5.0	5
	11.0	12.0	1.0	6.0	4.0	5.0	5
	54.0	0.0	1.4	6.0	5.5	5.0	7
	49.0	-11.0	1.3	6.0	4.0	5.0	7
	4.0	11.0	-0.7	5.0	1.5	5.0	8
	58.0	-10.0	-0.1	5.0	2.0	5.0	9

Table 1 Experiment Data

Figure 5 shows a plot of the residuals (from Table 2) versus the model output. No clear patterns are revealed. The first row of the data was removed from the analysis because standardized residuals showed this row to be an outlier.

Row	Observed	Predicted	Residual
2	2.00000	-4.70511	6.70511
3	18.00000	14.00258	3.99742
4	14.00000	30.57593	-16.57593
5	44.00000	46.29594	-2.29594
6	47.00000	34.83098	12.16902
7	23.00000	12.56487	10.43513
8	6.00000	9.14704	-3.14704
9	39.00000	35.31270	3.68730
10	82.00000	74.39116	7.60884
11	1.00000	3.51947	-2.51947
12	0.00000	-1.45349	1.45349
13	90.00000	80.53669	9.46331
14	63.00000	65.22723	-2.22723
15	83.00000	76.97873	6.02127
16	41.00000	42.82755	-1.82755
17	2.00000	-4.32410	6.32410
18	25.00000	37.23959	-12.23959
19	40.00000	34.83098	5.16902
20	66.00000	57.13543	8.86457
21	40.00000	54.38330	-14.38330
22	28.00000	31.05339	-3.05339
23	11.00000	21.92986	-10.92986
24	54.00000	44.94712	9.05288
25	49.00000	63.79825	-14.79825
26	4.00000	8.74858	-4.74858
27	58.00000	60.20539	-2.20539

Table 2 Residuals

DURBIN WATSON SERIAL CORRELATION = 2.29

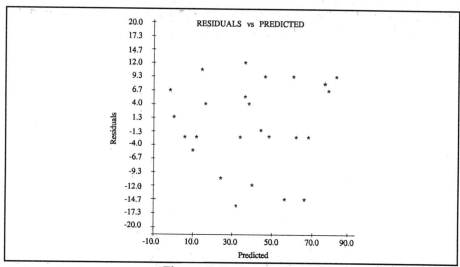

Figure 5 Residuals Plot
H Experiment - 26 Jan 89

Least clear differences are calculated to make block to block comparisons due to the unbalanced data set. LSD_1 is used for comparing blocks 1 and 2, LSD_2 for blocks 1 and 3, LSD_3 for 1 and 4, LSD_4 for 2 and 3, LSD_5 for 2 and 4, and LSD_6 for comparing 3 and 4.

$$t_{0.025}(27) = 2.052$$

$$s = \sqrt{\frac{4 \times 16.82^2 + 2 \times 20.60^2 + 9 \times 37.12^2 + 10 \times 20.33^2}{4 + 2 + 9 + 10}} = 27.21$$

LSD1 = 2.052 × 27.21 × sqrt (1/5 + 1/3) = 40.78

LSD2 = 2.052 × 27.21 × sqrt (1/5 + 1/9) = 31.14

LSD3 = 2.052 × 27.21 × sqrt (1/5 + 1/10) = 30.58

LSD4 = 2.052 × 27.21 × sqrt (1/3 + 1/9) = 37.22

LSD5 = 2.052 × 27.21 × sqrt (1/3 + 1/10) = 36.76

LSD6 = 2.052 × 27.21 × sqrt (1/9 + 1/10) = 28.83

The Bonferroni Limits are calculated such that BSD1 compares block 1 with block 2, etc., as is done with LSD above. The calculations use $t_{0.025/6}(27) = 2.848$.

$$BSD = 2.848 \times 27.21 \times \text{sqrt} (1/5 + 1/3) = 56.59$$
$$BSD = 2.848 \times 27.21 \times \text{sqrt} (1/5 + 1/9) = 43.22$$
$$BSD = 2.848 \times 27.21 \times \text{sqrt} (1/5 + 1/10) = 42.44$$
$$BSD = 2.848 \times 27.21 \times \text{sqrt} (1/3 + 1/9) = 51.66$$
$$BSD = 2.848 \times 27.21 \times \text{sqrt} (1/3 + 1/10) = 51.01$$
$$BSD = 2.848 \times 27.21 \times \text{sqrt} (1/9 + 1/10) = 35.61$$

The following table shows the results of both of the comparisons by block. Note that if zero is contained in an interval then no clear difference exists between the represented blocks.

	Block 2	Block 3	Block 4
Block 1	−2.73	−21.96	−14.90
LSD	(−43.51, 38.05)	(−53.10, 9.18)	(−45.48, 15.68)
BSD	(−59.32, 53.86)	(−65.18, 21.26)	(−57.34, 27.54)
Block 2		−19.23	−12.17
LSD		(−56.45, 17.99)	(−48.92, 23.59)
BSD		(−70.89, 32.43)	(−63.18, 38.84)
Block 3			7.06
LSD			(−21.77, 35.89)
BSD			(−28.55, 42.67)

It's clear that no differences exist between any of the blocks.

CATAPULTING STATISTICS INTO
ENGINEERING CURRICULA

Submitted by: S. E. Jones, PhD, University of Alabama
Stephen R. Schmidt, PhD, USAF Academy
Bruce Johnson, MS, HQ USAF/LE-RD

Today's rapidly changing technologies often result in products and processes that are not easily modelled using engineering theory alone. Other tools, such as statistics, can be used to supplement engineering knowledge and thereby speed up product and process development times. The use of a mechanical device such as the catapult (see Figure 1) discussed in this paper has been shown to be a very effective tool for demonstrating the power of blending engineering theory with statistically designed experiments.

Figure 1   Catapult

Analysis of the Catapult Using only Engineering Theory

The catapult is a device for launching a small projectile (ball) toward a predetermined, downrange impact site R (see Figure 2). The force F (see Figure 3), which rotates the extensible moment arm of the catapult, is provided by a stretched rubber band. The rubber band may be attached to several points on the superstructure of the device and may assume several stretch lengths for a given configuration. The force exerted by the rubber band on the moment arm of the catapult can be a complicated function of the stretch length. However, we assume that the stretched length is a function of the current position of the moment arm, $F = F(\theta)$

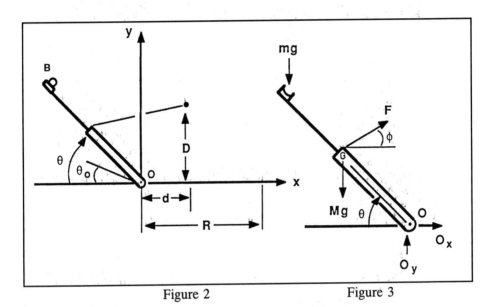

Figure 2   Figure 3

Taking moments about 0, and assuming that there are no applied moments at 0, we find

$$I_0 \ddot{\theta} = r_F F(\theta) \sin\theta \cos\theta - (Mg\, r_G + mg\, r_B) \sin\theta \tag{1}$$

where I_0 is the mass moment of inertia of the compound rod and ball assembly, r_F is the distance from 0 to the rubber band connection, r_G is the distance from 0 to the center of mass of the rod, r_B is the distance from 0 to the center of mass of the ball, M is the mass of the rod, m is the mass of the ball, and g is the local gravitational constant.

Since $F = F(\theta)$ and

$$\tan\theta = \frac{D - r_F \sin\theta}{d + r_F \cos\theta},$$ (2)

Equation (1) can be integrated once to give

$$\tfrac{1}{2} I_0 \; \dot\theta^2 \; = \; r_F \int_{\theta_0}^{\theta} F(\theta)\sin\theta\cos\theta\,d\theta - (Mg\,r_G + mg\,r_B)(\sin\theta - \sin\theta_0)$$ (3)

where we have assumed that the device is released from rest at $\theta = \theta_0$. This equation expresses the current angular rate $\dot\theta$ in terms of the current angular position θ (see Figure 2). When a stop position at $\theta = \theta_1$ is reached (see Figure 3), the angular speed is $\dot\theta_1$ and the ball will be launched with speed

$$v_B = r_B \; \dot\theta_1$$ (4)

toward the downrange target at R. We assume that the launch angle is normal to the rod position at $\theta = \theta_1$. From (4), $\dot\theta_1$ can be found, and the expression for it takes the form

$$\tfrac{1}{2} I_0 \; \dot\theta_1{}^2 \; = \; r_F \int_{\theta_0}^{\theta_1} F(\theta)\sin\theta\cos\theta\,d\theta - (Mg\,r_G + mg\,r_B)(\sin\theta_1 - \sin\theta_0).$$ (5)

Combining (5) with (4), we can find the initial velocity for the ball when the prescribed stop angle θ_1 is reached.

The small ball is launched with initial speed v_B in a direction that is normal to the catapult, as shown in Figure 4. The stop in the mechanism at $\theta = \theta_1$ provides for the initial launch angle, which is $\pi/2 - \theta_1$ radians. The initial launch position has coordinates $(-r_B \cos\theta_1, \; r_B \sin\theta_1)$ in the rectangular, cartesian coordinate system with origin at 0 (Figure 4).

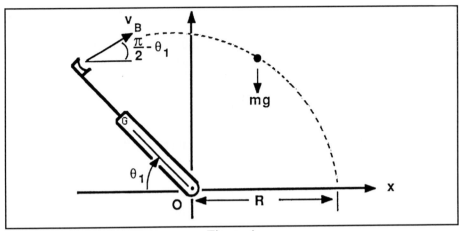

Figure 4

Neglecting the influence of atmospheric drag, the equations of motion for the ball are

$$\ddot{x} = 0, \quad \ddot{y} = 0 \tag{6}$$

where g is the local gravitational constant. These equations can be easily integrated, and the solution is

$$x = v_B \cos\left(\frac{\pi}{2} - \theta_1\right) t - \frac{1}{2} r_B \cos\theta_1 \tag{7}$$

and

$$y = r_B \sin\theta_1 + v_B \sin\left(\frac{\pi}{2} - \theta_1\right) t - \frac{1}{2} g t^2. \tag{8}$$

At impact, x=R and y=0 at the terminal time $t = \bar{t}$. Inserting these conditions into (7) and (8) and eliminating the terminal time between the resulting equations gives one relation involving v_B and θ_1.

$$r_B \sin\theta_1 + (R + r_B \cos\theta_1) \tan\left(\frac{\pi}{2} - \theta_1\right) - \frac{g}{2 v_B^2} \frac{(R + r_B \cos\theta_1)^2}{\cos^2\left(\frac{\pi}{2} - \theta_1\right)} = 0. \tag{9}$$

Equation (9) can be combined with (4) and (5) to form a nonlinear system for the unknowns θ_1, $\dot{\theta}_1$, and v_B when the initial launch angle θ_0 is given. We may also view this same system as one in which the unknowns are θ_0, $\dot{\theta}_1$, and v_B when the stop angle θ_1 is prescribed. When the constitutive properties of the rubber band are supplied, this system can be used to determine the launch conditions for a given, downrange site R.

The algebraic structure of the system of equations (4), (5) and (9) is such that they may be easily combined into one equation:

$$\frac{g I_0}{4 r_B} \frac{(R + r_B \cos\theta_1)^2}{\cos^2\left(\frac{\pi}{2} - \theta_1\right)\left[r_B \sin\theta_1 + (R + r_B \cos\theta_1)\tan\left(\frac{\pi}{2} - \theta_1\right)\right]} \tag{10}$$

$$= r_F \int_{\theta_0}^{\theta_1} F(\theta)\sin\theta\cos\phi \, d\theta - (M g \, r_G + m g \, r_B)(\sin\theta_1 - \sin\theta_0)$$

This equation is a complicated transcendental involving an integral. The integrand contains a force function $F = F(\theta)$ which is nonlinear for large stretches. The integrand also contains the cosine of the angle ϕ which is itself a complicated function of θ (see Equation (2)). There is no reasonable prospect for evaluation of this integral in terms of elementary functions. However, it can be dealt with numerically, and this will be the approach in subsequent work.

One comment must be made at this juncture. For a given value of R and a given force law $F = F(\theta)$, there may be no solution to (10) for any values of θ_0 and θ_1. On the other hand, when the force in the rubber band is sufficiently high, there will be solutions for a wide range of initial and stop angles θ_0 and θ_1.

There are several points of concern in the catapult analysis just presented. First, the force law for the rubber band has not been specified. Each rubber band that is used with the device must be tested and the force as a function of stretch must be determined. Rubber is generally a nonlinear elastic material. So, the force-extension law may be nonlinear.

We will assume that the stretch in the rubber band is uniform. However, for some configurations, the rubber band passes over pegs which can offer enough friction to make the stretch nonuniform. When these pegs are lubricated, this effect can be minimized.

A second concern is that the moment of inertia I_0 and the center of mass of the extensible moment arm r_G have not been specified. However, both of these quantities are easy to determine experimentally as functions of the rod extension length r_B. The moment of inertia can be computed for any given rod length by treating the compound moment arm as a pendulum and measuring the period of oscillation for various arm lengths. This will compute the moment of inertia, I_0, very conveniently and accurately. Similarly, the position of the mass center as a function of rod length can be determined from an elementary balance experiment.

Finally, the amount of negligible friction which is present in any device of this type is always a point of concern. The analysis does not account for any applied moments due to friction. We assume that the pin surface at 0 can be lubricated to eliminate any frictional moments.

However, an even greater source of difficulty is presented by the interaction between the ball (projectile) and the cup surface. If the ball is pressed into the cup, then friction is induced on the ball surface which may change the direction of the initial velocity vector. The ball will tend to roll out of the cup, as opposed to being launched cleanly at an angle of 90° to the moment arm. We can remedy this situation by either lubricating the cup or by taking care prior to launch not to press the ball into the cup.

In conclusion, a deterministic, engineering analysis for the catapult has been presented. The result of this analysis is a single equation (10) which related the initial and terminal angles, θ_0 and θ_1, to the applied force $F(\theta)$, the terminal distance R, and the various distances and weights in the system. In order to implement the analysis to select appropriate launch conditions, we must specify all of the input parameters from the problem and write a computer program to extract the information from equation (10). The input parameters consist of all of the distances r_B, r_F, and r_G, the moment of inertia I_0, the rubber band force characteristics, and the range R. Working with students, this project would take the better part of a semester to complete.

Blending Engineering with Statistics

Since it is going to be difficult at best to model the catapult using only an engineering approach, it is advisable to seek an alternative approximation technique to speed up the modeling process. Assume that we know the true functional relationship of distance with each of the variables contained in the catapult. This function could be approximated with a Taylor series model through the use of calculus. For the catapult, as well as today's complex products and processes, the problem is that this functional relationship is typically unknown or difficult to determine. A Taylor series model, however, can still be constructed using empirical data. For the catapult problem, a two level experimental design matrix can be used to generate the appropriate empirical data required to build a first order Taylor series approximation which can include desired 2-way linear interactions. Less than four hours of instruction are normally required to explain a simple approach to accomplish this type of modeling. Air Force Academy faculty and cadets attending an experimental design class recently used this approach for catapult modeling. They used engineering knowledge to brainstorm the four most likely variables that affect distance. These four variables were then included in an eight shot design matrix of highs and lows for each variable. The distance for these eight shots and the design matrix highs and lows were used to construct a first-order Taylor series model. Confirmation tests of the model indicated that $\pm$ 2-inch accuracy can be achieved over the 200 inch operational range of the device. After initial training, the entire experimental approach was accomplished in two hours.

The obvious conclusions are: (1) The Taylor series modeling approach is fast and accurate; (2) In today's competitive market environment, engineers can potentially increase their ability to reduce product and process development times through the use of statistically designed experiments blended with existing engineering theory; and (3) The use of training devices such as the catapult not only provides hands-on training, but also makes the learning process enjoyable.

Environmental Engineering
Operation of an Activated Sludge System

Submitted by: Majors James Brickell and Kenneth Knox
Department of Civil Engineering
United States Air Force Academy

Purpose

Demonstrate how experimental design can be used in environmental engineering. Specifically, show how operation of an activated sludge reactor can be better understood and optimized.

The Activated Sludge System

The activated sludge process is a biological treatment system used at many wastewater treatment plants. Briefly, biological organisms (termed sludge) within the reactor are used to aerobically convert incoming waste (influent) into additional biomass or innocuous carbon dioxide and water. The activated sludge reactor is followed immediately by a settling tank, called a secondary clarifier, where the liquid and solids (sludge) are separated. The sludge from the clarifier is then either wasted, or recycled back to the reactor to maintain an acceptable biomass population. Figure 1 shows a schematic of a typical completely-mixed activated sludge system.

The parameters shown in Figure 1 are defined as follows:

Q_o, Q_u, Q_w, Q_r = flow rate in influent, clarifier underflow, waste, and recycled flows, respectively (m^3 / d)

S_o, S = soluble food concentration (measured as BOD_5) in the fluent and reactor, respectively (mg / L)

X_o, X, X_e, X_u = biomass concentration in influent, reactor, effluent, and clarifier underflow, respectively (mg / L)

V = reactor volume (m^3)

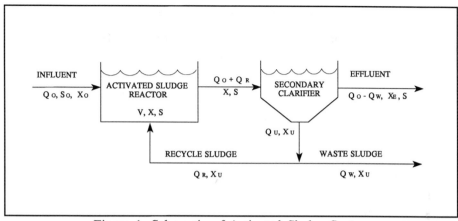

Figure 1   Schematic of Activated Sludge System

Performance of an activated sludge system is measured using the biochemical oxygen demand, BOD. The BOD is the amount of oxygen required to stabilize the decomposition matter in a water using aerobic biochemical action. In general, a lower BOD is indicative of a higher quality water. A typical municipal wastewater would have a BOD_5 concentration (BOD measured after 5 days at 20°C) of 110 - 400 mg/L. The EPA effluent standards for wastewater, on an average monthly basis, are 30 mg/L BOD_5 (Peavy et al., 1985).

A mass balance analysis of the system described in Figure 1, using the Monod equation to describe the rate of bacterial growth, yields the following analytical relationship:

$$(S_o - S) = \frac{(1 + k_d VX / Q_w X_u)}{(Q_o Y / Q_w X_u)} \tag{1}$$

where:

$S_o - S$ = amount of BOD removed by the system

k_d = biological growth rate constant (d^{-1})

Y = fraction of food (S) converted to biomass (X), or

$$Y = \frac{\text{kg X produced per day}}{\text{kg S consumed per day}}$$

Experimental Design

To demonstrate application of experimental design, a simulated activated sludge system was conceived. Key parameters of the system were then varied between expected ranges, and the results noted.

The parameters which affect the treatment efficiency for removing BOD $(S_o - S)$ from wastewater can be gleaned from the analytical equation describing the process (Equation (1)). Metcalf and Eddy, 1979, provide typical ranges for these parameters for an average municipal activated sludge system:

X,	reactor biomass concentration	3000 - 6000 mg / L
X_u,	clarifier biomass concentration	8000 - 12000 mg / L
k_d,	biological growth rate constant	0.040 - 0.075 d^{-1}
Y,	fraction of food to biomass	0.40 - 0.80 kg / kg

The system that will be used for this analysis originated from an example problem (10-1) presented by Metcalf and Eddy, 1979, where the reactor volume, V, is 4700 m^3, and the influent flow rate, Q_o, is 21,600 m^3 / d. Based on typical wasting rates, the waste sludge flow rate for this system was calculated to b 78.5 - 940.0 m^3 / d.

Initially, a series of experiments was designed using the five variables (X, X_u, k_d, Y, and Q_w). A 2-level design was used, with the low $(-)$ level at the lower limit of the typical range, and the high $(+)$ level at the higher limit for each variable. A full-factorial design would have required 32 runs (2^5), which may be excessive for an actual plant to perform. Therefore, a quarter-factorial design was used, requiring 8 runs (2^3). Aliasing and the results of each run (based on Equation (1)) are shown in Table 1.

Run	X	X_u	Q_w	$k_d = XX_u$	XQ_w	X_uQ_w	$Y = XX_uQ_w$	Response (mg/L)
1	+	+	+	+	+	+	+	775
2	+	+	−	+	−	−	−	354
3	+	−	+	−	+	−	−	1001
4	+	−	−	−	−	+	+	102
5	−	+	+	−	−	+	−	1371
6	−	+	−	−	+	−	+	87
7	−	−	+	+	−	−	+	496
8	−	−	−	+	+	+	−	195
Avg(+)	558	647	911	455	515	611	365	$\bar{y} = 548$
Avg(−)	537	448	184	640	581	485	730	
Δ	21	198	726	−185	−66	126	−365	
Δ/2	10	99	363	−93	−33	63	−183	

Table 1   Quarter-Factorial Design Matrix

Prediction equation (First-order Taylor series approximation):

$$S_o - S = 548 + 10\bar{X} + 99\bar{X}_u + 363\bar{Q}_w - 93\bar{k}_d - 33\bar{X}\bar{Q}_w + 63\bar{X}_u\bar{Q}_w - 183\bar{Y}$$

The bars over the variables indicate these are linearly "coded" variables, with the value of −1 at the lower limit, +1 at the higher limit, and 0 at the midrange value.   To evaluate the usefulness of the derived Taylor approximation of Equation (1), a series of 25 runs was made with the five variables allowed to randomly vary between their typical upper and lower limits.   The predicted response was then compared with the "correct" response from Equation (1), the results of which are shown in Figure 2.   The least squares regression line of the results has a respectable coefficient of determination (R_2) of 0.905.

An inspection of the prediction equation shows that the most important variables for controlling removal of BOD from wastewater seem to be Q_w and Y. A Pareto Diagram, Figure 3, more clearly demonstrates the contribution to BOD removal of each variable.   In this figure, X_u and k_d also seem to contribute significantly to the observed responses, while X and the tested interactions (XQ_w and X_uQ_w) appear less consequential.

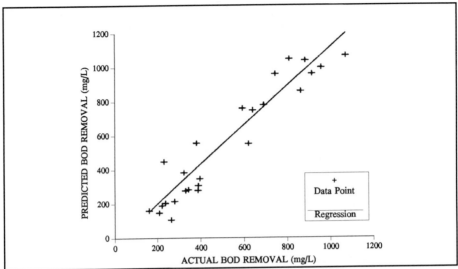

Figure 2   Comparison of Predicted versus Actual BOD Removal
Rates from the Quarter-Factorial Design Model

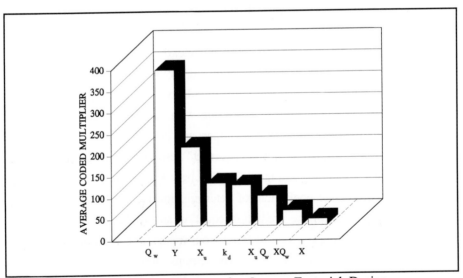

Figure 3   Pareto Diagram for Quarter-Factorial Design

Untested interactions and aliasing could materially impact the prediction. Aliasing cannot be evaluated without further testing. The untested interactions, however, were analyzed using 2-factor interaction graphs (Figures 4a-e) between the most significant main effects variables (Q_w, Y, X_u, and k_d). From Figure 4, the following observations can be made:

*   Since the interaction graphs are not parallel, significant interaction exists between Q_w and Y (Figure 4a); Q_w and k_d (Figure 4b); and between Y and k_d (Figure 4d).

*   Little to no interaction exists between Y and X_u (Figure 4c), and between X_u and k_d (Figure 4e).

Even though it may be impractical using a "real" activated sludge system, a 2-level full factorial model was tested on the computer using all five variables (32 runs) in order to more rigorously evaluate the desired experiments approach.

Figure 5 shows that the prediction equation does a good job of modeling the actual BOD removal equation, with an R^2 over 0.99. The corresponding Pareto Diagram, Figure 6, shows that with minor exceptions the activated sludge system accurately identified the key variables using the designed experiments approach with only eight runs.

Operational Evaluation of Results

Based on the results of the quarter-factorial design, to maximize BOD removal from wastewater one should maximize Q_w and X_u, while minimizing k_d and Y. X and the tested interactions are of lesser importance. The interactions between Q_w and Y, Q_w and k_d, and between Y and k_d appear to be significant, but further tests would be required to quantify these relationships.

Operationally, the only variables over which an operator has direct control (by turning a valve, for instance) are X, X_u, and Q_w. Y and k_d are biological rate constants which can only be controlled by relatively extreme measures such as changing the reactor water temperature or changing the type of microorganism used to stabilize the waste products. Therefore, a reasonable operational

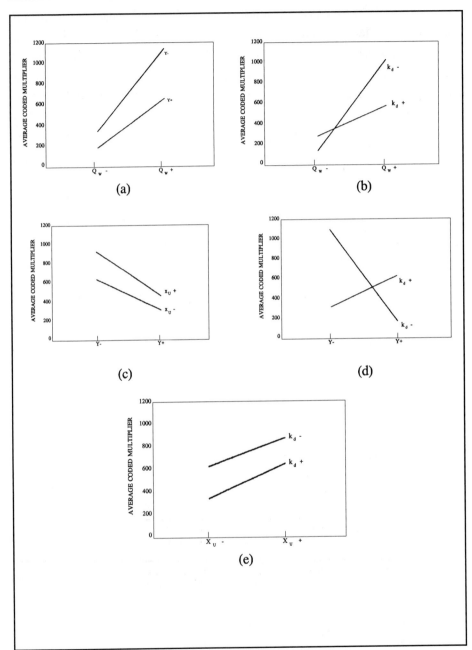

Figure 4 Two-Factor Interaction Graphs

recommendation based on the results of this study would be to maximize the flow rate of waste sludge (Q_w) and the clarifier biomass concentration (X_u) in order to maximize removal of BOD from wastewater. If the biological growth rate constant, k_d, proved to be unstable for some reason, Figure 4b suggests that adjusting Q_w to approximately 275 m^3 / d (the intersection of the two lines, accounting for the "coded" variable) would make the activated sludge system robust to k_d. The power of the designed experiments approach is demonstrated here by noting that these conclusions are certainly not obvious from a simple inspection of Equation (1).

An actual activated sludge process is extremely complex, however, and other considerations would come into play before undertaking changes to the system. One consideration is cost. For example, increasing Q_w will naturally increase the costs of disposing the waste sludge.

Figure 5 Comparison of Predicted versus Actual BOD Removal
Rates from the Full-Factorial Design Model

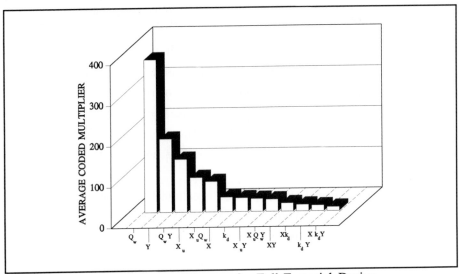

Figure 6 Pareto Diagram for Full-Factorial Design

Operational problems may also arise, such as the sludge in the clarifier may form nitrogen gas and float if the residence time in the settler becomes excessive while attempting to increase X_u.

This case study has attempted to demonstrate the power of the designed experiments approach. Using some simple statistical techniques coupled with engineering judgement, complex systems can be economically modeled, and their operation greatly simplified. As a result, adjustments and modifications to the system can be made, and the consequences of these changes predicted with some degree of confidence.

References for the Activated Sludge System Case Study

1. Metcalf & Eddy, Inc., *Wastewater Engineering: Treatment, Disposal, Reuse*, 2nd ed., McGraw-Hill, New York, 1979.

2. Howard S. Peavy, Donald R. Rowe, and George Tchobanoglous, *Environmental Engineering*, McGraw-Hill, New York, 1985.

Minimizing the Number of Large Scale Computer Simulation Runs:
The Testing, Validation, and Implementation of the Interagency Working Group (IWG) Acquired Immune Deficiency Syndrome (AIDS) Model

Submitted by:

Dr. Thomas F. Curry
LOGICON/RDA Colorado Springs

Captain John J. Tomick
Department of Mathematical Sciences
United States Air Force Academy

Captain Barbara A. Yost
Department of Mathematical Sciences
United States Air Force Academy

Dr. Stephen T. Dziuban
LOGICON/RDA Colorado Springs

Abstract

The IWG AIDS model requires over 600 lines of input data, most of which is uncertain. To determine the effect of the uncertain data, a sensitivity analysis is performed on 134 input variables. A Plackett-Burman experimental design is used to minimize the number of computer runs necessary to estimate the first-order approximation of the gradient of the local response surface of five dependent variables. For the input variables analyzed, this intermediate version of the model indicates that the six most important variables driving the gradient are the rate of progression from HIV to AIDS for adults, the percent of females age 15-44 that are married, the transmission rate of cocirculating sexually transmitted diseases, the rate of progression from AIDS to death for adults, the percent of men age 15-44 that are married, the percent of low risk married females that are HIV positive, and the number of contacts per month of high risk single females age 15-44. This suggests that intervention strategies should first consider methods to reduce risky behavior and control cocirculating sexually transmitted diseases.

Introduction and Background

In December 1988, researchers from the Department of Mathematical Sciences were asked to consult on the testing and validation of a simulation model designed to forecast, for most any nation, the effects of the AIDS epidemic. The request came from the State Department Interagency Working Group (IWG) which is chaired by the Deputy Assistant Secretary for the Environment. The IWG is a consortium of several agencies including the Subcommittee for AIDS Models and Methods, the US Agency for International Development, Health and Human Services, the National Institute of Health, the Center for Disease Control, the Department of Energy, and the Department of Defense. The AIDS model was developed during 1989, and this paper contains the experimental design methodology and the results of the initial testing and validation.

The AIDS model is a set of approximately 360 deterministic differential equations developed and programmed by scientists at Los Alamos National Laboratory (LANL) and the Merriam Laboratory at the University of Illinois at Urbana. The model uses risk-based methodology similar to models described in [3]. Data for the model is provided by the US Bureau of Census, Center for International Research (CIR).

The Problem: Uncertain Input Data and Lots of It

In spite of the excellent data collection procedures used by CIR, the AIDS data from Africa and the rest of the world are uncertain. Thus, the question arises as to the value of forecasts that are based on input parameters estimated from such data. The answer depends upon what one is asking the model to compute. Uncertainty in the model's output depends upon uncertainty in the input data as well as upon uncertainties in the model assumptions and formulation. We discuss the techniques used to analyze the sensitivity of the model to the input data, and how this sensitivity analysis can be used to help form decisions based on model results. The question of uncertainty due to incorrect model formulation is left to future studies.

The IWG AIDS model requires a large amount of input data. The input data file is over 600 lines long and contains demographic and epidemiological information including population distributions, fertility rates, migration rates, sexual risk behavior categories, sexual contact rates, cocirculating sexually transmitted diseases, condom use, percent of the population that is seropositive (has the HIV virus), risk of infection per sexual contact, rate of progression from infection to AIDS, and blood transfusion and blood screening. Most of these data are enumerated by sex, age, and marital status resulting in over 350 separate input variables that could be analyzed for their effect on the various response surfaces. Using "engineering knowledge," the team selected 134 input variables that were of interest.

Each run of the AIDS model requires approximately 15 minutes on a 386 based microcomputer. If "one-at-a-time" input data perturbation techniques were used, the required 2^{134} runs would take 10^{29} years. Clearly, an efficient design was needed to evaluate the model. In conferences at LANL, we determined that Plackett-Burman designs (using Hadamard matrices) were well suited for the initial first-order approximation of the gradient of the "local response surfaces."

Experimental Design

To study the relative sensitivity of the results to each of the 134 parameters in the input file, we investigated the response surfaces by varying the input data by small amounts around a baseline set of values for a "generic" country. Using the techniques described in [1,2,6,7], we developed a 136 by 136 Hadamard matrix to set up 136 runs. All parameters were varied simultaneously, and the effects of each were ranked. Since the parameters each take on their high and low values for half the runs, there is enough information to do t-tests to statistically compare the means of the outputs from the high value and low value runs.

Results

In our study of a recent version of the model, most inputs were varied about their baseline values by approximately 10% of their range. Using 10 year forecasts, the epidemic response surfaces were most sensitive to the rate of progression from HIV to AIDS for adults (expressed as the percent of the HIV+ population that converts to AIDS per year), and the percent of females age 15-44 that are married. Obtaining better estimates of these parameters appears to be the key to narrowing confidence bounds on the model results.

Frequently appearing variables key:

%MarFem15-44	=	Percent of females age 15-44 that are married
%MarMen15-44	=	Percent of males age 15-44 that are married
Inf Rate Men	=	Risk of infection per sexual contact for men
STD Trans	=	Unprotected transmission rate of cocirculating sexually transmitted diseases (STD's)
Cntct/mth HRSF 15-44	=	# of sexual contacts per month by high risk single females age 15-44
Rate AIDS to Death 5+	=	Rate of conversion from AIDS to Death for age 5 and older
Rate HIV to AIDS 5+	=	Rate of conversion from HIV to AIDS for age 5 and older
Months Dur STD Fem	=	Duration in months of and STD episode for females
%HIV+ LRMF 15-44	=	Percent of low risk married females that are HIV+

Response variable key:

1	=	Percent of population infected
2	=	Cumulative AIDS deaths
3	=	Percent of deaths due to AIDS
4	=	Number of infected females age 15-44
5	=	Percent of females age 15-44 infected

In all, five epidemic response surfaces were examined. The nine most important input variables driving each response are listed in Table 1.

Frequently appearing variables	Response Variable					
	1	2	3	4	5	Avg
Rate HIVtoAIDS5+	1	1	1	1	1	1.0
%MarFem15-44	2	2	2	2	2	2.0
STD Trans	3	10	9	4	3	5.8
Rate AIDStoDeath5+	4	11	11	5	4	7.0
%MarMen15-44	7	7	8	10	7	7.8
%HIV+ LRMF15-44	14	5	5	12	9	9.0
Cntct/mth HRSF 15-44	8	9	12	11	10	10.0
MonthsDur STD Fem	6	16	15	7	6	10.0
Inf Rate Men	5	18	23	6	5	11.4

Table 1   Most Important Input Variables

These results suggest the use of intervention strategies that reduce high risk behavior and control sexually transmitted diseases. When Ugandan President Yoweri K. Museveni was shown these model results, he immediately reversed a long-held position and urged Ugandans to use condoms [5].

The Value of the Experimental Design Results

Table 1 lists only the top nine variables controlling the response surfaces. When these results were briefed to the IWG, several health officials stated that this was the first time they had seen any information on what variables were important in driving the AIDS epidemic. They noted that the complete ranking of all 134 input parameters can be used to determine the "best" way to minimize the number of deaths due to AIDS. These results can also be used to guide data collection. Indeed, it is evident that accurate values for these variables must be determined before approximate confidence bounds on the forecasts can be calculated.

The experimental design runs also helped us determine model deficiencies. In other testing, by examining the pattern of the parameter settings in failed runs, we determined that when the number of men was increased by 10%, and the number of women was decreased by 10%, the model failed because there were insufficient numbers of women for the men to marry. This resulted in the programming of a variable marriage rate that can handle depletion of certain age groups in the population. This also suggests that important changes in societal structure and behavior may occur due to large segments of a population dying from AIDS.

The model was run with both 5% and 10% perturbations and the rank ordering of the independent variables remained the same indicating that the response surfaces are fairly smooth in the region of interest.

Lastly, the results of the main effects study have been used to screen the input parameters for analysis of 2-way interactions. Specifically, using a "two-to-the-seven-minus-one" experimental design, we examined seven input variables for pairwise interactions. From the marginal mean plots, we determined that (at 96% confidence) there was a statistically significant interaction between the number of men in the high risk group and the risk of infection per sexual contact for males. This supplies additional support for the need to accurately determine the risk of infection per sexual contact.

In summary, we found the use of experimental design techniques to be very valuable. They reduced the time required for model testing from years to weeks, provided us with valuable rankings of main effects, and gave direction to future data collection, and future model analyses.

References for the AIDS Case Study

1. J. Hadamard, "Resolution d'une Question Relative aux Determinants," *Bulletin des Sciences Mathematiques*, No.2, Vol. 17, Part 1, 1893, pp. 240-246 .

2. A. Hedayat and W.D. Wallis, "Hadamard Matrices and Their Applications," *The Annals of Statistics*, Vol. 6, No. 6, 1978, pp. 1184-1238.

3. J.M. Hyman and E.A. Stanley, "Using Mathematical Models to Understand the AIDS Epidemic," *Mathematical Biosciences*, Vol. 90, 1988, pp. 415-473 .

4. R.E.A.C. Paley, "On Orthogonal Matrices," *Journal of Mathematics and Physics*, Vol. 12, 1933, pp. 311-320.

5. J. Perlez, "Spread of AIDS is Worrying Uganda," *The New York Times*, January 30, 1991.

6. R.L. Plackett and J.P. Burman, "The Design of Optimum Multifactorial Experiments," *Biometrika*, Vol. 33, 1946, pp. 305-325.

7. S. R. Schmidt and R.G. Launsby, *Understanding Industrial Designed Experiments*, 3rd ed., Air Academy Press, Colorado Springs, CO 1992.

Karl Fischer Moisture Designed Experiment

Submitted by: Rita Whiteley and Robert Lawson
Cell Technology, Inc., Boulder, CO

Introduction

The Karl Fischer method of water determination, titrates water with iodine in the presence of sulfur dioxide, methanol, and a suitable base. See the following chemical reaction:

$$H_2O + I_2 + SO_2 + CH_3OH + 3RN \ ------> \ [RNH]SO_4CH_3 + 2[RNH]I$$

where RN = Base.

Use of the Mettler DL18 titrator allows the titration to be followed exactly through the use of a two-pin platinum electrode which has a current source applied to its poles. The voltage measured at the polarized electrode pins is used by the controls as an input signal. When the last traces of water have been titrated, voltage drops to virtually zero, the electrodes are then depolarized by the iodine now present.

The Karl Fischer titration permits determination of available water only. This means water must be brought into solution by suitable means before the titration can be carried out. Sample preparation for freeze-dried ImuVert is very critical. Since a portion of the moisture found in lyophilized ImuVert is tightly bound to the crystalline structure and difficult to make available for titration, an external extraction step must be performed prior to titration.

Objective

Determine how external extraction variables, such as solvents and shake time, effect the measured percent moisture titrated from lyophilized ImuVert.

The Factors and Levels chosen are shown in Table 1.

Factors	Levels
Methanol	0, 0.5 ml
Chloroform	0, 0.5 ml
Formamide	0.5, 1.0 ml
Shake Time	30 sec, 3630 sec

Table 1 Factors and Levels

Response: The response looked at was percent moisture with hopes to maximize the measured amount of moisture determined by the Karl Fischer titration.

Experimental Design

Defining Contrast: I = ABCD

$2^K - P$ Fractional Factorial

Yates Order

Run	TC	Methanol	Chloroform	Formamide	Shake Time (sec)	Response y_1	y_2
4	(1)	0	0	0.5 ml	30	1.2402	1.1005
2	a	0.5 ml	0	0.5 ml	3630	0.8562	0.9602
3	b	0	0.5 ml	0.5 ml	3630	0.9883	0.9907
1	ab	0.5 ml	0.5 ml	0.5 ml	30	1.0212	1.0845
7	c	0	0	1.0 ml	3630	0.9993	1.0068
8	ac	0.5 ml	0	1.0 ml	30	0.9625	1.0568
5	bc	0	0.5 ml	1.0 ml	30	0.8733	1.0253
6	abc	0.5 ml	0.5 ml	1.0 ml	3630	1.0151	0.9992

The experiment was performed in a randomized order.

Procedure

Prior to sample analysis, the titrator must be calibrated to ensure accurate mechanical operation. Calibration is completed by checking titrant concentration and instrument drift. Both parameters were within statistically acceptable ranges.

A blank vial determination (to determine moisture in the solvents as well as residual moisture attributed to an empty vial) is the next step. This is accomplished by using an empty vial which was processed exactly like lyophilized product; then, testing the blank vial with the same factor levels as will be done for the sample. For example, if run #2 required 0.5 ml formamide, 0.5 ml methanol, no chloroform, and a 3630 second shake time –the blank for that run would be done in the same manner.

Sample analysis for the designed experiment is the next stage. For each experimental run, two freeze-dried samples were tested with three to four repeat samplings per vial. Run #3 was an exception –one sampling/vial was done due to solvent volume. Solvents were added in a specific order as follows: formamide, methanol, and lastly chloroform. Swirling of the sample occurred after the formamide addition to expediate resolution of the freeze-dried product. The experimental design runs that received only the 30 second vigorous vortexing were tested immediately; while, other design runs after vortexing were allowed to shake continuously for one hour on a platform shaker before being tested.

In order for the titrator to determine the percentage moisture, several parameter values need to be entered by the operator. Sample volume to be tested was held constant at 0.4 ml. The blank vial's moisture content was entered based on the blank determination for that given run. Sample weight, which is based on the product's dry cake weight, solvent volume, and sample size varied due to the solvent volume used for the external extraction. The following sample weights were used: 2.0 ml solvent → 0.0073 g, 1.5 ml solvent → 0.0097 g, 1.0 ml solvent → 0.0146 g, and 0.5 ml solvent → 0.0292 g.

Data and Analysis

The results obtained from the designed experiment are presented in Table 2.

I = ABCD

Index	Total Observation	Sum of Square	Half Effect	Measures Average	DF	Mean Square	F Ratio
1	2.1057	16.3622	1.011				
2	1.8164	0.0025	0.013	$A^1=BCD$	1	0.0025	0.608
3	1.9790	0.0163	0.032	$B^2=ACD$	1	0.0163	3.904
4	2.3407	0.0188	0.034	AB=CD	1	0.0188	4.513
5	1.8986	0.0058	−0.019	$C^3=ABD$	1	0.0058	1.382
6	2.0143	0.0002	0.004	AC=BD	1	0.0002	0.048
7	2.0061	0.0051	−0.018	BC=AD	1	0.0051	1.219
8	2.0193	0.0355	−0.047	$ABC=D^4$	1	0.0355	8.518
	Error	0.0333			8	0.0042	
	Total	0.1174			15		

[1] Methanol
[2] Chloroform
[3] Formamide
[4] Shake Time

Table 2 ANOVA

Using a $F_{T0.05} = 5.31$, the only significant factor is D (shake time) or the 3-way interaction ABC (methanol, chloroform, and formamide). However, at the 90% confidence level, using a $F_{T0.10} = 3.46$, the B factor (chloroform) and the AB or CD interaction become significant.

Since the largest single effects (shake time and chloroform) are confounded with the ABC and ACD interactions, another experiment was needed to determine which was causing the significant effect.

The second designed experiment contained the factors and levels shown in Table 3.

Factors	Levels
Chloroform	0.5, 1.0 ml
Shake Time	30, 3630 sec

Table 3 Factors and Levels

Percent moisture was the response examined. Constants in the experiment were: methanol at 0.5 ml and formamide at 0.5 ml.

Experimental Design:

2^K Factorial

Yates Order

Run	TC	Chloroform	Shake Time	Response y_1	y_2
2	(1)	0.5 ml	30 sec	1.2264	1.2141
3	a	1.0 ml	30 sec	1.1330	1.1467
1	b	0.5 ml	3630 sec	1.0419	1.1271
4	ab	1.0 ml	3630 sec	1.1718	1.1533

The experiment was performed in a randomized order.

The data can be summarized in the ANOVA table (Table 4) and Figures 1, 2, 3 and 4.

I = AB

Index	Total Observations	Sum of Squares	Half Effect	Measures Average	DF	Mean Square	F Ratio
1	2.1690	10.6129	1.152				
2	2.4405	0.0126	0.040	A(CHCL$_3$)	1	0.0126	12.651
3	2.2797	0.0000	−0.001	B(Shake)	1	0.0000	0.003
4	2.3251	0.0064	−0.028	AB	1	0.0064	6.440
	Error	0.0040			4	0.0010	
	Total	0.0229			7		

Table 4 ANOVA

At the 95% confidence level only chloroform is an important effect; whereas, at the 90% confidence level the AB interaction becomes significant. In either case shake time alone has no significance, which means that the important effect noted in experiment one was a 3-factor interaction.

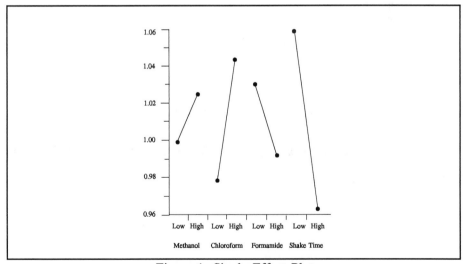

Figure 1 Single Effect Plot

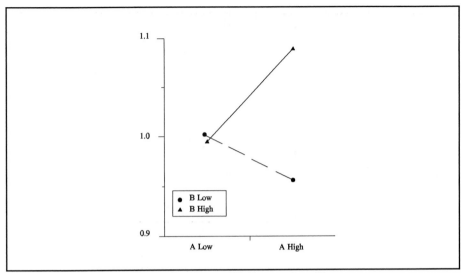

Figure 2 AB or CD Interaction Plot

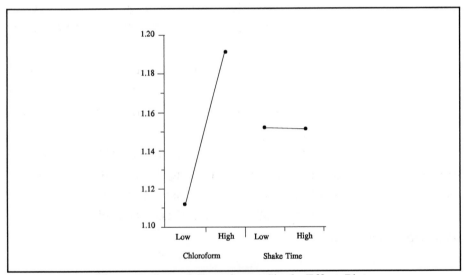

Figure 3 Second Experiment Single Effect Plots

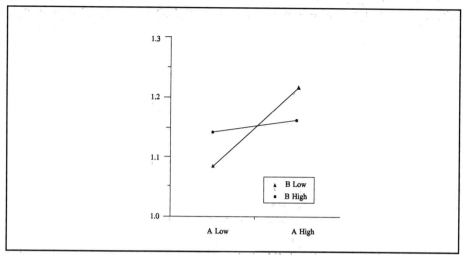

Figure 4   Second Experiment AB Interaction

Conclusions

The most significant effect on the external extraction process performed on ImuVert is the three solvent interaction. Based on the first experiments single effect plots, 0.5 ml of formamide, 0.5 ml of methanol, and 1.0 ml of chloroform are used for the external extraction.

Since shaking the product versus just vortexing the product was shown to have no effect in the designed experiment, for future testing the 30 second vortex and 60 minute shake will be used. This method did produce smaller coefficients of variation within sample testing per vial.

Attempts were made to remove the methanol to simplify the external extraction since it did not contribute significantly to the external extraction. However, when methanol was removed, two phases formed and measurements were harder to make. Therefore, all three external extraction solvent components were kept in the system.

Further optimization of this external extraction process could be done, including trying even less formamide. Also, the two factor interaction AB or CD which is significant at the 90% level may be important and a deconfounding experiment should be completed.

The End Count Method
Power with Small Samples

Submitted by: C. B. (Kip) Rogers
                     Digital Equipment Corporation, Marlboro, MA

In your company as in mine, we deal with IC's, SCR's, bearings, chemicals and a range of equipment in making a reliable and cost effective products. To keep things from crashing in our world we also need to make the right decisions often with limited samples and/or time. In this context End Count is a wonderful tool to add to your data analysis tool kit. Easy to learn, simple to apply, exact in its answers; it is a most useful supplement to one's existing statistical techniques. Now let's see what it is and how it works.

End Count

In the following paragraphs we will explore an increasingly popular technique for making sound decisions with very small samples. It is a technique which is of special use to those for whom:

   (a)   Samples are very expensive,

   (b)   Time is critical or,

   (c)   Both.

First let's look at how we make decisions in cases like these, then we'll do a couple of brief examples using the End Count method. Finally, for the statistically curious we'll show why the method works.

In decision making, we rarely can get all the data. Instead we rely on samples drawn from the entire data set (population). We then compare the results of these samples with other data and make a relative judgment on which we base our actions. This process involves some risk, which we hope to minimize, in return for the benefit of not having to measure the entire data set.

With End Count we are comparing two alternative solutions to a problem. Our goal is to decide whether the two alternatives are significantly different from each other. End Count is a simple technique for making this decision with very few assumptions or samples. It provides the exact probability of the observed outcome happening by chance (i.e., there is no difference between the two items or methods being tested). Then you, as a decision maker, can decide if the risk is worth taking.

Example 1

Two different styles of heat sinks are being evaluated. A standard test setup is designed and the heat sink results (on-chip temp) are as follows:

Style A: 110   113   108   111   (Results in degrees Centigrade)
Style B: 107   105   103   106

Looking at the data, your 'gut' reaction is that Style B, which yields a lower temperature, is better, but you wonder 'could this be a fluke?' — 'Should I do more testing?'. In this case the answers are No and No! Let's see why.

The first step is to rank order the data (keeping the data identified with its Style I.D.). From low to high we have:

103	105	106	107	108	110	111	113
B	B	B	B	A	A	A	A

$\Longleftarrow$ 4 $\Longrightarrow$ $\Longleftarrow$ 4 $\Longrightarrow$

By counting in the number of consecutive letters from each end until a letter change occurs and combining these "End Counts" we get 4 plus 4 equals 8. This is a measure of NON-homogeneity signifying that the two designs are really different. How likely is this degree of stratification to occur by chance if the designs were equal (which they aren't)? Table 1 has the answer. Letting style B = STD and style A = EXP we locate the 4/4 row and read across to the End Count = 8 column. The answer 2/70 or 2.86% is the likelihood of getting this degree of stratification from equal populations. In other words, we can be 97% certain that the two designs are really different.

Example 2

An adhesion test is performed to see if two flux types influence the bonding strength achieved with a new solder process. Noting that Flux Y had a much longer processing time, we take advantage of the fact that the End Count method does not require equal sample sizes. Results are:

Flux X:	435	440	428	432	437	(Data in grams)
Flux Y:	422	431	426	424		

Rank ordering we get:

422	424	426	428	431	432	435	437	440
Y	Y	Y	X	Y	X	X	X	X

$$| \quad \Leftarrow \quad 3 \quad \Rightarrow \quad | \qquad | \quad \Leftarrow \quad 4 \quad \Rightarrow \quad |$$

Once again the same nagging questions occur. Intuition says "X gives stronger bonds" but logic asks, "Could this be an accident?" and "Shouldn't I test more?" Referring to Table 1, we see that for 4 EXP and 5 STD samples with an End Count of $3 + 4 = 7$ we have only a 4/126 (3.17%) chance of this happening by accident and we conclude that type X really does give stronger bonds.

EXP Sample	STD Sample	END COUNT								
		10	9	8	7	6	5	4	3	2
5	5	2/252	N/A	4/252	8/252	16/252	32/252	60/252	100/252	140/252
5	4		2/126	N/A	4/126	8/126	16/126	30/126	50/126	70/126
4	5		2/126	N/A	4/126	8/126	16/126	30/126	50/126	70/126
4	4			2/70	N/A	4/70	8/70	16/70	28/70	40/70
4	3				2/35	N/A	4/35	8/35	14/35	20/35
3	4				2/35	N/A	4/35	8/35	14/35	20/35
3	3					2/20	N/A	4/20	8/20	12/20

Table 1 End Count Probabilities
Probability of Obtaining an End Count this high (or higher) when STD = EXP

Instructions for Using the End Count Method

1. Rank order data in a single row.

2. Write an E under each experimental value and an S under each standard value.*

3. Start at the left and count the number of consecutive letters which are the same. Go to the right end and repeat using the opposite letter.

4. Add the two counts. The total is your End Count. (If the same letter occurs at both ends then the End Count = 0 which simply means that the End Count method cannot detect a difference.)

5. Refer to the above table for the likelihood that an end count this high could have happened by chance.

6. Decide whether this likelihood is low enough for you to be comfortable stating that the observed difference is real.

For the Technically Curious

The End Count technique belongs to a class of statistics known as non-parametric methods. In these cases we don't assume a normal (or any other)

---

*If any data values are both E and S then flip a coin or put the letters in a hat to determine the assignment for those values.

distribution. The first published appearance of the techniques goes back to 1959, when John Tukey published his paper on what is referred to as the Tukey Quick Test. This procedure has been modified and popularized by a well known industrial consultant, Dorian Shainin.

Our End Count examples can be explained using a simple coin shuffling methodology. Starting with the combination of 4 pennies followed by 4 dimes we can by systematically moving the D's to the right generate all 70 possible combinations of coin flips (Table 2). Just as flips of an unbiased coin represent a 50/50 proposition so should test results from two identical processes fall equally above and below the median. In this case if EXP = STD, then each of the 70 combinations is equally likely! Given that the combinations are equally likely then let us first check to see if half the combinations begin with D and half with P. Thirty-five each—right!

Now let us consider the highly stratified cases with an End Count of 8. In this case there are only two such possible combinations out of all 70 orderings. Stated another way, the drawing 4 pennies followed by 4 dimes is very unlikely: 1/70 + 1/70 or 2.9%! Putting on our statistician's hat we conclude that this occurrence, though possible, is so unlikely that we reject the hypothesis that this ordering happened by chance. Transferring this concept to the real world, and our tests, we conclude that tests which produce End Counts with low probabilities in all likelihood come from populations having different averages.

NOTE: *The preceding examples were directed towards answering the question "Are STD and EXP different from each other?" In statistical terms we were performing a two-tailed test. Also worth considering is the (one-tailed) test case where we will only take action if the stratification is in one direction i.e., PPPPDDDD but not DDDDPPPP. This is often the case when a new process will be considered only if it is better (not just different from the standard). In this case all the End Count table values are halved (2/20 becomes 1/20) effectively doubling the sensitivity of the End Count method.*

N	Comb	End Count	N	Comb	End Count	N	Comb	End Count
1	PPPPDDDD	8	25	PDPDDDPP	0	49	DPDPDPPD	0
2	PPPDPDDD	6	26	PDDPPPDD	3	50	DPDPDPDP	2
3	PPPDDPDD	5	27	PDDPPPDD	2	51	DPDPDDPP	3
4	PPPDDDPD	4	28	PDDPPDDP	0	52	DPDDPPPD	0
5	PPPDDDDP	0	29	PDDPDPPD	2	53	DPDDPPDP	2
6	PPDPPDDD	5	30	PDDPDPDP	0	54	DPDDPDPP	3
7	PPDPDPDD	4	31	PDDPDDPP	0	55	DPDDDPPP	4
8	PPDPDDPD	3	32	PDDDPPPD	2	56	DDPPPPDD	0
9	PPDPDDDP	0	33	PDDDPPDP	0	57	DDPPPDPD	0
10	PPDDPPDD	4	34	PDDDPDPP	0	58	DDPPPDDP	3
11	PPDDPDPD	3	35	PDDDDPPP	0	59	DDPPDPPD	0
12	PPDDPDDP	0	36	DPPPPDDD	0	60	DDPPDPDP	3
13	PPDDDPPD	3	37	DPPPDPDD	0	61	DDPPDDPP	4
14	PPDDDPDP	0	38	DPPPDDPD	0	62	DDPDPPPD	0
15	PPDDDDPP	0	39	DPPPDDDP	2	63	DDPDPPDP	3
16	PDPPPDDD	4	40	DPPDPPDD	0	64	DDPDPDPP	4
17	PDPPDPDD	3	41	DPPDPDPD	0	65	DDPDDPPP	5
18	PDPPDDPD	2	42	DPPDPDDP	2	66	DDDPPPPD	0
19	PDPPDDDP	0	43	DPPDDPPD	0	67	DDDPPPDP	4
20	PDPDPPDD	3	44	DPPDDPDP	2	68	DDDPPDPP	5
21	PDPDPDPD	2	45	DPPDDDPP	3	69	DDDPDPPP	6
22	PDPDPDDP	0	46	DPDPPPDD	0	70	DDDDPPPP	8
23	PDPDDPPD	2	47	DPDPPDPD	0			
24	PDPDDPDP	0	48	DPDPPDDP	2			

Table 2 Generation of 4/4 Combinations by shifting "P's"

NOTE: *Using Table 2 as a guide, a simple software program can be developed which covers all practical sample sizes. The technique involves moving the rightmost P sequentially to the right; then popping it back as the next P moves one space to the right, ..., etc.*

A Designed Experiment to Reduce Circular Runout

Submitted by: Tom Tanis
Ampex Recording Systems

Background

The Ampex Precision Machining Business "Round Cell[*]" manufactured a machined lathe part that was used in a critical assembly (see Figure 1). Original production specifications required conformance to a .002" circular runout tolerance; however, experience showed that excessive runout caused complications during assembly. Achieving runouts of less that .0005" provided significant improvement at assembly operations. The question was whether we could consistently maintain a new tolerance of .0005" if we improved our process.

Figure 1

When we began our improvement efforts the process was generating parts with an average circular runout of .00055" and with a standard deviation

---

[*]Round Cell Team Members: Ray Atkins, Terry Bates, Tom Tanis, Dave Thiele, and Frank Williamson.

of .00024". With the tolerance set a .002" the resultant process capability (C_{pk}) was 2.01. If the tolerance was changed to .0005" a C_{pk} of -.069 would result, given no process improvements.

During the ten month period between November of 1992 and August of 1993, we attempted a number of single variable improvements to reduce runout. In August 1993, the process was generating parts with an average runout of .00031" and a standard deviation of .00013". With the runout tolerance set to .0005" the resultant C_{pk} was .487. This represented a substantial improvement over -.069. Nevertheless, the process capability was still unacceptable by industry standards. Unfortunately, despite further single variable improvement efforts, the process performance had remained relatively stationary for six months (from February 1993 until August 1993).

Definition

Runout is defined as "the composite deviation from the desired form of a part surface of revolution during full rotation (360°) of the part on a datum axis."[**] For this experiment, this means that the controlled surface must meet .0005" "total indicator reading" (with the new process tolerance) over the entire surface as it is rotated on datum -A-. The smaller the reading the better.

Experimental Objectives

The objective of this experiment was to create a process where the C_{pk} for a runout tolerance of .0005" was 1.4 or greater.

Procedures

Earlier attempts at process improvement by the Round Cell provided a detailed Ishikawa diagram. This diagram was analyzed to determine likely factors for experimentation. An experimental design was chosen. The experiment was

---

[**]Foster, L.W. Geo-Metrics II: The Application of Geometric Tolerancing Techniques. Reading: Addison-Wesley.

conducted by the researcher. The results were analyzed using regression analysis and confirmation runs were conducted to verify results.

Experimental Factors and Design

Analysis of the Ishikawa diagram, informed by process knowledge, led to the determination that there were two kinds of factors which could be major contributors to excessive circular runout: tool cutting pressure and the method of holding the part during machining. Tool cutting pressure could be influenced by a variety of factors, including the depth of the finish cut, the process speeds and feeds, and the number of finish cuts. The method of holding the part could be influenced by the physical geometry of the holding jaws and the amount of clamping pressure applied to the holding jaws.

After further consideration, the experimental factors were reduced to four: addition of a finish cut; two styles of chuck jaws; finish cut feed rate; and amount of chuck clamping pressure. Because the number of experimental factors was limited to four and because it was estimated that the effects of input variables would be generally linear, a two level full factorial design was chosen. Table 1 details the factors and settings chosen.

Experimental Factors			
Factor	Tool Pressure	High	Low
A	.0005" Finish cut	No	Yes
D	Feedrate	.003"/Rev.	.001"/Rev
	Holding Technique		
B	Physical Geometric of Jaws	New Style	Old Style
C	Clamping Pressure	35PSI	20PSI

Table 1

Except for the .0005" finish cut, a random number chart was used to randomize the experiment. Three replications were conducted to produce adequate sample sizes for the experiment. (For a detailed depiction of the experimental matrix see Addendum 1.)

Results

Regression analysis revealed two factors which significantly affected average runout. No factors significantly affected process variation. For a complete view of the experimental results see Addendum 1. Regression analyses are shown in Addendums 2 and 3. Addition of a finish cut (Factor A) and use of new style jaws (Factor B) both showed significant decreases in average runout. Regression analyses' predicted that adding a .0005" finish cut and using the new style jaw would yield a process with a mean runout of .000085" and a standard deviation of .000073".

A confirmation run of eighteen parts was conducted with the finish cut added and with the new style jaws. Chuck pressure was held at 35PSI and the feed rate was held at .003" per revolution. The confirmation run yielded a average runout of .000066" with a standard deviation of .000048". Both average runout and standard deviation were slightly better than the predictions.

With a mean value for runout of .000066" and a standard deviation of .000048", the process now has a C_{pk} of <u>3.01</u>. Since the goal of the experiment was to produce a process with a C_{pk} of 1.4 or greater, the experiment can be considered as successful.

Conclusion

Figure 2 gives graphic representation to the power of designed experiments. The chart shows 95% confidence intervals for average runout at various stages of our improvement process. Each line plotted in Figure 2 represents the performance of our process at a particular point in time and the true process mean at that time can be said to lie somewhere on the line. Because the process standard deviation is used to calculate the confidence intervals, the length of the line gives some sense of the process variation--the longer the line the more variation. (See Addendum 4 for descriptive statistics for each stage of improvement effort. Addendums 5 and 6 contain process control charts that provided the data used to calculate descriptive statistics.)

Figure 2

It can be seen that significant process improvement was attained between 11/11/92 and 9/16/93. These improvements were achieved through single variable modification. The process, though, had remained relatively stable for some six months.

The line representing process conditions after DOE shows a dramatic improvement in process performance. These results may have been eventually achieved through traditional single variable experimentation but DOE provided a means to discover answers to a number of questions in a highly efficient manner.

The DOE process provides the experimenter with two kinds of weapons. First, there is the statistical analysis. Statistical analysis gives the experimenter the tools with which to understand the impact of input variables with some degree of certainty. At the same time though, the DOE process requires the experimenter(s) to consider process characteristics in great detail (to think about what is happening). It is hard to say which of these weapons is more critical but, used together, they provide a powerful means for process improvement. In the case of this experiment, the goal of achieving a C_{pk} of greater than 1.4 was surpassed by the projected C_{pk} of 3.01.

Addendums

1. Experimental Matrix

Factor	A	B	C	D		Response Variable (Runout) Measured in Millionths of an Inch		
	.0005" Finish	Jaws	Chuck Pressure	Feed Rate				
Low (−)	Yes	Old	20 PSI	.001"/Rev.				
High (+)	No	New	35 PSI	.003"/Rev.				
Run	A	B	C	D		Y_1	Y_2	Y_3
1	−1	−1	−1	−1		50	300	170
2	−1	−1	−1	+1		180	220	100
3	−1	−1	+1	−1		140	180	300
4	−1	−1	+1	+1		140	200	200
5	−1	+1	−1	−1		40	60	20
6	−1	+1	−1	+1		200	50	100
7	−1	+1	+1	−1		120	60	60
8	−1	+1	+1	+1		100	80	80
9	+1	−1	−1	−1		280	400	450
10	+1	−1	−1	+1		320	390	390
11	+1	−1	+1	−1		220	290	300
12	+1	−1	+1	+1		500	350	350
13	+1	+1	−1	−1		180	260	340
14	+1	+1	−1	+1		120	380	140
15	+1	+1	+1	−1		200	370	210
16	+1	+1	+1	+1		300	500	210

2. Ŷ Model Prediction

Regression coefficients provided by *Q-EDGE* software. Factors A and B are significant at the .05 level and were included in the prediction equation. Interaction ACD appears close to significant but was not included in the prediction equation because a significant three way interaction was considered unlikely.

Effect	Coeff	p2t	Hi	Low	Expr
Const	220.8				
A	89.58	0.000	1.000	−1.000	−1.000
B	−46.6	0.001	1.000	−1.000	−1.000
ACD	22.08	0.078			
ABC	14.58	0.237			
ABD	−12.9	0.294			
D	12.50	0.310	1.000	−1.000	0.000
CD	10.83	.0377			
BC	10.00	0.415			
C	6.667	0.585	1.000	−1.000	0.000
AD	6.250	0.609			
BCD	−4.16	0.733			
AB	3.750	0.759			
ABCD	2.083	0.864			
BD	1.667	0.891			
AC	−0.41	0.973	Ypred	= 84.58	

3. Standard Deviation Prediction

Regression coefficients provided by *Q-EDGE* software. Sample sizes are too small to show statistical significance for factors' influence on process variation. No factors meet the "rule of thumb" for significance. So, best predictor of Standard Deviation is the constant .000073".

Effect	Coeff	t	tol	P(2 tail)
Const	73.3184	***	1.00	
A	17.4926	***	1.00	***
B	3.06632	***	1.00	***
AB	23.2510	***	1.00	***
C	−6.0537	***	1.00	***
AC	8.74810	***	1.00	***
BC	2.17356	***	1.00	***
ABC	−0.0812	***	1.00	***
D	2.15684	***	1.00	***
AD	12.0663	***	1.00	***
BD	16.7191	***	1.00	***
ABD	−1.5108	***	1.00	***
CD	0.88520	***	1.00	***
ACD	8.90565	***	1.00	***
BCD	−12.274	***	1.00	***
ABCD	−0.4270	***	1.00	***

4. Descriptive Sample Statistics at Various Stages of Improvement

The following table presents descriptive statistics for the "state of the process" at various stages of the improvement effort. Units are millionths of an inch. So, 200 represents 200 millionths of an inch, or, .0002".

Column A represents the state of the process preceding 11/11/92; before any real improvement efforts began.

Column B represents the state of the process between 11/11/92 and 2/4/93; after a single variable attempt at process improvement on 11/11/92.

Column C represents the state of the process between 2/4/93 and 9/16/93; after a second attempt at single variable process improvement on 2/4/93.

Column D represents the state of the process after Designed Experiment, 9/16/93. Calculations are based on confirmation run results.

	A	B	C	D
Data Elements (N):	98	98	98	18
Minimum:	200	200	100	10
Maximum:	1200	800	700	190
Arithmetic Mean:	552	511	314	66
95% Confidence Interval:	[504,599]	[479,543]	[289,340]	[42,90]
Standard Deviation:	238	159	127	48
Coefficient of Skewness:	0.312	-0.404	.655	1.08
Coefficient of Kurtosis:	2.192	2.11	3.23	3.60

5. $\overline{X}$ Control Charts

The control charts in Figure 3 represent the "State of the Process" average at various stages of our improvement effort.

NOTE: *Charts 4 and 5 have tighter control limits that were recalculated after improvement efforts on 2/4/93.*

6. $\overline{R}$ Control Charts

The control charts in Figure 4 represent the "State of the Process" dispersion at various stages of the improvement process.

NOTE: *Charts 4 and 5 have tighter control limits that were recalculated after process improvements.*

Figure 3 X̄ Charts

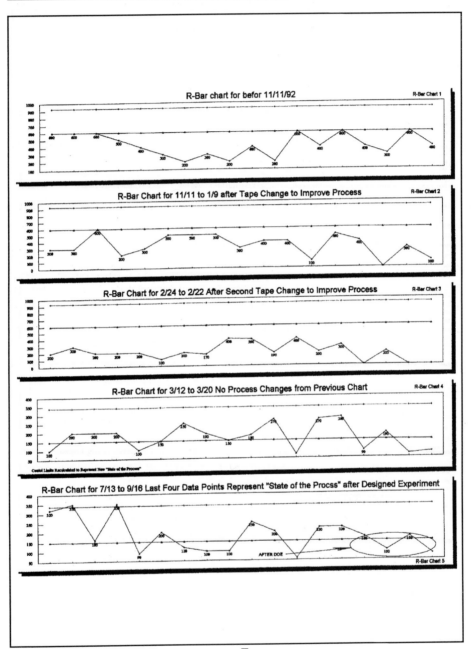

Figure 4 $\overline{R}$ Charts

Ball Weld Experiment

Submitted by: Kelly Johnson
Morton International

Description of the Problem

Process development engineering was evaluating several prototype welding processes for new product concepts in automobile air bag technologies. Traditional concepts use a chemical reaction involving sodium azide to generate gases to fill automobile air bags. Another concept showing promise involved filling a canister with an inert gas. This experiment involved the integrity of a seal in the automobile air bag inflator. When the inflator is produced, a canister is filled to high pressure through a small fill port, after which a BB-sized ball is placed over the hole and welded into place to seal the canister (see Figure 1). The purpose of this experiment was to maximize the strength of the weld as well as understand which variables have the largest influence on the strength of the weld.

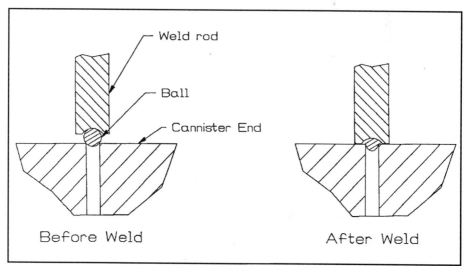

Figure 1   Canister Before and After the Weld

Several experiments were conducted involving a number of machine setup factors. In this particular study we will provide data and analysis from only the last experiment. The factors are shown in Table 1.

Weld Power	1, 2, 3
Hole Chamfer	1, 2, 3
Weld Tip Chamfer	1, 2, 3

Table 1 Machine Setup Factors

Response

Force required to push welded ball out of fill hole. A 12-run, D-optimal design with repetition was conducted (see Table 2).

Analysis

Analysis of the data was completed using *RS/Discover* software. The final model shown in Table 3 was obtained after conducting step-wise multiple regression.

Using the final regression model, residual analysis was conducted. *RS/Discover* recommends using studentized residuals which are computed as follows:

$$e^*_i = \frac{e_i}{S_i\sqrt{1 - h_i}}$$

where e^*_i = studentized residual

e_i = $Y_i - \hat{Y}_i$ (residual)

S_i = estimate of residual standard deviation based upon omitting the ith observation from the model

h_i = leverage of the ith data point, defined as the ith diagonal element of the matrix $H = X(X^T X)^{-1} X^T$

Run	Chamfer	Weld Power	Weld Tip Chamfer	Push Out
1	3	1	3	238
2	3	1	3	254
3	2	2	1	798
4	2	2	1	888
5	1	3	1	1461
6	1	3	1	1537
7	3	1	1	242
8	3	1	1	300
9	1	1	3	1262
10	1	1	3	1263
11	3	3	3	469
12	3	3	3	505
13	1	3	3	870
14	1	3	3	1310
15	1	1	1	1107
16	1	1	1	1228
17	3	3	1	420
18	3	3	1	540
19	2	1	2	752
20	2	1	2	903
21	1	2	2	1335
22	1	2	2	1357
23	2	2	3	946
24	2	2	3	1055
25	2	1	2	823

Table 2 12-Run D-optimal Design

Term		Coeff.	Std. Error	T-value	Signif.
1	1	846.745669	26.984354		
2	~C	-448.250481	31.498649	-14.23	0.001
3	~W	67.380895	30.815300		
4	~WT	-17.450000	29.746559		
5	~W×WT	-59.000000	33.257665	-1.77	0.0913

No. Cases = 25 R^2 = 0.9137 RMS Error = 133

Resid. df = 20 Adj. R^2 = 0.8964 Cond. No. = 1.183

~ indicates factors are transformed.

Table 3 Analysis

According to Draper and Smith [1] and [2], e^*_i is distributed approximately as a t-statistic. Therefore, if this assumption is correct, values for $|e^*_i|$ should be ≤ 3.0. Values greater than 3.0 suggest further evaluation may be required.

Two simple (but useful) graphics used to evaluate e^*_i are histograms and case order plots. See Figures 2 and 3. The histogram in Figure 2 suggests one unusual case ($|e^*_i| > 6$) and the case order plot in Figure 3 indicates Run 13 is the run with an unusual residual. A review of the raw data indicates Run 13 and Run 14 (repetitions of the same combination) have large variability compared to all the other replicates. The engineering group conducted repeated tests of this combination and confirmed the large variability for this combination. Further investigation revealed the following: After the canister is pressurized, the BB-sized ball is placed over the fill port and the weld rod is lowered on top of the ball (see Figure 1). A spring behind the weld rod maintains contact while the ball deforms during the welding process. If the weld rod comes too close to the end of the canister, it is possible for the weld current to arc across the gap. This reduces the energy passing through the ball, resulting in a weaker weld. The amount of strength degradation will be highly variable due to the nature of the arcing phenomenon. Therefore, process development engineering continued the analysis to maximize weld strength while avoiding a process setting where arcing is likely.

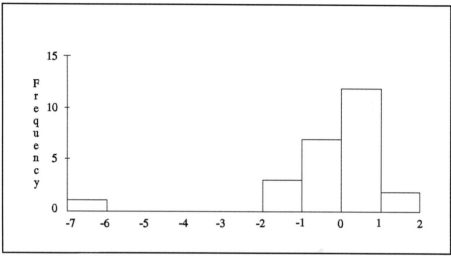

Figure 2 Histogram of Residuals Using Studentized Residuals

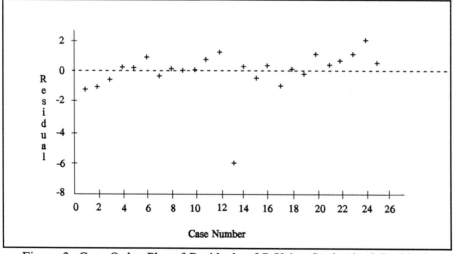

Figure 3 Case Order Plot of Residuals of P Using Studentized Residuals

Summary

Experimental design is a powerful tool for modeling a process. If we continue to torture the data through residual analysis, additional information can be gained. This example did not contain a large amount of data; however, blending experimental design with residual analysis and engineering knowledge resulted in a more complete understanding of the process which led to a better quality product.

References for the Ball Weld Case Study

1. Draper and Smith, *Applied Regression Analysis*.

2. *RS Explore Statistical Appendixes*, BBN Software Products, Cambridge, MA.

APPLICATION OF DOE TO AN ANALOG FILTER DESIGN

Submitted by:

David M. Fisk
Defense Systems & Electronics Group
Texas Instruments, Inc.

Abstract

A designed experiment was performed using a *SPICE* analog circuit simulator. Empirical models generated from this experiment were used to choose optimum values and tolerances for the components of an analog filter. Multiple responses were optimized simultaneously using response surface analysis. Performance defects per unit (DPU) were estimated and compared for the original and optimized circuits.

Introduction

Standard design practice calls for the designer to create a point design from the filter specifications, choose values and tolerances of components based on availability, and either build several to see if they work or run Monte Carlo simulations to predict DPU. If all responses satisfy their requirements you're a hero. If not, you must use engineering judgement, past experience, or considerable analytical effort to choose new values and tolerances that will work. This case study demonstrates the use of a designed experiment to choose component values and tolerances eliminating the need for the guess work. The result is a new design process incorporating statistical analysis and Six Sigma that can be applied to all types of analog designs. Using designed experiments in this way will result in higher quality designs and lower design cycle times.

Circuit Requirements

The objective is to design a passive analog filter circuit to be used in parallel video channels of a receiver. The most important characteristics of the filter are its pulse response, gain, and the relative delay between the two channels. Table 1 outlines the requirements. A bandwidth of 4 MHz is a goal, not a requirement. The filter is to be implemented in through-hole printed wire board

(PWB) technology using a minimum number of components at the lowest possible cost and DPU.

Response	Target	Tolerance
Rise Time	90 nsec	± 10 nsec
Delay Matching	0.0 nsec	± 5 nsec
Overshoot	0	+ 1%
Bandwidth	~ 4 MHz	−
DC Gain	0.50	± 0.05
DC Input Impedance	100 ohms	± 20 ohms

Table 1 Filter Requirements

Circuit Design

The goals and requirements shown above were used to define the architecture of the filter. Figure 1 shows a schematic of the ideal circuit. Component values were selected based on the bandwidth, gain, and input impedance goals.

Figure 1 Circuit Schematic

Table 2 shows the initial component values. A Bessel response was chosen to make it possible to meet the overshoot requirement. A three pole design was used to minimize part count. And, since inductor tolerances are typically wider than capacitor tolerances, a single inductor design was selected to minimize variation. The values of PWB stray and non-ideal component parameters were estimated and included in the simulation.

Component	Target	Tolerance
R1	50 ohms	—
C1	1753 pF	—
L1	1.93 uH	—
C2	268 pF	—
R2	50 ohms	—
R source	1 ohm	—
R etch	0.16 ohm	—
C etch	1 pF	—
L etch	5 nH	—
R capacitor	10 Meg ohms	—
C inductor	1 pF	—

Table 2 Initial Component Values

Performance Analysis

A single run was made in *SPICE* to determine if the values shown in Table 2 allow all responses to satisfy their requirements simultaneously. Table 3 shows the results of this run compared to the requirements.

Response	Result	Requirements	
Rise Time	87 nsec	90 nsec	$\pm$ 10 nsec
Delay Matching	0.0 nsec	0.0 nsec	$\pm$ 5 nsec
Overshoot	1.23%	0.0 nsec	+ 1%
DC Gain	0.50	0.50	$\pm$ 0.05
DC Input Impedance	100 ohms	100 ohms	$\pm$ 20 ohms

Table 3 Single *SPICE* Run Results

The results shown in Table 3 point out two problems. The most obvious is that the overshoot requirement is exceeded. A new set of component values must be selected which satisfies the overshoot specification as well as the requirements of the other responses. The second problem is that the single run provides no information regarding response variation. In order to predict the DPU for each response some information about variation must be obtained.

Variation was introduced in the simulation in two ways. Tolerances were assigned to the component values defining a range over which they could be varied in a Monte Carlo procedure. A uniform distribution between the tolerance limits was assumed for each component. Variation was also introduced by assigning tolerances to the estimated PWB effects and non-ideal component parameters. Uniform distributions between the tolerance limits were assumed. Table 4 shows the tolerances chosen for the components and the stray effects.

Table 5 shows the results of 200 Monte Carlo runs in *SPICE*. Only the four most critical responses were analyzed. The combined DPU for these four responses is 0.134 corresponding to a 3.32 sigma design, see reference [2]. The goal for each response is a C_p of 2.0 or greater and a C_{pk} of 1.5 or greater. Rise time and gain requirements are satisfied with considerable margin. Overshoot and delay performance must be improved.

Component	Target	Tolerance
R1	50 ohms	1%
C1	1753 pF	1%
L1	1.93 uH	5%
C2	268 pF	1%
R2	50 ohms	1%
R source	1 ohm	50%
R etch	0.16 ohm	62.5%
C etch	1 pF	100%
L etch	5 nH	100%
R capacitor	10 Meg ohms	33.3%
C inductor	1 pF	100%

Table 4 Initial Component Tolerances

Response	Estimated Mean	Estimated Standard Deviation	C_p/C_{pk}
Rise Time	89.4 nsec	1.06 nsec	3.15/2.96
Delay Matching	0.0 nsec	1.22 nsec	1.37/1.37
Overshoot	0.714%	0.257%	1.30/0.37
DC Gain	0.488	0.0036	4.63/3.52

Table 5 *SPICE* Monte Carlo Results

Design of Experiments (DOE) - Screening

DOE was used to determine ways to improve the circuit's overshoot and delay performance while maintaining rise time and gain capability. An iterative approach to DOE was used which first involved screening the circuit components to determine which have the greatest effect on the responses.

Screening DOE - CNX

The five primary circuit components were selected to be the controlled factors (X's) in the experiment. The rise time and voltage levels of the source were held constant (C's). And the remaining stray effects were chosen to be the noise factors (N's).

Screening DOE - Experimental Space

Table 6 shows the range over which each controlled factor was varied during the experiment. These ranges were selected to be wider than typical device tolerances so that a broad range of circuit options was explored. At this stage no attempt was made to choose standard component values. Table 4 defines the range over which the noise factors were allowed to vary during the experiment.

Component	Minimum	Maximum
R1	45 ohms	55 ohms
C1	1577 pF	1928 pF
L1	1.544 uH	2.316 uH
C2	241 pF	295 pF
R2	45 ohms	55 ohms

Table 6 Controlled Factor Ranges

NOTE: *Screening designs are typically used for ≥ 6 factors. Since this example will provide a generic DOE template for future users, the author elected to conduct the screening portion with only 5 factors, whereas we could have jumped straight to the modeling DOE.*

Screening DOE - Matrix Selection

A Taguchi L_{12} matrix was selected for this experiment; see reference [3]. A smaller matrix could have been used to screen five factors but this matrix has the unique property that the effects of interactions between factors are spread somewhat evenly across all columns. The factors were assigned to the first five columns of the matrix, resulting in the design shown in Table 7.

						Rise Time		Overshoot		Gain		Delay	
Run	R1	C1	L1	C2	R2	$\bar{y}$	s	$\bar{y}$	s	$\bar{y}$	s	$\bar{y}$	s
1	−	−	−	−	−	74.97	.407	1.44	.107	.489	.0035	58.03	.228
2	−	−	−	−	−	75.02	.436	1.47	.146	.488	.0037	58.29	.369
3	−	−	+	+	+	85.78	.445	1.78	.066	.537	.0017	69.86	.268
4	−	+	−	+	+	90.21	.442	.011	.024	.537	.0038	65.20	.372
5	−	+	+	−	+	100.9	.562	.508	.059	.536	.0023	74.69	.242
6	−	+	+	+	−	98.06	.303	2.80	.119	.489	.0038	76.93	.144
7	+	−	+	+	−	88.80	.419	4.36	.152	.442	.0032	72.56	.242
8	+	−	+	−	+	90.61	.552	1.54	.133	.488	.0029	70.72	.167
9	+	−	−	+	+	80.60	.460	.024	.041	.490	.0025	61.86	.375
10	+	+	+	−	−	102.4	.600	3.02	.143	.441	.0023	77.85	.306
11	+	+	−	+	−	91.39	.594	.006	.020	.443	.0018	67.67	.138
12	+	+	−	−	+	101.8	.745	.006	.018	.489	.0024	66.49	.233

Table 7 Taguchi L_{12} Screening Design and Response Statistics

Screening DOE - Experiment Execution

This 12-run experiment was repeated ten times (total of 120 runs). The large number of repetitions increased the probability of detecting small shifts in standard deviation; see reference [1]. For each run and repetition the noise factors were varied randomly according to uniform distributions. The controlled factors were set such that a − in the matrix indicates a minimum level (as shown in Table 6) and a + indicates a maximum level. No variation was applied to the control factor settings. Running the experiment this way allows estimation of how the controlled factors affect both the mean and variance of the responses. Response sensitivity to the noise factors can also be estimated.

Screening DOE - Analysis

Analysis of variance (ANOVA) was done for each response. Table 8 shows an ordering of the relative importance of each factor to both the mean and variance of the responses. All five controlled factors are important to more than one of the responses. The noise factors had little to no effect on the responses when compared to the effects of the controlled factors. These results tell us that all significant factors have been identified. Significant terms in the ANOVAs for standard deviation point to possible interactions or higher order terms which may be exploited to reduce variance. Empirical equations could be derived from these results and used for circuit optimization. However, if strong interactions and/or higher order terms do exist, these equations from a screening design may not be very accurate.

Response	Factor Order of Importance				
$\bar{y}$ (mean)	1	2	3	4	5
Rise Time	C1	L1	R1	R2	C2
Delay Matching	L1	C1	R1	C2	R2
Overshoot	L1	R2	C1		
DC Gain	R2	R1			
s (standard deviation)					
Rise Time	R2	L1	C2	C1	R1
Delay Matching	-				
Overshoot	C1	L1	C2		
DC Gain	R1				

Table 8 Factors Affecting Response $\bar{y}$ and s

Modeling DOE

The next step in this iterative approach involved the use of a higher resolution experimental matrix. This matrix allowed the analyst to estimate the effects due to factor interactions and second order terms. Empirical equations developed from regression analysis were used to determine optimum values and tolerances for the five controlled factors.

Modeling DOE - CNX

The results from the screening experiment confirmed that all five control factors are significant to circuit performance. The screening experiment also confirmed that the noise factors do not contribute significantly to circuit performance. Therefore, CNX for modeling remained the same as in screening.

Modeling DOE - Experimental Space

The experimental space also remained the same as in the screening experiment. Table 6 shows the ranges for the control factors and Table 4 shows the ranges for the noise factors.

Modeling DOE - Matrix Selection

A 5-factor Central Composite Design (CCD) was selected for this experiment. A half-fraction (16 runs) factorial section was used with seven center points and ten axial points (total of 33 runs). The number of center points is calculated based on the number of factors and the number of runs in the factorial section; see reference [1]. The control factors were arranged in the matrix so that a resolution V design was realized. This matrix allowed estimation of linear and quadratic effects, and linear 2-way interactions, for the control factors. Table 9 shows the design matrix.

					Rise Time (nsec)		Overshoot (nsec)		Gain		Delay (%)	
Run	R1	C1	L1	C2	R2	$\bar{y}$	s	$\bar{y}$	s	$\bar{y}$	s	$\bar{y}$	s
1	−	−	−	−	+	76.24	.327	57.68	.152	0.000	.000	.539	.0031
2	−	−	−	+	−	73.56	.423	59.36	.184	1.420	.028	.489	.0031
3	−	−	+	−	−	87.51	.240	69.09	.219	3.977	.038	.487	.0034
4	−	−	+	+	+	85.79	.376	70.03	.195	1.769	.029	.535	.0026
5	−	+	−	−	−	87.98	.407	64.06	.306	0.297	.034	.486	.0024
6	−	+	−	+	+	90.66	.680	65.27	.313	0.000	.000	.535	.0030
7	−	+	+	−	+	100.4	.341	74.71	.141	0.506	.040	.538	.0024
8	−	+	+	+	−	97.49	.375	76.76	.211	2.849	.042	.490	.0032
9	+	−	−	−	−	78.18	.486	60.02	.128	1.280	.062	.440	.0032
10	+	−	−	+	+	80.21	.681	61.77	.308	0.000	.000	.490	.0025
11	+	−	+	−	+	90.77	.251	70.69	.284	1.517	.050	.491	.0030
12	+	−	+	+	−	88.66	.242	72.53	.161	4.363	.058	.440	.0030
13	+	+	−	−	+	101.7	.423	66.57	.176	0.000	.000	.491	.0012
14	+	+	−	+	−	90.88	.177	67.45	.121	0.000	.000	.441	.0030
15	+	+	+	−	−	102.8	.255	78.16	.178	3.080	.093	.440	.0026
16	+	+	+	+	+	104.2	.716	79.16	.167	0.000	.000	.490	.0036
17	0	0	0	0	0	89.51	.645	68.67	.132	0.689	.047	.488	.0032
18	0	0	0	0	0	89.54	.108	68.80	.248	0.772	.085	.488	.0040
19	0	0	0	0	0	89.70	.544	68.70	.200	0.706	.034	.488	.0032
20	0	0	0	0	0	89.44	.433	68.70	.182	0.731	.065	.490	.0018
21	0	0	0	0	0	89.54	.377	68.71	.171	0.720	.038	.488	.0037
22	0	0	0	0	0	89.24	.352	68.72	.132	0.740	.028	.488	.0024
23	0	0	0	0	0	89.19	.468	68.73	.247	0.738	.039	.489	.0035
24	+2	0	0	0	0	93.54	.299	70.77	.185	0.396	.072	.443	.0008
25	−2	0	0	0	0	84.52	.555	66.02	.241	1.089	.036	.543	.0031
26	0	+2	0	0	0	104.4	.585	74.91	.094	0.000	.000	.487	.0021
27	0	−2	0	0	0	75.84	.313	61.90	.158	2.422	.058	.488	.0031
28	0	0	+2	0	0	100.3	.281	78.84	.171	3.076	.063	.491	.0031
29	0	0	−2	0	0	82.77	.679	56.75	.376	0.000	.000	.487	.0035
30	0	0	0	+2	0	87.82	.579	70.02	.121	0.541	.064	.490	.0030
31	0	0	0	−2	0	91.24	.512	67.15	.144	1.018	.065	.490	.0027
32	0	0	0	0	+2	91.82	.478	68.44	.231	0.000	.000	.535	.0030
33	0	0	0	0	−2	87.28	.318	68.88	.267	4.070	.075	.435	.0024

Table 9 Design Matrix and Response Statistics

Modeling DOE - Experimental Execution

This experiment was repeated four times (total of 132 runs). This number of repetitions follows the guideline for detecting shifts in standard deviation of 2.72 or greater; see reference [1]. For each run and repetition the noise factors were varied randomly according to uniform distributions. The controlled factors were set such that a − in the matrix indicates a minimum level (as shown in Table 9) and a + indicates a maximum level. The zeroes in the matrix correspond to midpoints between the minimum and maximum for each factor. Fewer center points could have been used since the experiment was repeated four times. The axial points represent factor values set to twice the difference between midpoint and the +/− levels. The factor of two is computed based on the number of runs in the factorial section; see reference [1]. No variation was applied to the control factor settings.

Modeling DOE - Analysis

Regression analysis was done to determine the significant terms and their coefficients for both y and s. The empirical equations derived for each response are shown below for coded values of the factors.

Rise Time (units are nsec)

$\hat{y} = 89.49$		$\hat{s} = 0.429$	
+7.176	C1	+0.080	C2R2
+4.715	L1	+0.063	R2
+2.325	R1	−0.067	L1
+1.331	R2		
−0.877	C2		
−0.852	L1R2		
+0.792	C1R2		
−0.682	C1L1		
+0.520	R1C1		
+0.471	L1^2		
−0.458	R1L1		
−0.329	C1C2		
−0.311	R1C2		
+0.211	L1C2		
−0.152	C2R2		
−0.149	R1^2		
+0.115	C1^2		

Delay (units are nsec)

$\hat{y} = 68.711$

+5.547	L1
+3.207	C1
+1.204	R1
+0.711	C2
−0.222	L1^2
−0.146	L1R2
+0.119	C1L1
+0.113	C2R2
+0.105	R1C1
−0.103	R2
+0.099	R1R2
−0.071	R1^2
−0.070	C1^2
−0.066	C1C2

$\hat{s} = 0.187$

+0.029	R1R2
−0.029	R1C1
+0.024	C2R2
−0.023	L1
+0.022	L1^2

Overshoot (units are percent)

$\hat{y} = 0.740$

−0.901	R2
+0.884	L1
−0.518	C1
−0.467	L1R2
+0.313	R2^2
−0.189	L1^2
−0.174	C1L1
−0.173	R1C2
+0.127	C1R2
−0.113	C1C2
+0.107	C1^2
−0.082	R1
−0.059	R1R2
−0.051	C2

$\hat{s} = 0.050$

−0.016	R2
+0.015	L1
+0.008	R1C2
−0.007	C1
−0.007	C1^2
−0.006	L1^2

Gain

$\hat{y} = 0.489$

+0.025	R2
−0.024	R1
+0.001	R1^2
−0.001	R2^2

$\hat{s} = 0.003$

Circuit Design Optimization

The coefficients of the terms in the ŝ models are small relative to the intercepts. In general, only those coefficients that are greater than half the value of the intercept would be considered significant; see reference [1]. This indicates that there is little to be gained by using the ŝ models to select values and tolerances of the control factors. The ŷ models were used to select optimum values and tolerances for the main factors.

A process of elimination was used to limit the number of factors involved in the response surface plots. According to the relative size of regression coefficients shown previously, C2 contributes the least to the four responses. R1 and R2 control the gain but play minor roles in the other responses. C1 and L1 drive three of the four responses, making them good choices for the response surface analysis.

A standard value and tolerance was chosen for C2 near the midpoint of its experimental range. Standard values and tolerances were selected for R1 and R2 based on the empirical equation for Gain. With these values fixed, response surfaces for rise time, delay, and overshoot were drawn in terms of C1 and L1. These three plots were overlayed and a region was defined where all three responses satisfied their requirements simultaneously (see Figure 2). Coded values

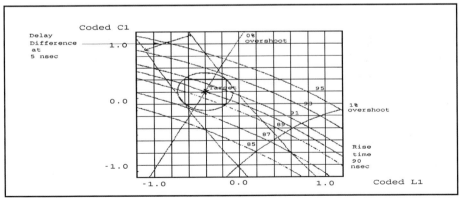

Figure 2 Overlayed Contour Plots to Find Optimal Settings

and tolerances for C1 and L1 were selected to fall within that region. Table 10 shows the final circuit actual values and tolerances.

Component	Target	Tolerance
R1	49.9 ohms	1%
C1	1800 pF	1%
L1	1.80 uH	5%
C2	270 pF	1%
R2	52.3 ohms	1%
R source	1 ohm	50%
R etch	0.16 ohm	62.5%
C etch	1 pF	100%
L etch	5 nH	100%
R capacitor	10 Meg ohms	33.3%
C inductor	1 pF	100 %

Table 10 Final Component Values, Tolerances

Another set of 200 Monte Carlo runs was made in *SPICE* using the values and tolerances shown in Table 10. Table 11 shows the resulting response statistics. Variance has been reduced for rise time, delay, and overshoot. Overshoot, rise time, and DC gain are all closer to their target values than in the original design. The combined DPU for the four responses has been reduced from 0.134 to 6.09E-6 corresponding to a 6.15 sigma design; see reference [2].

Response	Estimated Mean	Estimated Standard Deviation	C_p/Cpk
Rise Time	90.3 nsec	0.79 nsec	4.24/4.11
Delay Matching	0.0 nsec	1.14 nsec	1.46/1.46
Overshoot	0.000002 %	0.000016 %	> 10/ > 10
DC Gain	0.500	0.0036	4.63/4.63

Table 11 *SPICE* Monte Carlo Results

Conclusion

Statistical performance analysis was performed on an analog filter circuit using *SPICE*. Performance DPU for the circuit was estimated using Six Sigma analysis; see reference [2]. DOE was utilized to finalize the design of the circuit component values and tolerances. The process illustrated in this case study provides an efficient solution to a circuit design problem and insures the quality of that solution.

References for the Analog Filter Design Case Study

1.   Stephen R. Schmidt & Robert G. Launsby, *Understanding Industrial Designed Experiments*, Air Academy Press, Colorado Springs, CO, 1992.

2.   M.J. Harry & R.J. Lawson, "Six Sigma Producibility Analysis and Process Characterization," Motorola University, Schaumburg, IL, 1990.

3.   G. Taguchi & S. Konishi, *Taguchi Methods Orthogonal Arrays and Linear Graphs*, American Supplier Institute, Inc., Dearborn, MI, 1987.

A Robust Active Filter Design

Submitted by: Dan Warwick and Tom Cheek
Texas Instruments, Dallas, TX

Introduction

This study demonstrates a method of applying DOE techniques to board-level circuit design. The methodology consists of three parts: determining robust conditions, testing component tolerances, and specifying those tolerances. A commonly used operational amplifier (op amp) circuit will be employed for this demonstration.

Performance Specification

The chosen circuit for study is a low-pass filter. The requirements are that the filter have unity gain at low frequencies, deviate from unity by no more than 3dB until after 0.55kHz, to fall off to less than -20dB by 1.5kHz, and stay at less than -40dB beyond 15kHz. The four performance objectives chosen were

Figure 1 Low-Pass Filter Gain Requirements and Performance Objectives

taken directly from the specifications which appear in graphical form in Figure 1 as A) the -3dB frequency, B) the magnitude response at 1.5kHz, C) the maximum magnitude response beyond 15kHz, and D) the amount of peaking in the magnitude response prior to 0.55kHz. Further requirements were that the process capabilities (Cp's) for each performance objective be greater than 2, and that all parts should be from a standard parts list [1] if possible. Additionally, the number of components not on the standard parts list be kept to a minimum.

The constraint of using standard parts will introduce some of the variations inherent in any realizeable circuit design. Parts are usually not available in exactly the values derived by analytic calculations. In addition, they vary within their tolerance limits. These two effects cause much of the variation of performance objectives about their targeted means. How to deal with these problems was explored.

Initial Design Approach

A two-pole active filter of the type shown in the dashed box area of Figure 2 was chosen as the initial design architecture. Traditional designs for this type of filter involve looking up base values (e.g. C_1" and C_2" in Table 1) for the

Peak	ζ	C_1"	C_2"
2 dB	.4431	2.672 F	.5246 F

Table 1   Example Base Value Look-Up Table [2]

components in a table according to the characteristics of the frequency response. The actual component values are then derived from the corner frequency and resistor value (R), as shown in the following equations:

$$C_1 = \frac{C_1''}{2\pi f_c R} \qquad\qquad C_2 = \frac{C_2''}{2\pi f_c R}$$

This procedure, which is based on ideal components and ideal loads, then yields values which need to be "tweaked" for the particular circuit. The effects of non-ideal components and loads are reflected by that portion of Figure 2 which appears outside the dashed box, as well as by the fact that a detailed macro model of the op amp was used during simulations.

Figure 2 Schematic for the 2-Pole Low-Pass Filter to be Designed

Circuit Simulation

The circuit simulator used for the experiment was *PSPICE*, version 5.4 [3]. The circuit network listing appears in Figure 3; a goal function file and command file were used to automate the extraction of performance objectives. Concatenating the circuit files for each experimental condition (each row in the design matrix) allowed *PSPICE* to generate a table of performance objective values which could then be brought into *Q-EDGE*.

Experimental Design and Modeling

The methodology of the experiment is taken from the flow [4] presented in Figure 4. Three factors were determined to affect the performance objectives: the damping factor (ζ) of the filter, the quantity R used to derive the capacitor values, and the value used for resistors R1 and R2. Since the damping factor has a direct affect on the peaking in the magnitude response, peaking was chosen as a more direct way of specifying this control factor.

To simplify the example, resistors R1 and R2 were given the same value as the R used to derive capacitor values C1 and C2. Thus, the two experiment control factors chosen were the amount of peaking in decibels, P; and the resistor value used, R. The amount of peaking was used to look up the base values for C_1'' and C_2'' in the table. In turn, these base values were scaled by R and the filter corner frequency to obtain component values for C_1 and C_2. In this way, the four variables R_1, R_2, C_1, and C_2 were set by only two control factors.

```
* A low-pass active filter circuit *

* Op Amp Supplies
VPOS    VCC     0       DC      13.5v
VNEG    VEE     0       DC      -13.5v

* Signal Input Circuit
VSIG    1       0       AC1V
Rss     1       IN      73.1
Rps     IN      0       142.5
Cs      IN      0       .0026uF

* Load Circuit
Rload   OUT     0       4.32k
Cload   OUT     0       .01uF

* Filter
*R1     IN      FBCK    RMOD    10
R1      IN      FBCK    RMOD    13k
*R2     FBCK    PIN     RMOD    10
R2      FBCK    PIN     RMOD    13k
*C1     OUT     FBCK    CMOD    80.79uF
C1      OUT     FBCK    CMOD    69.07nF
*C2     PIN     0       CMOD    13.29uF
C2      PIN     0       CMOD    10.15nF
X741    PIN     OUT     VCC     VEE     OUT     UA741

* Modeling
.LIB MAC.MDL
.MODEL RMOD RES(R=1)
.MODEL CMOD CAP(C=1)

* Command Statements
.OPTIONS NOPAGE
.AC DEC 100 1Hz  1MegHz
.PROVE V([OUT]) V([IN])
.END
```

Figure 3 *PSPICE* Circuit List File Used to Perform the Experiments

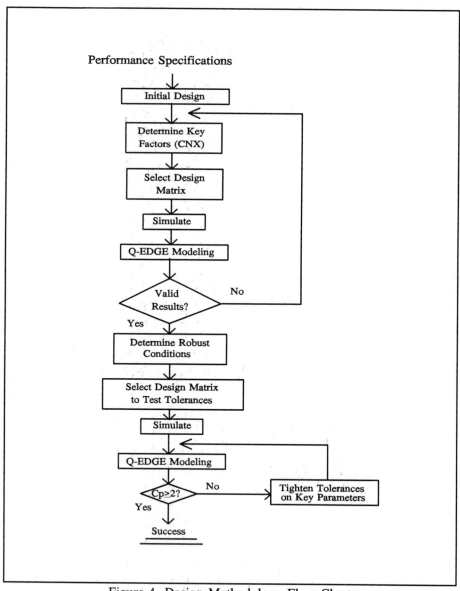

Figure 4 Design Methodology Flow Chart

R	P	$R_{1,2}$	C_1	C_2
−	−	10Ω	40.92μF	20.46μF
−	+	10Ω	80.79μF	13.19μF
+	−	13kΩ	31.48nF	15.74nF
+	+	13kΩ	69.07nF	10.15nF
0	0	6.495kΩ	109.2nF	25.08nF

Table 2   Initial Design Matrix and Component Values

A full-factorial design covers this experimental space and results in a design matrix and component values as shown in Table 2. The low and high settings for the R factor were chosen so as to provide a practical range of values for capacitors C_1 and C_2. Factor P high and low settings were 0dB and 3dB. The performance objective values obtained from the initial experiment appear in Table 3. The confirmation run at the center point shows both the *PSPICE* value and the value predicted by the *Q-EDGE* model. Clearly responses B and C show significant nonlinearities.

To model the nonlinearities, the experimental matrix was expanded from the original, full-factorial design to include center and axial points sufficient to create a central composite design (CCD). The circuit files were then reconstructed and the results again recorded in *Q-EDGE*. This time the model generated by *Q-EDGE* confirmed the *PSPICE* simulations. At this point, the flow proceeded to determining the region of feasibility and the robust conditions.

R P	Y_A,Hz	Y_B,dB	Y_C,dB	Y_D,dB
− −	549	−17.7	−6.43	0.00
− +	639	−20.4	−6.43	3.02
+ −	549	−27.5	−42.21	0.00
+ +	642	−19.9	−42.24	3.00
PSPICE 0 0	606	−19.8	−36.32	1.50
Q-EDGE 0 0	595	−18.9	−24.3	1.51
% err	−2	−5	−33	+1

Table 3 Initial Experiment Results

Determining Robust Conditions

A region of feasibility is a region of the control factor space which meets the design requirements (i.e., the range over which R and P can vary while still meeting the specifications). The robust conditions are then determined as being the values for R and P within this region which minimize variation in the performance objectives.

Q-EDGE allows the generation of contour plots for each of the objectives in terms of two control factors. Figure 5 is the *Q-EDGE* screen display of such a plot (Figures 7 and 8 are also *Q-EDGE* screen displays). Perhaps the easiest method of determining a region of feasibility is to sketch the specification limits for each objective on a piece of transparency plastic laid over each *Q-EDGE* contour plot. Figure 6 represents the result of overlaying the specification limit contours of all four performance objectives.

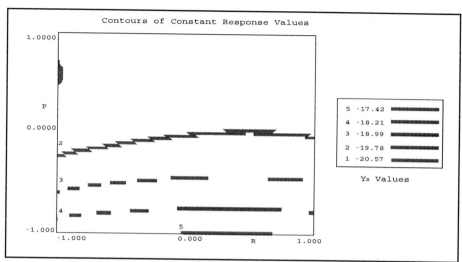

Figure 5 *Q-EDGE* Contour Plot for the Magnitude Response at 1.5kHz
(Output B)

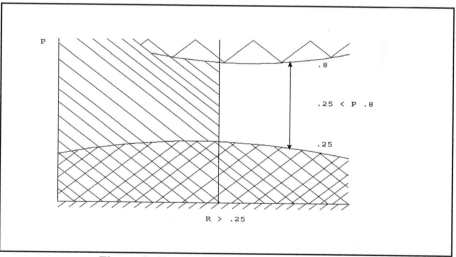

Figure 6 Sketch of the Region of Feasibility

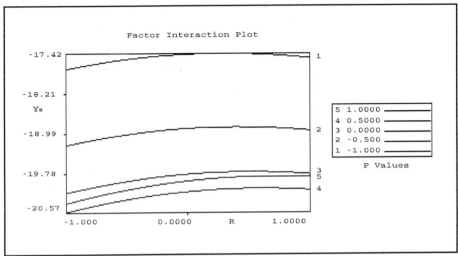

Figure 7 Response at 1.5kHz(y-axis) vs
R Values (x-axis) Over a Range of P Values

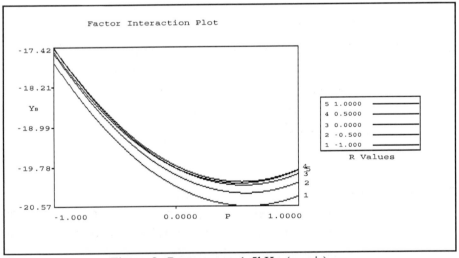

Figure 8 Response at 1.5kHz (y-axis) vs
P Value (x-axis) Over a Range of R Values

Determining the factor settings which maximize robustness is then a matter of choosing the values which minimize variation in the performance objectives due to variations in the factor values. Figures 7 and 8 show the *Q-EDGE* interaction plots of output B over a range of R and P settings. The most robust conditions then, for this performance objective, would be for coded R and P values of about +0.5 each. After considering the effects for each of the other objectives, a coded R value of +0.7 and a coded P value of +0.35 were selected, which translated to the component values in Table 4. These component values are not on the standard parts list and thus create the need to evaluate the impact of deviating from the robust factor settings.

R	P	C_1	C_2
11kΩ	2dB	70.29nF	13.8nF

Table 4   Robust Factor Settings and Component Values

Another designed experiment was conducted about the target parts values, and the set of standard values which gave the largest range of acceptable variation was chosen. This range was calculated as the difference between each of the nominal performance objective values and their specification limits. The details of this experiment are not included; however, the resulting standard component values were used in the following experiment to determine the effect of variation within their tolerances.

Testing Tolerances

In order to determine the effect of component values varying within their tolerances, an experiment using each component value as a control factor was necess ary. This resulted in a 16-run, full-factorial experiment which allowed R1, R2, C1, and C2 to vary within their widest available tolerances. The results of these simulations were then entered in *Q-EDGE*. Using the model predict capability, a model of the performance objectives in terms of individual component variations was created.

Range due to:						
	$R_1 \pm 1\%$	$R_2 \pm 1\%$	$C_1 \pm 10\%$	$C_2 \pm 10\%$	$R_1 * C_2$	TOTAL
Y_A	5.74	5.74	40.2	75.4	0	127.1
Y_B	.18	.18	1.76	1.86	0	3.98
Y_C	.14	0	0	.05	.05	.24
Y_D	0	0	.67	.66	0	1.33

Table 5 Effect of Component Variation on Performance Objectives

Setting Component Specifications

The *Q-EDGE* model was used to determine the range of variation in each performance objective due to variation in component values. Table 5 summarizes this information. This model was then used to determine the necessary specifications for the parts.

Performance objective Cp's of 2 or greater were determined to be achievable only by specifying tighter tolerances than standard for components C1 and C2. This result was arrived at through manipulation of the *Q-EDGE* model predict capability. Using this capability, the fact that the resistors used could vary in the widest range available for that part value was also determined. The final parts tolerances appear as Table 6, along with the estimates of Cp for each performance objective based on this component selection.

Part	Spec.	Performance Objective	Cp
$R_{1,2}$	$10.7 k\Omega \pm 1\%$	Y_A	3.58
C_1	$82nF \pm 1\%$	Y_B	2.83
C_2	$15nF \pm 1\%$	Y_C	3.35
		Y_D	11.25

Table 6   Final Parts Specifications and Performance Cp's

Conclusion

A method of applying DOE techniques to circuit design has been shown which is practical and which incorporates robustness to parts variations. The transition from initial design to final parts selection was shown along with results from the analyses used to make those selections. A robust design was achieved using only two parts (C_1 and C_2) with non-standard tolerances while allowing the remaining parts to have the widest tolerances available. Finally, this design met the requirement that all 4 performance objectives have Cp's of 2 or greater.

References for the Robust Active Filter Design Case Study

1.   Texas Instruments Inc., "Master Engineering Parts List, 2.14", Sept 1991.

2.   David Stout and Milton Kaufman, *Handbook of Operational Amplifier Circuit Design*, McGraw-Hill Book Company, 1976.

3.   *PSPICE*, MicroSim Corp., 1993.

4.   Dr. Steve Schmidt, "Basic Experimental Design", Air Academy Associates, June 1993.

Identifying Design Standards for Improved PWA Solderability

Submitted by: Cliff Polo, Tara Kroger, Maurice Berryman,
Jim Grayson, and Lisa Reagan
Texas Instruments

Abstract

As part of a strategy to apply Design for Manufacturability (DFM) principles to the definition of electronics hardware design standards, design of experiments (DOE) techniques were used to define and validate the criticality of selected printed wiring board (PWB) parameters impacting assembly solderability. A screening experiment and a follow-on higher resolution experiment were performed to analyze the effects and interactions of six PWB design and assembly process parameters on Mil Std 2000 solder defect rates. The results indicate that the design parameters of plane layer location, number of plane layers, external layer configuration and the interactions of these parameters were the most influential effects on PWB solderability. Additionally, flow solder process parameters were found to be benign factors when compared to the designed-in attributes. These results suggest that flow solder parameter selection cannot overcome board design parameters that are detrimental to good solderability.

Problem

With system packaging constraints requiring increasingly dense printed wiring assemblies (PWA's) and customer quality requirements moving to higher reliability levels, existing PWB design guidelines have failed to yield assemblies that meet both of these requirements. A case in point at Texas Instruments is the experience that a proprietary weapon system program had with the solderability of two mixed technology PWA's. The system packaging concept necessitated combining RF and analog/digital circuitry on both PWA's. Preferred RF design practices require high levels of noise reduction, which can result in high ground plane layer count and redundant ground pin terminations.

When the designs went to assembly, it was discovered that the PWA's had poor solderability. Solder defect rates were 25-30% and many of the defects could not be reworked or repaired to meet the quality requirements of Mil Std 2000. A quality improvement team (QIT) was formed to review the PWB design specification and electronic design automation (EDA) process flow and to generate corrective action plans to address high reliability solder specification compliance deficiencies inherent in the process. The testing described here is an outgrowth of this effort.

A series of designed experiments were ultimately run to determine the most significant factors affecting PWB solderability. The initial screening experiment will be discussed here.

Goals and Objectives of the Experiment

It was understood from the beginning that a series of experiments would be needed to gain a better understanding of the relationships between the PWB design parameters and assembly solderability. The goal of this initial experiment was to identify the significant board design and/or process parameters (within the limitations of the military standards) that would most affect producibility and process optimization. Follow-on modeling experiments were then conducted. The overall objective of the series of experiments was to generate information for making recommendations on design configurations to reduce solder defect rates.

Factors and Levels

To begin, brainstorming meetings were held resulting in the development of the fishbone diagram shown in Figure 1, which details the factors affecting board level solderability. The fishbone diagram includes both board design parameters as well as flow solder (process) parameters. Using this fishbone, other data gathered by the QIT and results from previous experimentation, six factors were ultimately chosen for study and are shown in Table 1 along with their levels.

Experimental Design

The team was unsure about what interactions might exist, so it was decided that an initial screening design (Resolution III) would be used. Since it was desired to study most of the factors at 3 levels, an L18 ("Taguchi") orthogonal array was selected and is shown in Table 2. This design also allows for the estimation of the interaction between the number of internal ground planes (IP) and the location of the internal ground planes (LOC), which was considered reasonable. A comparable 2-level design, the 12-run Plackett Burman design, would have allowed for only 2 levels for each factor. A 16-run design (2^{6-2}) would have resulted in a higher resolution (res IV) design but again only for 2-level factors.

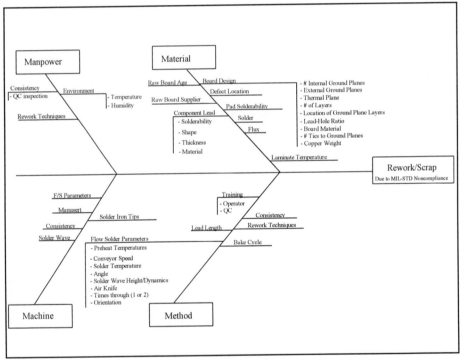

Figure 1 Fishbone Diagram

Factor	Name	Description	Levels		
			1	2	3
A	IP	# internal planes	2	4	-
B	LOC	location of internal planes	top	middle	bottom
C	EXT	# ties to external plane	0	2	4
D	INT	# ties to internal plane	2	4	-
E	TEMP	average topside temp (°F)	180	220	260
F	CONV	conveyor speed (ft/min)	2.6	2.9	3.2

Table 1 Factors and Levels

Run No.	A	B	C	D	E	F	y_1	y_2	y_3
1	1	1	1	1	1	1	68	51	59
2	1	1	2	2	2	2	267	165	72
3	1	1	3	3=2	3	3	287	85	41
4	1	2	1	1	2	2	52	47	54
5	1	2	2	2	3	3	33	57	31
6	1	2	3	3=2	1	1	54	77	56
7	1	3	1	2	1	3	48	39	60
8	1	3	2	3=2	2	1	47	61	54
9	1	3	3	1	3	2	36	61	70
10	2	1	1	3=2	3	2	247	156	114
11	2	1	2	1	1	3	220	304	84
12	2	1	3	2	2	1	252	282	56
13	2	2	1	2	3	1	46	63	67
14	2	2	2	3=2	1	2	217	57	57
15	2	2	3	1	2	3	56	74	69
16	2	3	1	3=2	2	3	53	38	61
17	2	3	2	1	3	1	37	51	64
18	2	3	3	2	1	2	74	76	54

Table 2 L18 Orthogonal Array

Data Collection and Sample Size

The PWB's used in the experiment were built at TI. Components were purchased from surplus TI stocks and were solderability tested to minimize component solderability variations. Thermal profiles were completed on the 18 board designs represented in the L18. The purpose of this was to determine the appropriate preheat temperature settings to achieve the desired topside laminate temperature. An in-process temperature monitoring device, MOLE (multichannel occurrent logger evaluator), was run with each board utilizing 5 thermocouples at predetermined board locations. These measurements were averaged to obtain the topside temperature. Boards designed with exterior ground planes were abraded at the thermocouple locations so that only the laminate temperature was measured. All process parameters not included in the L18 experiment were maintained as constants throughout the experiment.

Three boards were chosen to be run for each of the 18 experimental runs in accordance with the sample size rules of thumb in reference [1]. This sample size would be large enough to detect a change in the average defect rate as small as 1.5 σ with 95% confidence ($\alpha = .05$, $\beta = .05$). According to the rule of thumb,

$$ n_r = \frac{64}{n\Delta^2} = \frac{64}{14(1.5)^2} = 2.04 \cong 3 $$

where n = average number of $+$'s and $-$'s per column in the design = 14.

So, three boards were run for each of the 18 runs defined in the experimental design.

Quality Characteristics

The quality characteristic measured was the number of solder defects per board which was determined by visual examination under 10X magnification per Mil Std 2000 specifications. Defects were recorded by defect type (bridging, peaks, insufficient solder, etc.) but only the analysis for the total number of defects will be discussed here. One QC inspector inspected and documented all solder defects to minimize the effects of operator variability during the conduct of the experiment.

Results

The analysis of variance for the total number of defects, shown in Table 3, reveals that the most significant factors were the number of internal planes (IP) and the location of the internal plane layers (LOC). The level average graphs are shown in Figure 2. After IP and LOC, the IP X LOC interaction and the number of ties to external plane (EXT) are the next most significant effects, although much smaller.

Source	df	SS	MS	F	p-value
IP	1	14900.2	14900.2	4.21	0.0466
LOC	2	112355.8	56177.9	15.88	0.0001
EXT	2	9498.5	4749.2	1.34	0.2725
INT	1	3223.1	3223.1	0.91	0.3454
TEMP	2	1272.3	636.1	0.18	0.8361
CONV	2	5175.6	2587.8	0.73	0.4874
IP X LOC	2	9847.0	4923.5	1.39	0.2602
error	41	145053.6	3537.9		
Corr. Total	53	301326.1			

Table 3 Analysis of Variance Table

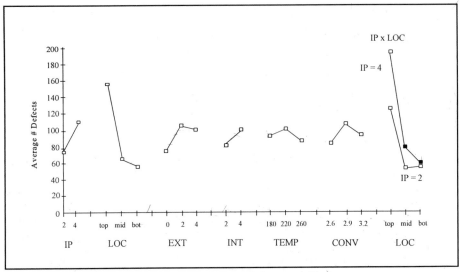

Figure 2 Factor Level Average Plots

From the graphs, we see that the location of the internal planes (LOC) at the top of the board stack (level 1) produces a much greater average defect rate than when placed in the middle (level 2) however, the difference between the middle (level 2) and the bottom location (level 3) is not as great. Also, 2 internal planes seem to result in significantly fewer defects per board.

Based on this analysis, settings were recommended for the significant factors which would minimize the total defect level. Other (non significant) factor levels were chosen based on minimizing the process cycle time and risk of board damage. The "optimum" (recommended) settings were identified as;

Factor	Level
IP	2
LOC	bottom
EXT	0 (pads only)
INT	2 way ties
TEMP	180
CONV	3.2 ft/min

Although a screening design is not generally used for modeling, a equation (using coded factor levels) for estimating the number of defects per board was generated as follows:

$$\hat{y}=64.83+(16.61*IP)-(50.72*LOC)-(16.33*IP*LOC)+(12.14*EXT)+(40.56*LOC^2)$$

The expected defect level (per board), using the prediction equation above was calculated as follows:

$$\hat{y}=64.83+16.61(-1)-50.72(+1)-16.33(-1)(+1)+12.14(-1)+40.56(+1)(+1)=42.2$$

This represents a defect rate of 5.2%. It is important to note that the process parameters examined in the experiment were found to be benign factors when compared to the board design characteristics.

It should be noted that a similar analysis was performed using standard deviation (of the defect measurements) as the quality characteristic. Only the location of the internal planes (LOC) significantly affected the standard deviation and the level which minimized variation was consistent with that recommended above.

Confirmation

A confirmation run was subsequently completed using 20 boards designed at the "optimized" levels shown above. The boards were flow soldered and inspected. An overall average defect level of 39.4 defects/board was found, corresponding to a 4.9% defect rate, which correlated very closely to the 5.2% predicted defect rate and was within a calculated prediction interval.

Therefore, the confirmation run validated the results obtained in the L18 screening experiment. This gave the team confidence in the experimental results and reinforced the recommended board design configurations which would result in reduced solder defect rates.

Follow-on Experimentation

To more closely define limits for interactions among the active PWB design parameters, a follow-on study is being pursued with the number of plane layers (IP), the location of the plane layers (LOC), and the condition of the external layers (EXT).

Summary

DOE methods were able to be successfully applied to define the effects of PWB design characteristics and solder process parameters on assembly solderability performance. The solder process parameters were seen as much less significant as compared with the board design parameters. That is, they can neither compensate for nor overcome designed-in PWB attributes. This emphasizes the need to consider assembly solderability during the PWB layout/routing cycles. DOE will play a continued and important role in helping to form design configuration recommendations in the future.

Assembly and PWB Fab Producibility Engineering have initiated efforts to update the Corporate PWB Design Specifications to reflect the design constraints discovered as a result of this testing. In addition, Assembly Producibility and EDA Support engineers are currently incorporating the results of this testing in a PWB design post processing routine. It is intended to identify and flag areas of assembly producibility risk early in the PWB design cycle to minimize system design cost and cycle time.

These results have only uncovered the tip of the iceberg in understanding the link between PWB design and assembly solderability. There are several other PWB design issues that are currently under investigation or planned for future analysis.

References for the PWA Solderability Case Study

1. Stephen R. Schmidt and Robert G. Launsby, *Understanding Industrial Designed Experiments*, 4th Edition, Air Academy Press, 1994.

2. Sammy G. Shina, "The Successful Use of the Taguchi Method to Increase Manufacturing Process Capability," *Quality Engineering*, Vol. 3, No. 3, 1991, pp. 333-349.

3. Beth Morgan and Lisa Reagan, "The Role of Experimental Design in Quality Improvement," *TI Technical Journal*, Vol. 7, No. 2, March-April 1990, pp. 55-61.

CHEMILUMINESCENCE:
CAN WE RULE OUT 3-WAY INTERACTIONS?

Submitted by: Karen Welch
Abbott Laboratories

Background

To produce light efficiently from a chemical reaction (chemiluminescence), you need to add peroxide to a basic solution with a pH of approximately 13. To date this reaction is produced using tablets that are dissolved in Sodium Hydroxide to create an **activator** solution. The inability to mass produce the tablets suggests evaluating the feasibility and optimization of a liquid formulation. The use of a liquid will also increase efficiency by decreasing the required preparation time. This experiment will examine the concentration of the reagents used in the liquid and examine the two different pump types utilized within the diagnostic instrument.

Experiment

The specific factors to be tested are:

A: DTPA - Reagent C - Binds to and thus inactivates metal ion contaminates which would react with Reagent A and interfere with the chemical reaction.

(Low Setting = 0.02% and High Setting = 1.0%)

B: Hydrogen Peroxide (H_2O_2) - Reagent A - Active ingredient for chemiluminescence.

(Low Setting = 0.4% and High Setting = 0.8%)

C: Sodium Hydroxide (NaOH) - Reagent B - Creates the basic environment.

(Low Setting = 0.4 Normal and High Setting = 0.6 Normal)

D: Pump Type   -   1) 50 microliter (ul)
                    -    2) 85 microliter (ul)

The response is the number of photons which is a measure of light intensity. The goal is to match or exceed the performance for the current tablet process. The experimental matrix shown in Table 1 consists of a 4-factor, 2-level, full factorial with 2 center points (one for each pump type). Although 8 replicates for each combination were obtained, the analysis is conducted only on the mean $(\bar{y})$ and standard deviation (s). Evaluation of the $\bar{y}$ ANOVA (see Table 2) indicates that the important terms are the B, C, and D linear effects, and the AB, BD and ABD interactions. The R^2 of 0.9796 is good although it is higher than the true R^2 because we are only analyzing the mean response $(\bar{y})$. The regression coefficients are shown in Table 3. The Pareto chart of effects in Figure 1 indicates that the largest effect is that of the ABD interaction.

Response: # of Photons

Run	A:DTPA	B:H$_2$O$_2$	C:NaOH	D:Pump Type	$\bar{y}$	s
1	0.02 (0)	0.6 (0)	0.5 (0)	50 (−)	44479	690.9
2	0.01 (−)	0.4 (−)	0.4 (−)	50 (−)	40488	961.0
3	0.03 (+)	0.4 (−)	0.4 (−)	50 (−)	40550	712.1
4	0.01 (−)	0.8 (+)	0.4 (−)	50 (−)	52047	1417.0
5	0.03 (+)	0.8 (+)	0.4 (−)	50 (−)	51478	805.8
6	0.01 (−)	0.4 (−)	0.6 (+)	50 (−)	34322	545.1
7	0.03 (+)	0.4 (−)	0.6 (+)	50 (−)	34898	615.5
8	0.01 (−)	0.8 (+)	0.6 (+)	50 (−)	45581	889.8
9	0.03 (+)	0.8 (+)	0.6 (+)	50 (−)	45484	1132.0
10	0.01 (−)	0.4 (−)	0.4 (−)	85 (+)	48788	1976.0
11	0.03 (+)	0.4 (−)	0.4 (−)	85 (+)	43385	1374.0
12	0.01 (−)	0.8 (+)	0.4 (−)	85 (+)	43227	1178.0
13	0.03 (+)	0.8 (+)	0.4 (−)	85 (+)	55486	1578.0
14	0.01 (−)	0.4 (−)	0.6 (+)	85 (+)	55235	2196.0
15	0.03 (+)	0.4 (−)	0.6 (+)	85 (+)	36600	1029.0
16	0.01 (−)	0.8 (+)	0.6 (+)	85 (+)	36968	702.0
17	0.03 (+)	0.8 (+)	0.6 (+)	85 (+)	48765	1438.0
18	0.02 (0)	0.6 (0)	0.5 (0)	85 (+)	48099	1564.0

Table 1 Experimental Matrix

Effect	SS	DF	MS	F-Ratio	P-Value
A: DTPA	210578	1	2.11E+05	0.04	0.8532
B: H_2O_2	125272056	1	1.25E+08	24083	0.0155
C: NaOH	88341201	1	8.83E+07	17.51	0.0249
D: Pump Type	41652843	1	4.17E+07	8.25	0.0639
AB	135831506	1	1.37E+08	27.12	0.0138
AC	10093329	1	1.01E+07	2.00	0.2522
AD	201694	1	2.02E+05	0.04	0.8563
BC	11029041	1	1.10E+07	2.19	0.2358
BD	120417702	1	1.20E+08	23.86	0.0164
CD	7507600	1	7.51E+06	1.49	0.3097
ABC	10126124	1	1.01E+07	2.01	0.2516
ABD	152510150	1	1.53E+08	30.22	0.0118
ACD	13468900	1	1.35E+07	2.67	0.2008
BCD	9000000	1	9.00E+06	1.78	0.2740
Total Error	15138512	3	5.05E+06		

Total (corr.)   741419830   17

$R^2 = .9796^*$ Adj. $R^2 = .8843$

*For all the data (16 replicates) the $R^2 = .9677$

Table 2 Analysis of Variance (ANOVA)

constant	=	44581.300
A:DTPA	=	-0.625
B:H$_2$O$_2$	=	2798.120
C:NaOH	=	-2349.700
D:Pump Type	=	1475.370
AB	=	2924.370
AC	=	-794.250
AD	=	2.875
BC	=	-830.250
BD	=	-2743.300
CD	=	685.000
ABC	=	795.500
ABD	=	3087.370
ACD	=	-917.500
BCD	=	-750.000
ABCD	=	800.750

Table 3 Coded Regression Coefficients for $\bar{y}$ Analysis

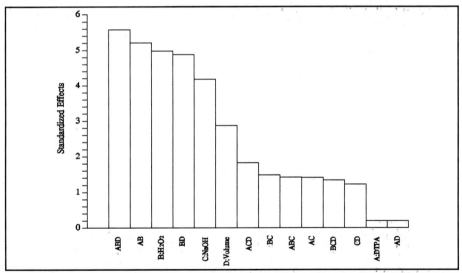

Figure 1 Pareto Chart for Mean Analysis Using Standardized Effects

The response surface plots (see Figure 2) indicate that when pump type is constant at 85ul, the strong interaction between H_2O_2 and DTPA is evident. At low H_2O_2 the response DECREASES as you increase DTPA concentration. At high H_2O_2, the response INCREASES as you increase DTPA. On the other hand, holding the pump at 50ul, (see Figure 3), you see no interaction. In other words, the 2-way interaction between H_2O_2 and DTPA depends on the third factor, pump type. This 3-way interaction is rare; yet the discovery of it provides important information towards optimizing the process.

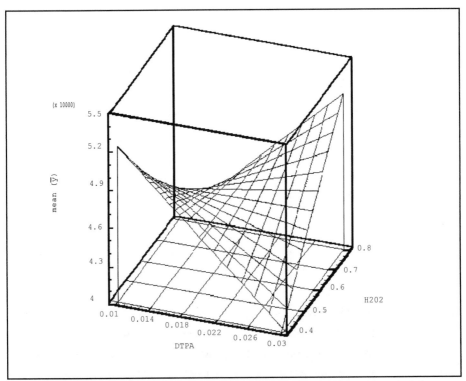

Figure 2 Estimated Response Function with Pump Constant at 85ul

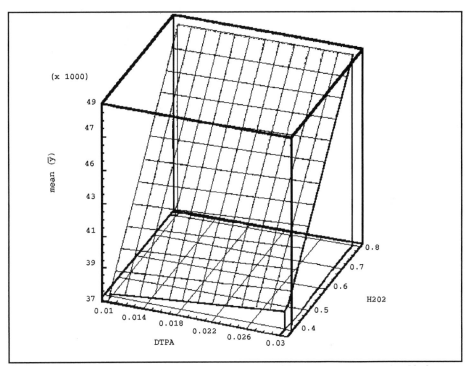

Figure 3 Estimated Response Function with Pump Constant at 50ul

The s regression analysis and Pareto diagram (see Table 4 and Figure 4) indicate that factor D: Pump Type was the most significant factor. Other important effects appear to be ABD, BD, and AB, similar to the interactions affecting $\overline{y}$. The end result is that changing the pump type from 85ul (+) to 50ul (−) will substantially decrease the standard deviation. (See Figures 5 and 6.)

Based on the previous findings, the decision to repeat the experiment holding the pump type constant at 50ul was made. This change in volume did eliminate numerous interactions and reduce the standard deviation. The $\overline{y}$ analysis was then used to adjust the reagent settings to match the tablet performance and set the proper tolerances for each reagent.

constant	=	1159.330
A:DTPA	=	-73.781
B:H$_2$O$_2$	=	-16.756
C:NaOH	=	-90.906
D:Pump Type	=	274.543
AB	=	169.656
AC	=	58.980
AD	=	-11.220
BC	=	-5.340
BD	=	-193.110
CD	=	-1.720
ABC	=	89.660
ABD	=	193.470
ACD	=	-87.610
BCD	=	89.690
ABCD	=	22.930

Table 4 Coded Regression Coefficients for s Analysis

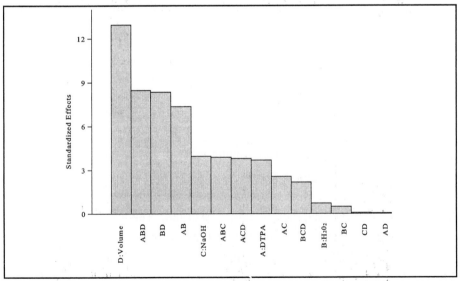

Figure 4 Pareto Chart for s Analysis of the Standardized Effects

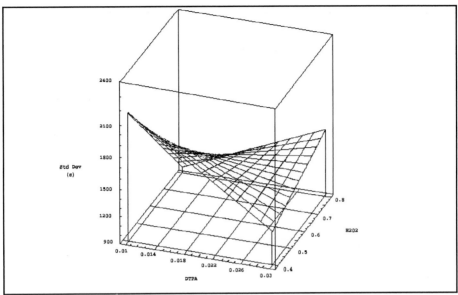

Figure 5 Estimated Response Function with Pump Constant at 85ul

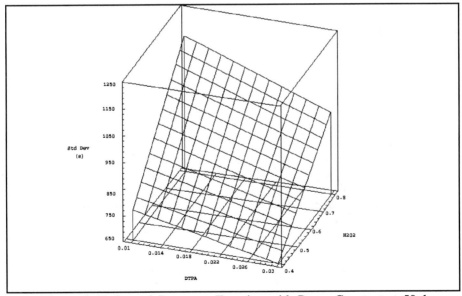

Figure 6 Estimated Response Function with Pump Constant at 50ul

Summary

Benefits gained from this DOE were:

- Identified a very significant 3-way interaction that would not have been found in a one-factor-at-a-time experiment.

- Using the 50ul pump type decreased the variance by half.

- Confirmed that the reagents do not interact with the 50ul pump.

- Determined reagent settings to match tablet performance.

- Scientifically set tolerances on the reagents.

Appendixes

Appendix A	**Table of Random Digits**
Appendix B	**Binomial Probability Distributions**
Appendix C	**Poisson Probability Distributions**
Appendix D	**Standardized Normal Distribution**
Appendix E	**Percentiles of the t Distribution**
Appendix F	**Percentiles of the F Distribution**
Appendix G	**Percentiles of the χ^2 Distribution**
Appendix H	**Studentized Ranges q_T**
Appendix I	**Power of the Analysis-of-Variance Test**
Appendix J	**Answers to Selected Problems**
Appendix K	**Time Line for Design of Experiment Techniques**
Appendix L	**List of Symbols and Abbreviations**
Appendix M	**Rules of Thumb**

Row	1-5	6-10	11-15	16-20	21-25	26-30	31-35
1	50527	38063	89093	22648	96867	34685	62753
2	40483	25946	46848	63507	86984	05724	73793
3	80462	82084	21308	94040	77165	29466	52079
4	90545	19515	78767	05365	88790	57080	41826
5	19169	99595	73714	19238	72251	05922	87552
6	52665	26336	40764	26694	19988	67596	36786
7	90799	50016	47988	52706	16346	31800	00971
8	45285	63834	42650	90117	15536	72206	44917
9	19441	42344	28192	99346	23158	19639	48645
10	37222	08805	23852	40256	85211	32672	64952
11	37065	26209	66719	27944	16944	81393	35310
12	48899	17553	47175	31199	26672	19188	38099
13	44622	87968	72342	43257	54973	60454	12025
14	73638	18241	52230	89038	67734	44774	55787
15	42221	90634	01822	75261	08195	60196	28252
16	89921	11815	52797	74868	34720	20249	95911
17	38555	79182	87003	42410	09189	29731	44611
18	26213	44234	84459	21895	01186	21979	70360
19	68455	81564	04625	69299	58598	41757	58267
20	91866	70002	70036	66806	00693	89570	22355
21	99347	68962	57338	30529	02209	82909	71087
22	82928	44796	04134	13362	69602	24334	86744
23	16242	87679	08615	30668	50641	63391	00152
24	71385	29319	42870	79158	67826	32664	21638
25	51420	85638	02017	31406	16928	66618	74314
26	07700	56663	07642	72029	15147	43861	48231
27	38514	82498	84649	38830	02403	07300	42212
28	39756	23408	97266	33946	58018	78874	25099
29	12796	42649	09878	25483	30689	39905	99100
30	93589	01789	63970	76873	33782	38460	87506
31	11213	10296	36181	92910	14471	21381	04219
32	61233	66409	94223	11695	36021	92241	85479
33	98421	75296	41506	71427	57443	64683	94253
34	04977	06872	34764	00911	95711	00664	08983
35	59298	13735	95985	56378	94783	54725	05701
36	15017	53836	00873	37602	85937	05872	74263
37	29311	95124	47110	83427	44799	68966	49907
38	82609	13408	68318	84181	36411	20749	83699
39	39952	93967	91401	50899	75843	47751	55027
40	09731	99395	53883	53855	88758	14556	83042
41	45821	84519	07493	12282	80953	13641	67947
42	84309	87119	70436	73462	98107	82617	35553
43	02929	44362	01024	67928	43415	79721	67653
44	42989	49536	64420	38132	58927	32576	51847
45	63063	49898	06019	98905	37156	84088	64254
46	57588	27694	01911	79714	87368	96074	90567
47	26093	88111	84279	85312	51340	23309	98691
48	73300	95180	55945	73516	31579	81353	15724
49	54085	91495	91575	46931	54054	18622	64911
50	76838	77175	69384	09011	69902	88684	81569

Column

Binomial Probabilities calculated from: $P(x) = \binom{n}{x} p^x (1-p)^{n-x}$

n	x	.01	.02	.03	.04	.05	.06	.07	.08	.09		
2	0	.9801	.9604	.9409	.9216	.9025	.8836	.8649	.8464	.8281	2	
	1	.0198	.0392	.0582	.0768	.0950	.1128	.1302	.1472	.1630	1	
	2	.0001	.0004	.0009	.0016	.0025	.0036	.0049	.0064	.0081	0	2
3	0	.9703	.9412	.9127	.8847	.8574	.8306	.8044	.7787	.7536	3	
	1	.0294	.0576	.0847	.1106	.1354	.1590	.1816	.2031	.2236	2	
	2	.0003	.0012	.0026	.0046	.0071	.0102	.0137	.0177	.0221	1	
	3	.0000	.0000	.0000	.0001	.0001	.0002	.0003	.0005	.0007	0	3
4	0	.9606	.9224	.8853	.8493	.8145	.7807	.7481	.7164	.6857	4	
	1	.0388	.0753	.1095	.1416	.1715	.1993	.2252	.2492	.2713	3	
	2	.0006	.0023	.0051	.0088	.0135	.0191	.0254	.0325	.0402	2	
	3	.0000	.0000	.0001	.0002	.0005	.0008	.0013	.0019	.0027	1	
	4	.0000	.0000	.0000	.0000	.0000	.0000	.0000	.0000	.0001	0	4
5	0	.9510	.9039	.8587	.8154	.7738	.7339	.6957	.6591	.6240	5	
	1	.0480	.0922	.1328	.1699	.2036	.2342	.2618	.2866	.3086	4	
	2	.0010	.0038	.0082	.0142	.0214	.0299	.0394	.0498	.0610	3	
	3	.0000	.0001	.0003	.0006	.0011	.0019	.0030	.0043	.0060	2	
	4	.0000	.0000	.0000	.0000	.0000	.0001	.0001	.0002	.0003	1	
	5	.0000	.0000	.0000	.0000	.0000	.0000	.0000	.0000	.0000	0	5
6	0	.9415	.8858	.8330	.7828	.7351	.6899	.6470	.6064	.5679	6	
	1	.0571	.1085	.1546	.1957	.2321	.2642	.2922	.3164	.3370	5	
	2	.0014	.0055	.0120	.0204	.0305	.0422	.0550	.0688	.0833	4	
	3	.0000	.0002	.0005	.0011	.0021	.0036	.0055	.0080	.0110	3	
	4	.0000	.0000	.0000	.0000	.0001	.0002	.0003	.0005	.0008	2	
	5	.0000	.0000	.0000	.0000	.0000	.0000	.0000	.0000	.0000	1	
	6	.0000	.0000	.0000	.0000	.0000	.0000	.0000	.0000	.0000	0	6
7	0	.9321	.8681	.8080	.7514	.6983	.6485	.6017	.5578	.5168	7	
	1	.0659	.1240	.1749	.2192	.2573	.2897	.3170	.3396	.3578	6	
	2	.0020	.0076	.0162	.0274	.0406	.0555	.0716	.0886	.1061	5	
	3	.0000	.0003	.0008	.0019	.0036	.0059	.0090	.0128	.0175	4	
	4	.0000	.0000	.0000	.0001	.0002	.0004	.0007	.0011	.0017	3	
	5	.0000	.0000	.0000	.0000	.0000	.0000	.0000	.0001	.0001	2	
	6	.0000	.0000	.0000	.0000	.0000	.0000	.0000	.0000	.0000	1	
	7	.0000	.0000	.0000	.0000	.0000	.0000	.0000	.0000	.0000	0	7
8	0	.9227	.8508	.7837	.7214	.6634	.6096	.5596	.5132	.4703	8	
	1	.0746	.1389	.1939	.2405	.2793	.3113	.3380	.3570	.3721	7	
	2	.0026	.0099	.0210	.0351	.0515	.0695	.0888	.1087	.1288	6	
	3	.0001	.0004	.0013	.0029	.0054	.0089	.0134	.0189	.0255	5	
	4	.0000	.0000	.0001	.0002	.0004	.0007	.0013	.0021	.0031	4	
	5	.0000	.0000	.0000	.0000	.0000	.0000	.0001	.0001	.0002	3	
	6	.0000	.0000	.0000	.0000	.0000	.0000	.0000	.0000	.0000	2	
	7	.0000	.0000	.0000	.0000	.0000	.0000	.0000	.0000	.0000	1	
	8	.0000	.0000	.0000	.0000	.0000	.0000	.0000	.0000	.0000	0	8
9	0	.9135	.8337	.7602	.6925	.6302	.5730	.5204	.4722	.4279	9	
	1	.0830	.1531	.2116	.2597	.2985	.3292	.3525	.3695	.3809	8	
	2	.0034	.0125	.0262	.0433	.0629	.0840	.1061	.1285	.1507	7	
	3	.0001	.0006	.0019	.0042	.0077	.0125	.0186	.0261	.0348	6	
	4	.0000	.0000	.0001	.0003	.0006	.0012	.0021	.0034	.0052	5	
	5	.0000	.0000	.0000	.0000	.0000	.0001	.0002	.0003	.0005	4	
	6	.0000	.0000	.0000	.0000	.0000	.0000	.0000	.0000	.0000	3	
	7	.0000	.0000	.0000	.0000	.0000	.0000	.0000	.0000	.0000	2	
	8	.0000	.0000	.0000	.0000	.0000	.0000	.0000	.0000	.0000	1	
	9										0	9
		.99	.98	.97	.96	.95	.94	.93	.92	.91	x	n

p

n	x	.10	.15	.20	.25	.30	.35	.40	.45	.50	x	n
2	0	.8100	.7225	.6400	.5625	.4900	.4225	.3600	.3025	.2500	2	
	1	.1800	.2550	.3200	.3750	.4200	.4550	.4800	.4950	.5000	1	
	2	.0100	.0225	.0400	.0625	.0900	.1225	.1600	.2025	.2500	0	2
3	0	.7290	.6141	.5120	.4219	.3430	.2746	.2160	.1664	.1250	3	
	1	.2430	.3251	.3840	.4219	.4410	.4436	.4320	.4084	.3750	2	
	2	.0270	.0574	.0960	.1406	.1890	.2389	.2880	.3341	.3750	1	
	3	.0010	.0034	.0080	.0156	.0270	.0429	.0640	.0911	.1250	0	3
4	0	.6561	.5220	.4096	.3164	.2401	.1785	.1296	.0915	.0625	4	
	1	.2916	.3685	.4096	.4219	.4116	.3845	.3456	.2995	.2500	3	
	2	.0486	.0975	.1536	.2109	.2646	.3105	.3456	.3675	.3750	2	
	3	.0036	.0115	.0256	.0469	.0756	.1115	.1536	.2005	.2500	1	
	4	.0001	.0005	.0016	.0039	.0081	.0150	.0256	.0410	.0625	0	4
5	0	.5905	.4437	.3277	.2373	.1681	.1160	.0778	.0503	.0312	5	
	1	.3280	.3915	.4096	.3955	.3601	.3124	.2592	.2059	.1562	4	
	2	.0729	.1382	.2048	.2637	.3087	.3364	.3456	.3369	.3125	3	
	3	.0081	.0244	.0512	.0879	.1323	.1811	.2304	.2757	.3125	2	
	4	.0004	.0022	.0064	.0146	.0283	.0488	.0768	.1128	.1562	1	
	5	.0000	.0001	.0003	.0010	.0024	.0053	.0102	.0185	.0312	0	5
6	0	.5314	.3771	.2621	.1780	.1176	.0754	.0467	.0277	.0156	6	
	1	.3543	.3993	.3932	.3560	.3025	.2437	.1866	.1359	.0938	5	
	2	.0984	.1762	.2458	.2966	.3241	.3280	.3110	.2780	.2344	4	
	3	.0146	.0415	.0819	.1318	.1852	.2355	.2765	.3032	.3125	3	
	4	.0012	.0055	.0154	.0330	.0595	.0951	.1382	.1861	.2344	2	
	5	.0001	.0004	.0015	.0044	.0102	.0205	.0369	.0609	.0938	1	
	6	.0000	.0000	.0001	.0002	.0007	.0018	.0041	.0083	.0156	0	6
7	0	.4783	.3206	.2097	.1335	.0824	.0490	.0280	.0152	.0078	7	
	1	.3720	.3960	.3670	.3115	.2471	.1848	.1306	.0872	.0547	6	
	2	.1240	.2097	.2753	.3115	.3177	.2985	.2613	.2140	.1641	5	
	3	.0230	.0617	.1147	.1730	.2269	.2679	.2903	.2918	.2734	4	
	4	.0026	.0109	.0287	.0577	.0972	.1442	.1935	.2388	.2734	3	
	5	.0002	.0012	.0043	.0115	.0250	.0466	.0774	.1172	.1641	2	
	6	.0000	.0001	.0004	.0013	.0036	.0084	.0172	.0320	.0547	1	
	7	.0000	.0000	.0000	.0001	.0002	.0006	.0016	.0037	.0078	0	7
8	0	.4305	.2725	.1678	.1001	.0576	.0319	.0168	.0084	.0039	8	
	1	.3826	.3847	.3355	.2670	.1977	.1373	.0896	.0548	.0312	7	
	2	.1488	.2376	.2936	.3115	.2965	.2587	.2090	.1569	.1094	6	
	3	.0331	.0839	.1468	.2076	.2541	.2786	.2787	.2568	.2188	5	
	4	.0046	.0185	.0459	.0865	.1361	.1875	.2322	.2627	.2734	4	
	5	.0004	.0026	.0092	.0231	.0467	.0808	.1239	.1719	.2188	3	
	6	.0000	.0002	.0011	.0038	.0100	.0217	.0413	.0703	.1094	2	
	7	.0000	.0000	.0001	.0004	.0012	.0033	.0079	.0164	.0312	1	
	8	.0000	.0000	.0000	.0000	.0001	.0002	.0007	.0017	.0039	0	8
9	0	.3874	.2316	.1342	.0751	.0404	.0207	.0101	.0046	.0020	9	
	1	.3874	.3679	.3020	.2253	.1556	.1004	.0605	.0339	.0176	8	
	2	.1722	.2597	.3020	.3003	.2668	.2162	.1612	.1110	.0703	7	
	3	.0446	.1069	.1762	.2336	.2668	.2716	.2508	.2119	.1641	6	
	4	.0074	.0283	.0661	.1168	.1715	.2194	.2508	.2600	.2461	5	
	5	.0008	.0050	.0165	.0389	.0735	.1181	.1672	.1928	.2461	4	
	6	.0001	.0006	.0028	.0087	.0210	.0424	.0743	.1160	.1641	3	
	7	.0000	.0000	.0003	.0012	.0039	.0098	.0212	.0407	.0703	2	
	8	.0000	.0000	.0000	.0001	.0004	.0013	.0035	.0083	.0176	1	
	9	.0000	.0000	.0000	.0000	.0000	.0001	.0003	.0008	.0020	0	9
		.90	.85	.80	.75	.70	.65	.60	.55	.50	x	n

					p						
n	x	.01	.02	.03	.04	.05	.06	.07	.08	.09	
10	0	.9044	.8171	.7374	.6648	.5987	.5386	.4840	.4344	.3894	10
	1	.0914	.1667	.2281	.2770	.3151	.3438	.3643	.3777	.3851	9
	2	.0042	.0153	.0317	.0519	.0746	.0988	.1234	.1478	.1714	8
	3	.0001	.0008	.0026	.0058	.0105	.0168	.0248	.0343	.0452	7
	4	.0000	.0000	.0001	.0004	.0010	.0019	.0033	.0052	.0078	6
	5	.0000	.0000	.0000	.0000	.0001	.0001	.0003	.0005	.0009	5
	6	.0000	.0000	.0000	.0000	.0000	.0000	.0000	.0000	.0001	4
	7	.0000	.0000	.0000	.0000	.0000	.0000	.0000	.0000	.0000	3
	8	.0000	.0000	.0000	.0000	.0000	.0000	.0000	.0000	.0000	2
	9	.0000	.0000	.0000	.0000	.0000	.0000	.0000	.0000	.0000	1
	10	.0000	.0000	.0000	.0000	.0000	.0000	.0000	.0000	.0000	0	10
12	0	.8864	.7847	.6938	.6127	.5404	.4759	.4186	.3677	.3225	12
	1	.1074	.1922	.2575	.3064	.3413	.3645	.3781	.3837	.3827	11
	2	.0060	.0216	.0438	.0702	.0988	.1280	.1565	.1835	.2082	10
	3	.0002	.0015	.0045	.0098	.0173	.0272	.0393	.0532	.0686	9
	4	.0000	.0001	.0003	.0009	.0021	.0039	.0067	.0104	.0153	8
	5	.0000	.0000	.0000	.0001	.0002	.0004	.0008	.0014	.0024	7
	6	.0000	.0000	.0000	.0000	.0000	.0000	.0001	.0001	.0003	6
	7	.0000	.0000	.0000	.0000	.0000	.0000	.0000	.0000	.0000	5
	8	.0000	.0000	.0000	.0000	.0000	.0000	.0000	.0000	.0000	4
	9	.0000	.0000	.0000	.0000	.0000	.0000	.0000	.0000	.0000	3
	10	.0000	.0000	.0000	.0000	.0000	.0000	.0000	.0000	.0000	2
	11	.0000	.0000	.0000	.0000	.0000	.0000	.0000	.0000	.0000	1
	12	.0000	.0000	.0000	.0000	.0000	.0000	.0000	.0000	.0000	0	12
15	0	.8601	.7386	.6333	.5421	.4633	.3953	.3367	.2863	.2430	15
	1	.1303	.2261	.2938	.3388	.3658	.3785	.3801	.3734	.3605	14
	2	.0092	.0323	.0636	.0988	.1348	.1691	.1979	.2273	.2496	13
	3	.0004	.0029	.0085	.0178	.0307	.0468	.0653	.0857	.1070	12
	4	.0000	.0002	.0008	.0022	.0049	.0090	.0148	.0223	.0317	11
	5	.0000	.0000	.0001	.0002	.0006	.0013	.0024	.0043	.0069	10
	6	.0000	.0000	.0000	.0000	.0000	.0001	.0003	.0006	.0011	9
	7	.0000	.0000	.0000	.0000	.0000	.0000	.0000	.0001	.0001	8
	8	.0000	.0000	.0000	.0000	.0000	.0000	.0000	.0000	.0000	7
	9	.0000	.0000	.0000	.0000	.0000	.0000	.0000	.0000	.0000	6
	10	.0000	.0000	.0000	.0000	.0000	.0000	.0000	.0000	.0000	5
	11	.0000	.0000	.0000	.0000	.0000	.0000	.0000	.0000	.0000	4
	12	.0000	.0000	.0000	.0000	.0000	.0000	.0000	.0000	.0000	3
	13	.0000	.0000	.0000	.0000	.0000	.0000	.0000	.0000	.0000	2
	14	.0000	.0000	.0000	.0000	.0000	.0000	.0000	.0000	.0000	1
	15	.0000	.0000	.0000	.0000	.0000	.0000	.0000	.0000	.0000	0	15
20	0	.8179	.6676	.5438	.4420	.3585	.2901	.2342	.1887	.1516	20
	1	.1652	.2725	.3364	.3683	.3774	.3703	.3526	.3282	.3000	19
	2	.0159	.0528	.0988	.1458	.1887	.2246	.2521	.2711	.2818	18
	3	.0010	.0065	.0183	.0364	.0596	.0860	.1139	.1414	.1672	17
	4	.0000	.0006	.0024	.0065	.0133	.0233	.0364	.0523	.0703	16
	5	.0000	.0000	.0002	.0009	.0022	.0048	.0088	.0145	.0222	15
	6	.0000	.0000	.0000	.0001	.0003	.0008	.0017	.0032	.0055	14
	7	.0000	.0000	.0000	.0000	.0000	.0001	.0002	.0005	.0011	13
	8	.0000	.0000	.0000	.0000	.0000	.0000	.0000	.0001	.0002	12
	9	.0000	.0000	.0000	.0000	.0000	.0000	.0000	.0000	.0000	11
	10	.0000	.0000	.0000	.0000	.0000	.0000	.0000	.0000	.0000	10
	11	.0000	.0000	.0000	.0000	.0000	.0000	.0000	.0000	.0000	9
	12	.0000	.0000	.0000	.0000	.0000	.0000	.0000	.0000	.0000	8
	13	.0000	.0000	.0000	.0000	.0000	.0000	.0000	.0000	.0000	7
	14	.0000	.0000	.0000	.0000	.0000	.0000	.0000	.0000	.0000	6
	15	.0000	.0000	.0000	.0000	.0000	.0000	.0000	.0000	.0000	5
	16	.0000	.0000	.0000	.0000	.0000	.0000	.0000	.0000	.0000	4
	17	.0000	.0000	.0000	.0000	.0000	.0000	.0000	.0000	.0000	3
	18	.0000	.0000	.0000	.0000	.0000	.0000	.0000	.0000	.0000	2
	19	.0000	.0000	.0000	.0000	.0000	.0000	.0000	.0000	.0000	1
	20	.0000	.0000	.0000	.0000	.0000	.0000	.0000	.0000	.0000	0	20
		.99	.98	.97	.96	.95	.94	.93	.92	.91	x	n
						p						

n	x	.10	.15	.20	.25	.30	.35	.40	.45	.50		
10	0	.3487	.1969	.1074	.0563	.0282	.0135	.0060	.0025	.0010	10	
	1	.3874	.3474	.2684	.1877	.1211	.0725	.0403	.0207	.0098	9	
	2	.1937	.2759	.3020	.2816	.2335	.1757	.1209	.0763	.0439	8	
	3	.0574	.1298	.2013	.2503	.2668	.2522	.2150	.1665	.1172	7	
	4	.0112	.0401	.0881	.1460	.2001	.2377	.2508	.2384	.2051	6	
	5	.0015	.0085	.0264	.0584	.1029	.1536	.2007	.2340	.2461	5	
	6	.0001	.0012	.0055	.0162	.0368	.0689	.1115	.1596	.2051	4	
	7	.0000	.0001	.0008	.0031	.0090	.0212	.0425	.0746	.1172	3	
	8	.0000	.0000	.0001	.0004	.0014	.0043	.0106	.0229	.0439	2	
	9	.0000	.0000	.0000	.0000	.0001	.0005	.0016	.0042	.0098	1	
	10	.0000	.0000	.0000	.0000	.0000	.0000	.0001	.0003	.0010	0	10
12	0	.2824	.1422	.0687	.0317	.0138	.0057	.0022	.0008	.0002	12	
	1	.3766	.3012	.2062	.1267	.0712	.0368	.0174	.0075	.0029	11	
	2	.2301	.2924	.2835	.2323	.1678	.1088	.0639	.0339	.0161	10	
	3	.0852	.1720	.2362	.2581	.2397	.1954	.1419	.0923	.0537	9	
	4	.0213	.0683	.1329	.1936	.2311	.2367	.2128	.1700	.1208	8	
	5	.0038	.0193	.0532	.1032	.1585	.2039	.2270	.2225	.1934	7	
	6	.0005	.0040	.0155	.0401	.0792	.1281	.1766	.2124	.2256	6	
	7	.0000	.0006	.0033	.0115	.0291	.0591	.1009	.1489	.1934	5	
	8	.0000	.0001	.0005	.0024	.0078	.0199	.0420	.0762	.1208	4	
	9	.0000	.0000	.0001	.0004	.0015	.0048	.0125	.0277	.0537	3	
	10	.0000	.0000	.0000	.0000	.0002	.0008	.0025	.0068	.0161	2	
	11	.0000	.0000	.0000	.0000	.0000	.0001	.0003	.0010	.0029	1	
	12	.0000	.0000	.0000	.0000	.0000	.0000	.0000	.0001	.0002	0	12
15	0	.2059	.0874	.0352	.0134	.0047	.0016	.0005	.0001	.0000	15	
	1	.3432	.2312	.1319	.0668	.0305	.0126	.0047	.0016	.0005	14	
	2	.2669	.2856	.2309	.1559	.0916	.0476	.0219	.0090	.0032	13	
	3	.1285	.2184	.2501	.2252	.1700	.1110	.0634	.0318	.0139	12	
	4	.0428	.1156	.1876	.2252	.2186	.1792	.1268	.0780	.0417	11	
	5	.0105	.0449	.1032	.1651	.2061	.2123	.1859	.1404	.0916	10	
	6	.0019	.0132	.0430	.0917	.1472	.1906	.2066	.1914	.1527	9	
	7	.0003	.0030	.0138	.0393	.0811	.1319	.1771	.2013	.1964	8	
	8	.0000	.0005	.0035	.0131	.0348	.0710	.1181	.1647	.1964	7	
	9	.0000	.0001	.0007	.0034	.0116	.0298	.0612	.1048	.1527	6	
	10	.0000	.0000	.0001	.0007	.0030	.0096	.0245	.0515	.0916	5	
	11	.0000	.0000	.0000	.0001	.0006	.0024	.0074	.0191	.0417	4	
	12	.0000	.0000	.0000	.0000	.0001	.0004	.0016	.0052	.0139	3	
	13	.0000	.0000	.0000	.0000	.0000	.0001	.0003	.0010	.0032	2	
	14	.0000	.0000	.0000	.0000	.0000	.0000	.0000	.0001	.0005	1	
	15	.0000	.0000	.0000	.0000	.0000	.0000	.0000	.0000	.0000	0	15
20	0	.1216	.0388	.0115	.0032	.0008	.0002	.0000	.0000	.0000	20	
	1	.2702	.1368	.0576	.0211	.0068	.0020	.0005	.0001	.0000	19	
	2	.2852	.2293	.1369	.0669	.0278	.0100	.0031	.0008	.0002	18	
	3	.1901	.2428	.2054	.1339	.0716	.0323	.0123	.0040	.0011	17	
	4	.0898	.1821	.2182	.1897	.1304	.0738	.0350	.0139	.0046	16	
	5	.0319	.1028	.1746	.2023	.1789	.1272	.0746	.0365	.0148	15	
	6	.0089	.0454	.1091	.1686	.1916	.1712	.1244	.0746	.0370	14	
	7	.0020	.0160	.0545	.1124	.1643	.1844	.1659	.1221	.0739	13	
	8	.0004	.0046	.0222	.0609	.1144	.1614	.1797	.1623	.1201	12	
	9	.0001	.0011	.0074	.0271	.0654	.1158	.1597	.1771	.1602	11	
	10	.0000	.0002	.0020	.0099	.0308	.0686	.1171	.1593	.1762	10	
	11	.0000	.0000	.0005	.0030	.0120	.0336	.0710	.1185	.1602	9	
	12	.0000	.0000	.0001	.0008	.0039	.0136	.0355	.0727	.1201	8	
	13	.0000	.0000	.0000	.0002	.0010	.0045	.0146	.0366	.0739	7	
	14	.0000	.0000	.0000	.0000	.0002	.0012	.0049	.0150	.0370	6	
	15	.0000	.0000	.0000	.0000	.0000	.0003	.0013	.0049	.0148	5	
	16	.0000	.0000	.0000	.0000	.0000	.0000	.0003	.0013	.0046	4	
	17	.0000	.0000	.0000	.0000	.0000	.0000	.0000	.0002	.0011	3	
	18	.0000	.0000	.0000	.0000	.0000	.0000	.0000	.0000	.0002	2	
	19	.0000	.0000	.0000	.0000	.0000	.0000	.0000	.0000	.0000	1	
	20	.0000	.0000	.0000	.0000	.0000	.0000	.0000	.0000	.0000	0	20
		.90	.85	.80	.75	.70	.65	.60	.55	.50	x	n

p (top); p

Poisson Probabilities calculated from: $P(x) = \dfrac{\lambda^x e^{-\lambda}}{x!}$

x	.1	.2	.3	.4	.5	.6	.7	.8	.9
0	.9048	.8187	.7408	.6703	.6065	.5488	.4966	.4493	.4066
1	.0905	.1637	.2222	.2681	.3033	.3293	.3476	.3595	.3659
2	.0045	.0164	.0333	.0536	.0758	.0988	.1217	.1438	.1647
3	.0002	.0011	.0033	.0072	.0126	.0198	.0284	.0383	.0494
4	.0000	.0001	.0003	.0007	.0016	.0030	.0050	.0077	.0111
5	.0000	.0000	.0000	.0001	.0002	.0004	.0007	.0012	.0020
6	.0000	.0000	.0000	.0000	.0000	.0000	.0001	.0002	.0003

x	1.0	1.5	2.0	2.5	3.0	3.5	4.0	4.5	5.0
0	.3679	.2231	.1353	.0821	.0498	.0302	.0183	.0111	.0067
1	.3679	.3347	.2707	.2052	.1494	.1057	.0733	.0500	.0337
2	.1839	.2510	.2707	.2565	.2240	.1850	.1465	.1125	.0842
3	.0613	.1255	.1804	.2138	.2240	.2158	.1954	.1687	.1404
4	.0153	.0471	.0902	.1336	.1680	.1888	.1954	.1898	.1755
5	.0031	.0141	.0361	.0668	.1008	.1322	.1563	.1708	.1755
6	.0005	.0035	.0120	.0278	.0504	.0771	.1042	.1281	.1462
7	.0001	.0008	.0034	.0099	.0216	.0385	.0595	.0824	.1044
8	.0000	.0001	.0009	.0031	.0081	.0169	.0298	.0463	.0653
9	.0000	.0000	.0002	.0009	.0027	.0066	.0132	.0232	.0363
10	.0000	.0000	.0000	.0002	.0008	.0023	.0053	.0104	.0181
11	.0000	.0000	.0000	.0000	.0002	.0007	.0019	.0043	.0082
12	.0000	.0000	.0000	.0000	.0001	.0002	.0006	.0016	.0034
13	.0000	.0000	.0000	.0000	.0000	.0001	.0002	.0006	.0013
14	.0000	.0000	.0000	.0000	.0000	.0000	.0001	.0002	.0005
15	.0000	.0000	.0000	.0000	.0000	.0000	.0000	.0001	.0002

x	5.5	6.0	6.5	7.0	7.5	8.0	9.0	10.0	11.0
0	.0041	.0025	.0015	.0009	.0006	.0003	.0001	.0000	.0000
1	.0225	.0149	.0098	.0064	.0041	.0027	.0011	.0005	.0002
2	.0618	.0446	.0318	.0223	.0156	.0107	.0050	.0023	.0010
3	.1133	.0892	.0688	.0521	.0389	.0286	.0150	.0076	.0037
4	.1558	.1339	.1118	.0912	.0729	.0573	.0337	.0189	.0102
5	.1714	.1606	.1454	.1277	.1094	.0916	.0607	.0378	.0224
6	.1571	.1606	.1575	.1490	.1367	.1221	.0911	.0631	.0411
7	.1234	.1377	.1462	.1490	.1465	.1396	.1171	.0901	.0646
8	.0849	.1033	.1188	.1304	.1373	.1396	.1318	.1126	.0888
9	.0519	.0688	.0858	.1014	.1144	.1241	.1318	.1251	.1085
10	.0285	.0413	.0558	.0710	.0858	.0993	.1186	.1251	.1194
11	.0143	.0225	.0330	.0452	.0585	.0722	.0970	.1137	.1194
12	.0065	.0113	.0179	.0263	.0366	.0481	.0728	.0948	.1094
13	.0028	.0052	.0089	.0142	.0211	.0296	.0504	.0729	.0926
14	.0011	.0022	.0041	.0071	.0113	.0169	.0324	.0521	.0728
15	.0004	.0009	.0018	.0033	.0057	.0090	.0194	.0347	.0534
16	.0001	.0003	.0007	.0014	.0026	.0045	.0109	.0217	.0367
17	.0000	.0001	.0003	.0006	.0012	.0021	.0058	.0128	.0237
18	.0000	.0000	.0001	.0002	.0005	.0009	.0029	.0071	.0145
19	.0000	.0000	.0000	.0001	.0002	.0004	.0014	.0037	.0084
20	.0000	.0000	.0000	.0000	.0001	.0002	.0006	.0019	.0046
21	.0000	.0000	.0000	.0000	.0000	.0001	.0003	.0009	.0024
22	.0000	.0000	.0000	.0000	.0000	.0000	.0001	.0004	.0012
23	.0000	.0000	.0000	.0000	.0000	.0000	.0000	.0002	.0006
24	.0000	.0000	.0000	.0000	.0000	.0000	.0000	.0001	.0003
25	.0000	.0000	.0000	.0000	.0000	.0000	.0000	.0000	.0001

x	λ								
	12	13	14	15	16	17	18	19	20
0	.0000	.0000	.0000	.0000	.0000	.0000	.0000	.0000	.0000
1	.0001	.0000	.0000	.0000	.0000	.0000	.0000	.0000	.0000
2	.0004	.0002	.0001	.0000	.0000	.0000	.0000	.0000	.0000
3	.0018	.0008	.0004	.0002	.0001	.0000	.0000	.0000	.0000
4	.0053	.0027	.0013	.0006	.0003	.0001	.0001	.0000	.0000
5	.0127	.0070	.0037	.0019	.0010	.0005	.0002	.0001	.0001
6	.0255	.0152	.0087	.0048	.0026	.0014	.0007	.0004	.0002
7	.0437	.0281	.0174	.0104	.0060	.0034	.0019	.0010	.0005
8	.0655	.0457	.0304	.0194	.0120	.0072	.0042	.0024	.0013
9	.0874	.0661	.0473	.0324	.0213	.0135	.0083	.0050	.0029
10	.1048	.0859	.0663	.0486	.0341	.0230	.0150	.0095	.0058
11	.1144	.1015	.0844	.0663	.0496	.0355	.0245	.0164	.0106
12	.1144	.1099	.0984	.0829	.0661	.0504	.0368	.0259	.0176
13	.1056	.1099	.1060	.0956	.0814	.0658	.0509	.0378	.0271
14	.0905	.1021	.1060	.1024	.0930	.0800	.0655	.0514	.0387
15	.0724	.0885	.0989	.1024	.0992	.0906	.0786	.0650	.0516
16	.0543	.0719	.0866	.0960	.0992	.0963	.0884	.0772	.0646
17	.0383	.0550	.0713	.0847	.0934	.0963	.0936	.0863	.0760
18	.0255	.0397	.0554	.0706	.0830	.0909	.0936	.0911	.0844
19	.0161	.0272	.0409	.0557	.0699	.0814	.0887	.0911	.0888
20	.0097	.0177	.0286	.0418	.0559	.0692	.0798	.0866	.0888
21	.0055	.0109	.0191	.0299	.0426	.0560	.0684	.0783	.0846
22	.0030	.0065	.0121	.0204	.0310	.0433	.0560	.0676	.0769
23	.0016	.0037	.0074	.0133	.0216	.0320	.0438	.0559	.0669
24	.0008	.0020	.0043	.0083	.0144	.0226	.0328	.0442	.0557
25	.0004	.0010	.0024	.0050	.0092	.0154	.0237	.0336	.0446
26	.0002	.0005	.0013	.0029	.0057	.0101	.0164	.0246	.0343
27	.0001	.0002	.0007	.0016	.0034	.0063	.0109	.0173	.0254
28	.0000	.0001	.0003	.0009	.0019	.0038	.0070	.0117	.0181
29	.0000	.0001	.0002	.0004	.0011	.0023	.0044	.0077	.0125
30	.0000	.0000	.0001	.0002	.0006	.0013	.0026	.0049	.0083
31	.0000	.0000	.0001	.0001	.0003	.0007	.0015	.0030	.0054
32	.0000	.0000	.0000	.0001	.0001	.0004	.0009	.0018	.0034
33	.0000	.0000	.0000	.0000	.0001	.0002	.0005	.0010	.0020
34	.0000	.0000	.0000	.0000	.0000	.0001	.0002	.0006	.0012
35	.0000	.0000	.0000	.0000	.0000	.0001	.0001	.0003	.0007
36	.0000	.0000	.0000	.0000	.0000	.0000	.0001	.0002	.0004
37	.0000	.0000	.0000	.0000	.0000	.0000	.0000	.0001	.0002
38	.0000	.0000	.0000	.0000	.0000	.0000	.0000	.0000	.0001
39	.0000	.0000	.0000	.0000	.0000	.0000	.0000	.0000	.0001

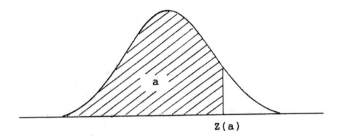

Z(a)

Cumulative Probabilities

Each 4-digit entry is the area "a" under the standard normal curve from $-\infty$ to Z(a). Reference the sketch above.

z	.00	.01	.02	.03	.04	.05	.06	.07	.08	.09
.0	.5000	.5040	.5080	.5120	.5160	.5199	.5239	.5279	.5319	.5359
.1	.5398	.5438	.5478	.5517	.5557	.5596	.5636	.5675	.5714	.5753
.2	.5793	.5832	.5871	.5910	.5948	.5987	.6026	.6064	.6103	.6141
.3	.6179	.6217	.6255	.6293	.6331	.6368	.6406	.6443	.6480	.6517
.4	.6554	.6591	.6628	.6664	.6700	.6736	.6772	.6808	.6844	.6879
.5	.6915	.6950	.6985	.7019	.7054	.7088	.7123	.7157	.7190	.7224
.6	.7257	.7291	.7324	.7357	.7389	.7422	.7454	.7486	.7517	.7549
.7	.7580	.7611	.7642	.7673	.7704	.7734	.7764	.7794	.7823	.7852
.8	.7881	.7910	.7939	.7967	.7995	.8023	.8051	.8078	.8106	.8133
.9	.8159	.8186	.8212	.8238	.8264	.8289	.8315	.8340	.8365	.8389
1.0	.8413	.8438	.8461	.8485	.8508	.8531	.8554	.8577	.8599	.8621
1.1	.8643	.8665	.8686	.8708	.8729	.8749	.8770	.8790	.8810	.8830
1.2	.8849	.8869	.8888	.8907	.8925	.8944	.8962	.8980	.8997	.9015
1.3	.9032	.9049	.9066	.9082	.9099	.9115	.9131	.9147	.9162	.9177
1.4	.9192	.9207	.9222	.9236	.9251	.9265	.9279	.9292	.9306	.9319
1.5	.9332	.9345	.9357	.9370	.9382	.9394	.9406	.9418	.9429	.9441
1.6	.9452	.9463	.9474	.9484	.9495	.9505	.9515	.9525	.9535	.9545
1.7	.9554	.9564	.9573	.9582	.9591	.9599	.9608	.9616	.9625	.9633
1.8	.9641	.9649	.9656	.9664	.9671	.9678	.9686	.9693	.9699	.9706
1.9	.9713	.9719	.9726	.9732	.9738	.9744	.9750	.9756	.9761	.9767
2.0	.9772	.9778	.9783	.9788	.9793	.9798	.9803	.9808	.9812	.9817
2.1	.9821	.9826	.9830	.9834	.9838	.9842	.9846	.9850	.9854	.9857
2.2	.9861	.9864	.9868	.9871	.9875	.9878	.9881	.9884	.9887	.9890
2.3	.9893	.9896	.9898	.9901	.9904	.9906	.9909	.9911	.9913	.9916
2.4	.9918	.9920	.9922	.9925	.9927	.9929	.9931	.9932	.9934	.9936

z	.00	.01	.02	.03	.04	.05	.06	.07	.08	.09
2.5	.9938	.9940	.9941	.9943	.9945	.9946	.9948	.9949	.9951	.9952
2.6	.9953	.9955	.9956	.9957	.9959	.9960	.9961	.9962	.9963	.9964
2.7	.9965	.9966	.9967	.9968	.9969	.9970	.9971	.9972	.9973	.9974
2.8	.9974	.9975	.9976	.9977	.9977	.9978	.9979	.9979	.9980	.9981
2.9	.9981	.9982	.9982	.9983	.9984	.9984	.9985	.9985	.9986	.9986
3.0	.9987	.9987	.9987	.9988	.9988	.9989	.9989	.9989	.9990	.9990
3.1	.9990	.9991	.9991	.9991	.9992	.9992	.9992	.9992	.9993	.9993
3.2	.9993	.9993	.9994	.9994	.9994	.9994	.9994	.9995	.9995	.9995
3.3	.9995	.9995	.9995	.9996	.9996	.9996	.9996	.9996	.9996	.9997
3.4	.9997	.9997	.9997	.9997	.9997	.9997	.9997	.9997	.9997	.9998

Selected Percentiles

Each entry is Z(a) where P[Z ≤ Z(a)] = a. For example,
P(Z ≤ 1.645) = .95 so Z(.95) = 1.645.

a	.10	.05	.025	.02	.01	.005	.001
Z(a)	-1.282	-1.645	-1.960	-2.054	-2.326	-2.576	-3.090

a	.90	.95	.975	.98	.99	.995	.999
Z(a)	1.282	1.645	1.960	2.054	2.326	2.576	3.090

t(a,v)

df				a			
ν	.75	.90	.95	.975	.99	.995	.9995
1	1.000	3.078	6.314	12.706	31.821	63.657	636.619
2	0.816	1.886	2.920	4.303	6.965	9.925	31.599
3	0.765	1.638	2.353	3.182	4.541	5.841	12.924
4	0.741	1.533	2.132	2.776	3.747	4.604	8.610
5	0.727	1.476	2.015	2.571	3.365	4.032	6.869
6	0.718	1.440	1.943	2.447	3.143	3.707	5.959
7	0.711	1.415	1.895	2.365	2.998	3.499	5.408
8	0.706	1.397	1.860	2.306	2.896	3.355	5.041
9	0.703	1.383	1.833	2.262	2.821	3.250	4.781
10	0.700	1.372	1.812	2.228	2.764	3.169	4.587
11	0.697	1.363	1.796	2.201	2.718	3.106	4.437
12	0.695	1.356	1.782	2.179	2.681	3.055	4.318
13	0.694	1.350	1.771	2.160	2.650	3.012	4.221
14	0.692	1.345	1.761	2.145	2.624	2.977	4.140
15	0.691	1.341	1.753	2.131	2.602	2.947	4.073
16	0.690	1.337	1.746	2.120	2.583	2.921	4.015
17	0.689	1.333	1.740	2.110	2.567	2.898	3.965
18	0.688	1.330	1.734	2.101	2.552	2.878	3.922
19	0.688	1.328	1.729	2.093	2.539	2.861	3.883
20	0.687	1.325	1.725	2.086	2.528	2.845	3.850
21	0.686	1.323	1.721	2.080	2.518	2.831	3.819
22	0.686	1.321	1.717	2.074	2.508	2.819	3.792
23	0.685	1.319	1.714	2.069	2.500	2.807	3.768
24	0.685	1.318	1.711	2.064	2.492	2.797	3.745
25	0.684	1.316	1.708	2.060	2.485	2.787	3.725
26	0.684	1.315	1.706	2.056	2.479	2.779	3.707
27	0.684	1.314	1.703	2.052	2.473	2.771	3.690
28	0.683	1.313	1.701	2.048	2.467	2.763	3.674
29	0.683	1.311	1.699	2.045	2.462	2.756	3.659
30	0.683	1.310	1.697	2.042	2.457	2.750	3.646
40	0.681	1.303	1.684	2.021	2.423	2.704	3.551
60	0.679	1.296	1.671	2.000	2.390	2.660	3.460
120	0.677	1.289	1.658	1.980	2.358	2.617	3.373
∞	0.674	1.282	1.645	1.960	2.326	2.576	3.291

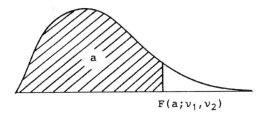

$$F(a;\nu_1,\nu_2)$$

The F distribution is skewed to the right and has two parameters:

ν_1 = numerator degrees of freedom, and

ν_2 = denominator degrees of freedom.

The random variable F is denoted by $F(\nu_1,\nu_2)$. Entries in the tables denote the positive values $F(a;\nu_1,\nu_2)$ such that

$$P[F(\nu_1,\nu_2) \leq F(a;\nu_1,\nu_2)] = a \quad \text{(reference the above sketch)}.$$

The superscripted entries, like 4053^2, should be read as 405300.

Because the F distribution has two parameters, any set of tables containing even a few percentiles will be extensive. We have included the following percentiles in the pages that follow: 50th, 75th, 90th, 95th, 97.5^{th}, 99th, and 99.9^{th}. This set of seven percentiles is given for each of 20 different values that ν_1 and ν_2 can assume (i.e., 400 possible combinations of ν_1 and ν_2).

Percentiles less than the 50th can be found by using the following relationship:

$$F(a;\nu_1,\nu_2) = \frac{1}{F(1-a;\nu_2,\nu_1)}$$

Please note that the order of the parameters is reversed in the two F expressions. For example, if a = .05, ν_1 = 8, and ν_2 = 30, then

$$F(.05;8,30) = \frac{1}{F(.95;30,8)} = \frac{1}{3.08} = .325$$

						v_1					
v_2	a	1	2	3	4	5	6	7	8	9	10
1	.50	1.00	1.50	1.71	1.82	1.89	1.94	1.98	2.00	2.03	2.04
	.75	5.83	7.50	8.20	8.58	8.82	8.98	9.10	9.19	9.26	9.32
	.90	39.9	49.5	53.6	55.8	57.2	58.2	58.9	59.4	59.9	60.2
	.95	161	200	216	225	230	234	237	239	241	242
	.975	648	800	864	900	922	937	948	957	963	969
	.99	4052	5000	5403	5625	5764	5859	5928	5982	6022	6056
	.999	4053^2	5000^2	5404^2	5625^2	5764^2	5859^2	5929^2	5981^2	6023^2	6056^2
2	.50	.667	1.00	1.13	1.21	1.25	1.28	1.30	1.32	1.33	1.34
	.75	2.57	3.00	3.15	3.23	3.28	3.31	3.34	3.35	3.37	3.38
	.90	8.53	9.00	9.16	9.24	9.29	9.33	9.35	9.37	9.38	9.39
	.95	18.5	19.0	19.2	19.2	19.3	19.3	19.4	19.4	19.4	19.4
	.975	38.5	39.0	39.2	39.2	39.3	39.3	39.4	39.4	39.4	39.4
	.99	98.5	99.0	99.2	99.2	99.3	99.3	99.4	99.4	99.4	99.4
	.999	998	999	999	999	999	999	999	999	999	999
3	.50	.585	.881	1.00	1.06	1.10	1.13	1.15	1.16	1.17	1.18
	.75	2.02	2.28	2.36	2.39	2.41	2.42	2.43	2.44	2.44	2.44
	.90	5.54	5.46	5.39	5.34	5.31	5.28	5.27	5.25	5.24	5.23
	.95	10.1	9.55	9.28	9.12	9.01	8.94	8.89	8.85	8.81	8.79
	.975	17.4	16.0	15.4	15.1	14.9	14.7	14.6	14.5	14.5	14.4
	.99	34.1	30.8	29.5	28.7	28.2	27.9	27.7	27.5	27.3	27.2
	.999	167	149	141	137	135	133	132	131	130	129
4	.50	.549	.828	.941	1.00	1.04	1.06	1.08	1.09	1.10	1.11
	.75	1.81	2.00	2.05	2.06	2.07	2.08	2.08	2.08	2.08	2.08
	.90	4.54	4.32	4.19	4.11	4.05	3.98	3.95	3.94	3.92	3.92
	.95	7.71	6.94	6.59	6.39	6.26	6.16	6.09	6.04	6.00	5.96
	.975	12.2	10.6	9.98	9.60	9.36	9.20	9.07	8.98	8.90	8.84
	.99	21.2	18.0	16.7	16.0	15.5	15.2	15.0	14.8	14.7	14.5
	.999	74.1	61.2	56.2	53.4	51.7	50.5	49.7	49.0	48.5	48.0
5	.50	.528	.799	.907	.965	1.00	1.02	1.04	1.05	1.06	1.07
	.75	1.69	1.85	1.88	1.89	1.89	1.89	1.89	1.89	1.89	1.89
	.90	4.06	3.78	3.62	3.52	3.45	3.40	3.37	3.34	3.32	3.30
	.95	6.61	5.79	5.41	5.19	5.05	4.95	4.88	4.82	4.77	4.74
	.975	10.0	8.43	7.76	7.39	7.15	6.98	6.85	6.76	6.68	6.62
	.99	16.3	13.3	12.1	11.4	11.0	10.7	10.5	10.3	10.2	10.1
	.999	47.2	37.1	33.2	31.1	29.7	28.8	28.2	27.6	27.2	26.9
6	.50	.515	.780	.886	.942	.977	1.00	1.02	1.03	1.04	1.05
	.75	1.62	1.76	1.78	1.79	1.79	1.78	1.78	1.78	1.77	1.77
	.90	3.78	3.46	3.29	3.18	3.11	3.05	3.01	2.98	2.96	2.94
	.95	5.99	5.14	4.76	4.53	4.39	4.28	4.21	4.15	4.10	4.06
	.975	8.81	7.26	6.60	6.23	5.99	5.82	5.70	5.60	5.52	5.46
	.99	13.7	10.9	9.78	9.15	8.75	8.47	8.26	8.10	7.98	7.87
	.999	35.5	27.0	23.7	21.9	20.8	20.0	19.5	19.0	18.7	18.4
7	.50	.506	.767	.871	.926	.960	.983	1.00	1.01	1.02	1.03
	.75	1.57	1.70	1.72	1.72	1.71	1.71	1.70	1.70	1.69	1.69
	.90	3.59	3.26	3.07	2.96	2.88	2.83	2.78	2.75	2.72	2.70
	.95	5.59	4.74	4.35	4.12	3.97	3.87	3.79	3.73	3.68	3.64
	.975	8.07	6.54	5.89	5.52	5.29	5.12	4.99	4.90	4.82	4.76
	.99	12.2	9.55	8.45	7.85	7.46	7.19	6.99	6.84	6.72	6.62
	.999	29.2	21.7	18.8	17.2	16.2	15.5	15.0	14.6	14.3	14.1

ν_2	a	ν_1 11	12	15	20	24	30	40	60	120	∞
1	.50	2.05	2.07	2.09	2.12	2.13	2.15	2.16	2.17	2.18	2.20
	.75	9.36	9.41	9.49	9.58	9.63	9.67	9.71	9.76	9.80	9.85
	.90	60.5	60.7	61.2	61.7	62.0	62.3	62.5	62.8	63.1	63.3
	.95	243	244	246	248	249	250	251	252	253	254
	.975	973	977	985	993	997	1000	1010	1010	1010	1020
	.99	6080	6110	6160	6210	6230	6260	6290	6310	6340	6370
	.999	6090^2	6110^2	6160^2	6210^2	6230^2	6260^2	6290^2	6310^2	6340^2	6370^2
2	.50	1.35	1.36	1.38	1.39	1.40	1.41	1.42	1.43	1.43	1.44
	.75	3.39	3.39	3.41	3.43	3.43	3.44	3.45	3.46	3.47	3.48
	.90	9.40	9.41	9.42	9.44	9.45	9.46	9.47	9.47	9.48	9.49
	.95	19.4	19.4	19.4	19.4	19.5	19.5	19.5	19.5	19.5	19.5
	.975	39.4	39.4	39.4	39.4	39.5	39.5	39.5	39.5	39.5	39.5
	.99	99.4	99.4	99.4	99.4	99.5	99.5	99.5	99.5	99.5	99.5
	.999	999	999	999	999	999	999	999	999	999	999
3	.50	1.19	1.20	1.21	1.23	1.23	1.24	1.25	1.25	1.26	1.27
	.75	2.45	2.45	2.46	2.46	2.46	2.47	2.47	2.47	2.47	2.47
	.90	5.22	5.22	5.20	5.18	5.18	5.17	5.16	5.15	5.14	5.13
	.95	8.76	8.74	8.70	8.66	8.63	8.62	8.59	8.57	8.55	8.53
	.975	14.4	14.3	14.3	14.2	14.1	14.1	14.0	14.0	13.9	13.9
	.99	27.1	27.1	26.9	26.7	26.6	26.5	26.4	26.3	26.2	26.1
	.999	129	128	127	126	126	125	125	124	124	123
4	.50	1.12	1.13	1.14	1.15	1.16	1.16	1.17	1.18	1.18	1.19
	.75	2.08	2.08	2.08	2.08	2.08	2.08	2.08	2.08	2.08	2.08
	.90	3.91	3.90	3.87	3.84	3.83	3.82	3.80	3.79	3.78	3.76
	.95	5.94	5.91	5.86	5.80	5.77	5.75	5.72	5.69	5.66	5.63
	.975	8.79	8.75	8.66	8.56	8.51	8.46	8.41	8.36	8.31	8.26
	.99	14.4	14.4	14.2	14.0	13.9	13.8	13.7	13.7	13.6	13.5
	.999	47.7	47.4	46.8	46.1	45.8	45.4	45.1	44.7	44.4	44.0
5	.50	1.08	1.09	1.10	1.11	1.12	1.12	1.13	1.14	1.14	1.15
	.75	1.89	1.89	1.89	1.88	1.88	1.88	1.88	1.87	1.87	1.87
	.90	3.28	3.27	3.24	3.21	3.19	3.17	3.16	3.14	3.12	3.10
	.95	4.71	4.68	4.62	4.56	4.53	4.50	4.46	4.43	4.40	4.36
	.975	6.57	6.52	6.43	6.33	6.28	6.23	6.18	6.12	6.07	6.02
	.99	9.96	9.89	9.72	9.55	9.47	9.38	9.29	9.20	9.11	9.02
	.999	26.6	26.4	25.9	25.4	25.1	24.9	24.6	24.3	24.1	23.8
6	.50	1.05	1.06	1.07	1.08	1.09	1.10	1.10	1.11	1.12	1.12
	.75	1.77	1.77	1.76	1.76	1.75	1.75	1.75	1.74	1.74	1.74
	.90	2.92	2.90	2.87	2.84	2.82	2.80	2.78	2.76	2.74	2.72
	.95	4.03	4.00	3.94	3.87	3.84	3.81	3.77	3.74	3.70	3.67
	.975	5.41	5.37	5.27	5.17	5.12	5.07	5.01	4.96	4.90	4.85
	.99	7.79	7.72	7.56	7.40	7.31	7.23	7.14	7.06	6.97	6.88
	.999	18.2	18.0	17.6	17.1	16.9	16.7	16.4	16.2	16.0	15.7
7	.50	1.04	1.04	1.05	1.07	1.07	1.08	1.08	1.09	1.10	1.10
	.75	1.69	1.68	1.68	1.67	1.67	1.66	1.66	1.65	1.65	1.65
	.90	2.68	2.67	2.63	2.59	2.58	2.56	2.54	2.51	2.49	2.47
	.95	3.60	3.57	3.51	3.44	3.41	3.38	3.34	3.30	3.27	3.23
	.975	4.71	4.67	4.57	4.47	4.42	4.36	4.31	4.25	4.20	4.14
	.99	6.54	6.47	6.31	6.16	6.07	5.99	5.91	5.82	5.74	5.65
	.999	13.9	13.7	13.3	12.9	12.7	12.5	12.3	12.1	11.9	11.7

ν_2	a	\(\nu_1\) 1	2	3	4	5	6	7	8	9	10
8	.50	.499	.757	.860	.915	.948	.971	.988	1.00	1.01	1.02
	.75	1.54	1.66	1.67	1.66	1.66	1.65	1.64	1.64	1.64	1.63
	.90	3.46	3.11	2.92	2.81	2.73	2.67	2.62	2.59	2.56	2.54
	.95	5.32	4.46	4.07	3.84	3.69	3.58	3.50	3.44	3.39	3.35
	.975	7.57	6.06	5.42	5.05	4.82	4.65	4.53	4.43	4.36	4.30
	.99	11.3	8.65	7.59	7.01	6.63	6.37	6.18	6.03	5.91	5.81
	.999	25.4	18.5	15.8	14.4	13.5	12.9	12.4	12.0	11.8	11.5
9	.50	.494	.749	.852	.906	.939	.962	.978	.990	1.00	1.01
	.75	1.51	1.62	1.63	1.63	1.62	1.61	1.60	1.60	1.59	1.59
	.90	3.36	3.01	2.81	2.69	2.61	2.55	2.51	2.47	2.44	2.42
	.95	5.12	4.26	3.86	3.63	3.48	3.37	3.29	3.23	3.18	3.14
	.975	7.21	5.71	5.08	4.72	4.48	4.32	4.20	4.10	4.03	3.96
	.99	10.6	8.02	6.99	6.42	6.06	5.80	5.61	5.47	5.35	5.26
	.999	22.9	16.4	13.9	12.6	11.7	11.1	10.7	10.4	10.1	9.89
10	.50	.490	.743	.845	.899	.932	.954	.971	.983	.992	1.00
	.75	1.49	1.60	1.60	1.59	1.59	1.58	1.57	1.56	1.56	1.55
	.90	3.28	2.92	2.73	2.61	2.52	2.46	2.41	2.38	2.35	2.32
	.95	4.96	4.10	3.71	3.48	3.33	3.22	3.14	3.07	3.02	2.98
	.975	6.94	5.46	4.83	4.47	4.24	4.07	3.95	3.85	3.78	3.72
	.99	10.0	7.56	6.55	5.99	5.64	5.39	5.20	5.06	4.94	4.85
	.999	21.0	14.9	12.6	11.3	10.5	9.92	9.52	9.20	8.96	8.75
11	.50	.486	.739	.840	.893	.926	.948	.964	.977	.986	.994
	.75	1.47	1.58	1.58	1.57	1.56	1.55	1.54	1.53	1.53	1.52
	.90	3.23	2.86	2.66	2.54	2.45	2.39	2.34	2.30	2.27	2.25
	.95	4.84	3.98	3.59	3.36	3.20	3.09	3.01	2.95	2.90	2.85
	.975	6.72	5.26	4.63	4.28	4.04	3.88	3.76	3.66	3.59	3.53
	.99	9.65	7.21	6.22	5.67	5.32	5.07	4.89	4.74	4.63	4.54
	.999	19.7	13.8	11.6	10.3	9.58	9.05	8.66	8.35	8.12	7.92
12	.50	.484	.735	.835	.888	.921	.943	.959	.972	.981	.989
	.75	1.46	1.56	1.56	1.55	1.54	1.53	1.52	1.51	1.51	1.50
	.90	3.18	2.81	2.61	2.48	2.39	2.33	2.28	2.24	2.21	2.19
	.95	4.75	3.89	3.49	3.26	3.11	3.00	2.91	2.85	2.80	2.75
	.975	6.55	5.10	4.47	4.12	3.89	3.73	3.61	3.51	3.44	3.37
	.99	9.33	6.93	5.95	5.41	5.06	4.82	4.64	4.50	4.39	4.30
	.999	18.6	13.0	10.8	9.63	8.89	8.38	8.00	7.71	7.48	7.29
15	.50	.478	.726	.826	.878	.911	.933	.948	.960	.970	.977
	.75	1.43	1.52	1.52	1.51	1.49	1.48	1.47	1.46	1.46	1.45
	.90	3.07	2.70	2.49	2.36	2.27	2.21	2.16	2.12	2.09	2.06
	.95	4.54	3.68	3.29	3.06	2.90	2.79	2.71	2.64	2.59	2.54
	.975	6.20	4.76	4.15	3.80	3.58	3.41	3.29	3.20	3.12	3.06
	.99	8.68	6.36	5.42	4.89	4.56	4.32	4.14	4.00	3.89	3.80
	.999	16.6	11.3	9.34	8.25	7.57	7.09	6.74	6.47	6.26	6.08
20	.50	.472	.718	.816	.868	.900	.922	.938	.950	.959	.966
	.75	1.40	1.49	1.48	1.47	1.45	1.44	1.43	1.42	1.41	1.40
	.90	2.97	2.59	2.38	2.25	2.16	2.09	2.04	2.00	1.96	1.94
	.95	4.35	3.49	3.10	2.87	2.71	2.60	2.51	2.45	2.39	2.35
	.975	5.87	4.46	3.86	3.51	3.29	3.13	3.01	2.91	2.84	2.77
	.99	8.10	5.85	4.94	4.43	4.10	3.87	3.70	3.56	3.46	3.37
	.999	14.8	9.95	8.10	7.10	6.46	6.02	5.69	5.44	5.24	5.08

v_2	a	11	12	15	20	24	30	40	60	120	∞
8	.50	1.02	1.03	1.04	1.05	1.06	1.07	1.07	1.08	1.08	1.09
	.75	1.63	1.62	1.62	1.61	1.60	1.60	1.59	1.59	1.58	1.58
	.90	2.52	2.50	2.46	2.42	2.40	2.38	2.36	2.34	2.32	2.29
	.95	3.31	3.28	3.22	3.15	3.12	3.08	3.04	3.01	2.97	2.93
	.975	4.24	4.20	4.10	4.00	3.95	3.89	3.84	3.78	3.73	3.67
	.99	5.73	5.67	5.52	5.36	5.28	5.20	5.12	5.03	4.95	4.86
	.999	11.4	11.2	10.8	10.5	10.3	10.1	9.92	9.73	9.54	9.34
9	.50	1.01	1.02	1.03	1.04	1.05	1.05	1.06	1.07	1.07	1.08
	.75	1.58	1.58	1.57	1.56	1.56	1.55	1.55	1.54	1.53	1.53
	.90	2.40	2.38	2.34	2.30	2.28	2.25	2.23	2.21	2.18	2.16
	.95	3.10	3.07	3.01	2.94	2.90	2.86	2.83	2.79	2.75	2.71
	.975	3.91	3.87	3.77	3.67	3.61	3.56	3.51	3.45	3.39	3.33
	.99	5.18	5.11	4.96	4.81	4.73	4.65	4.57	4.48	4.40	4.31
	.999	9.71	9.57	9.24	8.90	8.72	8.55	8.37	8.19	8.00	7.81
10	.50	1.01	1.01	1.02	1.03	1.04	1.05	1.05	1.06	1.06	1.07
	.75	1.55	1.54	1.53	1.52	1.52	1.51	1.51	1.50	1.49	1.48
	.90	2.30	2.28	2.24	2.20	2.18	2.16	2.13	2.11	2.08	2.06
	.95	2.94	2.91	2.85	2.77	2.74	2.70	2.66	2.62	2.58	2.54
	.975	3.66	3.62	3.52	3.42	3.37	3.31	3.26	3.20	3.14	3.08
	.99	4.77	4.71	4.56	4.41	4.33	4.25	4.17	4.08	4.00	3.91
	.999	8.58	8.44	8.13	7.80	7.64	7.47	7.30	7.12	6.94	6.76
11	.50	1.00	1.01	1.02	1.03	1.03	1.04	1.05	1.05	1.06	1.06
	.75	1.52	1.51	1.50	1.49	1.49	1.48	1.47	1.47	1.46	1.45
	.90	2.23	2.21	2.17	2.12	2.10	2.08	2.05	2.03	2.00	1.97
	.95	2.82	2.79	2.72	2.65	2.61	2.57	2.53	2.49	2.45	2.40
	.975	3.47	3.43	3.33	3.23	3.17	3.12	3.06	3.00	2.94	2.88
	.99	4.46	4.40	4.25	4.10	4.02	3.94	3.86	3.78	3.69	3.60
	.999	7.76	7.62	7.32	7.01	6.85	6.68	6.52	6.35	6.17	6.00
12	.50	.995	1.00	1.01	1.02	1.03	1.03	1.04	1.05	1.05	1.06
	.75	1.50	1.49	1.48	1.47	1.46	1.45	1.45	1.44	1.43	1.42
	.90	2.17	2.15	2.11	2.06	2.04	2.01	1.99	1.96	1.93	1.90
	.95	2.72	2.69	2.62	2.54	2.51	2.47	2.43	2.38	2.34	2.30
	.975	3.32	3.28	3.18	3.07	3.02	2.96	2.91	2.85	2.79	2.72
	.99	4.22	4.16	4.01	3.86	3.78	3.70	3.62	3.54	3.45	3.36
	.999	7.14	7.01	6.71	6.40	6.25	6.09	5.93	5.76	5.59	5.42
15	.50	.984	.989	1.00	1.01	1.02	1.02	1.03	1.03	1.04	1.05
	.75	1.44	1.44	1.43	1.41	1.41	1.40	1.39	1.38	1.37	1.36
	.90	2.04	2.02	1.97	1.92	1.90	1.87	1.85	1.82	1.79	1.76
	.95	2.51	2.48	2.40	2.33	2.39	2.25	2.20	2.16	2.11	2.07
	.975	3.01	2.96	2.86	2.76	2.70	2.64	2.59	2.52	2.46	2.40
	.99	3.73	3.67	3.52	3.37	3.29	3.21	3.13	3.05	2.96	2.87
	.999	5.93	5.81	5.54	5.25	5.10	4.95	4.80	4.64	4.47	4.31
20	.50	.972	.977	.989	1.00	1.01	1.01	1.02	1.02	1.03	1.03
	.75	1.39	1.39	1.37	1.36	1.35	1.34	1.33	1.32	1.31	1.29
	.90	1.91	1.89	1.84	1.79	1.77	1.74	1.71	1.68	1.64	1.61
	.95	2.31	2.28	2.20	2.12	2.08	2.04	1.99	1.95	1.90	1.84
	.975	2.72	2.68	2.57	2.46	2.41	2.35	2.29	2.22	2.16	2.09
	.99	3.29	3.23	3.09	2.94	2.86	2.78	2.69	2.61	2.52	2.42
	.999	4.94	4.82	4.56	4.29	4.15	4.01	3.86	3.70	3.54	3.38

						ν_1					
ν_2	a	1	2	3	4	5	6	7	8	9	10
24	.50	.469	.714	.812	.863	.895	.917	.932	.944	.953	.961
	.75	1.39	1.47	1.46	1.44	1.43	1.41	1.40	1.39	1.38	1.38
	.90	2.93	2.54	2.33	2.19	2.10	2.04	1.98	1.94	1.91	1.88
	.95	4.26	3.40	3.01	2.78	2.62	2.51	2.42	2.36	2.30	2.25
	.975	5.72	4.32	3.72	3.38	3.15	2.99	2.87	2.78	2.70	2.64
	.99	7.82	5.61	4.72	4.22	3.90	3.67	3.50	3.36	3.26	3.17
	.999	14.0	9.34	7.55	6.59	5.98	5.55	5.23	4.99	4.80	4.64
30	.50	.466	.709	.807	.858	.890	.912	.927	.939	.948	.955
	.75	1.38	1.45	1.44	1.42	1.41	1.39	1.38	1.37	1.36	1.35
	.90	2.88	2.49	2.28	2.14	2.05	1.98	1.93	1.88	1.85	1.82
	.95	4.17	3.32	2.92	2.69	2.53	2.42	2.33	2.27	2.21	2.16
	.975	5.57	4.18	3.59	3.25	3.03	2.87	2.75	2.65	2.57	2.51
	.99	7.56	5.39	4.51	4.02	3.70	3.47	3.30	3.17	3.07	2.98
	.999	13.3	8.77	7.05	6.12	5.53	5.12	4.82	4.58	4.39	4.24
40	.50	.463	.705	.802	.854	.885	.907	.922	.934	.943	.950
	.75	1.36	1.44	1.42	1.40	1.39	1.37	1.36	1.35	1.34	1.33
	.90	2.84	2.44	2.23	2.09	2.00	1.93	1.87	1.83	1.79	1.76
	.95	4.08	3.23	2.84	2.61	2.45	2.34	2.25	2.18	2.12	2.08
	.975	5.42	4.05	3.46	3.13	2.90	2.74	2.62	2.53	2.45	2.39
	.99	7.31	5.18	4.31	3.83	3.51	3.29	3.12	2.99	2.89	2.80
	.999	12.6	8.25	6.60	5.70	5.13	4.73	4.44	4.21	4.02	3.87
60	.50	.461	.701	.798	.849	.880	.901	.917	.928	.937	.945
	.75	1.35	1.42	1.41	1.38	1.37	1.35	1.33	1.32	1.31	1.30
	.90	2.79	2.39	2.18	2.04	1.95	1.87	1.82	1.77	1.74	1.71
	.95	4.00	3.15	2.76	2.53	2.37	2.25	2.17	2.10	2.04	1.99
	.975	5.29	3.93	3.34	3.01	2.79	2.63	2.51	2.41	2.33	2.27
	.99	7.08	4.98	4.13	3.65	3.34	3.12	2.95	2.82	2.72	2.63
	.999	12.0	7.76	6.17	5.31	4.76	4.37	4.09	3.87	3.69	3.54
120	.50	.458	.697	.793	.844	.875	.896	.912	.923	.932	.939
	.75	1.34	1.40	1.39	1.37	1.35	1.33	1.31	1.30	1.29	1.28
	.90	2.75	2.35	2.13	1.99	1.90	1.82	1.77	1.72	1.68	1.65
	.95	3.92	3.07	2.68	2.45	2.29	2.18	2.09	2.02	1.96	1.91
	.975	5.15	3.80	3.23	2.89	2.67	2.52	2.39	2.30	2.22	2.16
	.99	6.85	4.79	3.95	3.48	3.17	2.96	2.79	2.66	2.56	2.47
	.999	11.4	7.32	5.79	4.95	4.42	4.04	3.77	3.55	3.38	3.24
∞	.50	.455	.693	.789	.839	.870	.891	.907	.918	.927	.934
	.75	1.32	1.39	1.37	1.35	1.33	1.31	1.29	1.28	1.27	1.25
	.90	2.71	2.30	2.08	1.94	1.85	1.77	1.72	1.67	1.63	1.60
	.95	3.84	3.00	2.60	2.37	2.21	2.10	2.01	1.94	1.88	1.83
	.975	5.02	3.69	3.12	2.79	2.57	2.41	2.29	2.19	2.11	2.05
	.99	6.63	4.61	3.78	3.32	3.02	2.80	2.64	2.51	2.41	2.32
	.999	10.8	6.91	5.42	4.62	4.10	3.74	3.47	3.27	3.10	2.96

ν_2	a	ν_1 11	12	15	20	24	30	40	60	120	∞
24	.50	.967	.972	.983	.994	1.00	1.01	1.01	1.02	1.02	1.03
	.75	1.37	1.36	1.35	1.33	1.32	1.31	1.30	1.29	1.28	1.26
	.90	1.85	1.83	1.78	1.73	1.70	1.67	1.64	1.61	1.57	1.53
	.95	2.21	2.18	2.11	2.03	1.98	1.94	1.89	1.84	1.79	1.73
	.975	2.59	2.54	2.44	2.33	2.27	2.21	2.15	2.08	2.01	1.94
	.99	3.09	3.03	2.89	2.74	2.66	2.58	2.49	2.40	2.31	2.21
	.999	4.50	4.39	4.14	3.87	3.74	3.59	3.45	3.29	3.14	2.97
30	.50	.961	.966	.978	.989	.994	1.00	1.01	1.01	1.02	1.02
	.75	1.35	1.34	1.32	1.30	1.29	1.28	1.27	1.26	1.24	1.23
	.90	1.79	1.77	1.72	1.67	1.64	1.61	1.57	1.54	1.50	1.46
	.95	2.13	2.09	2.01	1.93	1.89	1.84	1.79	1.74	1.68	1.62
	.975	2.46	2.41	2.31	2.20	2.14	2.07	2.01	1.94	1.87	1.79
	.99	2.91	2.84	2.70	2.55	2.47	2.39	2.30	2.21	2.11	2.01
	.999	4.11	4.00	3.75	3.49	3.36	3.22	3.07	2.92	2.76	2.59
40	.50	.956	.961	.972	.983	.989	.994	1.00	1.01	1.01	1.02
	.75	1.32	1.31	1.30	1.28	1.26	1.25	1.24	1.22	1.21	1.19
	.90	1.73	1.71	1.66	1.61	1.57	1.54	1.51	1.47	1.42	1.38
	.95	2.04	2.00	1.92	1.84	1.79	1.74	1.69	1.64	1.58	1.51
	.975	2.33	2.29	2.18	2.07	2.01	1.94	1.88	1.80	1.72	1.64
	.99	2.73	2.66	2.52	2.37	2.29	2.20	2.11	2.02	1.92	1.80
	.999	3.75	3.64	3.40	3.15	3.01	2.87	2.73	2.57	2.41	2.23
60	.50	.951	.956	.967	.978	.983	.989	.994	1.00	1.01	1.01
	.75	1.29	1.29	1.27	1.25	1.24	1.22	1.21	1.19	1.17	1.15
	.90	1.68	1.66	1.60	1.54	1.51	1.48	1.44	1.40	1.35	1.29
	.95	1.95	1.92	1.84	1.75	1.70	1.65	1.59	1.53	1.47	1.39
	.975	2.22	2.17	2.06	1.94	1.88	1.82	1.74	1.67	1.58	1.48
	.99	2.56	2.50	2.35	2.20	2.12	2.03	1.94	1.84	1.73	1.60
	.999	3.43	3.31	3.08	2.83	2.69	2.56	2.41	2.25	2.09	1.89
120	.50	.945	.950	.961	.972	.978	.983	.989	.994	1.00	1.01
	.75	1.27	1.26	1.24	1.22	1.21	1.19	1.18	1.16	1.13	1.10
	.90	1.62	1.60	1.55	1.48	1.45	1.41	1.37	1.32	1.26	1.19
	.95	1.87	1.83	1.75	1.66	1.61	1.55	1.50	1.43	1.35	1.25
	.975	2.10	2.05	1.95	1.82	1.76	1.69	1.61	1.53	1.43	1.31
	.99	2.40	2.34	2.19	2.03	1.95	1.86	1.76	1.66	1.53	1.38
	.999	3.12	3.02	2.78	2.53	2.40	2.26	2.11	1.95	1.76	1.54
∞	.50	.939	.945	.956	.967	.972	.978	.983	.989	.994	1.00
	.75	1.24	1.24	1.22	1.19	1.18	1.16	1.14	1.12	1.08	1.00
	.90	1.57	1.55	1.49	1.42	1.38	1.34	1.30	1.24	1.17	1.00
	.95	1.79	1.75	1.67	1.57	1.52	1.46	1.39	1.32	1.22	1.00
	.975	1.99	1.94	1.83	1.71	1.64	1.57	1.48	1.39	1.27	1.00
	.99	2.25	2.18	2.04	1.88	1.79	1.70	1.59	1.47	1.32	1.00
	.999	2.84	2.74	2.51	2.27	2.13	1.99	1.84	1.66	1.45	1.00

Entry is $\chi^2(a,v)$, where

$P[\chi^2(v) \leq \chi^2(a,v)] = a.$

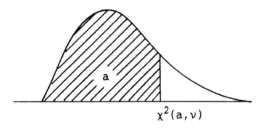

$\chi^2(a,v)$

df	a									
v	.005	.010	.025	.050	.100	.900	.950	.975	.990	.995
1	$.0^4 39$	$.0^3 16$	$.0^3 98$	$.0^2 39$	.0158	2.71	3.84	5.02	6.63	7.88
2	.0100	.0201	.0506	.103	.211	4.61	5.99	7.38	9.21	10.60
3	.072	.115	.216	.352	.584	6.25	7.81	9.35	11.34	12.84
4	.207	.0297	.484	.711	1.064	7.78	9.49	11.14	13.28	14.86
5	.412	.554	.831	1.145	1.61	9.24	11.07	12.83	15.09	16.75
6	.676	.872	1.24	1.64	2.20	10.64	12.59	14.45	16.81	18.55
7	.989	1.24	1.69	2.17	2.83	12.02	14.07	16.01	18.48	20.28
8	1.34	1.65	2.18	2.73	3.49	13.36	15.51	17.53	20.09	21.96
9	1.73	2.09	2.70	3.33	4.17	14.68	16.92	19.02	21.67	23.59
10	2.16	2.56	3.25	3.94	4.87	15.99	18.31	20.48	23.21	25.19
11	2.60	3.05	3.82	4.57	5.58	17.28	19.68	21.92	24.73	26.76
12	3.07	3.57	4.40	5.23	6.30	18.55	21.03	23.34	26.22	28.30
13	3.57	4.11	5.01	5.89	7.04	19.81	22.36	24.74	27.69	29.82
14	4.07	4.66	5.63	6.57	7.79	21.06	23.68	26.12	29.14	31.32
15	4.60	5.23	6.26	7.26	8.55	22.31	25.00	27.49	30.58	32.80
16	5.14	5.81	6.91	7.96	9.31	23.54	26.30	28.85	32.00	34.27
17	5.70	6.41	7.56	8.67	10.09	24.77	27.59	30.19	33.41	35.72
18	6.26	7.01	8.23	9.39	10.86	25.99	28.87	31.53	34.81	37.16
19	6.84	7.63	8.91	10.12	11.65	27.20	30.14	32.85	36.19	38.58
20	7.43	8.26	9.59	10.85	12.44	28.41	31.41	34.17	37.57	40.00
21	8.03	8.90	10.28	11.59	13.24	29.62	32.67	35.48	38.93	41.40
22	8.64	9.54	10.98	12.34	14.04	30.81	33.92	36.78	40.29	42.80
23	9.26	10.20	11.69	13.09	14.85	32.01	35.17	38.08	41.64	44.18
24	9.89	10.86	12.40	13.85	15.66	33.20	36.42	39.36	42.98	45.56
25	10.52	11.52	13.12	14.61	16.46	34.38	37.65	40.65	44.31	46.93
26	11.16	12.20	13.84	15.38	17.29	35.56	38.89	41.92	45.64	48.29
27	11.81	12.88	14.57	16.15	18.11	36.74	40.11	43.19	46.96	49.64
28	12.46	13.56	15.31	16.93	18.94	37.92	41.34	44.46	48.28	50.99
29	13.12	14.26	16.05	17.71	19.77	39.09	42.56	45.72	49.59	52.34
30	13.79	14.95	16.79	18.49	20.60	40.26	43.77	46.98	50.89	53.67
40	20.71	22.16	24.43	26.51	29.05	51.81	55.76	59.34	63.69	66.77
50	27.99	29.71	32.36	34.76	37.69	63.17	67.50	71.42	76.15	79.49
60	35.53	37.48	40.48	43.19	46.46	74.40	79.08	83.30	88.38	91.95
80	51.17	53.54	57.15	60.39	64.28	96.58	101.9	106.6	112.3	116.3
100	67.33	70.06	74.22	77.93	82.36	118.5	124.3	129.6	135.8	140.2

$$\alpha = 0.05$$

					p				
v	2	3	4	5	6	7	8	9	10
1	17.97	17.97	17.97	17.97	17.97	17.97	17.97	17.97	17.97
2	6.085	6.085	6.085	6.085	6.085	6.085	6.085	6.085	6.085
3	4.501	4.516	4.516	4.516	4.516	4.516	4.516	4.516	4.516
4	3.927	4.013	4.033	4.033	4.033	4.033	4.033	4.033	4.033
5	3.635	3.749	3.797	3.814	3.814	3.814	3.814	3.814	3.814
6	3.461	3.587	3.649	3.680	3.694	3.697	3.697	3.697	3.697
7	3.344	3.477	3.548	3.588	3.611	3.622	3.626	3.626	3.626
8	3.261	3.399	3.475	3.521	3.549	3.566	3.575	3.579	3.579
9	3.199	3.339	3.420	3.470	3.502	3.523	3.536	3.544	3.547
10	3.151	3.293	3.376	3.430	3.465	3.489	3.505	3.516	3.522
11	3.113	3.256	3.342	3.397	3.435	3.462	3.480	3.493	3.501
12	3.082	3.225	3.313	3.370	3.410	3.439	3.459	3.474	3.484
13	3.055	3.200	3.289	3.348	3.389	3.419	3.442	3.458	3.470
14	3.033	3.178	3.268	3.329	3.372	3.403	3.426	3.444	3.457
15	3.014	3.160	3.250	3.312	3.356	3.389	3.413	3.432	3.446
16	2.998	3.144	3.235	3.298	3.343	3.376	3.402	3.422	3.437
17	2.984	3.130	3.222	3.285	3.331	3.366	3.392	3.412	3.429
18	2.971	3.118	3.210	3.274	3.321	3.356	3.383	3.405	3.421
19	2.960	3.107	3.199	3.264	3.311	3.347	3.375	3.397	3.415
20	2.950	3.097	3.190	3.255	3.303	3.339	3.368	3.391	3.409
24	2.919	3.066	3.160	3.226	3.276	3.315	3.345	3.370	3.390
30	2.888	3.035	3.131	3.199	3.250	3.290	3.322	3.349	3.371
40	2.858	3.006	3.102	3.171	3.224	3.266	3.300	3.328	3.352
60	2.829	2.976	3.073	3.143	3.198	3.241	3.277	3.307	3.333
120	2.800	2.947	3.045	3.116	3.172	3.217	3.254	3.287	3.314
∞	2.772	2.918	3.017	3.089	3.146	3.193	3.232	3.265	3.294

Copied with permission from *Probability and Statistics for Engineers and Scientists*. Walpole, R.E., Myers, R.H., McMillan, 1985

$$\alpha = 0.01$$

v	2	3	4	5	6	7	8	9	10
					p				
1	90.03	90.03	90.03	90.03	90.03	90.03	90.03	90.03	90.03
2	14.04	14.04	14.04	14.04	14.04	14.04	14.04	14.04	14.04
3	8.261	8.321	8.321	8.321	8.321	8.321	8.321	8.321	8.321
4	6.512	6.677	6.740	6.756	6.756	6.756	6.756	6.756	6.756
5	5.702	5.893	5.989	6.040	6.065	6.074	6.074	6.074	6.074
6	5.243	5.439	5.549	5.614	5.655	5.680	5.694	5.701	5.703
7	4.949	5.145	5.260	5.334	5.383	5.416	5.439	5.454	5.464
8	4.746	4.939	5.057	5.135	5.189	5.227	5.256	5.276	5.291
9	4.596	4.787	4.906	4.986	5.043	5.086	5.118	5.142	5.160
10	4.482	4.671	4.790	4.871	4.931	4.975	5.010	5.037	5.058
11	4.392	4.579	4.697	4.780	4.841	4.887	4.924	4.952	4.975
12	4.320	4.504	4.622	4.706	4.767	4.815	4.852	4.883	4.907
13	4.260	4.442	4.560	4.644	4.706	4.755	4.793	4.824	4.850
14	4.210	4.391	4.508	4.591	4.654	4.704	4.743	4.775	4.802
15	4.168	4.347	4.463	4.547	4.610	4.660	4.700	4.733	4.760
16	4.131	4.309	4.425	4.509	4.572	4.622	4.663	4.696	4.724
17	4.099	4.275	4.391	4.475	4.539	4.589	4.630	4.664	4.693
18	4.071	4.246	4.362	4.445	4.509	4.560	4.601	4.635	4.664
19	4.046	4.220	4.335	4.419	4.483	4.534	4.575	4.610	4.639
20	4.024	4.197	4.312	4.395	4.459	4.510	4.552	4.587	4.617
24	3.956	4.126	4.239	4.322	4.386	4.437	4.480	4.516	4.546
30	3.889	4.056	4.168	4.250	4.314	4.366	4.409	4.445	4.477
40	3.825	3.988	4.098	4.180	4.244	4.296	4.339	4.376	4.408
60	3.762	3.922	4.031	4.111	4.174	4.226	4.270	4.307	4.340
120	3.702	3.858	3.965	4.044	4.107	4.158	4.202	4.239	4.272
∞	3.643	3.796	3.900	3.978	4.040	4.091	4.135	4.172	4.205

Copied with permission from *Probability and Statistics for Engineers and Scientists.* Walpole, R.E., Myers, R.H., McMillan, 1985

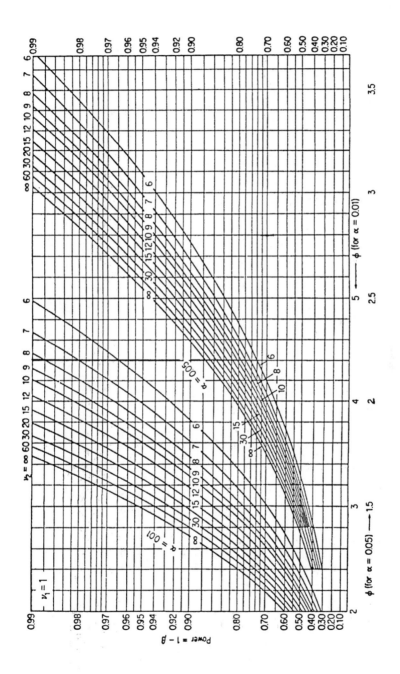

Copied with permission from *Probability and Statistics for Engineers and Scientist.* Walpole, R.E, Myers, R.H., McMillan, 1985

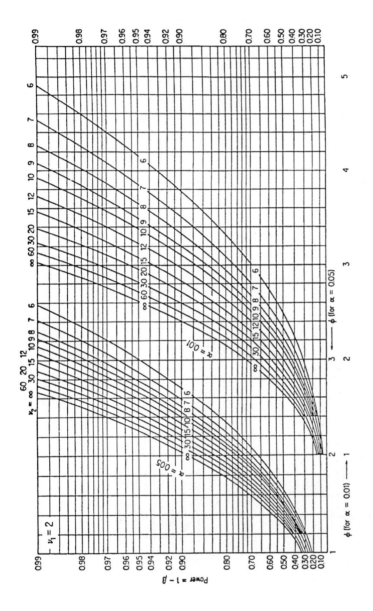

Copied with permission from *Probability and Statistics for Engineers and Scientist*, Walpole, R.E. Myers, R.H., McMillan, 1985

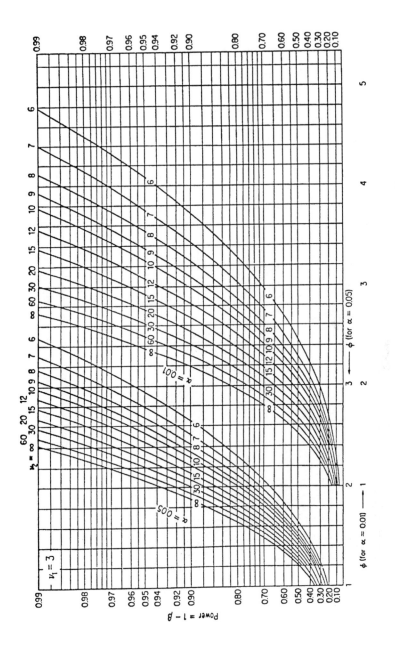

Copied with permission from *Probability and Statistics for Engineers and Scientist*, Walpole, R.E, Myers, R.H., McMillan, 1985

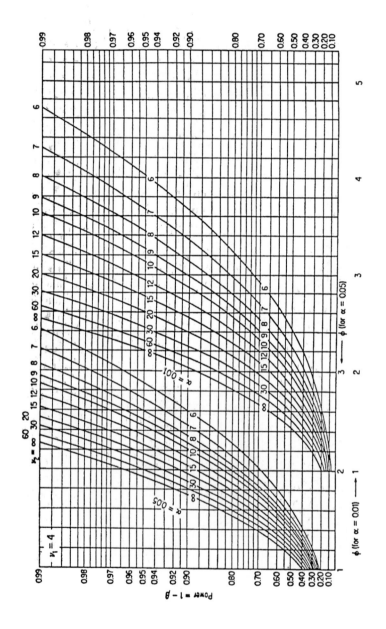

Copied with permission from *Probability and Statistics for Engineers and Scientist*. Walpole, R.E, Myers, R.H., McMillan, 1985

ANSWERS TO SELECTED PROBLEMS

Chapter 1

7. C_p = (Upper Spec − Lower Spec) / 6σ
     = (2100 − 1900) / 6(20)
     = 200 / 120
     = 1.67

C_{pk} = minimum (Upper Spec − $\bar{y}$, $\bar{y}$ − Lower Spec) / 3σ
     = minimum (2100 - 2040, 2040 - 1900) / 3(20)
     = 60 / 60
     = 1.0

Just Looking at C_p, this is very desirable, producing a minimum number of defects. However, C_{pk} indicates that the process is not as good as it appears, mostly because the average is not centered on the target value.

Chapter 2

1.   b)

	RPM	Feed Rate	Tool Radius
1.	−	−	−
2.	−	−	+
3.	−	+	−
4.	−	+	+
5.	+	−	−
6.	+	−	+
7.	+	+	−
8.	+	+	+

2. a)

	(A) RPM	(B) Feed Rate	(C) Radius	AB	AC	BC	ABC
Avg($-$)	131.9	66.9	107.2	110.0	134.1	119.1	108.4
Avg($+$)	81.3	146.2	105.9	103.1	79.1	94.1	104.7
Δ	-50.6	79.3	-1.3	-6.9	-55.0	-25.0	-3.7
$\Delta/2$	-25.3	39.7	-0.6	-3.5	-27.5	-12.5	-1.9

b) The Pareto chart implies B, AC, A, and perhaps BC are important.

Chapter 3

5. By folding over the R_{III} design, you can get a R_{IV} design.

6. I = ABCDF = BCDEG = AEFG, which makes this a R_{IV} design.

13. a) I = ACD = BCE = ABDE, which makes this a R_{IV} design.

b) A   : CD = ABDE = BDE
    B   : ABCD = CE = ADE
    C   : AD = BE = ABCDE
    D   : AC = BCDE = ABE
    E   : ACDE = BC = ABD

Yes, B is aliased with CE, and C is aliased with AD

16. a) α distance $= (n_F)^{1/4} = (8)^{1/4} = 1.68$

b) Factor B: 10 to 80% (Assume this represents the maximum allowable range for the factors.)

Center $= 45$

design values	$-\alpha$	-1	0	1	α
real values	10	24.17	45	65.83	80

c) $n_c = 4\sqrt{n_F + 1} - 2k = 6$

d) Total number of runs: 8(factorial) + 6(centerpoints) + 6(axial) = 20

17. $R^2_{AB} = 4^2 / 8^2 = .25 \rightarrow$ there is a 25% overlap between the two variables.

18. $_8C_2 = \dfrac{8!}{2! \ 6!} = 28$

Chapter 4

1. MSE = .0415

a) 95% C.I. = $(4.10 - 3.86) \pm 2.306\sqrt{.0415\left(\dfrac{1}{5} + \dfrac{1}{5}\right)}$

$= [-.0571, .5371]$

Since zero is contained in the interval, there is no difference.

b) $t_0 = \dfrac{(4.1 - 3.86)}{\sqrt{.0415\left(\dfrac{1}{5} + \dfrac{1}{5}\right)}} = 1.862$

$t_{critical} = 2.306$

Since $t_0 < t_{crit}$, fail to reject H_0, there is no difference.

c) $F_{crit}(0.05, 1, 8) = 5.32$

Source	SS	df	MS	F_0	P
Between	.1440	1	.1440	3.47	>0.05
Error	.3320	8	.0415		
Total	.4760	9			

Since $F_0 < F_{crit}$, retain H_0, there is no difference.

2. $\Phi = .56\sqrt{n'}$

$\alpha = 0.05$, goal $(1 - \beta) \geq 0.90$

n'	df = k(n' − 1)	Φ	$1 - \beta$	Comments
15	28	2.17	0.820	too small
17	32	2.31	0.870	too small
18	34	2.38	0.890	close
19	36	2.44	0.915	correct

Use n = 19 for each group of bearings.

3.   a)   MSB = 17843
          MSE = 50.64
          $F_0 = 352.3$
          $F_{crit}(0.01, 4, 30) = 4.02$

Since $F_0 >> F_{crit}$, reject H_0 with 99% confidence, i.e., there is a significant difference in the means.

The next step is to complete the Tukey pairwise comparison test, where dt=11.43.

b) A_5 is the best choice for maximization.

4. a) $\bar{x} = 15.8$, $\bar{y} = 9.6$
       $b_1 = .9415$ (slope)
       $b_0 = -5.314$ (intercept)

   b) $R = .968$
      $R^2 = .938$ or 93.8% of the variability is explained

5. a)

Source	SS	df	MS	F_0
Temp (A)	1.439	2	.7195	7195
H_2SO_4	.0345	2	.0173	173
A × B	.1940	4	.0485	485
Error	8.9×10^{-4}	9	.0001	
Total	1.669	17		

Chapter 5

1. a)

Run #	D = ABC	$\bar{y}$	s	ln s
1	−	60.0	14.14	2.65
2	+	50.5	4.95	1.60
3	+	39.5	6.36	1.85
4	−	62.5	12.02	2.49
5	+	55.0	2.83	1.04
6	−	45.5	3.54	1.26
7	−	40.5	2.12	0.75
8	+	58.5	2.12	0.75

   b) $Y = 51.5 - 1.25B + 2.75C + 7.50BC$

c) To minimize the response, the settings should be B−, and C+. It is important that (−BC) = +, so BC = −. To minimize the variance, set A at the high setting.

2. a) The prediction equation for the mean is:

$$y = 3.512 + 0.255(A) + 1.008(B) + .044(AB)$$

3. a)

	A	B	C	D	E	F	G
Avg (−)	72.26	77.77	77.39	79.80	78.18	77.30	77.22
Avg (+)	82.89	77.37	77.76	75.35	76.97	77.84	77.93
Δ	10.63	−.40	.37	−4.45	−1.21	.54	.71

b) To maximize, set A at high and D at low.
   $\hat{y} = 77.57 + 5.32 + 2.23 = 85.12$

5.

ANALYSIS OF MEANS

	A	B	C	D
Avg(−)	23.50	25.08	24.17	15.08
Avg(+)	26.17	24.58	25.50	34.58
Δ	2.67	−.50	1.33	19.50
SS	42.77	1.50	10.61	2281.50
F	1.09	0.04	0.27	58.20

$F_{crit}(0.05, 1, 16) = 4.49$ so D is the only significant factor.

Chapter 6

4. Using S/N (nominal is best), the averages are:

	D	A	E	B
Avg(−)	109.9	107.9	105.9	104.9
Avg(+)	107.9	109.9	111.9	112.9
Δ	−2.0	2.0	6.0	8.0

5. Using S/N (larger is better), the averages are:

	A	B	C	D	D
Avg(−)	25.79	26.68	26.36	22.97	22.97
Avg(+)	27.41	26.52	26.84	30.23	30.23
Δ	1.62	−.16	.48	7.26	7.26

The best settings appear to be:
   D at +1   A at +1   B at −1

Chapter 7

1. You should use response surface methodology when you think the optimum is outside the experimental region. The indicator for this condition when using ANOVA is a non-significant quadratic term.

2. The gradient vector is the first derivative of the model with respect to each factor. It indicates which direction to travel to find the optimum.

4. Coded values allow us to avoid multicollinearity problems and inflated p values in the analysis of the experimental data.

Historical Perspective on DOE

Early Paper on Latin Squares (Meziriac)	1st Statistical Hypothesis Test (Gauss)	Least Squares (Gauss) (Legendre)	Regression Concepts (Galton, Pearson)	Orthogonal Matrices (Hadamard)	t-test (Gosset)
1624	1812	1830	1888	1890	1908
Design of Experiments Developed (Fisher)	Hypothesis Testing Formalized (Neyman, Pearson)	Fisher's texts on DOE Blocking, Randomizing, Factorial Design, Anova (Fisher)	Incomplete Randomized Block Designs (Yates)	2^k Factorial Designs (Yates)	
1920	1933	1925-35	1936	1937	
$L_{27}(2^{13})$ (Fisher)	Fractional Factorials (Finney, Rao)	Plackett-Burman Designs (Plackett-Burman)	$L_{18}(2^1x3^7)$ (Burman)	1st text on DOE Applications (Brown Lee)	
1945	1945	1945	1946	1948	
Taguchi Develops his Approach to DOE (Taguchi)	Multiple Regression (?)	Central Composite Design (Box, Wilson)	Optimal Designs (Elfving)	Multiple Comparisons (Tukey, Scheffe)	$L_{36}(2^3x3^{13})$ (Seiden)
1950	1950	1951	1952	1954	1954
RSM EVOP (Box)	Tukey Quick Test (Tukey)	Mixture Experiments (Scheffe)	Half Normal Plots (C. Daniel)	Optimal Designs (Kiefer, Wolfowitz)	Box-Behnken Designs (Box-Behnken)
1957	1958	1958	1959	1960	1960
Addelman Designs 2&3 Levels (Addelman)	2^{k-p} Fractional Factorials, Resolution (Box, Hunter)	Formalized Optimal Design Approach (Fedorov)	Detmax D-Optimal Algorithm (Mitchell)	1st text on RSM (R. Myers)	Taguchi version of Loss Function & Robust Design (Taguchi)
1961	1961	1972	1974	1976	1980
K-Exchange Algorithm for D-Optimal (Johnson, Nachtsheim)	Residuals to Find Variance Reducing Factors (Box, Meyer)	1st text to Blend Taguchi and Classical (Schmidt/ Launsby)	D-Optimal Blocking Techniques (Cox, Nachtsheim)	$\ln \dfrac{S^2+}{S^2-} \sim Z$ (Montgomery)	Sample Size Rules of Thumb for § Models (Schmidt/ Launsby)
1983	1986	1988	1989	1990	1991

Symbols and Abbreviations

$AB = A \times B$	2-way interaction of factors A and B
ANOVA	Analysis of Variance
CCD	Central Composite Design
COPQ	Cost of Poor Quality
C_p	Process capability index (assumes process is centered on the target)
C_{pk}	Process capability index (does not assume process in centered)
CPI	Continuous Process Improvement
df	Degrees of freedom
DOE	Design of Experiments
dpm	Number of defects per million opportunities
e_i	Error of prediction for the ith data point
$F_{critical}$	Tabled F value
F_0	Calculated F value
H_0	Null hypothesis
H_1	Alternative hypothesis
L	Loss
LCL	Lower control limit (for control charts)
LSL	Lower specification limit
MSB	Mean square between
MSE	Mean square error
n_a	# of combinations associated with the axial points in a CCD
n_c	# of centerpoints in a CCD
n_f	# of combinations in the factorial portion of a CCD
NGT	Nominal Group Technique
PAT	Process Action Teams
QFD	Quality Function Deployment
R	Correlation coefficient
R^2	Squared correlation coefficient (coefficient of determination)

S/N	Signal-to-noise ratio
SPC	Statistical Process Control
SSB	Sum-of-squares due to between group differences
SSE	Sum-of-squares due to error
SSR	Sum-of-squares due to regression
SST	Sum-of-squares total
T	Target or nominal specification value
TQM	Total Quality Management
t_0	Calculated t values
t_c	Tabled t values
UCL	Upper control limit (for control charts)
USL	Upper specification limit
VRP	Variability Reduction Process
y	Some quality characteristic (response)
$\bar{y}$	Average response
$\hat{y}$	Predicted response
α	Axial point distance from the center of a CCD (Chapter 3) or Probability of a Type I error (Chapter 4)
Δ	Effect
$\Delta/2$	Half effect
σ	Population standard deviation
$\hat{\sigma} = s$	Standard deviation calculated from a sample
σ^2	Population variance
$\hat{\sigma}^2 = s^2$	Variance calculated from a sample
Σ	Mathematical summation (indicates "add them all up")
μ	Population mean
0	Usually represents the middle level of a factor
$+ = +1$	Usually represents the high level of a factor
$- = -1$	Usually represents the low level of a factor

RULES OF THUMB

The following Rules of Thumb are based on sound experimental logic; however, they are not necessarily precise from a statistical point of view. The reader should only use these temporarily as a quick and easy substitute for some of the rigorous methods discussed in the actual text. The Rules of Thumb in this appendix are categorized as follows:

1) Sample size per experimental condition or run (often referred to as the number of replicates and/or repetitions per experimental condition).

2) Quick test for detecting a significant shift in the average.

3) Quick test for detecting a significant shift in the standard deviation.

4) Two-level and three-level design choices when the number of factors (k) equal 2, 3, 4, 5, and greater than 5.

5) Determining statistical significance for terms in the $\hat{y}$ model.

6) Determining statistical significance for terms in the $\hat{s}$ (or ln $\hat{s}$) model.

7) Interpreting R^2, Adj R^2, and reasons for low R^2.

8) Interpreting tolerance values from statistical output.

Rule of Thumb (1)	Sample Size per Experimental Condition

Historically, Rules of Thumb on sample size have been based on the amount of data required for only the $\hat{y}$ model. Since shifts in the variance (or standard deviation) are as important (if not more so) than shifts in the average, the following Rules of Thumb are based on the amount of data to adequately model $\hat{s}$ or ln $\hat{s}$. As a follow-on Rule of Thumb, it takes approximately twice as much data to estimate a standard deviation as it does for an average. Therefore, sample sizes for $\hat{y}$ models would only be half those indicated in the table provided.

The sample size information contained in this Rule of Thumb is statistically derived and can be verified in *Engineering Statistics* [1]. The following Rules of Thumb on sample size also assume that the experiment is performed off-line (i.e., factor lows and highs are chosen to be relatively far apart). For on-line experiments (i.e., factor lows and highs are relatively close together) you will want to multiply all tabled sample sizes by at least 10.

Percent confidence that a term identified as significant, truly does belong in ŝ (α = false detection rate)	Percent chance of finding a significant term if one actually exists (β = missed detection rate)	Total number of (−) and (+) values in an experimental matrix column (n)**				
		2	4	8	12	16
		Number of Replications or Sample Size per Experimental Condition (n_r)*				
95% (α = .05)	40% (β = .60)	5	3	2	N/A	N/A
95% (α = .05)	75% (β = .25)	9	5	3	2	2
95% (α = .05)	90% (β = .10)	13	7	4	3	N/A
95% (α = .05)	95% (β = .05)	17	9	5	4	3
95% (α = .05)	99% (β = .01)	21	11	6	5	4
99% (α = .01)	15% (β = .85)	5	3	2	N/A	N/A
99% (α = .01)	50% (β = .50)	9	5	3	N/A	2
99% (α = .01)	70% (β = .30)	13	7	4	3	N/A
99% (α = .01)	90% (β = .10)	17	9	5	4	3
99% (α = .01)	96% (β = .04)	21	11	6	5	4

* The sample sizes shown are for evaluating the standard deviation. Divide these numbers by 2 to determine sample sizes for evaluating the mean while using the same tabled confidence level.

** Number of runs in a 2-level design versus just the number of (-) and (+) values in a 3-level design.

Example 1:

Assume you are testing only 1 factor A at a low and high level. The matrix would appear as shown below:

Run	A	y_1 y_2 ...	... y_{n_r}	$\bar{y}$	s
1	Low $(-)$				
2	High $(+)$				

Since the matrix has a total of 2 $(-)$ and $(+)$ values, you enter the table with n = 2. Thus, the sample size required to have a 95% chance in detecting a standard deviation shifting effect (i.e., a .05 missed detection rate) is 17 for a statistical test based on a .05 false detection rate.

Example 2:

Consider a 16-run fractional factorial for the number of factors (k) = 5. Since each column in the 16-run matrix has a total of 16 $(-)$ and $(+)$ values, the ideal sample size per experimental condition (or run) is n_r = 3. This sample size produces 95% confidence in terms identified as significant and provides a 95% chance of finding a standard deviation shifting effect if one exists. If we wanted to be 99% confident that a term identified as significant truly does belong in $\hat{s}$, there would be a 90% chance of finding an actual significant term.

Example 3:

For an L_{12} screening design, you would need to have 2 replicates for testing the <u>average</u> with a .05 missed detection rate and a .05 false detection rate.

A general approach to sample size determination for 2-level designs is described next. (Source: David LeBlond.)

1. Let σ = experimental error

2. Let your false detection rate = 0.05. (i.e., α = .05)

3. Let n = number of occurrences of $\pm$ 1 per column (i.e., the number of runs in a 2-level design).

To detect a change in ...	... as small as ...	... with a missed detection rate of ... (i.e., a β value)	... run this many reps per trial (n_r).
Mean	$\Delta\sigma$ $(1 < \Delta < 2)$	0.05	$n_r = \dfrac{64}{n\Delta^2}$
		0.10	$n_r = \dfrac{54}{n\Delta^2}$
		0.25	$n_r = \dfrac{36}{n\Delta^2}$
SD	λ fold $(2 < \lambda < 4)$	0.05	$n_r = \dfrac{242}{n\lambda^2} + 1$
		0.10	$n_r = \dfrac{190}{n\lambda^2} + 1$
		0.25	$n_r = \dfrac{128}{n\lambda^2} + 1$

In other words, if you want to know the sample size needed for detecting a 1.5σ shift in averages with a .05 missed detection rate using a 2-level design of n = 8 runs, then Δ = 1.5 and n = 8, indicating $n_r = \dfrac{64}{8(1.5)^2} = 3.56$ or 4 replicates.

Similarly, if you want to detect a standard deviation shift of $\dfrac{\sigma_i}{\sigma_j} \geq (\lambda = 2.0)$ with a .10 missed detection rate while using a Taguchi L_{12} design (i.e., n = 12), then

$$n_r = \frac{190}{12\,(2^2)} + 1 = 4.96 \text{ or 5 replicates.}$$

All of the previous sample sizes have been for continuous outputs. If you have an attribute output, the Rule of Thumb is a sample size large enough to see at least 5 bad ones. For example, if your yield is 98%, it will take a sample size of 250 to see at least 5 bad ones.

Rule of Thumb (2)	**Quick Test for Detecting a Significant Shift in the Average**

To determine if a significant average shift has occurred, a quick test developed by John Tukey in 1959, *Technometrics* [2], and popularized by Dorian Shainin, *World Class Quality* [3], is called the Tukey Quick Test or End Count Technique. To perform this test, arrange all of the data on a scale such that each of the two groups has a different label. Starting from the left-most data value, count the number of similar symbols until you reach an opposite symbol. Starting from the right, count the number of similar symbols until you reach an opposite symbol. Sum the two counts which is your end count. If both ends start with the same symbol, your end count is zero.

The significance of an end count can be determined using the next table:

End Count	Significance	% Confidence there exists a significant shift in average
≥ 6	$\leq .10$	$\geq 90\%$
≥ 7	$\leq .05$	$\geq 95\%$
≥ 10	$\leq .01$	$\geq 99\%$
≥ 13	$\leq .001$	$\geq 99.9\%$

An example demonstrating the simplicity of the End Count Technique is shown below on the same data from Chapter 4 where a t-test was used to determine whether or not a significant shift in the average occurred.

Strength Measurements											
Temp	y_1	y_2	y_3	y_4	y_5	y_6	y_7	y_8	y_9	$\bar{y}$	s^2
200°	2.8	3.6	6.1	4.2	5.2	4.0	6.3	5.5	4.5	4.6889	1.3761
300°	7.0	4.1	5.7	6.4	7.3	4.7	6.6	5.9	5.1	5.8667	1.1575

Composite Material Data (Low (L) = 200°, High (H) = 300°)

The hypotheses to be tested are:
$$H_0: \quad \mu_L = \mu_H$$
$$H_1: \quad \mu_L \neq \mu_H$$

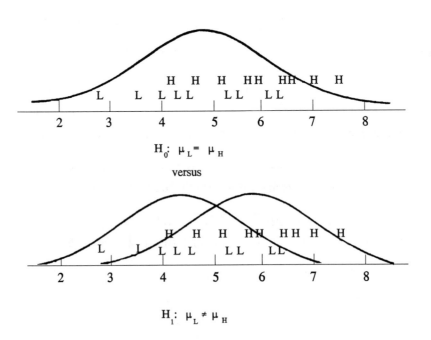

Starting from the left, the number of L symbols before the first H is 3. Starting from the right, the number of H symbols before the first L is 4. Thus, the overall End Count is 7. Using the previous table with End Count values and associated significance values, our data indicates that we are approximately 95% confident in concluding H_1: $\mu_L \neq \mu_H$. This conclusion is identical to that of the t-test performed in Chapter 4.

Rule of Thumb (3)	**Quick Test for Detecting a Significant Shift in the Standard Deviation**

To detect a significant shift in standard deviation the proper statistical test is based on the ratio of sample variances, i.e., $F = \dfrac{s_-^2}{s_+^2}$. In the text, *Design and Analysis of Experiments* [4], the author states that $\ln \dfrac{s_-^2}{s_+^2}$ is approximately the standard normal, i.e., $\ln \dfrac{s_-^2}{s_+^2} \sim Z(0,1)$. Consequently, a quick test for H_0: $\sigma_1 = \sigma_2$ vs H_1: $\sigma_1 \neq \sigma_2$ can be developed as follows:

a) Find the maximum of $\dfrac{s_-}{s_+}$ and $\dfrac{s_+}{s_-}$.

b) If the maximum $\left[\dfrac{s_-}{s_+}, \dfrac{s_+}{s_-}\right] > 2.72$ you are approximately 95% confidence that the two conditions have significantly different standard deviations, (i.e., a significant shift in standard deviation is highly probable).

Note 1: This test is best performed if $[n_- = n_+] \geq 16$.

Note 2: The value 2.72 is not a hard and fine line. You should use good judgement or more precise tests when max $\left[\dfrac{s_-}{s_+}, \dfrac{s_+}{s_-}\right]$ produces a value close to 2.72. Values between 1.5 and 2.72 constitute a gray zone, and the alternative Rule of Thumb shown on the next page could be applied in this case.

As an example for Rule of Thumb (3), consider the following:

$$n_- = n_+ = 16$$

$$s_- = 1.9 \qquad \max \left[\frac{1.9}{5.8}, \frac{5.8}{1.9} \right] = 3.053$$

$$s_+ = 5.8$$

Therefore, we are at least 95% confident that $\sigma_- \neq \sigma_+$.

Because of the limitations of the approach above, an alternative Rule of Thumb follows:

a)   Determine $\dfrac{S^2_{Largest}}{S^2_{Smallest}} \sqrt{\dfrac{n_1 + n_2}{2}}$ where n_1 is group 1 sample size, and n_2 is group 2 sample size (assume $\dfrac{n_1 + n_2}{2} \leq 60$; else refer to statistical F test.)

b)   if $\dfrac{S^2_{Largest}}{S^2_{Smallest}} \sqrt{\dfrac{n_1 + n_2}{2}} > 10$ you are approximately 95% confident in

$$H_1: \sigma_1 \neq \sigma_2.$$

Using the same example previously discussed,

$$\frac{S^2_{Largest}}{S^2_{Smallest}} \sqrt{\frac{n_1 + n_2}{2}} = \frac{(5.8)^2}{(1.9)^2} \sqrt{\frac{16 + 16}{2}} = 9.32\,(4) = 37.27$$

which is substantially greater than 10. Therefore, we can be much more than 95% confident that $H_1: \sigma_- \neq \sigma_+$ is true.

Note: The alternative Rule of Thumb is very close to an F test for $\dfrac{n_1 + n_2}{2} \leq 60.$

Rule of Thumb (4)	Two-Level and Three-Level Design Choices for the Number of Factors (k) = 2, 3, 4, 5, and greater than 5

Two-Level Designs

a) $k = 2$: The only option is a full factorial where $n = 2^2 = 4$ experimental conditions (runs). According to Rule of Thumb (1), the ideal sample size per condition (run) is 9. Thus, the total resources to conduct the ideal experiment of 2 factors at two levels is 36.

Run	A	B	AB	y_1	y_2	....	y_9	$\bar{y}$	s
1	−1	−1	1						
2	−1	1	−1						
3	1	−1	−1						
4	1	1	1						

b) $k = 3$: Option 1 is to run a full factorial where $n = 2^3 = 8$ with 5 samples per condition (run). Thus, the total resources required would be 40.

			Factors and Interactions							
Run	A	B	AB	C	AC	BC	ABC	$y_1 \dots y_5$	$\bar{y}$	s
1	−1	−1	1	−1	1	1	−1			
2	−1	−1	1	1	−1	−1	1			
3	−1	1	−1	−1	1	−1	1			
4	−1	1	−1	1	−1	1	−1			
5	1	−1	−1	−1	−1	1	1			
6	1	−1	−1	1	1	−1	−1			
7	1	1	1	−1	−1	−1	−1			
8	1	1	1	1	1	1	1			

Option 2 is a R_{III} half fraction where $n = 2^{3-1} = 4$ with 9 samples per condition (run). The total resources required is 36.

Run	A	B	C=AB	y_1 y_2 $\ldots$ y_9	$\bar{y}$	s
1	-1	-1	1			
2	-1	1	-1			
3	1	-1	-1			
4	1	1	1			

ALIAS PATTERN for a 2^{3-1} Design: A = B·C   B = A·C   C = A·B

Note that Option 2 only saves four resources, yet would result in loss of all interactions. Therefore, most experimenters would prefer Option 1.

c) k = 4 : Option 1, the full factorial, will have n = 2^4 = 16 runs, each with a sample size of 3. Total resources will be 48.

	1	2	3	4	5	6	7	8	9	10	11	12	13	14	15			
Run	A	B	AB	C	AC	BC	ABC	D	AD	BD	ABD	CD	ACD	BCD	ABCD	y_1 y_2 y_3	$\bar{y}$	s
1	-1	-1	1	-1	1	1	-1	-1	1	1	-1	1	-1	-1	1			
2	-1	-1	1	-1	1	1	-1	1	-1	-1	1	-1	1	1	-1			
3	-1	-1	1	1	-1	-1	1	-1	1	1	-1	-1	1	1	-1			
4	-1	-1	1	1	-1	-1	1	1	-1	-1	1	1	-1	-1	1			
5	-1	1	-1	-1	1	-1	1	-1	1	-1	1	1	-1	1	-1			
6	-1	1	-1	-1	1	-1	1	1	-1	1	-1	-1	1	-1	1			
7	-1	1	-1	1	-1	1	-1	-1	1	-1	1	-1	1	-1	1			
8	-1	1	-1	1	-1	1	-1	1	-1	1	-1	1	-1	1	-1			
9	1	-1	-1	-1	-1	1	1	-1	-1	1	1	1	1	-1	-1			
10	1	-1	-1	-1	-1	1	1	1	1	-1	-1	-1	-1	1	1			
11	1	-1	-1	1	1	-1	-1	-1	-1	1	1	-1	-1	1	1			
12	1	-1	-1	1	1	-1	-1	1	1	-1	-1	1	1	-1	-1			
13	1	1	1	-1	-1	-1	-1	-1	-1	-1	-1	1	1	1	1			
14	1	1	1	-1	-1	-1	-1	1	1	1	1	-1	-1	-1	-1			
15	1	1	1	1	1	1	1	-1	-1	-1	-1	-1	-1	-1	-1			
16	1	1	1	1	1	1	1	1	1	1	1	1	1	1	1			

Option 2 is a R_{IV} ½ fraction where $n = 2^{4-1} = 8$ and the sample size per run is 5. Total resources is 40.

Run	1 A	2 B	3 AB	4 C	5 AC	6 BC	7 D=ABC	y_1 y_2 y_3 y_4 y_5	$\bar{y}$	s
1	-1	-1	1	-1	1	1	-1			
2	-1	-1	1	1	-1	-1	1			
3	-1	1	-1	-1	1	-1	1			
4	-1	1	-1	1	-1	1	-1			
5	1	-1	-1	-1	-1	1	1			
6	1	-1	-1	1	1	-1	-1			
7	1	1	1	-1	-1	-1	-1			
8	1	1	1	1	1	1	1			

ALIAS PATTERN FOR 2^{4-1} DESIGN

$$A = BCD \qquad AB = CD$$
$$B = ACD \qquad AC = BD$$
$$C = ABD \qquad BC = AD$$
$$D = ABC$$

Note that Option 2 will save us 8 resources compared to Option 1. However, of the 6 total 2-way interactions, Option 1 will evaluate them all and Option 2 will evaluate half of them. Which option to choose will typically depend on the cost per resource and the number of anticipated interactions.

d) $k = 5$: Option 1 is a full factorial with $n = 2^5 = 32$ runs, each replicated 2 times for a total of 64 resources.

Option 2 is a R_V ½ fraction where $n = 2^{5-1} = 16$ runs, each with 3 replications. Total resources will be 48.

Note that Option 2 is usually the preferred choice because 16 resources are saved and all 2-way interactions are still evaluated.

	1	2	3	4	5	6	7	8	9	10	11	12	13	14	15	y₁ y₂ y₃	ȳ s
Run	A	B	AB	C	AC	BC	ABC DE	D	AD	BD	ABD CE	CD	ACD BE	BCD AE	E ABCD	$y_1\ y_2\ y_3$	$\bar{y}$ s
1	-1	-1	1	-1	1	1	-1	-1	1	1	-1	1	-1	-1	1		
2	-1	-1	1	-1	1	1	-1	1	-1	-1	1	-1	1	1	-1		
3	-1	-1	1	1	-1	-1	1	-1	1	1	-1	-1	1	1	-1		
4	-1	-1	1	1	-1	-1	1	1	-1	-1	1	1	-1	-1	1		
5	-1	1	-1	-1	1	-1	1	-1	1	-1	1	1	-1	1	-1		
6	-1	1	-1	-1	1	-1	1	1	-1	1	-1	-1	1	-1	1		
7	-1	1	-1	1	-1	1	-1	-1	1	-1	1	-1	1	-1	1		
8	-1	1	-1	1	-1	1	-1	1	-1	1	-1	1	-1	1	-1		
9	1	-1	-1	-1	-1	1	1	-1	-1	1	1	1	1	-1	-1		
10	1	-1	-1	-1	-1	1	1	1	1	-1	-1	-1	-1	1	1		
11	1	-1	-1	1	1	-1	-1	-1	-1	1	1	-1	-1	1	1		
12	1	-1	-1	1	1	-1	-1	1	1	-1	-1	1	1	-1	-1		
13	1	1	1	-1	-1	-1	-1	-1	-1	-1	-1	1	1	1	1		
14	1	1	1	-1	-1	-1	-1	1	1	1	1	-1	-1	-1	-1		
15	1	1	1	1	1	1	1	-1	-1	-1	-1	-1	-1	-1	-1		
16	1	1	1	1	1	1	1	1	1	1	1	1	1	1	1		

e) $k \geq 6$: In this case, the experimenter will usually conduct a screening experiment followed by a subsequent experiment on the important factors. Two-level design types for screening and ideal sample size are given below.

k	Design Type	Sample Size	Comments
6, 7	Taguchi L_8 or 2^{k-q}	5	2 ways aliased with mains
6 - 11	Taguchi L_{12} or PB_{12}	4	2 ways partially confounded with mains; sensitive to large numbers of interactions
6 - 15	Taguchi L_{16} or 2^{k-q}	3	for $k \leq 8$, 2 ways unconfounded with mains; for $k \geq 9$, 2 ways are perfectly confounded with mains
12 - 19	PB_{20}	3	2 ways partially confounded with mains

Note that partial confounding causes less confusion than perfect confounding. Thus, to simplify the screening design choice, use the L_{12} for up to 11 factors.

2 Level Screening Designs

Run	1	2	3	4	5	6	7	y_1	y_2	y_3	y_4	y_5	$\bar{y}$	s
				L_8 Design										
1	−1	−1	−1	−1	−1	−1	−1							
2	−1	−1	−1	1	1	1	1							
3	−1	1	1	−1	−1	1	1							
4	−1	1	1	1	1	−1	−1							
5	1	−1	1	−1	1	−1	1							
6	1	−1	1	1	−1	1	−1							
7	1	1	−1	−1	1	1	−1							
8	1	1	−1	1	−1	−1	1							

Run	1	2	3	4	5	6	7	8	9	10	11	y_1	y_2	y_3	y_4	$\bar{y}$	s
					L_{12} Design												
1	−1	−1	−1	−1	−1	−1	−1	−1	−1	−1	−1						
2	−1	−1	−1	−1	−1	1	1	1	1	1	1						
3	−1	−1	1	1	1	−1	−1	−1	1	1	1						
4	−1	1	−1	1	1	−1	1	1	−1	−1	1						
5	−1	1	1	−1	1	1	−1	1	−1	1	−1						
6	−1	1	1	1	−1	1	1	−1	1	−1	−1						
7	1	−1	1	1	−1	−1	1	1	−1	1	−1						
8	1	−1	1	−1	1	1	1	−1	−1	−1	1						
9	1	−1	−1	1	1	1	−1	1	1	−1	−1						
10	1	1	1	−1	−1	−1	−1	1	1	−1	1						
11	1	1	−1	1	−1	1	−1	−1	−1	1	1						
12	1	1	−1	−1	1	−1	1	−1	1	1	−1						

Run	L_{16} Design															y_1	y_2	y_3	$\bar{y}$	s
	1	2	3	4	5	6	7	8	9	10	11	12	13	14	15					
1	−1	−1	−1	−1	−1	−1	−1	−1	−1	−1	−1	−1	−1	−1	−1					
2	−1	−1	−1	−1	−1	−1	−1	1	1	1	1	1	1	1	1					
3	−1	−1	−1	1	1	1	1	−1	−1	−1	−1	1	1	1	1					
4	−1	−1	−1	1	1	1	1	1	1	1	1	−1	−1	−1	−1					
5	−1	1	1	−1	−1	1	1	−1	−1	1	1	−1	−1	1	1					
6	−1	1	1	−1	−1	1	1	1	1	−1	−1	1	1	−1	−1					
7	−1	1	1	1	1	−1	−1	−1	−1	1	1	1	1	−1	−1					
8	−1	1	1	1	1	−1	−1	1	1	−1	−1	−1	−1	1	1					
9	1	−1	1	−1	1	−1	1	−1	1	−1	1	−1	1	−1	1					
10	1	−1	1	−1	1	−1	1	1	−1	1	−1	1	−1	1	−1					
11	1	−1	1	1	−1	1	−1	−1	1	−1	1	1	−1	1	−1					
12	1	−1	1	1	−1	1	−1	1	−1	1	−1	−1	1	−1	1					
13	1	1	−1	−1	1	1	−1	−1	1	1	−1	−1	1	1	−1					
14	1	1	−1	−1	1	1	−1	1	−1	−1	1	1	−1	−1	1					
15	1	1	−1	1	−1	−1	1	−1	1	1	−1	1	−1	−1	1					
16	1	1	−1	1	−1	−1	1	1	−1	−1	1	−1	1	1	−1					

Three-Level Designs

a) For $k \le 3$ qualitative factors use a full factorial, i.e.,

   $k = 2$ will have $n = 9$ with 7 replications
   $k = 3$ will have $n = 27$ with 3 replications
   See Chapter 3.

b) For $k \ge 4$ qualitative (or some qualitative and some quantitative) factors use the L_{18} for screening up to 7 factors at 3 levels with 1 factor at 2 levels. The minimum recommended number of replicates is 4. See Chapter 3.

c)   For $k \leq 5$ quantitative factors use either a Box-Behnken (BB) or Central Composite Design (CCD) as shown below:

$k = 2 \ldots n_{reps} \geq 9$ (CCD)

$k = 3 \ldots n_{reps} \geq 5$ (CCD or BB)

$k = 4 \ldots n_{reps} \geq 3$ (CCD or BB)

$k = 5 \ldots n_{reps} \geq 3$ (CCD)

See Chapter 3.

Rule of Thumb (5)	**Determining Statistical Significance for Terms in the ŷ Model**

a.   Analysis of Variance (ANOVA) produces a F value for each potential term in the ŷ model. Based upon the sample sizes discussed in Rule of Thumb (1), the critical F value will typically be between 4.0 and 6.0, assuming a confidence level of 95%. Therefore, a simple Rule of Thumb is:

F < 4.0 implies insignificance (leave term out of ŷ model),

F > 6.0 implies significance (leave term in ŷ model),

$4.0 \leq F \leq 6.0$ is a gray zone;

Note: It is usually better to make an error placing the term in ŷ as opposed to leaving it out.

Therefore, use F > 4.0 as a Rule of Thumb criteria, but remember it is just a Rule of Thumb and not an absolute decision criteria.

b.   Multiple Regression usually results in a t value for each term where $t^2 = F$. Therefore, the same Rules of Thumb from ANOVA could apply to t^2. However, most regression software also produces a P(2 Tail) value which represents the probability of placing an insignificant term or effect in the ŷ model. Thus, the Rule of Thumb for P(2 Tail) values is usually as follows:

P(2 Tail) < .05 implies term is significant; place in ŷ model.

P(2 Tail) > .10 implies term is insignificant; leave out of ŷ model.

.05 ≤ P(2 Tail) ≤ .10 is a gray zone where most experimenters will typically decide to place the term in ŷ.

Note: (1 – P(2 Tail)) * 100% is your confidence that the term belongs in the model.

Again if there is any doubt, it is better to place a term in the model rather than to leave it out. Therefore, use P < .10 as a Rule of Thumb.

It should be noted that these Rules of Thumb do not represent a precise set of decision criteria. Neither Rules of Thumb nor rigorous statistical tests should ever replace common sense, size of effects, and prior knowledge. Ideally, a blend of all these approaches works best.

Rule of Thumb (6)	**Determining Statistical Significance for Terms in the ŝ or ln ŝ Model**

- ŝ models from 2-level designs: $|\frac{\Delta}{2}|$ for any term should be ≥ $\frac{\bar{s}}{2}$ (i.e., the |term coefficient| should be ≥ ½ of the constant).

As previously stated with the ŷ model Rules of Thumb, the reader should use several complementary inputs such as common sense, prior knowledge, and the above Rules of Thumb when making the final decision for significance. For example, any main factors with $\left|\frac{\Delta}{2}\right|$ close to $\frac{\bar{s}}{2}$ would be considered significant; whereas, a $\left|\frac{\Delta}{2}\right|$ close to $\frac{\bar{s}}{2}$ for a 3-way interaction is highly unlikely to be significant. Note that 3-level designs will provide a P(2 Tail) for ŝ models which can be used similar to that described in Rule of Thumb (5).

Rule of Thumb (7)	Interpreting R^2, Adj R^2, and Reasons for Low R^2

- R^2 measures strength of the model, $0 \leq R^2 \leq 1$

- Adj $R^2 = 1 - \dfrac{(n-1)}{(n-p)}(1 - R^2)$

> n = total number of observations
>
> p = total number of terms in the model (including the constant)

Adj R^2 adjusts R^2 down for small sample size and/or for too many insignificant terms in the model.

Reasons for Low R^2

1. Poor PF/CE/CNX/SOPs
2. Low and High x settings close together
3. x's not related to y
4. wrong model, i.e., linear fit when the true relationship is non-linear

Rule of Thumb (8)	Interpreting Tolerance Values from Statistical Output

- Tolerance is the proportion of orthogonality for each effect

- (1 - Tolerance) is the proportion of partial confounding for each effect

- All tolerance values should be $\geq$ .5 for meaningful coefficients

Appendix M - Bibliography

1.   Bowker, A.H. and Lieberman (1972), *Engineering Statistics*, Prentice-Hall Inc., Englewood Cliffs, New Jersey.

2.   Tukey, J.W. (1959), *Technometrics*, Vol. 1, pp. 31-48.

3.   Bhote, K.R. (1988), *World Class Quality*, American Management Association, New York.

4.   Montgomery, D.C. (1990), *Design and Analysis of Experiments*, 3rd Edition, John Wiley and Sons, Inc., New York.

DOE KISS Student Version

Software User's Guide

This appendix contains information
on installing and running the software
accompanying *Understanding
Industrial Designed Experiments:*

DOE KISS Student Version

AirSim

FactSim

(719) 531-0777

License Agreement

This book cannot be returned for credit or refund if the seal on the disk envelope has been broken or tampered with in any way. If you do not accept the terms of this license, you must return the book with the disk seal unbroken immediately to the party from whom you received it.

License Grant

This license is granted to the Licensee to possess, use, and make limited copies of the software licensed hereunder only on the terms and conditions specifically set out in this document. The Licensee evidences its acceptance of these terms and conditions by opening the product package or by any use of the licensed product.

The Licensee may:

• Use one copy of the Software on a single computer ("Dedicated Computer");

• Make one copy of the Software for archival purposes, or copy the Software onto the hard disk of your computer and retain the original for archival purposes;

• Transfer the entire Product on a permanent basis to another person or entity, provided you retain no copies of the Product and the transferee agrees to the terms of this agreement;

• Access the Software from a hard disk, over a network, or any other method you choose, so long as you otherwise comply with this agreement;

• Use on a network, however, only one computer may be using the program for each copy of the program owned.

Disclaimer of Warranty

This Software and Manual are sold "AS IS" and without warranties as to performance or merchantability. The seller's salespersons may have made statements about this software. Any such statements do not constitute warranties and shall not be relied on by the buyer in deciding whether to purchase this Program.

The Program is sold without any express or implied warranties whatsoever Because of the diversity of conditions and hardware under which this Program may be used, no warranty of fitness for a particular purpose is offered. The user must assume the entire risk of using the Program. Any liability of seller or manufacturer will be limited exclusively to Product replacement or refund of the purchase price.

Limited Warranty

This Product is warranted to be free of defects in materials and workmanship for a period of 90 days from your receipt of this Product. Workmanship is defined as the media, documentation, and associated packaging. If the Product fails to comply with the warranty set forth herein, the entire liability and your exclusive remedy will be replacement of the disk or, at our option, reasonable effort to make the Product meet the warranty set forth. This limited warranty applies only if you return all copies of the Product, along with a copy of your paid invoice, to the seller, shipping prepaid.

Limit of Liability

In no event shall Digital Computations, Inc. (DC) or Air Academy Associates,LLC (AAA), or its suppliers be liable for any damages whatsoever (including, but not limited to, damages for loss of profits, business interruption, loss of information, or other pecuniary loss) arising out of the use of or inability to use this Product, even if DC has been advised of the possibility of such damages.

Digital Computations, Inc. and Air Academy Associates, LLC, disclaim all other warranties, either express or implied, including but not limited to implied warranties of merchantability and fitness for a particular purpose, with respect to the Product.

General

This agreement constitutes the entire agreement between you and DC and AAA and supersedes any prior agreement concerning the contents of this package. It shall not be modified except by written agreement dated subsequent to the date of this agreement and signed by an authorized DC representative. DC is not bound by any provision of any purchase order, receipt, acceptance, confirmation, correspondence, or otherwise, unless DC specifically agrees to the provision in writing. This agreement is governed by the Federal laws of the United States of America.

Save This License For Future Reference.

Overview

The software disk that accompanies *Understanding Industrial Designed Experiments*, Air Academy Associates, LLC, contains three programs. Each is described briefly on the following pages. Note that replicated data created in AirSim can be analyzed in DOE KISS Student Version.

INSTALLATION

System Requirements

To install, you need a system running Microsoft® Windows 3.1 or later. DOE KISS Student Version requires a system running Microsoft Excel 5.0 or later.

Installing the Software

To install the software, insert the installation disk into your floppy drive and double-click on the Setup.exe icon. (For Windows 3.1x, run A:\Setup.exe.) Follow the instructions on the screen.

Running the Software

When the installation is complete, you will see a window that contains the software icons. To run a program, double-click on its icon.

You can also run the programs from the Start menu in Windows 95 or Windows NT. The default folder is DOE KISS Student Version Software.

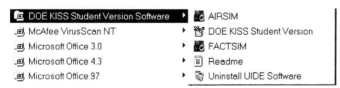

When you select the DOE KISS icon, you launch Microsoft Excel and DOE KISS Student Version. See the DOE KISS Student Version Description (following) for more information on how DOE KISS works.

Adding In DOE KISS Student Version

If you would like to have access to DOE KISS Student Version at all times from Excel, you can manually add-in DOE KISS Student Version to Excel's list of add-ins. To do this, follow these directions:

1. Start Excel using the Excel icon.

    Very Important: **Do not start Excel using the DOE KISS Student Version icon for this procedure! Be sure you start Excel using the normal Excel icon or from the Start menu at Microsoft Excel. Do not launch any program or icon titled DOE KISS Student Version at this time!**

2. From the Excel menu bar, select "Tools" then Add-Ins".

3. Use the Browse button to locate the directory in which DOE KISS Student Version is installed.

4. Select the file Doekis97.xla, then OK.

5. DOE KISS Student Version is now listed in the Add-Ins available window. Select OK to add-in DOE KISS Student Version and return to Excel.

DOE KISS Student Version will now be available every time Excel is run.

Note: You should not start Excel from the DOE KISS Student Version icon after you add in DOE KISS. We recommend that you delete the DOE KISS Student Version icon to avoid future confusion.

TECHNICAL SUPPORT

Before you call technical support, check the Readme file that came on your distribution disk for up-to-date information. You may find the answer to your question there. (The Readme file appears in the same window or Start menu level as the UIDE Software after installation.)

For technical support, contact Air Academy Associates, LLC at (719) 531-0777 or FAX (719) 531-0778.

Please have ready the version numbers of Microsoft Excel that you are running, and your network type, if applicable.

DOE KISS Student Version Description

Running DOE KISS

After DOE KISS Student Version is installed, you will be able to run it by selecting the DOE KISS menu item from Excel. The operation of DOE KISS is very simple and straightforward. First you select a design, then you input and analyze your data.

Note: The DOE KISS Student Version is limited in functionality to creating the following: an 8 run design; 2 factor Central Composite Design; Taguchi L12; and Historical Data with 4 main effects and a maximum of 30 rows. The full version of DOE KISS can be obtained from Air Academy Associates, LLC, at (800) 748-1277.

Design Selection

Your first step in design selection is to start DOE KISS. This is done from the DOE KISS menu item on Excel's menu bar. Select "Create Design", then either "Computer Aided…" or "Non-Computer Aided…".

Computer Aided

The computer aided option follows the Rules of Thumb developed by Air Academy Associates to help you choose the right design.

Once you start DOE KISS and select "Create Design - Computer Aided…", you will see a window allowing you to input your design specifications and choose your design. You may be prompted for the number of center points and the alpha value. You will be asked to name each factor and specify a low and high value, and finally choose the number of replications. Depending on the complexity of your design, you may be asked if your data is composed of quantitative or qualitative factors.

After the design has been selected and the factors named, DOE KISS creates a worksheet that contains the design and a place for data entry (a yellow block to the right of the design matrix). When the data has been entered in the data entry area, you are ready for analysis.

Non-Computer Aided

Begin the non-computer aided design selection process by selecting "Create Design - Non-Computer Aided…" from the DOE KISS menu. The non-computer aided option allows you to directly select the design of your choice. You will specify a two or three level design, then choose your design from a list. If applicable, set the number of factors in your design, give each factor a name, and set the low and high values. Choosing the number of replications finishes this selection process.

DOE KISS creates a new worksheet containing your design and a yellow block to the right of your design for entry of your replicated data. Once data entry is complete, you are ready for analysis.

Historical Data

Historical data is accessed by selecting "Historical Data /Custom Designs - Create Design Matrix..." from the menu. This option allows advanced users to analyze data that has been created in the past (probably non-coded). In one easy step, you set the number of factors and the number of runs in your design. A new sheet will be created with a place for the design matrix to be typed in.

After you have entered your data, select "Historical Data /Custom Designs - Interactions/Coding/Reps..." to select the interactions you want to use with this design. You will see the following interaction selection window.

Once you have chosen your interactions, you will be prompted to choose how you want to standardize or code your data. After you have made those choices, the option chosen for each column will appear below that column. If no standardizing or coding was chosen, an "N" appears below the column. If the column was standardized, an "S" appears below it. If it was coded, a "C" appears.

Design Analysis

After you have chosen your design and input your data, you can use any of the analysis functions under the DOE KISS menu item. These include multiple regression, marginal means plots, optimization, plotting and residuals. Make your selections from the "Analyze Design" or "Graphs, Optimize, Residuals" menu choices.

Confirmation Capability

Predicted Capability Sheet

Find the predicted capability of your process by choosing "Confirmation/Capability - Add Predicted Capability Sheet". The predicted Ybar and S values on the selected regression sheet are used to compute the process capability.

Creating and Analyzing Confirmation Runs

Analysis of confirmation runs is a three step process:

1. Create a new worksheet by selecting "Confirmation/Capability - Create Confirmation Sheet".
2. Enter your confirmation runs on the confirmation sheet.
3. Choose "Confirmation/Capability - Analyze Confirmation Runs...". Select the area on the confirmation sheet that contains the confirmation runs, and provide the specification limits (optional).

A new worksheet will be added to your workbook with process capability statistics. You may change the specification limits after the sheet has been created; however, if you change the confirmation runs after the analysis sheet has been created, those changes will not be reflected on the analysis worksheet. You will need to analyze the confirmation runs again.

Printing

You can print any part of your design or output by selecting one of the "Print" options from the DOE KISS menu item. These options include printing the design, the Y-hat or S models, the prediction area, and the residuals.

When you select a print option, you will immediately be shown Excel's print preview window. This allows you to see exactly what will be printed.

AirSim Description

Overview

AirSim is an animated simulation of an airline's departures. The program shows the number of minutes late for each departure. The outcome is controlled by adjusting the factor settings on the left of the screen with the mouse. You can view the animations, the DOE output, or the X bar R chart of the departure data. The goal is to minimize the number of minutes late of each departure.

Running AirSim

To run AirSim, double-click on the AirSim icon. Select an animation graphic to use. You can adjust any of the factor settings, then select Continuous Output or Single Output to run the simulation. Select Change View to change to the DOE view or see the X bar R chart output.

Using AirSim data with DOE KISS Student Version

Data created in AirSim can be analyzed in DOE KISS Student Version. Use the following steps to create replicated data in AirSim.
1. Run AirSim; change to the DOE view.
2. Choose an eight run matrix, and select the factors to include in the design.
3. Press the "Fill Matrix" button to create replicated data sets.
3. Select "Save Data" and choose ASCII format. (Give your data a name of no more than eight characters, such as AIRDATA). Your file is saved as a .txt file in the same directory in which DOE KISS Student Version Software is installed.
4. Exit AirSim and launch DOE KISS Student Version.

These directions show you how to open your AirSim data and analyze it using DOE KISS Student Version.
1. Open the ASCII file you just created using "File - Open..." from Excel's menu. At the bottom of the Open window, change the Files of Type box to read Text Files or All Files to see the name of the ASCII .txt file in the list of files you can open. Excel's Text Import Wizard converts the .txt ASCII file to .xls to be read by Excel.

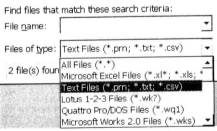

2. Now run DOE KISS and create an eight run design (to match the design you created in AirSim). Set the number of factors and number of runs to match your AirSim data, and give each factor a name and low and high value.
3. From the AirSim sheet, select the Y data (replicated data) with the mouse, choose "Edit - Copy" from the menu, then paste the data into the data entry area of your DOE KISS design sheet.
4. You are now ready to analyze your data using DOE KISS.

FactSim Description

Overview

FactSim is an animated simulation of a factory that produces nuts (as in nuts and bolts). This program shows the nuts coming off the factory line, and provides an Xbar and R chart of the process. It is an excellent tool for demonstrating the concepts of Statistical Process Control and Design of Experiments and how the two can fit together. The goal of the factory is to produce nuts with a consistent target thickness of 0.5 inches.

Running FactSim

To run FactSim, double-click on the FactSim icon. The only keys you will need to use are the arrow keys and the Enter key. The up and down arrows position the "hand" selector, the left and right arrows change the settings in the dials, and the Enter key starts and stops the factory. Press Esc or End to exit FactSim when the factory is stopped.

Use the up and down arrows to move between the dials, and the left and right arrows to adjust the factor settings.

Move to the mode button with the up and down arrows. Select "cont" or "one" for continuous or single output with the left and right arrows. Press Enter to start or stop the factory.

GLOSSARY

α **(alpha) risk** the probability of concluding the alternative hypothesis (H_1) when the null hypothesis (H_0) is true.

β **(Beta) risk** the probability of concluding the null hypothesis (H_0) when the alternative (H_1) is true.

2-level design an experiment where all factors are set at one of two levels, denoted as low and high (-1, 1 or 1, 2).

2-tailed test also known as a two-sided test, it is a hypothesis test with a two-sided alternative hypothesis. That is, one could possibly err on either side of the center.

3-level design an experiment where all factors are set at one of three levels, denoted as low, medium, and high (-1, 0, 1 or 1, 2, 3).

aliasing when two factors or interaction terms are set at identical levels throughout the entire experiment (i.e., the two columns are 100% correlated).

alternative hypothesis the hypothesis to be accepted if the null hypothesis is rejected. It is denoted by H_1.

analysis of variance (ANOVA) a procedure for partitioning the total variation. It is often used to compare more than two population means.

x̄ and R charts for variables data: control charts for the average and range of subgroups of data.

assignable cause (of variation) significant, identifiable change in a response which is caused by some specific variable from the cause and effect diagram.

attributes data (quality) data coming basically from GO/NO-GO, pass/fail determinations of whether units conform to standards. Also includes noting presence or absence of a quality characteristic.

average (of a statistical sample) $(\overline{x})$ also called the sample mean, it is the arithmetic average value of all of the sample values. It is calculated by adding all of the sample values together and dividing by the number of elements (n) in the sample.

balanced design a 2-level experimental design is balanced if each factor is run the same number of times at the high and low levels.

bar chart a graphical method which depicts how data fall into different categories.

bell-shaped curve a curve or distribution showing a central peak and tapering off smoothly and symmetrically to "tails" on either side. A normal (Gaussian) curve is bell-shaped.

bias (in measurement) systematic error which leads to a difference between the average result of a population of measurements and the true, accepted value of the quantity being measured.

blocking variable a variable (factor) which cannot be randomized. The experiment is usually run in blocks for each level of the blocking variable and randomization is performed within blocks.

Box-Behnken design a 3-level design used for quantitative factors and designed to estimate all linear, quadratic, and 2-way interaction effects.

brainstorming a group activity which generates a list of possible factors and levels, and the method by which the results may be evaluated.

calibration (of instrument) adjusting an instrument using a reference standard to reduce the difference between the average reading of the instrument and the "true" value of the standard being measured, i.e., to reduce measurement bias.

capability (of process) a measure of quality for a process usually expressed as sigma capability, C_{pk}, or defects per million (dpm). It is obtained by comparing the actual process with the specification limit(s).

causality the assertion that changes to an input factor will directly result in a specified change in an output.

cause-and-effect diagram a pictorial diagram showing possible causes (process inputs) for a given effect (process output).

center points experimental runs with all factor levels set halfway between the low and high settings.

central composite design a 3-level design that starts with a 2-level fractional factorial and some center points. If needed, axial points can be tested to complete quadratic terms. Used typically for quantitative factors and designed to estimate all linear effects plus desired quadratics and 2-way interactions.

central tendency a measure of the point about which a group of values is clustered; some measures of central tendency are mean, mode, and median.

characteristic a process output which can be measured and monitored for control and capability.

chi-square distribution the distribution of chi-square statistics.

chi-square the test statistic used when testing the null hypothesis of independence in a contingency table or when testing the null hypothesis of a set of data following a prescribed distribution.

classical methods statistical experimental design thoughts and processes as originally developed by Fisher and others as early as the 1920's. Uses ANOVA as the primary analysis tool, along with orthogonal designs such as fractional factorials, latin squares, Plackett-Burman, Box-Behnken, central composite, and D-optimal.

coefficient of variation the ratio of the standard deviation to the mean. It is a standardized method of looking at variation.

coefficient of determination (R^2) the square of the sample correlation coefficient; it represents the strength of a model. $(1 - R^2)*100\%$ is the percentage of noise in the data not accounted for by the model.

common causes of variation those sources of variability in a process which are truly random, i.e., inherent in the process itself.

confidence interval range within which a parameter of a population (e.g., mean, standard deviation, etc.) may be expected to fall, on the basis of measurement, with some specified confidence level or confidence coefficient.

confidence limits the upper and lower boundaries of a confidence interval.

confounded two factors are considered to be confounded when their test profiles contain the same pattern of test settings, thus making it impossible to evaluate the two factors independently.

control (of process) a process is said to be in a state of statistical control if the process exhibits only random variations (as opposed to systematic variations and/or variations with known sources). When monitoring control with control charts, a state of control is exhibited when all points remain between set control limits without any abnormal (non-random) patterns.

control chart the basic tool of statistical process control. It consists of a run chart, together with statistically determined upper and lower control limits and a centerline.

control limits upper and lower bounds in a control chart that are determined by the process itself. They can be used to detect special causes of variation. They are usually set at $\pm$ 3 standard deviations from the centerline.

correlation coefficient (R) a measure of the linear relationship between two random variables. Not as useful as R^2, the coefficient of determination.

controllable factors factors the experimenter has control of during all phases, i.e., experimental, production, and operational phases.

C_p during process capability studies, C_p is a capability index which shows the process capability potential but does not consider how centered the process is. C_p may range in value from 0 to infinity, with a large value indicating greater potential capability. A value of 1.33 or greater is usually desired.

C_{pk} during process capability studies, C_{pk} is an index used to compare the natural tolerance of a process with the specification limits. C_{pk} has a value equal to C_p if the process is centered on the nominal; if C_{pk} is negative, the process mean is outside the specification limits; if C_{pk} is between 0 and 1 then the natural tolerances of the process fall outside the spec limits. If C_{pk} is larger than 1, the natural tolerances fall completely within the spec limits. A value of 1.33 or greater is usually desired.

D-optimal design an experimental design in which the minimum number of runs is based on degrees of freedom needed to analyze the desired effects. Not necessarily orthogonal or balanced, it does minimize the correlation (confounding) between factors.

defect departure of a quality characteristic from its acceptable level or state, i.e., the measured value of the characteristic is outside of specification. Also referred to as non-conformance to requirements.

defective unit a sample (part) which contains one or more defects, making the sample unacceptable for its intended, normal usage.

defining relationship a statement of one or more factor word equalities used to determine the aliasing structure in a fractional factorial design.

defining words factor word equalities in building a defining relationship.

degrees of freedom a parameter in the t, F, and χ^2 distributions. It is a measure of the amount of independent information available for estimating the population variance, σ^2. It is the number of independent observations minus the number of parameters estimated.

design array (also referred to as design matrix) an array representing the experimental settings. Usually contains values ranging from -1 to $+1$, but could be wider if using a CCD. The rows represent the runs and the columns represent the factors.

deviation the difference between an observed value and the mean or average of all observed values.

discrete random variable a variable that is based on count data (number of defects, number of births, number of deaths, etc.). It supposedly has a countable number of possible outcomes.

dispersion (of a statistical sample) the tendency of the values of the elements in a sample to differ from each other. Dispersion is commonly expressed in terms of the range of the sample (difference between the lowest and highest values) or by the standard deviation.

estimation an approach to making inferences about population parameters. This includes both point estimates and interval estimates (confidence intervals).

experimental design purposeful changes to the inputs (factors) to a process (or activity) in order to observe corresponding changes in the outputs (responses). A means of extracting knowledge from a process (or activity).

exponential distribution a probability distribution mathematically described by an exponential function. Used to describe the probability that a product survives a length of time t in service, under the assumption that the probability of a product failing in any small time interval is independent of time.

F table provides a means for determining the significance of a factor to a specified level of confidence by comparing calculated F ratios to those from the F distribution (see Appendix F). If the F ratio is greater than the table value, there is a significant effect.

F-distribution distribution of F-statistics.

F-ratio a ratio of two independent estimates of experimental error. If there isn't an effect, the ratio should be close to 1.

factor an input to a process which can be manipulated during experimentation.

failure rate the average number of failures per unit time. Used for assessing reliability of a product in service.

fault tree analysis (FTA) a technique for evaluating the possible causes which might lead to the failure of a product. For each possible failure, the possible causes of the failure are determined; then the situations leading to those causes are determined; and so forth, until all paths leading to possible failures have been traced. The result is a flow chart for the failure process. Plans to deal with each path can then be made.

feedback using the results of a process to control it. The feedback principle has wide application. An example would be using control charts to keep production personnel informed on the results of a process. This allows them to make suitable adjustments to the process. Some form of feedback on the results of a process is essential in order to keep the process under control.

fishbone diagram see cause-and-effect diagram.

flow chart or diagram (for programs, decision making, process development) a pictorial representation of a process indicating the main steps, branches, and eventual outcomes of the process.

foldover design a way to obtain a resolution four (R_{IV}) design based on two designs of R_{III}. Used when the confirmation runs from a resolution III design differ substantially from their prediction and the experimenter desires to de-alias the 2-way interactions from the main effects.

fractional factorials instead of using a full factorial, a subset or fraction of it can be used if the experimenter can assume some interactions will not occur.

freehand regression line the best-fit line drawn by "eyeballing."

full factorial all possible combinations of the factors and levels. Given k factors, all with two levels, there will be 2^k runs. If the factors have three levels, then 3^k runs would be needed.

Gaussian distribution the engineering name for the normal distribution.

gradient the slope at a point on a surface.

grand average overall average of data.

histogram a bar chart that depicts the frequencies of numerical or measurement data.

hypothesis test a procedure whereby one of two mutually exclusive and exhaustive statements about a population parameter is concluded. Information from a sample is used to infer something about a population from which the sample was drawn.

inner array a Taguchi term used in parameter design to identify the combinations of controllable factors to be studied in a designed experiment. Also called a "design array" or "design matrix."

interaction 2 factors (input variables) are said to interact if one factor's effect on the response is dependent upon the level of the other factor.

Ishikawa diagram see cause-and-effect diagram.

just-in-time (JIT) manufacturing a strategy that coordinates scheduling, inventory, and production to move away from the batch mode of production in order to improve quality and reduce inventories.

latin squares a classical method used to generate orthogonal designs which are a subset of the full factorial. Typically used for screening strategies.

LCL (lower control limit) for control charts: the limit above which the process subgroup statistics ($\overline{X}$, R, etc.) must remain when the process is in control. Typically 3 standard deviations below the center line.

least squares a method of curve-fitting that defines the "best" fit as the one that minimizes the sum of the squared deviations of the data points from the fitted curve.

level a setting or testing value of a factor.

level of significance a measure of the outcome of a hypothesis test. It is the P-value or probability of making a Type I error.

linear graph a tool used by Taguchi to identify sets of interacting columns in orthogonal arrays.

loss function a technique for quantifying loss due to product deviations from target values.

lower confidence limit the smaller of the two numbers that form a confidence interval.

LSL (lower specification limit) the lowest value of a product dimension or measurement which is acceptable.

main effect the influence a single factor has on the response when it is changed from one level to another. Often used to represent the "linear effect" associated with a factor.

mean square error a weighted average of the variances for each run.

mean the average of a set of values. We usually use $\bar{x}$ or $\bar{y}$ to denote a sample mean, whereby we use the Greek letter μ to denote a population (or true) mean.

measures of central tendency numerical measures that depict the center of a data set. The most commonly used measures are the mean and the median.

median the middle value of a data set when the values are arranged in either ascending or descending order.

modeling designs designed experiments used to build a linear and/or non-linear relationship of the output(s) with a reasonable number (≤ 5) of inputs.

MTBF (mean time between failures) mean time between successive failures of a repairable product. This is a measure of product reliability.

multicollinearity the existence of strong correlations between input factors or independent variables. High levels of multicollinearity will prevent terms from being evaluated independently.

multiple regression a model where several independent variables are used to predict one dependent variable.

natural tolerances (of a process) 3 standard deviations on either side of the center point (mean value). In a normally distributed process, the natural tolerances encompass 99.73% of all measurements.

noise unexplained variability in the response. Typically, due to variables which are not controlled.

nominal the desired mean value for a particular product dimension; the target value.

nonconforming unit a sample (part) which has one or more nonconformities, making the sample unacceptable for its intended use.

normal distribution the distribution characterized by the smooth, bell-shaped curve.

null hypothesis (H_0) the conclusion that typically includes equality, i.e., $H_0: \mu_1 = \mu_2$ or $H_0: \sigma_1 = \sigma_2$.

one-at-a-time approach a popular, but inefficient way to conduct a designed experiment.

orthogonal design an experimental design is said to be orthogonal if each factor is tested in such a way that it can be evaluated independently of the other factors.

outer array a Taguchi term used in parameter design to identify the combinations of noise factors to be studied in a robust designed experiment.

out of control (of a process) a process is said to be out of control if it exhibits variations larger than its control limits, or shows a systematic pattern of variation.

P(2 Tail) the probability that a term does not belong in a regression model. Typically, Rules of Thumb will state that P(2 Tail) $\leq$ .1 indicate the term should be placed in the model.

P-value the probability of making a Type I error. This value comes from the data itself. It also provides the exact level of significance of a hypothesis test.

Pareto chart a bar chart for attribute (or categorical) data that is presented in descending order of frequency.

percent defective for acceptance sampling: the percentage of units in a lot which are defective, i.e., of unacceptable quality.

population a set or collection of objects or individuals. It can also be the corresponding set of values which measure a certain characteristic of a set of objects or individuals.

predicted value of y a value of the response (dependent) variable y that is computed from the prediction equation for some particular value of the input (independent) variable x. It is labeled $\hat{y}$ and is computed from $\hat{y} = b_0 + b_1 x$.

primary reference standard for measurements: a standard maintained by the National Institute of Standards and Technology (NIST) for a particular measuring unit. The primary reference standard duplicates as nearly as possible the international standard and is used to calibrate other (transfer) standards, which in turn are used to calibrate measuring instruments for industrial use.

probability a measure of the likelihood of a given event occurring. It is a measure that takes on values between 0 and 1 inclusive, with 1 being the certain event and 0 meaning that there is relatively no chance at all of the event occurring. How probabilities are assigned is another matter. The relative frequency approach to assigning probabilities is one of the most common.

process capability comparing actual process performance with process specification limits. There are various measures of process capability, such as C_{pk}, σ-level, and dpm (defects per million).

process control a process is said to be in control or it is a stable, predictable process if all special causes of variation have been removed. Only common causes or natural variation remains in the process.

quality assurance the function of assuring that a product or service will satisfy given needs. The function includes necessary verification, audits, and evaluations of quality factors affecting the intended usage and customer satisfaction. This function is normally the responsibility of one or more upper management individuals overseeing the quality assurance program.

quality characteristic a particular aspect of a product which relates to its ability to perform its intended function.

quality function the function of maintaining product quality levels; i.e., the execution of quality control.

quality specifications particular specifications of the limits within which each quality characteristic of a product is to be maintained in order to meet the minimum functional requirements of the customer.

quality control the process of maintaining an acceptable level of product quality.

random sample a sample selected from a population in such a way that every element of the population had an equally likely chance of being selected.

random varying with no discernable pattern.

range a measure of the variability in a data set. It is a value, namely the difference between the largest and smallest values in a data set.

regression analysis a statistical technique for determining the mathematical relation between a measured quantity and the variables it depends on.

regression line the line that is fit to a set of data points by using the method of least squares.

reliability the probability that a product will function properly for some specified period of time, under specified conditions.

repeatability (of a measurement) the extent to which repeated measurements of a particular object with a particular instrument produce the same value.

reproducibility the variation between individual people taking the same measurement and using the same gaging.

residual the difference between an observed value and a predicted value: residual $= y - \hat{y}$.

resolution (of a design) is a value such as III, IV, V, etc., that represents the type of confounding for that design.

resolution (of a measuring instrument) the smallest unit of measure which an instrument is capable of indicating.

Rule of Thumb (ROT) a simplified, practical procedure that can be used in place of a formal statistical test that will produce approximately the same result.

run chart a basic graphical tool that charts a process over time, recording either individual readings or averages over time.

sample size the number of elements, or units, in a sample.

sample a set of values or items selected from some population.

sampling variation the variation of a sample's properties from the properties of the population from which it was drawn.

sampling the process of selecting a sample of a population and determining the properties of the sample. The sample is chosen in such a way that its properties are representative of the population.

scatterplot a 2-dimensional plot for displaying bivariate data.

screening designs designed experiments used to identify the "vital few" factors from a large number of factors (≥ 6) to be tested.

short-run SPC a set of techniques used for SPC in low-volume, short duration manufacturing or service.

sigma limits for histograms: lines marked on the histogram showing the points n standard deviations above and below the mean.

sigma (σ) the standard deviation of a statistical population.

Sigma capability a commonly used measure of process capability that represents the number of standard deviations between the center of a process and the closest specification limit.

signal-to-noise (s/n) a comparison of the influence of controllable factors (signals) to the influence of noise factors. The higher the s/n value, the better.

significance level see level of significance.

simple linear regression a model where one independent variable is used to predict one dependent variable.

simulation (modeling) using a mathematical model of a system or process to predict the performance of the real system. The model consists of a set of equations or logic rules which operate on numerical values representing the operating parameters of the system. The result of the equation is a prediction of the system's output.

skewed distribution a distribution that (graphically) has a longer tail on the right than it does on the left (or vice versa), i.e., it is not symmetric about its center point. Numerically, a distribution is skewed if the mean and the median are not the same.

slope the term b_1 in the prediction equation $\hat{y} = b_0 + b_1 x$.

special causes of variation those nonrandom causes of variation that can be detected by the use of control charts and good process documentation.

specification (spec) limits the bounds of acceptable values for a given product or process. They should be customer driven.

SSE sum of squares due to error, or sum of squared residuals.

SSR sum of squares due to regression, or the sum of squared deviations of the predicted values from the mean.

SST total sum of squares, or the sum of squared deviations of the observed values from the mean.

stability (of a process) a process is said to be stable if it shows no recognizable pattern of change.

standard (measurement) a reference item providing a known value of a quantity to be measured. Standards may be primary—i.e., the standard essentially defines the unit of measure—or secondary (transfer) standards, which have been compared to the primary standard (directly or by way of an intermediate transfer standard). Standards are used to calibrate instruments which are then employed to make routine measurements.

standard deviation one of the most common measures of variability in a data set or in a population.

standardized normal distribution a normal distribution or random variable having a mean and standard deviation of 0 and 1, respectively. It is denoted by the symbol Z and is also called the Z distribution.

stationary point the corollary in single variable calculus would be a critical point. A stationary point is a point where the gradient is zero, i.e., no slope at the point.

statistic a value calculated from a random sample which is used to estimate a population parameter.

statistical inference the process of drawing conclusions about a population on the basis of statistics.

statistical quality control (SQC) the application of statistical methods for measuring and improving the quality of processes. SPC is one method included in SQC.

statistical control (of a process) a process is said to be in a state of statistical control when it exhibits only random variation and is otherwise stable and predictable.

statistical process control (SPC) the use of basic graphical and statistical methods for analyzing and controlling the variation of a process, and thus continuously improving the process.

statistics numerical measures obtained from a sample (as opposed to parameters, which are numerical measures of a population).

system design the selection of materials, parts, products, factors, equipment, and process parameters.

systematic variation (of a process) variation which exhibit a predictable pattern. The pattern may be cyclic (i.e., a recurring pattern) or may progress linearly (i.e., a trend).

t-distribution a symmetric, bell-shaped distribution that resembles the standardized normal (or Z) distribution, but it typically has more area in its tails than does the Z distribution. That is, it has greater variability than the Z distribution.

t-test a hypothesis test of population means when small samples are involved.

Taguchi methods experimental design thoughts and processes as developed by Taguchi. Based on the philosophy of the loss function, he uses signal-to-noise ratios as the primary analysis tool. Orthogonal design matrices are tabled and originate from fractional factorials, Plackett-Burman, and latin square designs.

test statistic a single value which combines the evidence obtained from sample data. The P-value in a hypothesis test is directly related to this value.

tolerance design the specification of appropriate tolerances, product parameters, and process factors.

total quality management (TQM) a management philosophy of integrated controls, including engineering, purchasing, financial administration, marketing and manufacturing, to ensure customer satisfaction and economical cost of quality.

trend a gradual, systematic change with time or some other variable.

Tukey test a statistical test to measure the difference between several mean values and tell the user which ones are statistically different from the rest.

Type II error concluding H_0 when H_1 is really true.

Type I error concluding H_1 (or rejecting H_0) when H_0 is really true.

UCL (upper control limit) for control charts: the upper limit below which a process statistic ($\bar{X}$, R, etc.) must remain to be in control. Typically this value is 3 standard deviations above the center line.

uncontrollable factors factors that are difficult, undesirable, or impossible to change. Also called "noise factors." The uncontrollable factors included in a design matrix must be controllable in the experimental phase.

uniform distribution a distribution in which all outcomes are equally likely.

upper confidence limit the larger of the two numbers that form a confidence interval.

USL (upper specification limit) the highest value of a product dimension or measurement which is acceptable.

variability the property of exhibiting variation, i.e., changes or differences, in key measurements of a process.

variables data concerning the values of a variable, or measurement data, as opposed to attribute data. A dimensional value can be recorded and is only limited in value by the resolution of the measurement system.

variables quantities which are subject to change or variability.

variance a measure of variability in a data set or population. It is the square of the standard deviation.

y-intercept the term b_0 in the prediction equation $\hat{y} = b_0 + b_1 x$.

z-value a standardized value formed by subtracting the mean and then dividing this difference by the standard deviation.

Index

A

Adjusted R², 4-48

Aliasing, 1-45, 2-25, 2-28, 3-4,
  3-10, 3-11

Analysis of Variance (ANOVA), 4-1
  through 4-20, 4-52, 4-54, 4-65, 5-9

Answers to selected homework
  problems, Appendix J

Average, 1-36, 1-48, 1-49, 4-2

B

Barker, 1-24

Binomial probability distribution table,
  Appendix B

Blocking variable, 3-20

Box-Behnken design, 3-30, 3-33,
  4-43, 5-58

Box-Meyer approach, 5-28 through
  5-31

Box-Wilson design, (see Central
  Composite)

Brainstorming, 1-24, 1-25

C

Capability, 8-220

Catapult, 8-181

Causality, 1-34

Cause-and-Effect diagrams, 1-43,
  5-34, 8-162, 8-264

Central Composite Design, 3-30,
  3-32, 3-34, 5-41, 5-52, 7-12
  through 7-15, 8-58, 8-145

Chi Square distribution table,
  Appendix G

Coding of input factors, 1-32

Comparison of:
  Dispersion factor identification
    methods, 5-16 through 5-27
  Three-level designs, 5-47
    through 5-61
  Classical and Taguchi, 6-19
    through 6-22
  ANOVA, t-tests, confidence
    intervals, and regression, 4-64
    through 4-68

Component swapping, 3-26

Conducting the experiment, 1-33

Confidence intervals, 4-64

Confirmation test, 2-22, 2-36, 5-12

Confounding, 3-22, 3-28

Contour plot, 5-44, 5-46, 7-5,
  8-52, 8-65 through 8-71, 8-150
  through 8-153

Control chart, 1-41, 8-227, 8-228
Controllable factors, 5-33, 6-9
Correlation coefficient, 4-26
C_p, 1-20 , 6-7, 8-260
C_{pk}, 1-20, 1-41, 2-35, 6-7
Crosby, 1-15
Crossvalidation, 4-48

D

D-optimal designs, 3-24, 3-26, 4-33,
  5-56, 8,156, 8-231
De-alias, 3-12, 3-14
Defects per million (dpm), 1-21
Defining relation, 3-9
Defining word, 3-11, 3-14
Degrees of freedom, 4-7
Deming, 1-13
Design generator, 3-11
Design matrix, 1-30, 1-32, 1-44, 3-1
Dispersion effects, 5-16 through 5-28
  5-65 through 5-73
Dot product, 3-3, 3-4

E

Effect, 1-48, 2-8, 2-9, 2-13, 2-38,
  4-12, 5-3
Effects plot, 1-48, 1-49, 2-16, 2-31
End Count, 8-212 through 8-217,
  Appendix M
Experimental Design
  Definition, 1-2   Steps, 1-26, 1-27

Experimental error, 2-8
Extrapolation, 4-33
Eigenvalues, 7-16

F

F distribution table, Appendix F
F ratio, 4-7
Factor types, 2-3, 5-17
Fishbone diagrams, (see Cause-and-
  Effect)
Flow chart, (see Process flow
  chart)
Foldover design, 3-20
Foldover Koshal, 3-24, 3-25
Fractional factorial designs, 1-11,
  1-45, 2-20, 2-22, 2-24, 2-25,
  3-19, 3-29
Full factorial designs, 1-11, 1-30,
  1-44, 2-2, 2-5, 2-24, 2-28, 3-40,
  4-18, 5-25, 5-49

G

Generators for fractional factorials,
  3-11
Gradient vector, 7-9

H

Hadamard matrix, 3-21, 3-42
Hierarchy rule, 2-20
Histogram, 1-40

History of experimental design,
  Appendix K
House of Quality, 1-59
Hypothesis test, 4-3 through 4-17,
  4-29

I

Interactions, 1-10, 1-49
  2-11 through 2-14, 3-5, 4-12,
  4-13, 4-14, 5-13, 8-272

J

Juran, 1-15, 1-17

K

Koshal design, 3-25, 3-26

L

Ln s, 5-23, 5-36
Lack of fit, 4-40
Latin Square designs, 3-28
Linear graph, 8-114, 8-118
Loss function, 1-17, 1-22, 1-49, 5-28,
  6-3 through 6-6

M

Mean Square Between (MSB), 4-7, 5-8
Mean Square Error (MSE), 4-6, 4-16,
  4-18, 4-28, 4-30, 4-39, 5-8

Measurements of quality, 1-19
Method of least squares, 4-24
Missing data, 2-50 through 2-52
Modeling $\hat{y}$ and $\hat{s}$, 2-19, 2-34
Multicollinearity, 4-42
Multiple regression, 4-36 through
  4-48
Multi-variate chart, 5-37 through
  5-40

N

Noise factors, 1-52, 5-33, 6-9
Normal distribution, 1-40
Normal probability plots, 5-6
  through 5-8
Nuisance variables, 1-34

O

One-factor-at-a-time, 1-9
Optimizing experiments, 2-34, 5-38,
  5-45 through 5-47, 7-16, 8-153,
  8-247
Orthogonal, 1-44, 3-3
Orthogonal coding, 1-32, 2-6,
  2-14, 3-3, 4-18

P

Parameter designs, 1-59, 6-8
  through 6-11

Partially confounded, 3-17

Pareto diagram, 1-28, 2-17, 2-32, 2-38, 5-5, 5-39

Percent contribution, 5-10

Pick the winner, 5-3

Plackett-Burman designs, 3-21

Poisson probability distripution table, Appendix C

Polynomial regression, 4-33 through 4-36

Power charts for ANOVA, Appendix I

Prediction equation, 2-19, 2-28, 4-17, 5-3, 5-12, 5-27, 5-54

Prediction interval, 4-30

Press residual, 4-49

Process action teams, 1-54

Process definition, 1-3

Process flow charts, 1-37

Process model, 1-3 through 1-6

Pure error, 4-40, 7-8

Q

QFD, 1-51, 1-54, 1-59

Qualitative factors, 3-29

Quality

  Definition, 1-1

  Old philosophy, 1-16

  Measurements of, 1-19

  New philosophy, 1-17

Quality improvement steps, 1-54, 1-55

Quantitative factors, 3-29

R

R & M 2000, 1-21, 1-57 through 1-71

Random digits table, Appendix A

Random balanced design, 3-24, 3-25

Randomization, 1-34, 2-7

Reasons for not confirming, 5-12

Regression:

  Multiple, 4-36 through 4-48, 5-11

  Polynomial, 4-33 through 4-36

  Significance tests, 4-29 through 4-43

  Simple linear, 4-21 through 4-33

Relative press, 4-49

Repetition, 2-7

Replication, 2-7

Residual, 4-22, 4-39, 5-28

Residual analysis, 8-229 through 8-233

Resolution, 3-14

Response selection, 1-29

Response surface methods, 7-1 through 7-15

Robustness, 1-52

Robust designs, 1-11, through 1-42, 1-52, 1-59, 1-62, 5-32 through 5-36, 6-12, 8-156, 8-250

Rotatability, 3-35

Rules of Thumb, Appendix M, and inside front cover

Runout, 8-219

Run chart, 1-38

S

Sample size, 4-61 through 4-63, 5-23, Appendix M

Saturated designs, 3-15

Saturation, 6-11

Scatter diagram, 4-22

Screening designs, 2-36, 5-1 through 5-11

Selecting a design, 2-5, Appendix M

Shainin, 3-26, 5-33

Signal-to-Noise, 1-2, 5-18 through 5-21, 5-65 through 5-71, 6-12

Simple linear regression, 4-21 through 4-33

Standarized normal distribution table, Appendix D

Standardization of input factor settings, 1-32

Standard deviation
Analysis of, 1-48, 1-50
Definition, 1-20, 1-39

Statistical Process Control (SPC), 1-30, 1-33, 1-58

Statistical significance, Appendix M

Steps for experimental design, 1-26

Steps before collecting data, 2-1

Studentized ranges, Appendix H

Sum of squares error, 4-23

Sum of squares regression, 4-23

Sum of squares total, 4-23

Supersaturated, 3-24

System design, 6-8

T

t distribution table, Appendix E

t-test, 4-11, 4-65

Taguchi designs, 1-2, 1-52, 3-29, 3-40, 3-42, 4-17, 6-1 through 6-26

Taylor Series, 2-20, 7-7

Time line history for DOE, Appendix K

Tolerance design, 6-8, 8-259

Total Quality Management (TQM), 1-53, 1-54, 1-56

Transformations, 5-10

Transmitted variance, 6-10

Tukey pairwise comparisons, 4-53

Tukey Quick Test, (see End Count)

Type I and Type II errors, 4-5

U

Uncontrollable factors, (see Noise factors)

Unreplicated designs, 5-28 through 5-32

Unsaturated designs, 3-15

V

Variance, 1-20, 1-38, 4-2, 5-5

Variance-covariance matrix, 4-39

Variability Reduction Process
   (VRP), 1-45

Variation, 1-16